*The Moths and
Butterflies of
Great Britain and
Ireland*

Volume 2

Volume 2

Cossidae–Heliodinidae

Editors: John Heath
A Maitland Emmet

Associate Editors:
D S Fletcher, E C Pelham-Clinton
B Skinner, W G Tremewan

Artists: Brian Hargreaves
Timothy Freed
Brenda Jarman

THE MOTHS AND BUTTERFLIES OF GREAT BRITAIN AND IRELAND

Harley Books

Harley Books (B. H. & A. Harley Ltd.),
Martins, Great Horkesley,
Colchester, Essex, CO6 4AH, England

Text set in Linotron 202 Plantin and made up by
Rowland Phototypesetting Ltd, Bury St Edmunds, Suffolk
Text printed by St Edmundsbury Press,
   Bury St Edmunds, Suffolk.
Colour plates printed at Westerham Press, Kent.
Bound by Hunter & Foulis, Edinburgh, Scotland.

Zygaenidae, by W. G. Tremewan
© British Museum (Natural History), London, 1985
*The Moths and Butterflies of Great Britain
and Ireland* Volume 2
© John Heath; Harley Books, 1985

ISBN 0 946589 02 X   Volume 2

Volume 2: Contents

Plate A faces page 9; Plate B faces
page 25

Preface

A number of name changes in the families dealt with in this volume result from the check-lists mentioned in the preface to Volume 10, an *Index* to Bradley & Fletcher (1979) by Hall-Smith (1983) which contains addenda and corrigenda by the same authors, and, in the Tineidae, Hodges R. W. *et al.* 1983, *Check List of the Lepidoptera of America North of Mexico*. Changes have also been made, by the author, in the Psychidae which bring the classification of that family up-to-date.

Dot distribution maps have been provided by the Institute of Terrestrial Ecology, Biological Records Centre, for the Macrolepidoptera with the exception of the Sesiidae, where the data was inadequate, and for which, therefore, vice-county maps are given. On the dot distribution maps solid dots represent post-1960 records and open circles earlier records. Vice-county maps are provided for the Microlepidoptera.

Species presumed not to be resident in the British Isles and of casual or reputed occurrence are represented by a brief paragraph in the text without a map.

We wish to acknowledge the continuing help of the Trustees of the British Museum (Natural History) and, in particular, Dr L. Mound, Keeper of Entomology for making available the collections and library of that department and for permitting the reproduction of figures from the Bulletin of the British Museum (Natural History), Entomology Series. These are acknowledged specifically in the Tineidae. We wish especially to thank D. J. Carter, Dr G. S. Robinson, Dr M. G. Fitton, Mr N. P. Wyatt and Miss Pamela Gilbert.

Thanks are also due to Dr R. H. Davis, Wye College, Professor C. M. Naumann, Bielefeld University, Dr G. Reiss, Stuttgart, Dr M. R. Shaw, Royal Scottish Museum and Dr G. M. Tarmann, Tyrolian Museum, Innsbruck, for assistance with the Zygaenidae. C. F. Threadgall, British Museum (Natural History), prepared the colour figures of the zygaenid larvae and Dr M.-D. Crapon de Caprona, Bielefeld University, the colour plate of zygaenid imagines. Dr M. R. Shaw also gave valuable advice on the leaf-mining families.

Special thanks are due to A. Rodger Waterston who has read both the manuscript and proofs, to Roy Swash who has read the proofs and prepared the index, and to Joan Heath who entered the text onto computer files. We are most grateful to Basil Harley for the skilful way in which he has executed the layout of the text and figures.

Timothy Freed has joined the team of illustrators and is responsible for the line-drawings of wing-venation and genitalia.

John Heath,   A. Maitland Emmet,
St. Ives, Huntingdon.   Saffron Walden, Essex.
February 1985

The clearwing *Pennisetia hylaeiformis* (Laspeyres) showing its wasp-like attitude while resting ($\times$ 6).   Photo Paolo Ragazzini

Miriam Rothschild

BRITISH APOSEMATIC LEPIDOPTERA

I. Introduction

Violent and bitter controversy still accompanies such topics as Darwinian evolution and kin selection, but the theory of warning coloration and self-advertisement seems something of an exception for it appears to be generally accepted as a reality rather than a working hypothesis. Krebs & Davies (1981), for instance, state baldly that 'some insects are warningly coloured, often with conspicuous yellow and red stripes'. In 1940 when Cott published his superlative book on adaptive coloration, there were still dissenting voices and non-believers, but today there are few, if any, who would seriously question the views which that author expressed on the basic aposematic defence mechanisms.

Firstly, we ourselves are animals that depend on our eyes for survival, and we react more positively to what we can see and assess visually than to any argument, however subtle and persuasive. Like birds we learn to respond rapidly to alerting colours. Secondly, in relatively recent times, new techniques have enabled us to identify minute quantities of toxic chemicals found in insects' haemolymph and defensive secretions. We can now work with a few specimens instead of a nightmare ton of material. Thus the purely theoretical association of bright colours and harmful or disagreeable qualities which was assumed by the older naturalists (Swynnerton, 1919; Marshall, 1902; 1909; Poulton, 1890; and others) has now been placed on a sound factual basis.

We may still speculate, however, on the origins of warning colour and self-advertisement, both in time and space. It is not unreasonable to suppose that they arose independently many times in the sea and on land, for carotenoids are the basis of colour vision both in water and in air. Nor need the appearance of self-advertising patterns have lagged far behind the evolution of the eye, for some of the most striking warning hues are still, to-day, due to carotenoids. They are to be seen, for example, in sponges, or starfish, or saturnid caterpillars, or the eggs of certain red mites, or sea-anemones, or goldfinches (Rothschild, 1975). In the case of insects their carotenoids are sequestered directly from plants and concentrated in the appropriate tissues where they are visible externally.

Were the huge Carboniferous dragonflies (Whalley, 1978) endowed with colour vision? One suspects they were. We know that some fossil beetles, for instance buprestids from the Lower Cretaceous, had strong patterns on the elytra – long before the flowering plants had put in an appearance on earth. Were they already metabolizing toxic or distasteful substances? Marine crustacea such as *Canadaspis*

possessed stalked eyes in the Middle Cambrian, far antedating the advent of fishes. The eye (not eye-spots) is possibly an arthropod invention, and it seems likely that, on land, colour vision was a novelty evolved by insects for insects before the evolution of birds. Millions of years later the avian eye (probably aided and abetted by that of reptiles) must have taken command, and imposed the basic colour patterns which now characterize almost all aposematic insect species and their mimics. The self-advertising life-style is, of course, not confined to the Insecta and must have evolved independently over and over again. It cuts right across the classification of both the animal and plant kingdoms and is found, for example, in frogs, lizards, snakes, fish, coelenterates, molluscs, plant-bugs, hornets, spiders, beetles, butterflies, lilies and legumes – all have recourse to the same type of alerting colours and patterns. We ourselves employ them for letter-boxes, road-crossings, sleeping-pills and so forth.

The general account of adaptive coloration, including aposematic attributes such as the types of pattern and the principal colours involved, scents, sound behaviour and learning by predators, has been so well described by Cott (1940) that I have deliberately avoided a repetition of the basic principles which he propounded.★ In the following account emphasis has been placed on more recent developments, particularly the chemical nature of the defence mechanisms (figure 1) of highly-coloured Lepidoptera. In this connection the invaluable work on the chemical defences of arthropods (Blum, 1981) which provides a catalogue of all important compounds with the relevant references, should be consulted together with the same author's chapter on arthropod alkaloids (Jones & Blum, 1983) in Pelletier's *Alkaloids*, volume 1.

Although a discussion of mimicry itself is outside the scope of this paper, it should be remembered that Batesian mimics are often warningly coloured – even if the warning conveyed is a false one – and therefore cannot be ignored.

The types of defence mechanisms and deterrents met with in the Lepidoptera – even if they can be fitted into a general framework or classification – are just as varied as the morphological details of the group. No two wings of even closely related species are exactly the same shape, nor are the secretions of two defensive exocrine glands precisely alike, either in taste or scent or chemistry.

II. Different Aposematic Strategies illustrated by British Lepidoptera†

A. *Butterflies*

It is generally agreed that although butterflies are eaten by birds, especially in the absence of alternative prey or when fledglings are still being fed in the nest, they are not a particularly popular article of diet with the medium-sized or small passerines characteristic of our insectivorous avifauna.‡ The few comparative experimental trials carried out with butterflies and wild birds confirm this impression. Thus Jones

★ The term 'aposematic' was first used as a synonym for 'warningly coloured' but has now been expanded to include warning scents, sounds and behaviour as well as colour and pattern.

† This is not a complete list of aposematic British Lepidoptera (for instance we have omitted the mother of pearl (*Pleuroptya ruralis* (Scopoli)) and the small magpie (*Eurrhypara hortulata* (Linnaeus)) (Pyralidae), but it covers the principal examples among the larger species.

‡ There is an amusing observation of Collenette & Talbot (1928) who saw a tyrant flycatcher (Tyrannidae sp.) in Brazil sitting among a group of butterflies on wet sand but catching and eating other insects.

(**a**) Calactin-Calotropin

(**b**) Calotropogenin

(**c**) Calotoxin

(**d**) Uzarigenin

(**e**) Integerrinine

(**f**) Senecionine

(**m**) Ingenol ester

(**g**) Seneciphylline

(**h**) Seneciphylline N-oxide

(**n**) Ingol ester

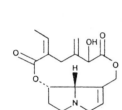

(**i**) Jacobine

(**j**) Jacoline

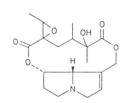

(**k**) Jacozine

(**l**) Jaconine

$$CH_2=CH-CH_2-C\overset{SGlc}{\underset{NOSO_3}{}} \xrightarrow{\text{myrosinase}} CH_2=CH-CH_2-N=C=S$$

(**o**) Sinigrin

(**p**) Allyl isothiocyanate

Figure 1 Defensive secondary plant compounds sequestered and stored by aposematic insects:
(**a–d**) cardiac glycosides from Asclepiadaceae
(**e–l**) pyrrolizidine alkaloids from Compositae
(**m, n**) diterpenes from Euphorbiaceae
(**o, p**) glucosinolates from Cruciferae

(1932) gave the butterflies he tested an average acceptability rating of 37.3, whereas hawk-moths received 93.3 and noctuids 78.3. They are obviously on the whole less popular than moths. It is evident that no insect could expose itself so blatantly as a typical adult nymphalid, say feeding on Michaelmas daisies, without some form of protection. The obvious deduction is that – even if they are not toxic – butterflies in general have a disagreeable taste, are too leathery in texture, and also prove difficult to catch and handle. Birds are loath to chase insects that involve a considerable expenditure of energy (Swynnerton, 1919; Jones, 1932). There are, however a number of species that are unquestionably aposematic.

1. The white butterflies (Pieridae)

The white butterflies or Pieridae form a world-wide Mullerian mimicry complex: eleven species are recorded from Britain, of which six are residents. These six butterflies are on the wing intermittently from April to October, and display several characteristics typical of aposematic insects.

(a) All are coloured white or yellow on the upper surface and white, yellow or mottled green on the undersides (the male orange-tip (*Anthocharis cardamines* (Linnaeus)), the resident species, has bright orange markings), and in Britain all, with the exception of the brimstone (*Gonepteryx rhamni* (Linnaeus)), have a slow, flapping flight.

(b) They display a true so-called acceptability or palatability spectrum (see also pp. 44, 47) which is contained within the Mullerian complex (Pocock, 1911; Marsh & Rothschild, 1974; Aplin *et al.*, 1975). Here we are not involved in shifting individual or group *variation* (p. 47) (due to such factors as different toxic secondary plant substances among the foodplants) but well-defined interspecific differences in repellent and toxic qualities.

(c) The foodplants of the Pierinae (whites) in Britain are various Cruciferae but they also feed on *Reseda* and *Tropaeolum* spp., all of which are characterized by the presence in their tissues of mustard oils which function for these butterflies as oviposition cues, larval feeding stimulants and avian deterrents if sequestered and stored. About 80 natural glucosinolates are known (Van Etten & Tookey, 1979).

The largest of the British whites, the large white (*Pieris brassicae* (Linnaeus)), is the most toxic of the pierid species so far investigated and it is known that:

(i) it sequesters and stores mustard oils derived from its foodplant (Aplin *et al.*, 1975);

(ii) it secretes a toxic substance, probably a protein, which is lethal to the laboratory mouse, rabbits, guinea pigs, etc. if injected (Marsh & Rothschild, 1974);

(iii) it lays brightly coloured yellow eggs in batches, which also contain mustard oils (Aplin *et al.*, 1975);

(iv) it has gregarious, conspicuously coloured larvae with a repellent odour, and pupae (cabbage-reared) with an abhorrent taste for at least one bird predator (Marsh & Rothschild, 1974: 110[*]) and probably for many others (Pocock, 1911);

[*] Reared on artificial diet pupae were eaten avidly by a tame magpie, but reared on cabbage they were rejected on taste alone.

(v) its larva, in the later instars, responds to imitation bird whistles by synchronized to-and-fro lashings of the anterior end of the body (pers. obs.; see also Minnich, 1925);

(vi) the green element in the coloration of its larva is not due to the usual physical mixture of carotenoids and bile pigments (the latter are only synthesized during the last instars) but to the contents of the gut seen from without (pers.obs.).

The majority of Mullerian complexes of this type contain a number of species which are more toxic than others (see (b) above) – the most poisonous species usually serving as a sort of model within the complex – since all are easily confused one with another in flight. This role is fulfilled by the large white in Britain – the other species being sheltered to some extent by its larger size and longer flight period, two attributes which are important for predator learning (Swynnerton, 1915; Bullini et al., 1969; Bullini & Sbordoni, 1970; 1971; Waldbauer & Sheldon, 1971; Rothschild, 1979; Sbordoni et al., 1979).

The pupa of the small white (*Pieris rapae* (Linnaeus)) but not the adult, also stores mustard oils (Aplin et al., 1975). Its larva also synthesizes a lethal protein passed on to the pupa and imago (Dempster, 1967) and no doubt other members of the complex, not yet examined, will be found to do the same.

The white butterflies are broadly speaking unpalatable to birds, the large white more so than the other related British species. There are, however, various widely disseminated records of these insects at all stages of their life-cycle serving as food for birds as well as spiders, wasps and mice. A tame fox, at liberty, has been seen catching the adult butterflies on the wing and eating them (pers.obs.). Poulton *in* Pocock (1911) considered that the British species with mottled green undersides were less unpalatable and more cryptic, when resting, than the species with all-white hindwings (Marshall, 1909; Pocock, 1911; Poulton *in* Pocock, 1911; Jones, 1932; Witherby et al., 1938; Cott, 1940; Lane, 1957a; Frazer & Rothschild, 1961; Baker, 1970; Feltwell & Rothschild, 1974; Marsh & Rothschild, 1974; Rothschild et al., 1977a).

It is not impossible that the sequestration of glucosinolates from the foodplants was originally based on their anti-bacterial properties and that the protection from bird predation was of secondary importance. Similarly the plants' co-evolution with bacteria and fungi probably preceded that with the Lepidoptera and may well have shaped these events. We previously drew attention to this possibility in the case of cardiac glycoside storage by danaids (Reichstein et al., 1968) for these heart poisons are also bacteriocidal. We also pointed out that aristolochic acids stored by the pharmacophagus swallowtails exert an effect on bacteria (von Euw et al., 1968; Urzua et al., 1983). This important aspect of the storage of toxic secondary plant substances has been somewhat neglected.

2. The danaids (Danaidae)

The milkweed or monarch butterfly (*Danaus plexippus* (Linnaeus)) is a rare, non-breeding vagrant in Britain. It provides the following different aspects of aposematic defences:

(a) Classical warning coloration and warning pattern combined, consisting of contrasting white circular spots on a black or very dark background in conjunction with bright, tawny areas of the wings. In this case the butterfly's underside as well as the upper surface is also unequivocally aposematic.

(b) A tough, thick, pliable integument which withstands mutilation – one of the basic features of the 'tenacity of life' which so greatly impressed the early naturalists.

(c) Longevity of the adult stage, a characteristic which favours 'group selection' (Blest, 1963; Turner, 1971 in relation to long-lived heliconids) or 'kin selection' (but see Grant, 1978) since it results in more than one generation being on the wing together – adult monarchs can live at least nine months.

(d) A sailing, leisurely flight; not flapping (Urquhart, 1960).

(e) Larval ability to sequester and store emetic cardiac glycosides and other toxic secondary plant substances such as pyrrolizidine alkaloids (Reichstein *et al.*, 1968; Rothschild & Edgar, 1978) from the foodplant. The cardenolides are present in the adult (Rothschild, 1967), eggs, larvae and pupae (Reichstein *et al.*, 1968), wings (Parsons, 1965), haemolymph, scales, integument and fat-body (Brower & Glazier, 1975; Blum, 1981: 452). The danaids do not respond to cardiac glycosides as oviposition or feeding cues. Their chemical ties to these toxic secondary plant substances appear not to be obligatory.

(f) Imaginal ability to sequester and store pyrrolizidine alkaloids, obtained from wilting vegetation or specific nectar, which are believed to enhance their chemical defences (Edgar *et al.*, 1974; Edgar, 1975; Edgar & Culvenor, 1975; Edgar *et al.*, 1976; Rothschild & Reichstein, 1976; Rothschild & Marsh, 1978).

(g) Selection, conversion, concentration and differential storage of the plant cardenolides. Such changes often occur during the pupal stages (Dixon *et al.*, 1978).

(h) The synthesis of substances which mimic the action of the cardiac glycosides on the mammalian heart (Rothschild *et al.*, 1978).

(i) The secretion or sequestration of pyrazines which produce the powerful warning scent characteristic of a large range of aposematic insects (Rothschild *et al.*, 1984) and some irritant plants (*e.g.*, coccinellids, pyrgomorphid grasshoppers, ants, tiger moths, nettles, poppies, etc.).

(j) Communal night roosting during migration from the north of America towards its massive winter hibernating congregations (Urquhart, 1960; 1976; pers. obs.).

(k) Aposematic larvae but cryptic pupae (Ackery & Vane-Wright, 1984).

(l) The larvae feeding directly on the latex of the plant if it oozes from a broken stem or cut leaf (see below, p. 29).

(m) Foodplants: Asclepiadaceae but only selected species. Many apocynaceous plants are rejected (Ackery & Vane-Wright, 1984).

The monarch and its relatives in Africa, *Danaus chrysippus* (Linnaeus) and *Amauris* spp., rank as some of the most distasteful butterflies known and which produce taste aversion (Brower *et al.*, 1967; Nicolaus, 1984), although there are various birds which have broken through their defences and consume them with relish [see references below]. The larvae and pupae are also usually rejected by birds. Wasps, however, attack and eat the larvae (pers.obs.) and mice eat the adults (Marshall, 1902; Eltringham, 1910; Swynnerton, 1915; Brower, 1958; Urquhart, 1960; Reichstein *et al.*, 1968; Fink & Brower, 1981). The degree of distastefulness, repellency and taste aversion produced, may be altered with the species of larval foodplant.

3. Enigmatical species of butterflies (Satyridae; Nymphalidae; Papilionidae)

It is not always possible to decide to which category of defence strategies a species subscribes. Frazer & Rothschild (unpublished observations) investigated the female of the cryptic meadow brown (*Maniola jurtina* (Linnaeus)), spurred on by the fact that a tame shama (*Copsychus* [as *Kittacincla*] *malabaricus*) would eat only the males! To their great surprise they found it contained appreciable quantities of a substance with histamine-like activity lacking in the male. Many butterflies, even Mullerian mimics, have cryptic undersides. No one, for instance, can deny that the dorsal surface of the red admiral (*Vanessa atalanta* (Linnaeus)) is warningly coloured. On catching sight of it for the first time my captive tit uttered a warning chatter (Lane & Rothschild, 1965). The red admiral also 'clicks' when disturbed, but there, apparently, the aposematic features lapse – unless its integument is brittle or strong enough to qualify. It appears to depend mainly on speed and strength to elude birds.

The adult peacock butterfly (*Inachis io* (Linnaeus)) has eye-spots on a grand scale, and an unquestionably aposematic larva – coal black with branched spines and the gregarious habit. Extracts of this larva have no effect on the laboratory mouse if injected via the intraperitoneal route (Marsh & Rothschild, 1974). Both it and the red admiral, as well as three other British nymphalids, feed on nettle (*Urtica dioica*), which provides a well-protected foodplant niche (Rothschild, 1964a). The nettle synthesizes histamine, ACh and 5HT, associated with its irritant hairs (Emmelin & Feltberg, 1947a,b; Chesher & Collier, 1955). In Australia a related species of the genus *Laportea* also contains highly toxic proteinaceous material (Everist, 1981). Is it possible that a similar or related substance, though less dangerous, occurs in our nettle, but has hitherto been overlooked? The nettle-feeders may obtain a chemical deterrent from this plant, not only physical protection. The Camberwell beauty (*Nymphalis antiopa* (Linnaeus)) can also be included here. Poulton *in* Pocock (1911) considered the small tortoiseshell (*Aglais urticae* (Linnaeus)) and the peacock to be somewhat unpalatable. He and Pocock (1911) concluded from a few feeding experiments and from the appearance and life-style of the marsh fritillary (*Eurodryas* [as *Euphydras*] *aurinia* (Rottemburg)) that it was an aposematic and distasteful species. This may well be so, for its foodplant, the devil's-bit scabious (*Succisa pratensis*), although not toxic, is rich in secondary plant substances. The heath fritillary (*Mellicta athalia* (Rottemburg)) feeds occasionally on foxglove (*Digitalis purpurea*) which is also suggestive. Bowers (1980) has shown that a related North American fritillary, *Euphydryas phaeton* (Drury), is distasteful and causes vomiting in predators if the larva feeds up on its favoured foodplant, *Chelone glabra* (Scrophulariaceae). One may assume it stores and sequesters the secondary chemical concerned, a fact which was not determined by Bowers.

If the fauna of this country is considered without reference to the Continent, the marbled white (*Melanargia galathea* (Linnaeus)) seems a particularly isolated butterfly, but in Europe it forms part of a goodly crowd of black and white satyrids, confusing to most entomologists, let alone bird predators. They must form a Mullerian complex which dovetails with the whites. Wilson has shown that the marbled white sequesters and stores flavonoids from its foodplants, but it may also harbour some other toxic substance as yet undetected, since flavonoids are also sequestered in much larger amounts by the blues (Feltwell & Valadon, 1970; Wilson, 1983) which are not toxic or aposematic. Pocock (1911) has demonstrated that the marbled white is fairly unpalatable to birds. Lane (1957a) rated it 'uneatable' for his tame shama.

The adult swallowtail butterfly (*Papilio machaon* (Linnaeus)), though not toxic, is somewhat distasteful (Wicklund, in prep.) but it relies on speed, toughness, size and an *aide mémoire* type of coloration. It is too large for most of our birds to tackle. The caterpillar, however, is frankly aposematic, provided with a fleshy, orange-coloured, eversible glandular spray or osmaterium (Eisner & Meinwald, 1965), which ejects a stream of isobutyric acid and α-methylbutyric acid at an aggressor. The colour of this organ is due to plant-sequestered and stored carotenoids. On a carotenoid-free diet, *Papilio* caterpillars are blue, not green, but the osmaterium is then white, pterobilins being excluded from this organ.

B. *Moths*

There is a much greater variety of aposematic species among the British moths than among the butterflies. One characteristic is paramount: the most toxic species are diurnal. Even among the tigers, those which do not feed as adults and are nocturnal, such as the garden tiger (*Arctia caja* (Linnaeus)) (figure 2), are more readily eaten by vertebrate predators than the scarlet tiger (*Callimorpha dominula* (Linnaeus)) or the

Figure 2 Aposematic display of the garden tiger (*Arctia caja* (Linnaeus)): note the cervical glands beneath the raised patagia

burnets, both of which imbibe nectar from flowers by day. This stark, aposematic, diurnal life-style may well have been forced on to some of the tigers by the sequestration of toxic secondary plant substances. The burnets and foresters, however, must have come into the sunlight by another route, for apart from pyrazines which are probably sequestered from legumes (for the pyrazines in pea-pods and zygaenids smell remarkably alike), they do not acquire their toxins, but synthesize them. Many aposematic moths which fly at dusk, or in the night, are conspicuous because they are white, not brightly coloured.

Collins & Watson (1983) noted that in Venezuela the aposematic geometers are, on the whole, more distasteful to bird predators than moths of other families, although warningly coloured geometers are not so numerous as warningly coloured noctuids. It would seem that once a species has become sufficiently poisonous to be snatched out of the strait-jacket of the cryptic life-style, it evolves extreme specializations in toxic defence mechanisms, and strikingly vivid patterns. In the British fauna there are several species such as the July highflyer (*Hydriomena furcata* (Thunberg)) and the spinach (*Eulithis* [as *Lygris*] *mellinata* (Fabricius)) which, although unacceptable to birds and bats, being distasteful if not toxic, still retain the cryptic life-style (Rothschild, 1967). In order to brave the daylight and withstand the full impact of diurnal bird predation, a small insect must be highly toxic. There is a dearth of Batesian mimics in the British butterfly fauna. This is hardly surprising as we lack the number of species – alternative prey – without which this type of mimicry simply cannot work. Even in Africa in the dry season, when insects are relatively scarce, mimics are scarce too (Marshall, 1902).

1. The burnets and foresters (Zygaenidae)

The burnets (Zygaeninae) and foresters (Procridinae) (ten British species) are either brilliant red and burnished blue-black, or burnished green moths, almost iridescent, displaying various characteristics associated with the aposematic life-style, some of which are not found in the pierids or danaids.

(a) They live in colonial aggregations at all stages of the life-cycle and, as adults, participate in a Mullerian mimicry 'ring'. This includes more than one species of burnet and other red and black insects such as ladybirds, plant-bugs and tiger beetles, feeding on flower heads and plants in the same habitat.

(b) Cyanide is released from crushed tissues at all stages of the life-cycle including the eggs (Jones *et al.*, 1962). Both as larva and adult they synthesize the cyanoglucosides which are the source of their HCN (Davis & Nahrstedt, 1979; Nahrstedt & Davis, 1983). This is probably a good taxonomic character (Jordan, 1907-08) unlike many chemical self-secretions of Lepidoptera, which can have little relevance to accepted classifications.

(c) Presumably because they are relatively insensitive to HCN, some species of *Zygaena* (but not *Adscita* [as *Procris*]) feed as larvae on cyanogenic plants such as common bird's-foot trefoil (*Lotus corniculatus*) (Rothschild *et al.*, 1970) and are thus provided with a protected niche.*

---

* Our lycaenids which feed on cyanogenic plants are not storers, and detoxify HCN by means of the enzyme rhodanese (Parsons & Rothschild, 1964). The known foodplants of *Adscita* in Britain are not cyanogenic; this genus, therefore, synthesizes its own HCN.

Figure 3 (a) (left) Colourless pungent fluid oozing out round the mouth parts of the six-spot burnet (*Zygaena filipendulae* (Linnaeus)) when the moth is molested. The fluid smells strongly of pyrazines but lacks HCN. **(b)** (right) Bright yellow haemolymph oozing from the 5th tarsal segment when the leg of the six-spot burnet was grasped with forceps. This fluid smells strongly of HCN

 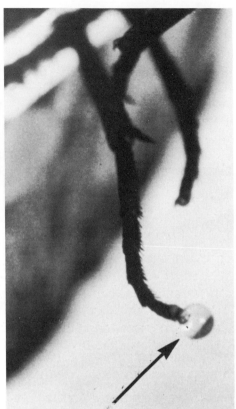

(d) The female secretes an unusually toxic protein (zygènine) in the eggs and ovaries (Rocci, 1916; Lane, 1959; Marsh & Rothschild, 1974).

(e) The cuticle is exceptionally soft, elastic and pliable and the insect can recuperate following massive injuries inflicted by a predator or with forceps (pers.obs.).

(f) Defensive bleeding spots include the thoracic intersegmental membranes and the tarsi (figure 3b) (Rothschild, 1961a).

(g) Pungent transparent watery, colourless fluid which contains pyrazines flows out round the base of the mouth parts of the adult moth (figure 3a).

(h) Both sexes also secrete histamine and acetylcholine (Morley & Schachter, 1963; Rothschild, 1972a).

(i) Although the five-spot burnet (*Zygaena trifolii* (Esper)) is the most toxic species of Lepidoptera so far found in Britain (Marsh & Rothschild, 1974), none of its toxins or defensive substances (apart, possibly, from pyrazines) appear to be plant-derived.

(j) The caterpillar is yellow with black markings.

It is interesting that these moths and various plants synthesize similar defensive substances. The insect with a gut full of a toxic foliage, in a sense, becomes an extension of the plant (Blum, 1981).

The zygaenids are recognized as a group of moths rejected by birds and various other predators (Poulton *in* Pocock, 1911; Pocock, 1911; Rocci, 1914, 1916; Lane, 1959; Frazer & Rothschild, 1961; Bullini *et al.*, 1969; Bullini & Sbordoni, 1971; Marsh & Rothschild, 1974).

Apart from the amines, pyrazines, HCN and toxic proteins found in these moths, Ford (1941) reported the presence of flavonoids in the subfamily Chalcosiinae (also said to be distasteful by Poulton (*in* Pocock, 1911)). It will be extremely interesting to see if these pigments are also found in the Zygaeninae.

2. The tiger moths and footmen (Arctiidae)

These comprise an impressive group of aposematic Lepidoptera. In recent classification (Hodges *et al.*, 1983) this includes the subfamilies Lithosiinae, Arctiinae, Pericopinae and Ctenuchinae which contain 32 British species. They form part of a world-wide Mullerian mimicry complex.

The Compositae are the favourite foodplants for the majority of species (Rothschild *et al.*, 1979b), but many are widely polyphagous (Table I); and the garden tiger *Arctia caja* (Linnaeus) has been dubbed a poison plant specialist. Unlike the other subfamilies, the Arctiinae are largely nocturnal and pair after dark. Many do not feed as adults and their bright colours are mainly directed at birds which disturb them at rest, hence there is usually a contrast between fore- and hindwing colour patterns. White (in our species) is used extensively with red and yellow. Colour schemes are more contrasting, garish and crude than in most day-flying butterflies. This family illustrates certain aposematic attributes more strikingly than the previous examples:

(a) 'Stockpiling' of deterrents, especially varied amines.

Table I: Larval Foodplants of 282 Species* of Arctiidae (from varied sources in the literature)

Plant family*	Total number of species†	Old World (Africa, India, Europe)	New World (North, South & Central America)	Australia
Compositae	82	47	34	3
Leguminosae Papilionaceae	42	21	18	3
Gramineae	34	21	12	
Plantaginaceae	26	13	12	1
Rosaceae	25	12	13	1
Asclepiadaceae Apocynaceae	23	7	15	1
Solanaceae	18	7	11	1
Salicaceae	15	8	8	
Boraginaceae	14	11	2	2
Polygonaceae	13	9	4	
Euphorbiaceae	12	5	6	1
Convolvulaceae	11	8	2	
Moraceae	11	7	4	2
Ulmaceae	11	3	8	

* There are 1–10 records from 96 other plant families not included here.
† Some species are found in both Old World and New World.

Figure 4 Oscillograph traces made from a sound recording of the defensive 'rattle' of the garden tiger (*Arctia caja* (Linnaeus)), audible to the human ear. This is accompanied by a wing-flapping display, when the resting moth is disturbed

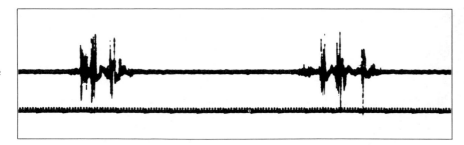

(b) Warning sound (figure 4) and scent production (Rothschild, 1961b; Treat *in* Busnel, 1963; Blest, 1964; Rothschild & Haskell, 1966) including sounds to alert hunting bats of their toxic attributes (Roeder, 1967).

(c) Production of defensive froths – sometimes accompanied by hissing sounds – mixed with haemolymph (figure 5).

(d) Synthesis of toxic proteins of both high and low molecular weight* principally in the female but probably present in small amounts in the males (Rothschild *et al.*, 1979c; Hsaio *et al.*, 1980).

(e) The fact that some arctiids such as *Diacrisia* spp. contain enlarged glands in the tarsi which may secrete a toxic substance. (The tarsi often detach if the insect is caught in a net.) This gland is quite prominent in the garden tiger (see figures 6a,b,c).

(f) Sequestration by the larvae and storage by pupae and adults of a variety of secondary plant substances (more than one of them by the same species) principally pyrrolizidine alkaloids but also cardenolides (Rothschild, 1972a; Rothschild *et al.*, 1977b).

* The term cajin probably covers both.

Figure 5 Blood cells in the secretion from the cervical glands of the garden tiger (*Arctia caja* (Linnaeus))

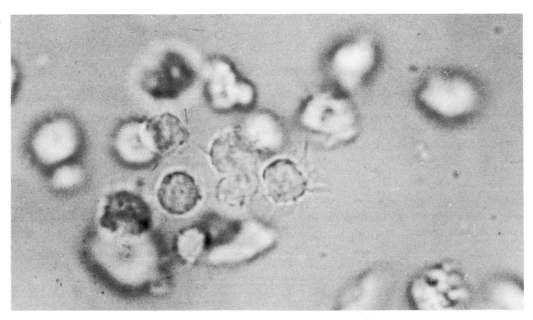

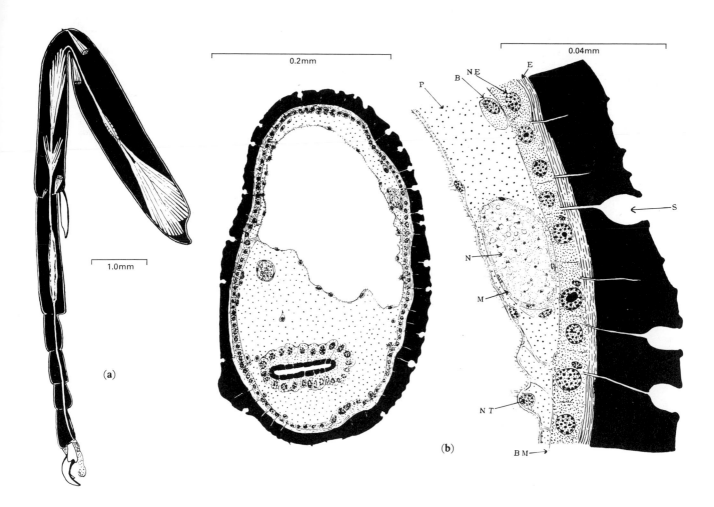

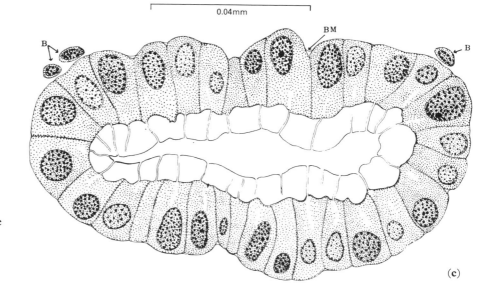

KEY.
B	haemocyte
BM	basement membrane
E	endocuticle
M	myelin sheath
N	nerve
NE	nucleus of epidermal cell
NT	nodule of trachea
S	duct of dermal gland

Figure 6 Dissection drawings of the garden tiger (*Arctia caja* (Linnaeus))
(**a**) Dissection of the hind leg to show the position of the tarsal gland
(**b**) Transverse section through the 1st tarsal segment showing position of nerve and suspension of the trachea
(**c**) Section through the tarsal gland

(g) Static displays (Rothschild, 1961b; Blest, 1964;) (figure 2, p. 16) which reveal the bright colours of underwings, etc., apertures of defensive glands and faked bleeding spots: in the case of *A. caja*, a specific display which includes stridulation audible to the human ear (Rothschild & Haskell, 1966).

(h) Moderate resistance to HCN in some species, for example the crimson speckled (*Utetheisa pulchella* (Linnaeus)) (Rothschild, 1961a).

(i) Occasionally vast congregations of aestivating adults (figures 7a,b) during the summer from May to October, for example the Jersey tiger (*Euplagia quadripunctaria* (Poda)) in the Valley of the Butterflies, Rhodes (Walker, 1966).

(j) Hirsute larvae with long histamine-charged hairs (Frazer, 1965) which protect them from attack by wasps (pers.obs.); larvae of the scarlet tiger and Jersey tiger yellow with black markings; larvae of the cinnabar (*Tyria jacobaeae* (Linnaeus)) with yellow and black alternating transverse stripes, gregarious habit and highly repellent elements in the integument (Windecker, 1939). The latter species demonstrates a dual signal strategy: cryptic at a distance feeding among ragwort flowers, but aposematic at close quarters (Wood, 1869; Poulton, 1908; Rothschild, 1975). It lacks pterobilins (Rothschild & Bois-Choussy, in prep.) like several other aposematic larvae. The eggs may contain pyrrolizidine alkaloids (Rothschild, 1972a; van Zoelen, pers.comm.) and high concentrations of amines.

The beautiful footman (*Utetheisa bella* (Linnaeus))[*] which has recently been re-split into two closely related species, *U. bella* and *U. ornatrix* (Linnaeus) is a very rare vagrant in Britain. I myself have accidentally released this species near Oxford. It is an extremely beautiful moth, remarkable for its pink hindwings and the relatively huge bubbles of yellow froth oozed out of the cervical glands if the insect is seized or disturbed, thus adding to the aposematic effect (Pl.B, fig.3). This froth contains the same amine as that of the garden tiger, $\beta\beta$-dimethylacrylylcholine and 77 µg/g of carotenoids (Mummery & Rothschild, unpublished) and almost certainly pyrrolizidine alkaloids as well (Table II, p. 24). It is worthy of special note because Connor *et al.* (1981) have shown by experiments with various predators, such as spiders and birds, that the pyrrolizidine alkaloids sequestered from the foodplant and stored (Rothschild & Aplin, 1971; Rothschild, 1972a) act as defensive and protective substances – long suspected (Rothschild & Reichstein, 1976) but never proved. They also function as precursors of the male sex pheromone which are distributed via coremata (Pliske & Eisner, 1969). It should be noted that not all arctiids which store pyrrolizidine alkaloids have coremata. Thus, for example, the white ermine (*Spilosoma lubricipeda* (Linnaeus)) has these organs very well developed (figure 8), but they are altogether lacking in the buff ermine (*S. luteum* (Hufnagel)) (Rothschild *et al.*, 1979b; Birch, 1979). In its native haunts the beautiful footman consorts with a magnificent assembly of aposematic insects, all associated with *Crotalaria*, but its colour is too unusual for it to participate in a Mullerian ring.

A somewhat less rare migrant visitor to Britain is the crimson speckled which feeds on borage (*Borago officinalis*) and forget-me-not (*Myosotis* spp.). Unfortunately we have never had enough specimens to test, but one may assume it is a storer of pyrrolizidine alkaloids.

[*] The English name is very misleading as this moth is not a lithosiine.

Figure 7 (**a**) (above) Aestivating congregation of adult Jersey tigers (*Euplagia quadripunctaria* (Poda)) in the Valley of the Butterflies, Rhodes. One or two specimens rest with wings open, displaying their red hindwings. The monarch butterflies in their winter congregations do likewise

(**b**) (below) Close-up of aestivating Jersey tigers

Table II: Choline Esters in *Arctia caja* (Linnaeus)
(Adapted from Bissett *et al.*, 1960; Morley & Schachter, 1963; Frazer, 1965)

Organ investigated	Acetylcholine μg/g dry weight
Cervical defensive glands (imago ♂ ♀)	3000–6000 μg/g dry weight (ββ-dimethylacrylylcholine)
Abdomens of imagines (♂ ♀)	200–400 μg/g (freeze dried)
Ejaculatory ducts (imago ♂)	2500–5500 μg/g (freeze dried)
Accessory sex glands (imago ♂)	500–1600 μg/g (freeze dried)
Testis and seminal vesicle (imago ♂)	Not detectable (< 5 μg/g)
Malpighian tubules, trachae and fat body (imago ♂)	Not detectable (< 5 μg/g)
Ovaries (imago ♀)	5–250 μg/g (freeze dried)
Eggs	5–400 μg/g (freeze dried) / 330 μg/g (extracted in hydrochloric acid, pH 3–4, heated to 98°C) / 300 μg/g (extracted in 95% ethyl alcohol at room temperature)
Bursa copulatrix (imago ♀)	4 specimens < 5 μg/g
Silk glands of larvae (hypertrophied: 20 specimens pooled)	3000–4500 μg/g
Cocoon silk	300 μg/g
Isolated larval hairs	Acetylcholine-like activity

The nine-spotted (*Syntomis phegea* (Linnaeus)) (Ctenuchinae) deserves special attention. This moth is sometimes shipped or flown to Britain in crates of fruit or flowers. It is not a resident, nor does it arrive 'under its own steam'. The subfamily is involved in a world-wide mimicry complex with clearwing moths (Sesiidae). Both are found together sitting on flower-heads in the sunshine (Rothschild, 1979). Bullini has made *S. phegea* into a famous species, for he has shown that it is responsible for 'capturing' a normally coloured red and blue-black burnet that coincides in part of its distribution range with *S. phegea* which, abandoning its classical coloration, mimics this yellow and black ctenuchid (Sbordoni *et al.*, 1979). The chemistry of this moth has not been investigated although it smells strongly of pyrazines (pers.obs.) but Moore (unpublished) has identified these substances in a related Australian species. However, by injection into mice and by feeding experiments, it has been shown to be mildly toxic (Rothschild *et al.*, 1972; Marsh & Rothschild, 1974). This moth demonstrates an interesting principle connected with the evolution of mimicry. It can become the model in a pair of ecologically associating species even if it is only mildly toxic, providing its flight period is longer and its numerical superiority is maintained. The males of *S. phegea* are more toxic than the females, which is the exact reverse of the situation found in the burnets. There is, relatively, so little HCN in the males of *Zygaena* species that some authors (Davis & Nahrstedt, 1979) have failed to find it, but they now agree it is present (pers.comm.). Generally, females are the more toxic sex. This is evident, for example, in large whites, the garden tiger, the yellow-tail (*Euproctis similis* (Fuessly)) (Lymantriidae) and white ermine. Male danaids, however, may store more pyrrolizidine alkaloids than females [see below p. 40].

Syntomis phegea is unusually easy to rear and it would be delightful if this moth could be introduced into Britain.

1

2

3

1. Eyed hawk-moth (*Smerinthus ocellata* (Linnaeus)) displaying. Note the raised end of the abdomen which thus resembles a snout, and the 'corners' of the eye-spots where they join the body ($\times$ 2¾). **2.** Cinnabar (*Tyria jacobaeae* (Linnaeus) var. *coneyi* Watson), the only British species with red fore- and hindwings, bred at Ashton by the author from stock obtained from the late Bernard Kettlewell. This dominant variety contains only traces of callimorphine ($\times$ 1½). **3.** The beautiful footman (*Utetheisa bella* (Linnaeus)). Secretion of the cervical glands, which contain $\beta\beta$-dimethylacrylylcholine as well as pyrazines, *Crotalaria* alkaloids and carotenoids, bubbling out when the moth is seized with forceps ($\times$ 10). Photos Miriam Rothschild

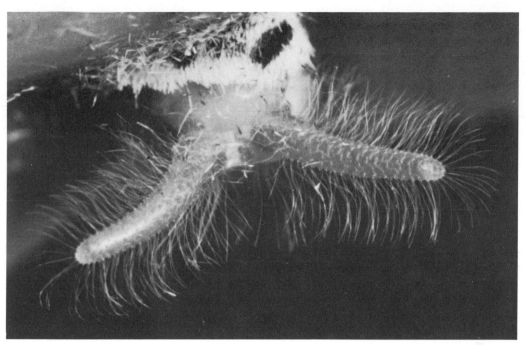

Figure 8 The coremata of the white ermine (*Spilosoma lubricipeda* (Linnaeus)), greatly magnified, everted and inflated under pressure

The great majority of arctiids are rejected by birds: the most repellent species in Britain so far tested are the cinnabar (Mostler, 1935; Windecker, 1939; Frazer & Rothschild, 1961) and the scarlet tiger. The principal deterrent in both these moths is unknown. Neither the histamine nor pyrrolizidine alkaloids present in the cinnabar can account for their unusually repellent qualities. Like the garden tiger and the white ermine, despite the sequestered and stored pyrrolizidine alkaloids, it goes through a 'latent' period during the pupal stage when it is eaten by predators (Windecker, 1939; pers.obs.). It seems likely that the adult cinnabar contains in the haemolymph a toxic protein which, like cajin, develops during the last days before metamorphosis.

Sometimes 'stings' have been experienced by those handling the garden tiger and the ruby tiger (*Phragmatobia fuliginosa* (Linnaeus)) but the mechanism is not understood (Rothschild *et al.*, 1979b).

3. The hawk-moths (Sphingidae)

This fascinating family of moths, of which nine species are British residents and an additional eight migrant visitors, relies chiefly on speed, aerobatics, flight and crypsis for defence, but a few larvae which sequester toxic secondary plant substances such as that of the spurge hawk-moth (*Hyles euphorbiae* (Linnaeus)) (Marsh *et al.*, 1984) are highly aposematic. This caterpillar is black in the first and second instars and then acquires a bright red head and appendages, with rows of brilliantly eye-catching, round, yellow spots along the sides. The horn is red.

The few species which fall into the aposematic or near aposematic category are so specialized that they must be described in greater detail. It is not possible to list general attributes as I have done for the tiger moths and danaids.

Furthermore, various sphingid species as adults adopt warning devices such as eye-spots, or the colour, appearance and behaviour of Hymenoptera, whilst as larvae they resemble snakes. Often they are not mimics in the usual sense, but demonstrate the *aide mémoire* type of mimicry (Rothschild, 1984) in which some facet of their general appearance reminds the potential predator of an unpleasant experience or warns him of impending danger. The adult eyed hawk-moth (*Smerinthus ocellata* (Linnaeus)) is an excellent example of this strategy. The forewings are cryptic, but if disturbed the moth displays its pink hindwings adorned with large realistic eye-spots. The abdomen is curled upwards and the eye-spots with dark 'corners' where the wings meet the body complete the illusion of a mammalian head, with a snout and huge glaring eyes (Pl.B, fig.1, facing p. 25). Warning colour and behaviour are here perfectly synchronized.

Aposematic insects which are themselves highly toxic like the arctiids, have no need to mimic the Hymenoptera or produce frightening eye-spots: they use garish colours to advertise their own adequate chemical defences, not the dangerous attributes of other species.

As I have indicated in the introduction, the self-advertising colours of mimics must be considered here, even if the warning they convey is misleading or deceptive.

Among the British hawk-moths there are two species, narrow-bordered bee hawk-moth (*Hemaris tityus* (Linnaeus)) and broad-bordered bee hawk-moth (*H. fuciformis* (Linnaeus)), which as adults bear an amazing resemblance to bumble-bees, having transparent wings and yellowish furry bodies. One feeding experiment was carried out by Jones (1934) with a related and somewhat similar North American species of bee hawk-moth, the common clearwing (*H. thysbe* (Fabricius)). The insect remained on the feeding tray for three and a half hours, while birds cleared 37 smaller Lepidoptera, and though mauled by a visiting bird (or birds) was rejected. This suggests the bee hawks may be Mullerian rather than Batesian-type mimics. However, one female *H. tityus*, when extracted and injected into the laboratory mouse, produced no ill effects (Marsh & Rothschild, 1974). The foodplants are not toxic (Rothschild & Jordan, 1903; South, 1961).

One of the most interesting and unusual examples of the warning life-style, a form of so-called aggressive mimicry, is illustrated by the death's-head hawk-moth (*Acherontia atropos* (Linnaeus)). This species may well be, like the tobacco hornworm (*Manduca sexta* (Johansson)) (Rothschild *et al.*, 1979a), an opportunistic sequesterer and storer of toxins found in the Solanaceae. In the Canary Isles and other parts of Africa it feeds frequently on *Datura* spp. which contain hyoscine, atropine and hyoscyamine (Henry, 1949). Its habits are unusual, for it frequently enters hives, by day as well as by night, to rob the bees of honey (figure 9a). It has developed two different lines of defence whereby it attempts to ward off or inhibit attacks by the occupants of the hives. The first is the production of a squeaking, rasping sound via the proboscis (Prell, 1920; Haskell, 1961; Busnel, 1963), said to resemble the 'piping' of the queen bee (see figures 10a–c), while the second is the pale, bee-like 'face' developed on the black or dark brown background of the thorax (figures 9b,c*

---

* Entomologists are so anthropomorphic in their approach; they see a human skull on the moth's thorax, rather than a bee's face!

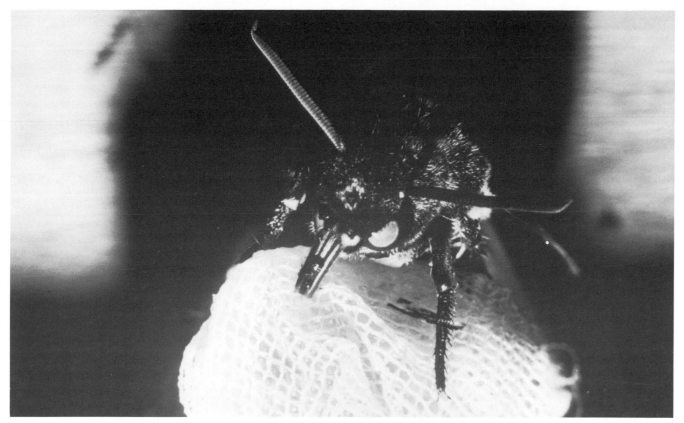

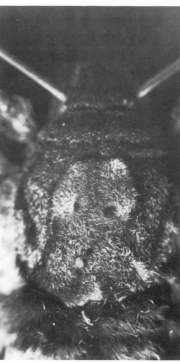

Figure 9 (**a**) (above) Death's-head hawk-moth (*Acherontia atropos* (Linnaeus)) stealing honey. The short, stout proboscis is probably adapted for this purpose as well as for sound production (**b**) (far left) The bee-like 'face' on the thorax was photographed in daylight while the moth was alive (**c**) (centre left) The bee-like 'face' illustrated here was photographed from a preserved specimen under U.V. light

and *MBGBI* **9**: Pl.2, fig.3). In the obscurity of the hive, and seen from above by the guard bees, this combination must be quite impressive, for the 'face' set immediately above the brown and yellow striped segmented abdomen topped by the bee's own antennae could give the impression of a huge gravid queen bee: a super model, if ever there was one.

In Britain the spurge hawk-moth is the outstanding example of an aposematic larva although larvae of its close relations, the bedstraw hawk-moth (*Hyles gallii* (Rottemburg)) and the striped hawk-moth (*H. lineata* (Fabricius)), also veer towards warning coloration. The spurge hawk-moth larva is most arresting when its ground colour is black. This form lacks bile pigments (pterobilins) (Rothschild & Bois-Choussy, in prep.). The green form is less aposematic and tones in well with the plant background, the warning nature of the coloration becoming obvious only on closer scrutiny. Unlike the black form it synthesizes blue bile pigments which are associated in the tissues with carotenoids, and mix in the eye of the beholder to produce the classical green larval ground coloration. This hawk-moth can 'ring the changes' and both larva and adult are sometimes cryptic and sometimes warningly coloured, according to circumstances. At rest the moth can be well concealed, for instance among the flowers of pink valerian on which it feeds, but if disturbed it produces a

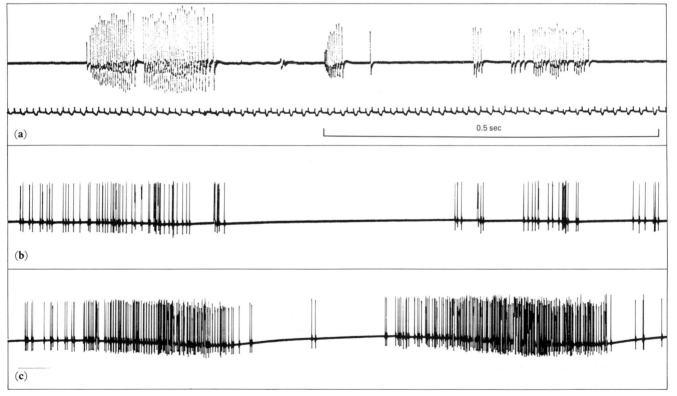

Figure 10 (**a**) (above) Oscillograph traces made from sound recording of 'squeaking' sound made by a death's-head hawk-moth (*Acherontia atropos* (Linnaeus)) when molested
(**b**) and (**c**) (centre and below) Oscillograph traces made from sound recordings of a queen honey-bee (*Apis mellifera* Linnaeus). The effectiveness of sound mimicry would depend on the relative importance of fundamental frequency, pulse length and pulse repetition frequency. These parameters are all variable in both insects

buzzing and stinging display reminiscent of any large Hymenoptera, which must prove terrifying to a small predator (Rothschild, 1984). So far the secondary plant substances which have been sequestered by the spurge hawk-moth and identified are diterpenes – ingols and ingenols (Marsh *et al.*, 1984) – but only relatively low concentrations are found in the adult moth. The caterpillars eject copious green fluid from the mouth if molested, and there is also a voluminous secretion of bright green fluid from the lateral glands. So far nothing is known about the reaction of predators to this species in nature. It is striking that both this moth and the monarch butterfly (*Danaus plexippus*) feed on plants with a milky latex. Danaids occasionally use species of Moraceae as alternative foodplants, several genera of which contain cardiac glycosides. These plants also have a milky latex. It seems likely that this emulsified fluid protects the plant from the effects of its own toxins. Perhaps the larvae can use it for a similar purpose.

There are two other hawk-moth larvae that deserve mention. The first is that of the oleander hawk-moth (*Daphnis nerii* (Linnaeus)) which is not a storer of cardenolides (Rothschild, 1972a). It is the most spectacular, cryptic species on the British list. In the early instars it is green and mimics a young twig of the oleander bush. In the third and fourth instars it develops white spots along the lateral sides of its body which resemble the cast skins of *Aphis nerii* (Fonscolombe) (Hemiptera Homoptera) which abound in the same region of the plant which it then frequents. Its yellow horn is truncated and both in colour and shape is reminiscent of an adult *Aphis nerii*. Not infrequently these aphids crawl on to the caterpillar which restricts them to the rear end of its body, round the horn, by snapping at them like a dog snapping at flies! In the fifth instar the caterpillar turns dull pink and is then found among the pink flowers of the oleander bush, hiding between the petals. When it prepares to pupate, it turns brown and descends to the soil below the bush. At this stage the concealed eye-spots (see below) are replaced by darker brown, square marks on the side of its head, which are the shape and colour of its own frass lying on the ground beneath the bush on which it has been feeding.

Its right to figure among warningly coloured larvae is due to the enormous eye-spots which develop during the last instars and are placed laterally on the anterior end of the body, usually concealed, but which can be flashed into view when the caterpillar ducks its head. At this moment it must present a terrifying aspect – 'glaring' with bright blue and black eyes from between the petals of the pink inflorescence.

The larva of the elephant hawk-moth (*Deilephila elpenor* (Linnaeus)), although it is coloured brown or green, can, with greater justification, be described as aposematic – not merely a case of flash coloration. It has eye markings on the side of its head and by withdrawing (or perhaps even inflating) this end of the body and its trunk-like snout the eye-spots are stretched and distended (figure 11b) – the whole giving a very fair impression of a small bright-eyed snake.

In other countries there are various hawk-moth larvae which have become snake mimics. Professor Lincoln Brower once showed me a photograph of one of these, the gaudy sphinx (*Eumorpha labruscae* (Linnaeus)), and I noticed it possessed an incredible refinement – the outline of a forked tongue on the front of its head (figure 11a).

Figure 11 (**a**) Caterpillar of the gaudy sphinx (*Eumorpha labruscae* (Linnaeus)), a hawk-moth from Central America: the acme of a snake mimic. Note the pseudo forked tongue
(**b**) (inset) Caterpillar of elephant hawk-moth (*Deilephila elpenor* (Linnaeus)) mimicking a small snake. Note the inflated anterior end

Some adult African hawk-moths are armed with formidable femoral spurs which can inflict wounds if the moth is handled. One of these, the widow sphinx (*Acanthosphinx guessfeldtii* (Dewitz)) smells strongly of pyrazines, while *Lophostethus demolini* (Angas) produces a powerful odour of burnt chocolate. So far, I have not detected these warning odours in any of our native species, although there is one report of the death's-head hawk-moth emitting a powerful scent while squeaking (Swinton *in* Birch, *MBGBI* **9**: 9). This species lacks long spurs.

The hawk-moths, except in the case of the few aposematic larvae and the *Hemaris* spp. mentioned above, are generally highly palatable and much sought after by birds and bats. As we have pointed out their only defences are crypsis and strong flight. Various authors (Jones, 1932; Lane, 1964; Carcasson, pers.comm.) and many scattered references designate them as the most favoured of all lepidopterous prey.

4. The clearwings (Sesiidae)

The Hymenoptera attract more mimics than any other group of insects, and there can be no doubt that the most striking examples of this life-style in Britain are the two hornet moths (hornet clearwings) (*Sesia apiformis* (Clerck) and *S. bembeciformis* (Hübner)) which are astonishing mimics of the larger wasps, with their bright yellow and brown bodies and transparent wings; and, perhaps most impressive of all, their hornet-like attitudes when at rest (see also Pl. A). It has been suggested that the larger hornet moth (*S. apiformis*) mimics the sound produced by the hornet in flight (figures 12a,b).

Unfortunately nothing is known about their chemistry and we cannot even say whether some or all of the clearwings, of which there are 15 species in Britain, are Mullerian or Batesian mimics. On the one hand the hornet moth (*Sesia apiformis* (Clerck)) and lunar hornet moth (*S. bembeciformis* (Hübner)) are apparently too rare vis-à-vis the common wasps to pertain to the former, but their behaviour is so sluggish I doubt whether any naïve bird which tries them would find them palatable.★

5. The red underwings and Clifden nonpareil (*Catocala* (Noctuidae: Catocalinae))

The five British species in this genus are beautiful moths and large for our fauna, have highly cryptic forewings with markings, resembling the bark of trees, but bright crimson or violet areas, outlined in black, on the hindwings. These species provide a different example of a startle device from the 'eyes' of the oleander hawk-moth caterpillar. Despite their scarlet markings, they are not aposematic in the conventional sense. If the moth is disturbed at rest its brilliantly coloured wings suddenly flash into view and the startled predator momentarily loses track of the insect. Meanwhile it re-settles abruptly, instantly concealing the crimson hindwings beneath the cryptic forewings, and is lost to sight.

★ Thanks to David Lees' remarkable success in breeding hornet clearwings (*Sesia apiformis*) we were able to feed them to two naïve hand-reared magpies. In our opinion, for this species, the moth functions as a Mullerian rather than a Batesian mimic, since it proved distasteful to both birds. It elicited beak-wiping and total rejection in the case of one of the magpies and the rejection of parts of the body by the second bird. Both magpies had accepted and eaten the hornet (*Vespa crabro* Linnaeus) with sting removed.

We were also able to confirm the distastefulness of the broad-bordered bee hawk-moth (*Hemaris fuciformis*).

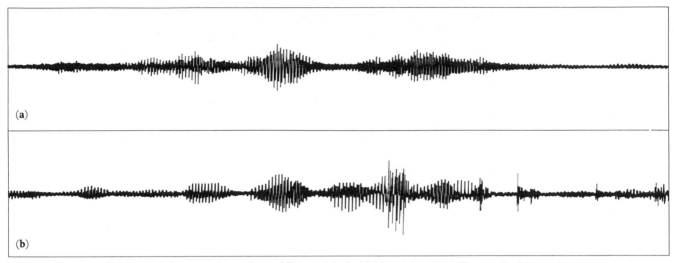

Figure 12 Oscillograph traces made from sound recordings of flight activity in (**a**) (above) the hornet (*Vespa crabro* Linnaeus) and (**b**) (below) the hornet clearwing (*Sesia apiformis* (Clerck)). The fundamental frequency is very similar in the two insects

The six British yellow underwing species (*Noctua*: Noctuinae, Noctuidae) display, on a smaller scale and in a less flamboyant manner, a similar strategy. Most cryptic noctuids are relished by birds and this group is only slightly less attractive.

6. The magpie moths (*Abraxas* (Geometridae: Ennominae))

Of the two British species, the magpie moth (*Abraxas grossulariata* (Linnaeus)) is known to be exceptionally distasteful to a wide variety of predators – birds, bats, spiders, frogs, lizards and man! (Poulton, 1890; Marshall, 1902; Pocock, 1911; Cott, 1940, 1946; and others). It is the only British species in which all four stages of the life-cycle are aposematic. Moreover the colour scheme of caterpillar and imago are remarkably similar. The larva spins a net-like cocoon with sufficient spaces to allow the black- and yellow-banded pupa to be clearly visible from without. An aposematic pupa is very rare indeed among the Lepidoptera, yet the toxic or distasteful substances present in this geometer, which must be unusually potent, have never been investigated beyond establishing the presence of histamine-like activity in extracts of whole moths (Frazer & Rothschild, 1961). Whatever they may be, the chemicals in question should provide excellent avian deterrents. Like many aposematic species, the larvae can feed on plants, like spindle (*Euonymus europaeus*), which contain toxic secondary plant substances. It is not known if it can store the cardiac glycosides present in the foliage of the spindle tree, although related species in other parts of the world also favour this family of plants which is suggestive (Sato & Nakajima, 1975). It does, however, sequester unusually large amounts of carotenoids (675.40 µg/g) (Feltwell & Rothschild, 1974). The related British species, the clouded magpie (*Abraxas sylvata* (Scopoli)), is also probably distasteful but it does not feed on toxic plants. Nothing is known about its chemical defences. Various foreign species are involved in mimetic relationships. One of these, *A. paucinoltala* Warren, greatly resembles our clouded magpie. It is closely mimicked by a distasteful Chalcosiinae (Rothschild, 1979: pl.1, figs 2,4).

7. The tussock moths (Lymantriidae) and eggars (Lasiocampidae)

Brightly coloured gregarious larvae, some spinning communal webs, are characteristic of both families, represented in the British Isles by 11 species in each. As far as we know they do not sequester toxins from plants but themselves secrete irritant or toxic substances – probably proteinaceous compounds – associated with glandular and spiny hairs (Quiroz *in* Bettini, 1978: Chap. 20). This is a form of defence characteristic of moths rather than butterflies. Reared on oleander, for instance, the pupa and cryptic adult of *Orgyia dubia judaea* Staudinger are negative for cardenolides, but the brightly coloured caterpillars form part of a Mullerian assemblage in Israel, which includes lygaeid bugs and oleander aphids (pers.obs.). The toxic hairs of the lymantriid caterpillars are frequently attached to the moth's cocoon and in some cases are collected by the emerging female and eventually spread over her eggs as a protection.

The adult yellow-tail (*Euproctis similis* (Fuessly)), however, is clearly aposematic, being white with a brush of yellow hairs at the end of the abdomen (displayed if the moth is disturbed), and is rejected by many birds (Frazer & Rothschild, 1961). The larva is black, red and white with histamine-loaded hairs, frequently the cause of 'nettle rash' in children who handle them. It feeds on various non-poisonous plants. The adult female, even with the brush of terminal setae removed, contains seven or eight times (100 µg/g) more histamine than the male. She also spreads larval hairs and spicules over her eggs. The caterpillars of the brown-tail (*E. chrysorrhoea* (Linnaeus)) live in webs and are more toxic and irritant than the larvae of the previous species, although utilizing the same foodplants. Protease, esterase and phospholipase A_2 activity has been recorded in extracts from their spicules (de Jong & Bleumink, 1977). Both larva and adult of the white satin moth (*Leucoma salicis* (Linnaeus)) and the black arches (*Lymantria monacha* (Linnaeus)) are aposematic. The latter smells strongly of pyrazines, but their chemistry has not been investigated.

The cryptic adult eggar moths are heavily predated by birds (Kettlewell, 1973). In the case of the lackey (*Malacosoma neustria* (Linnaeus)), the caterpillars, whenever feeding in groups on laurel, neither sequester nor secrete the HCN present in the plant (Parsons & Rothschild, 1964). Their coloration is exceptional since the bright blue marks along the back are due to Tyndall blue, not pigments. They proved highly distasteful to caged predators although the adult is eaten with relish. Similarly the caterpillar of the drinker (*Philudoria potatoria* (Linnaeus)) is rejected by birds (Pocock, 1911).

8. Aposematic larvae of other species of moths

(a) The alder moth (*Acronicta alni* (Linnaeus)) (Noctuidae)

One of the most peculiar aposematic larvae in Britain is that of the alder moth. It is warningly coloured only in the last two instars. In the first three instars the larva mimics a bird-dropping, which it resembles to an astonishing degree – even to a surface shine simulating a wet and recent defecation. At the fourth moult an extraordinary change occurs and a larva with the yellow and black transverse stripes emerges. In addition it is adorned with curious long clubbed hairs. Does the change indicate a sudden biochemical switch or is it a demonstration that size can play a vital role determining crypsis★ and aposematism? Unfortunately the moth is rather rare, and I have never fed either the adult or the larva to birds but anticipate that it would prove distasteful despite its range of innocuous foodplants. This larva poses a question to which at present we have no answer.

(b) The prominents (Notodontidae)

The adult moths of this family, represented in Britain by 25 species, are marvellously cryptic and never cease to astonish us with their refined adaptations of colour, form and movement. A few of their larvae are aposematic. Thus the larva of the puss moth (*Cerura vinula* (Linnaeus)) adopts a threatening attitude if disturbed, lashing and waving the filamentous whip-like extensions of the hind claspers and releasing a cervical gland secretion which contains formic acid. In some related foreign species the secretion also contains ketones and acetates, irritants which probably enhance the effect of formic acid against arthropod predators (Blum, 1981).

The buff-tip larva (*Phalera bucephala* (Linnaeus)) is gregarious and polyphagous, favouring tall trees such as ash, as well as various shrubs and bushes. The colour is greenish yellow and black, and it is said to have an obnoxious smell (Sheppard, 1967: 139); we presume it advertises its own secretions, not sequestered toxins. The moth, which in the daytime frequently rests on the ground beneath trees, achieves the acme of crypsis resembling a birch twig broken at both ends. Curiously enough it is refused by caged and wild garden birds such as sparrows, and therefore some toxic carry-over from the larva via the pupa is probable. It squirts a well-aimed stream of excreta if handled or molested. At night on the wing it looks white, a conspicuous object, and is probably usually avoided by bats and owls (Rothschild, 1981). Moths which appear white in flight at dusk or at night can be regarded as aposematic, although they are not always sufficiently toxic to have been forced out of the cryptic diurnal life-style, when bird predation is more intense. Other species of the family in other parts of its range are more toxic, among them the processionary caterpillars.

Scattered among the various families of moths there are larvae with a somewhat similar type of convergent aposematic coloration, bluish grey and yellow with black markings. The adults are cryptic. Two examples of these caterpillars are the figure of eight moth (*Diloba caeruleocephala* (Linnaeus)) and the mullein moth (*Cucullia verbasci* (Linnaeus) (Noctuidae)). The latter feeds gregariously, in the last instars among the brilliant yellow flower-heads of mullein, when it is, like the cinnabar among ragwort, cryptic at a distance.† It also has a marvellous 'shine' on the anterior end, to match the glittering secretions of the mullein flowers. But it is unquestionably aposematic at close quarters, and is said to be distasteful (Poulton, 1908). The foodplant contains toxic secondary substances.

(c) The goat moth and leopard moth (Cossidae)

Maybe one should add a last word about the larvae of the goat moth (*Cossus cossus*

★ An African swallowtail larva passes the first two or three instars mimicking a bird-dropping, but then changes to a different type of crypsis – disruptive green and brown markings. Observers have assumed this change is related to its increase in size (Edmunds, 1974). There are a number of other species which resemble bird-droppings only in the early instars.

† First noted by Wood (1869); restated by Poulton (1908).

(Linnaeus)) and leopard moth (*Zeuzera pyrina* (Linnaeus)), representatives of the Cossidae of which there are only three British species. They are aposematic only with regard to their vile odour. Pavan & Dazzini (1976) located the source of the smell in the mandibular glands which secrete polyphenols – highly distinctive alcohols not encountered otherwise in lepidopterous larval defence secretions which they named cossine 1, 2 and 3 ($C_{16}H_{24}O_2$) (Trave *et al.*, 1960a,b, 1966). Both species feed in the solid wood of various trees. The goat moth larva is distasteful to birds (Pocock, 1911).

III. The Sequestration of Carotenoids

These yellow pigments must be considered apart from the other secondary plant substances, since they are present in all Lepidoptera so far examined. However, in the aposematic species they take on special roles, participating in warning colour schemes and mixing with the secretions from exocrine glands (Pl.B, fig.3). It is repeatedly stated in the literature that Lepidoptera sequester carotenoids unselectively, merely mirroring those present in the foodplant but this is incorrect. Although Lepidoptera only very rarely metabolize or convert carotenoids (Valadon & Mummery, 1978), they do store these pigments selectively. Thus the large white (*Pieris brassicae*) stores 14 different carotenoids, and an Asiatic and Australian species of swallowtail (*Graphium sarpedon* (Linnaeus)) only one. The difference is striking in the case of the large and small whites fed on shared cabbage leaves (half a leaf each, repeated) when the large white has 332 μg/g of pupal carotenoids and the small white only 41 μg/g. Furthermore there are different concentrations in different parts of the insect's body (Feltwell & Rothschild, 1974), suggesting the carotenoids perform specialized functions in these areas. These are characteristic features of sequestration and storage of secondary plant substances *sensu stricto* by Lepidoptera in general, since *all* the different cardiac glycosides or glucosinolates or aristolochic acids present in the foodplants are not sequestered. There is rigorous selection, and the types stored or converted depend, on the one hand, on the species of insect concerned and, on the other, on the type of mustard oil or heart poison present in the species of plant on which it is feeding.

In monarch larvae, carotenoids, mixing with the blue bile pigments also present in the larval tissues, produce the well-known green coloration. In the pupae, they are responsible for the golden diadem and golden flecks. As we have seen (p. 16), they are also the cause of the orange colour of the osmaterium of the larval swallowtail.

The bright yellow aposematic spots along the sides of the spurge hawk-moth larva and the yellow colour of the eggs of the large white are due to the presence of carotenoids. Reared on a carotenoid-free diet, the large white butterfly lays eggs that are pure white: the yellow colour of its larva, however, is due to the presence of sepiapterins.

It is now well understood that carotenoids play a decisive role in insect vision and in various aspects of larval and pupal coloration. Their presence in the testes and ovaries of many Lepidoptera is probably a protective device (Kayser, 1982). It is possible they have another subtle role in directing the distribution of bile pigments (pterobilins) within the tissues (Rothschild *et al.*, 1975b; Allyn *et al.*, 1982) but this is at present only a matter for speculation. Nevertheless butterflies can be reared successfully on artificial diet lacking carotenoids.

It is interesting that in conjunction with toxic secondary plant substances carotenoids assist in forcing insects into the risky role of self-advertisement, but in their absence the yellow carotenoid pigments are one of the principal factors in evolving crypsis.

IV. The Defences of British Aposematic Lepidoptera

The defences of arthropods are intensely complex and incredibly varied and we therefore cling gratefully to such a relatively simple aspect as red or white or yellow-and-black coloration (for here we are supplied with straightforward evidence) a peg on which to hang different less well-known and less well-understood aposematic attributes. Are there any other cues which are as widely distributed and as easily recognizable as these?

Before we go on to describe the alerting devices there is an odd feature of aposematic Lepidoptera worth considering. There is only one British moth that has completely red forewings as well as hindwings – var. *coneyi* Watson, an all-red variety of the cinnabar (Pl.B, fig.2). Breeding experiments (Watson, 1975) have proved that this coloration is dominant to the normal pattern, yet the variety has not spread through the population. In fact, uniformly red forewings are practically unknown among moths the world over. It would seem that few can afford to be constantly and violently conspicuous. Among British Lepidoptera only the diurnal monarch and a few of the white butterflies are sufficiently well protected to achieve this distinction. Dual signalling is almost universal whether it involves disruptive colours of the forewings or crypsis-at-a-distance; or is concerned with ragwort flowers and striped caterpillars; or butterflies flying in the dappled sunlight penetrating different levels of the neotropical forest (Papageorgis, 1975). Nevertheless *at close quarters* gregariousness, with massing of red coloration, etc., increases the effectiveness of aposematic signals. This is striking on a warm, sunny July afternoon – even in Britain – when an umbrella of warning scents, which includes that of the pyrazines, seems to hang over the remnants of our flowering meadows.

The Alerting Devices

Red is an alerting colour. It is recognized instantly by all vertebrates endowed with normal colour vision. It warns birds if potential prey is dangerous, and attracts them to edible fruit and berries. It appears to be a signal learned more rapidly than others (Bisping *et al.*, 1974). For animals which hunt by scent rather than sight, pyrazines (substituted pyrazines are among the most powerful odorous substances known) fulfil a similar warning function, although so far they have been investigated only in insects and plants (Rothschild, 1961b; Seifert *et al.*, 1970; Moore & Brown, 1981; Jones & Blum, 1983). Here, like the colour red, their presence cuts across all accepted classifications. In the British fauna and flora they are present in the exocrine glands, or haemolymph, or juices at reflex bleeding-points, or crushed tissues, variously in tiger moths, burnets, monarchs, ladybirds, tiger beetles, ants and plant-bugs; and in nettles, poppies and many other plant species. The pyrazines produce a powerful, evocative and at the same time quizzical odour (Lane, 1957b) in the ruby tiger (*Phragmatobia fuliginosa* (Linnaeus)) and, although not in themselves highly objectionable, they seem to enhance or elevate the other odours present; but above all they are memory-stimulating. Just as mimics have adopted red colour as part of their deception so have they accepted the storage and release of pyrazines as another facet of mimicry (Moore & Brown, 1981; Rothschild, 1961b).

Relatively high concentrations of the biogenic amines in non-nervous tissues is another widespread feature of both aposematic animals and plants (Guggenheim, 1951; Smith, 1977a,b). Such concentrations are often associated with stinging or biting apparatus or are mixed with defensive sprays or froths, and are also found in the reproductive ducts of insects and their eggs. They occur in such diverse groups as jellyfish, molluscs, snakes, Hymenoptera, nettles and cacti (Kaiser & Michl, 1958) as well as in the aposematic British Lepidoptera. Originally I referred to them loosely as 'boosters' of toxic qualities, probably also pain-producing, inflammatory or functioning as spreading factors or releasers of other substances. Although they may well play all these roles (especially since they are so often associated with offensive weapons) I now consider that the most important function of the neurotransmitters such as serotonin and acetylcholine is the amplification of neurotransmission in predators. They would then also act as memory-boosters and learning aids. Recent work on the increase in brain acetylcholine which may in some cases improve memory in human patients (Goodman Gilman *et al.*, 1980; Wurtman, 1982) provides a new line of approach for us.

It has been suggested by Evans & Bella (1980) that in some cases in plants the amines themselves are the principal feeding deterrents. This may be true, just as certain alerting yellow pigments are themselves toxic. But their distribution is too widespread and too general for such a specific function. Among British Lepidoptera the highest concentrations of choline and histamine are found in the garden tiger (*Arctia caja* (Linnaeus)), the six-spot burnet (*Zygaena filipendulae* (Linnaeus)) and the cinnabar (*Tyria jacobaeae* (Linnaeus)). In the two former species they are associated with the highly toxic proteinaceous cajin and zygènine. Do they combine with the colour red and the pyrazines to produce an indelible mark on the senses of the predator? Do they initiate a type of 'instant imprinting'? This suggests a most fascinating, almost limitless vista of experiments with prey and predators for, curiously enough, the memory-stimulating, evocative qualities of defensive secretions have, up to date, attracted little if any attention.

I would like to include stridulation or sound-production as a fourth alerting device since I believe it plays a general role in the defence mechanisms of many adult and pupal Lepidoptera (Bourgogne, 1951; Downey & Allyn, 1973, 1978) – not only in those insects which buzz but also in the garden tiger and many other arctiids (Blest, 1964) which stridulate warningly, both by day and by night (Rothschild & Haskell, 1966; Roeder, 1967). But at present too little is known about sound-production of British moths and butterflies.

Finally we can ask ourselves whether we should classify repellent *taste* among the alerting signals or among the fundamentally defensive chemicals. This is really a matter of opinion. Swynnerton (1915) and Marshall (1902) both maintained, and quite rightly, that a nasty taste by itself was, ultimately, inadequate as a basis for immunity from attack and unacceptability of prey and added: 'Indigestibility is the real defence.'

It is also true that without the link with more fundamentally pleasant or unpleasant attributes, taste is not important. The same, as we know, applies to the colour red. But, like the smell of pyrazines and scarlet hues, it cuts across classification, for most alkaloids taste bitter and so do cardiac glycosides and a host of other unrelated toxins. It is also significant that Moore & Brown (1981) found bitter-tasting 1-methyl-2-quinolone and 3-phenylpropanamide associated with isopropylpyrazine in the repellent secretions of a lycid beetle. Rhoades (1979), although apparently unaware of the relevant literature (see also Rothschild, 1961b), came to the right conclusion when he stated that 'Mullerian convergence of properties of plant defensive substances to animal receptors could benefit both plants and herbivores . . . we are led to the interesting possibility that . . . defensive substances of unapparent plants have diverged with respect to toxicity but have converged in conjunction with animals receptors . . .'

This is exactly what has happened, but it should be noted that all alerting signals are not truly comparable, since colour, odour and noise function at a distance, while the enigmatical amines and taste receptors require some degree of contact with, or ingestion of the prey concerned.

Despite the fact that the alerting signals are of a more universal character, they are not single signals but limited to several striking and easily memorized effects. Just as – in addition to red – yellow, black and white are essentially alerting colours, so there are probably about the same number of basic alerting scents, amines and tastes, which in due course will be recognized and identified. Notwithstanding, individual alerting signals themselves vary; no two reds are exactly alike to our eyes, and the smell of pyrazines, unmistakable though they are, are never identical. Probably they are mixed with other substances or, as with yellow carotenoids, variations exist on the central theme. In any case, the smell of nettle pyrazines, or those of crushed poppy-stalks, or ladybirds, or monarch butterflies, or black arches moths are all subtly different.* There are also geographical variations in the pyrazine scent of at least one species of moth. Nevertheless it is evident that the alerting signals are useless as phylogenetic characters, however valuable as zoological indicators; if the chemical signals are to give us a clue to relationship this must be sought in the complex defensive substances of related species (see also Harborne, 1977). Thus Jordan (*in* Seitz, 1907–08) eighty years ago, with his inimitable insight and flair, was the first to do so when he selected the resistance to HCN as a taxonomic character of the Zygaeninae.

The Punishing Devices

I have yet to find a frankly aposematic British butterfly or moth (no suspected local Batesian mimics being available for testing) which is as acceptable to our bird

---

* To my nose only two British moths have an indistinguishable pyrazine smell – the garden tiger and cream-spot tiger (*Arctia villica* (Linnaeus)).

predators as a cryptic species of the same size. Distastefulness is unquestionably the most widespread defence, whether we think of it functioning as an alerting signal or as a facet of an underlying and more fundamental chemical toxicity or inedibility. This is most evident in those species that have adopted diurnal crypsis but which may be aposematic after dark. If, on the contrary, the insect is warningly coloured and flies by day, mere distastefulness appears inadequate; indigestibility or some deterrent which produces pain, discomfort or distress is required. Once sequestration and storage of secondary toxic plant substances is established, a cycle of stockpiling of alerting and actively defensive signals is initiated. It is very significant that cryptic species feeding on *Senecio* spp. do not store pyrrolizidine alkaloids and that the species of ladybird which contains no toxic alkaloids (in this case synthesized by these beetles) is not brightly coloured (Pasteels, 1976). We have yet to find the exception which proves this rule – namely a *cryptic* insect storer of cardiac glycosides.* No doubt this insect will eventually turn up! Nor is it necessary that such a cycle always originates or evolves in a similar manner. Colour may precede or follow storage of toxic material.

Pasteels *et al.* (1983) remark that 'with very few exceptions defensive secretions are remarkably complex mixtures'. To give one example: over 40 volatile compounds have been identified in Dufour's gland of formicine ants (Bergstrom & Lofqvist, 1971). In this way they differ from the purely alerting signals. Bergstrom (1979) suggests that each component may be involved in more than one functional aspect. Blum (1981) lists over 300 empirical formulae for arthropod defensive chemicals. (In 1967, Weatherstone recorded only 25!) These include hydrocarbons, alcohols, aldehydes, ketones, acids, terpenes, quinones, esters, lactones, phenols, steroids, alkaloids, amines, peptides and various proteinaceous venoms. Dazzini & Finzi (1974) list almost 380 chemically known constituents of such secretions. Wasp venoms which have been better investigated than any from Lepidoptera, include a mixture of toxic proteins, histamine, serotonin, dopamine, adrenaline, noradrenaline, acetylcholine, acid phosphatase, phospholipase A, hyaluronidase, kinins, histidine, decarboxylase and a mast cell degranulating peptide.

Moths and butterflies also synthesize non-exocrine defensive compounds which are often present in specific organs but also distributed in the haemolymph and oozed or secreted on to the surface via gland openings, thin ruptured membranes or by reflex bleeding at joints. If a drop of this usually yellow fluid is spread on a glass slide the amazing array of crystals which form – quite apart from various blood cells – give one a little insight into the complex nature of the exudate (Rothschild, 1970). It is interesting that non-toxic species do not usually have warningly coloured haemolymph while the chemically protected Lepidoptera do.

It is probable that the exocrine secretions or even the haemolymph may contain fractions which are selectively toxic against different predators, such as mammals or ants or birds as the case may be. This has been demonstrated to be so in the venom of scorpions and snakes (Zlotkin, 1973; Zlotkin *et al.*, 1975).

---

* It is a puzzle why honey-bees lack warning colour. Is there more significance to the buzz of these insects than we can appreciate? Certainly the buzz of *Chrysis ignita* (Linnaeus) is very effective – but so are its brilliant colours.

But not a single exocrine gland nor the haemolymph of a lepidopteron has as yet been subjected to a thorough analytical scrutiny! Usually a major deterrent, such as a heart poison, has been discovered and identified, and attention then centred on that particular component to the exclusion of any others which may be present. Thus specimens of the monarch butterfly (*Danaus plexippus* (Linnaeus)) which lacked cardenolides have been dubbed 'automimics' (Brower *et al.*, 1970) without regard to the pyrrolizidine alkaloids and pyrazines and other defensive substances (Rothschild *et al.*, 1978) which would also be present.

Pavan & Dazzini (1976) listed 76 species of Lepidoptera with chemical defences of which 20 are British (and a number of additional non-British species could now be added), with an additional four, toxic only in the larval stages.

For the sake of convenience, their chemical defences can be roughly divided into three categories:

(a) Toxic secondary plant substances (figure 1) sequestered and stored at various stages of the life-cycle, including the adult (Table III);

(b) Substances synthesized or secreted by the moth or butterfly (or its larva or pupa) and carried through to the adult (Table IV). These are more numerous but less well known;

(c) Substances synthesized and secreted only during the larval stages and not carried through to subsequent stages (Table V).

In the first category, among British species, one must include six major groups: cardenolides (for example, from *Digitalis* and probably *Convallaria*, *Euonymus*, etc.); pyrrolizidine alkaloids (from *Senecio*, *Symphytum* and probably certain nectars); glucosinolates (from Cruciferae, *Tropaeolum* and *Reseda*); diterpenes (from *Euphorbia*); pyrazines (from various plants); and flavonoids (from Gramineae).

It should be noted that callimorphine is produced by metabolism of different pyrrolizidine alkaloids by the garden tiger, the cinnabar and the scarlet tiger (*Callimorpha dominula* (Linnaeus)).

In the second category (Table IV) are included HCN synthesized by zygaenids, the amines found in a variety of species, and the toxic proteinaceous substances such as cajin (from the garden tiger), pierin (from the large and small whites), and zygènine (from burnets). The pyrazines are also almost certainly synthesized as well as sequestered.

In the third category we can include formic acid, present in the larval cervical glands of the puss moth, and isobutyric acid in the osmaterium of the swallowtail larva; also cossine, from the mandibular glands of the goat moth and leopard moth, and protease, esterase and phosolipase activity from spicules of lymantriid caterpillars.

Very often the female proves more toxic than the male. Comparisons are generally based on extracts made from the entire insect and the presence of toxic eggs (in some cases loaded with amines) in the female can account for the difference. But this is not always the case. If the larva of the monarch butterfly is fed with pyrrolizidine alkaloids the male stores twice as much as the female (Rothschild & Edgar, 1978). Since these alkaloids play a part in the synthesis of the male sex pheromones released

Table III: Toxic Secondary Plant Substances sequestered and stored by British Macro-Lepidoptera

Adult species	Toxins	Where found	Author
Pieris brassicae (L.)	Glucosinolates (sequestered from cabbage and other plants)	Pupa, eggs, adult	Aplin *et al.*, 1975
Pieris rapae (L.)	Glucosinolates (sequestered from cabbage and other plants)	Pupa only tested	Aplin *et al.*, 1975
Danaus plexippus (L.)*	Various cardiac glycosides (sequestered as larvae from asclepiad plant species)	Extract of eggs, adult, pupa and larva (stored selectively)	Reichstein *et al.*, 1968
	Pyrrolizidine alkaloids (sequestered as adults from *Senecio*, *Heliotropium* and other, unknown, sources)	Extract of whole insect	Edgar *et al.*, 1976
	Pyrazines‡	Extract of whole insect (haemolymph?)	Rothschild *et al.*, 1984
Zygaena filipendulae (L.)	Pyrazines‡	Odorous secretions	Rothschild (pers. obs.), 1961b
Zygaena lonicerae (Schev.)	Pyrazines‡	Extract of whole insect (haemolymph?)	Rothschild *et al.*, 1984
Hyles euphorbiae (L.)	Diterpenes: ingenol and ingol (sequestered as larvae from *Euphorbia*)	Whole moth, pupa and larva	Marsh *et al.*, 1984
Utetheisa bella (L.)†	Pyrrolizidine alkaloids (sequestered from *Crotalaria*)	Extract of whole insect	Rothschild & Aplin, 1971
	Pyrazines‡ (estimated)	Cervical gland secretion	Rothschild (pers. obs.)
Arctia caja (L.)	Cardiac glycosides (sequestered from *Digitalis*)	Extract of whole adult	Aplin & Rothschild, 1972
	Pyrrolizidine alkaloids (sequestered from *Senecio vulgaris* and stored selectively)	Extract of whole adult	Aplin & Rothschild, 1972
	Cannabidiols (sequestered from *Cannabis sativa*) (in the laboratory)	Stored only by larva	Rothschild & Fairbairn, 1980
	Pyrazines‡ (estimated)	Cervical gland secretion	Rothschild, 1961b
Arctia villica (L.)	Pyrazines‡ (estimated)	Cervical gland secretion	Rothschild, 1961b
Spilosoma luteum (Hufn.)	Pyrrolizidine alkaloids (sequestered selectively from *Senecio vulgaris* & *S. jacobaea*)	Extract of whole adult and pupa	Rothschild *et al.*, 1979b
	Pyrazines‡ (estimated)	Cervical gland secretion	Rothschild, 1961b
Spilosoma lubricipeda (L.)	Pyrrolizidine alkaloids (sequestered selectively from *Senecio vulgaris* & *S. jacobaea*)	Extract of pupa	Rothschild *et al.*, 1979b
Diaphora mendica (Clerck)	Pyrrolizidine alkaloids (sequestered from *Senecio vulgaris* and stored selectively)	Extract of pupa and adult	Rothschild *et al.*, 1979b
	Pyrazines‡ (estimated)	Cervical gland secretion	Rothschild (pers. obs.)
Phragmatobia fuliginosa (L.)	Pyrazines‡ (estimated)	Cervical gland secretion	Rothschild (pers. obs.); Lane, 1957b
Callimorpha dominula (L.)	Pyrrolizidine alkaloids (sequestered selectively from *Symphytum*)	Extract of whole body	Edgar *et al.*, 1980
	Pyrazines‡ (estimated)	Cervical gland secretion	Rothschild, 1961b
Tyria jacobaeae (L.)	Pyrrolizidine alkaloids (sequestered selectively from *Senecio jacobaea* & *S. vulgaris*)	Extract of eggs, pupa and adult	Aplin & Rothschild, 1972
	Pyrazines‡ (estimated)	Cervical gland secretion	Rothschild, 1961b

* Scarce migrant
† Scarce vagrant or migrant
‡ It is not known if these are sequestered only from the food plant or whether some are synthesized.

Table IV: Defensive Substances synthesized by British Lepidoptera and present in the Imago

Species	Amine	Where found	Author
Pieris brassicae (L.)	Pierin (proteinaceous toxin)	Extract from adult, larva and pupa	Marsh *et al.*, 1984 Marsh & Rothschild, 1974
Pieris rapae (L.)	Proteinaceous toxin	Extract of larva	Dempster, 1967
Danaus plexippus (L.)	Acetylcholine absent	Extract of whole insect	Parsons & Rothschild, unpublished
	Histamine-like activity (slight)	Extract of whole insect	Kellett & Rothschild, unpublished
	Cardioactive substance	Extract of whole insect	Rothschild *et al.*, 1978
Adscita (= *Procris*) *geryon* (Hüb.)	HCN (synthesized)	In crushed tissues of moth	Jones *et al.*, 1962
Zygaena filipendulae (L.)	HCN (synthesized)	Crushed tissues of moth	Jones *et al.*, 1962
	Probably zygènine: protein-aceous toxin (synthesized)	Extract of whole insect	Rocci, 1916
	15.25 mg/g acetylcholine	♂ accessory sex gland	Morley & Schachter, 1963
	50–60 mg/g acetylcholine	♂ ejaculatory duct	Morley & Schachter, 1963
Zygaena trifolii (Esper)	HCN (synthesized)	Crushed tissues of moth and eggs	Jones *et al.*, 1962
	Zygènine: proteinaceous toxin (synthesized)	Adult female (traces in male?) also eggs	Marsh & Rothschild, 1974
Zygaena lonicerae (Schev.)	HCN (synthesized)	Crushed tissues of all stages of life-cycle	Jones *et al.*, 1962
	Probably zygènine: protein-aceous toxin (synthesized)	Extract of whole insect	Rocci, 1916
	300–500 μg/g acetylcholine	♀ ♂ bodies	Bisset *et al.*, 1960
	1.5–5 mg/g acetylcholine	♂ accessory sex gland	Morley & Schachter, 1963
	4–20 mg/g acetylcholine	♂ ejaculatory duct	Morley & Schachter, 1963
	5 μg/g acetylcholine	Ovaries	Morley & Schachter, 1963
	5 μg/g acetylcholine	Gut and spent abdomen	Morley & Schachter, 1963
	250 μg/g histamine	♀ ♂ bodies	Bisset *et al.*, 1960
Abraxas grossulariata (L.)	Histamine	Extract of insect	Frazer & Rothschild, 1961
Smerinthus ocellatus (L.)	75 μg/g histamine	Whole insect	Bisset *et al.*, 1960 Morley & Schachter, 1963
Laothöe populi (L.)	< 10 μg/g acetylcholine	♂ accessory gland and ejaculatory duct	Morley & Schachter, 1963
	100 μg/g histamine	Whole insect	Bisset *et al.*, 1960
Euproctis chrysorrhoea (L.)	Histamine	Terminal hairs of adult	Frazer, 1965
Euproctis similis (Fuessly)	100 μg/g histamine	Extract of whole adult	Frazer & Rothschild, 1961
Utetheisa bella (L.)	ββ-dimethylacrylylcholine	Extract of cervical glands	Rothschild & Haskell, 1966
Arctia caja (L.)	ββ-dimethylacrylylcholine	Extract of cervical glands	Bisset *et al.*, 1960
(see Table II)	Choline esters	Thorax & abdomen	Morley & Schachter, 1963
	Histamine: trace only		Morley & Schachter, 1963
	Cajin: proteinaceous substance synthesized during the pupal stage and carried through to the adult		Marsh & Rothschild, 1974 Rothschild *et al.*, 1979c Hsiao *et al.*, 1980
Arctia villica (L.)	Acetylcholine activity	Whole adult and abdomen	Frazer & Rothschild, 1961
	400–700 μg/g acetylcholine	♂ accessory gland and ejaculatory ducts	Morley & Schachter, 1963
	Adrenaline-like depressor		Frazer & Rothschild, 1961
Spilosoma lubricipeda (L.)	Acetylcholine-like activity	Pooled abdomens	Bisset *et al.*, 1960
	700 μg/g histamine	Adult bodies	Bisset *et al.*, 1960
	175 μg/g histamine	Head and thorax	Bisset *et al.*, 1960
	20 μg/g histamine	Wings	Morley & Schachter, 1963

Table IV – *continued*

Species	Amine	Where found	Author
Spilosoma luteum (Hufn.)	Acetylcholine-like activity	Pooled abdomens	Morley & Schachter, 1963
	< 5 µg/g histamine	Whole insect	Bisset *et al.*, 1960
Phragmatobia fuliginosa (L.)	Histamine-like activity Suspected toxic protein	Extract of whole moth	Frazer & Rothschild, 1961
Callimorpha dominula (L.)	< 5 µg/g histamine	Whole insect, head and thorax, wings, legs	Bisset *et al.*, 1960
	< 10 µg/g histamine	♂ accessory gland and ejaculatory duct	Morley & Schachter, 1963
Tyria jacobaeae (L.)	< 10 µg/g acetylcholine	♂ accessory gland and ejaculatory duct	Morley & Schachter, 1963
	< 30–80 µg/g acetylcholine	Ovaries	Morley & Schachter, 1963
	750 µg/g histamine	Body	Bisset *et al.*, 1960

from the hairpencils or coremata as well as in defence, this is suggestive.[*] Extracts of the male of *Syntomis phegea*, when injected into mice are more toxic than female extracts. On the other hand male burnets secrete less HCN than females. They also lack zygènine except possibly in very small amounts. Apart from the pyrrolizidine alkaloids which have been investigated by Boppré (1977) and Edgar *et al.* (1979), it is not known whether the chemical defence substances, or fractions therefore, are also used for sexual signals within the species.

Altogether, whether we are considering the volatile or the non-volatile chemical characteristics of the insects' protective devices, the functional/evolutionary relationship between the various signals and the responses they elicit is obscure. This, as Bergstrom (1979) suggests, is because we know 'only the bare bones'.

Table V: British Larvae with Chemical Defences† (not found in adult)
(Compiled from Blum, 1981)

Species	Where located	Chemical defence
Cossus cossus (L.)	Mandibular glands	Cossine ($C_{16}H_{24}O_2$)
Zeuzera pyrina (L.)	Mandibular glands	Zeuzerine ($C_{14}H_{26}O_2$) n-Dodecyl acetate ($C_{14}H_{28}O_2$)
Cerura vinula (L.)	Cervical gland	Formic acid (CH_2O_2)
Papilio machaon (L.)	Osmaterium	Isobutyric acid ($C_4H_8O_2$) $\propto$-methylbutyric acid ($C_5H_{10}O_2$)

[*] Professor Keith Brown (pers.comm.) notes that many females he has examined contain more pyrrolizidine alkaloids than the males which use them as pheromones.

† Species with setae with histamine-like activity or spicules with protease activity, etc., are not included here. We have also excluded the small white (*Pieris rapae* (Linnaeus)) in which mustard oils are present in the larva and pupa but are not carried through to the adult.

V. Bird Predators*

Swynnerton between 1909 and 1919 made detailed observations on the behaviour of insectivorous bird predators in Africa, both wild and captive birds. His generalizations apply as well today as they did at the time. In Britain (apart from cuckoos) we lack the specialist feeders like bee-eaters, hornbills, ashy wood-swallows and bulbuls which concentrated on specific toxic or dangerous prey, or could, at any rate, eat them in large quantities – 50 *Danaus chrysippus* in quick succession – without suffering disagreeable consequences. An interesting point which has not received enough attention is the fact that many brightly coloured or conspicuous birds, such as bulbuls and wood hoopoes and drongos which are utterly repugnant to carnivores† (Swynnerton, 1915; 1919; see also Carpenter, 1941–42), are among the species which can feed on toxic insects with greatest impunity.

Our insectivorous birds fall into the middle ranges although some species like the corvids appear much more sensitive to insect toxins than, for instance, the flycatchers or the ground-feeding pheasants, or quails, (Rothschild & Kellett, 1972) although the latter birds come into contact with Lepidoptera only relatively infrequently. Nor can we boast a pigmy falcon (*Polihierax semitorquatus*) which lines its nest with the wings of the butterflies and dragonflies that it eats.

Although we have never had a field observer in Britain of the calibre of Swynnerton, our anecdotal literature is nevertheless stuffed with records of birds capturing Lepidoptera, from the kestrel on Ballard Down (Marshall, 1909: 335) to the wheatear which regularly fed cinnabar moths to its nestlings (Turner, *in* Poulton, 1932; see also Collenette, 1935 and Carpenter, 1941 for beak marks on butterflies' wings). Studying what is now usually referred to as the palatability‡ spectrum he came to the following conclusions: although different species of birds had different orders of preference, the likes and dislikes (excluding the few possible poison 'specialists') were fairly similar. In his region of Africa the danaids and acraeas (*Danaus, Amauris* and *Acraea* (Nymphalidae)) were the least acceptable of all the butterflies available, and the brown cryptic moth *Sphingomorpha* (Noctuidae) was an all-round favourite. The same can be said of our aposematic Lepidoptera. I know of no experienced local bird here except the cuckoo (Rothschild, 1981) which will eat

---

* The reader is advised to consult Cott's (1940) copious bibliography for records of British butterflies and moths captured by birds, especially the papers of Carpenter, Poulton and of many others.

† Cott (1946) showed that hornets also preferred the flesh of cryptic birds and generally found brightly coloured species like hoopoes unattractive as food.

‡ Blum (1981: 454) rightly points out that some authors have confused palatability (i.e. acceptance or refusal on the grounds of taste) with emetic or indigestible qualities (i.e. refusal based on previous experience of subsequent illness or disagreeable after-effects). We use the term palatability in Blum's and Swynnerton's sense – *nice or nasty taste* (which elicits an immediate contact reaction). It is understood, however, that once a disagreeable experience or an illness is associated with specific taste or smell these can themselves rapidly *become* repellent and unpalatable. I distinctly recall that on the first occasion I was dosed with castor oil I wondered why my contemporaries thought it so awful. In due course I developed a life-long shuddering dislike, not only of the taste, smell and consistency of castor oil, but also of the excellent cognac with which my mother attempted to mask the flavour! Thus there is a difference between the palatability spectrum considered from the point of view of a naïve or an experienced bird. In the latter case 'secondary unpalatability' may be added to the spectrum.

any of the arctiids unless it is extremely hungry, whereas the brindled beauty (*Lycia hirtaria* (Clerck)) is always an all-round favourite.

Although within this broad framework each species of bird was found on the whole to have exceedingly consistent preferences,* circumstances could vary the order. Thus when birds are hard-pressed when feeding young, they will capture species ignored at other times. (The spotted flycatcher hawking for small whites is a familiar sight at Ashton just before the nestlings leave the nest.) 'Unless through sheer impossible hardness, size, etc., there is practically no such thing as "inedibility". In the early morning, or after the ejection of a pellet, a bird may quite readily eat Acraeninae or even Danainae. As it fills up somewhat it refuses such very low-grade prey, but still eats other species, which in turn it rejects when slightly fuller – and so to repletion-point. A bird will often accept eagerly the most nauseous insect when hungry enough, or reject a favourite species when replete' (Swynnerton, 1915).

It is quite exceptional to find a bird (like Professor Brower's blue jays) which, after vomiting from the effects of cardiac glycosides in their prey, will not eat the same species of butterfly even if they are dying of hunger (Brower *et al.*, 1967). Nicolaus *et al.* (1983) have shown that crows also develop acute taste aversion after experiencing non-lethal but indigestible toxins. This would appear to be a reaction highly developed in the Corvidae, for crows are very intelligent birds with an unexpectedly large number of brain cells (Walker, 1983), but Nicolaus has stressed (pers.comm.) that the tendency to form a conditioned taste aversion to an illness-producing food is a *fundamental* feature of the neurology of all vertebrates and has nothing at all to do with intelligence since the mechanisms which produce the effect are known to reside in the medulla.

Swynnerton also stressed the point that birds tend to concentrate their attention on what prey at that time is most *readily attainable in quantity*. He wrote: 'but there are two other kinds of preference – the choice of the largest, and the turning of the whole attention to the commonest, insect that they are at the time hungry enough for. Absorption in the search or watch for one particular kind of insect is, I believe, fairly frequent, and it probably pays the bird [see also Sheppard, 1967: 140; de Ruiter, 1952]; it certainly accounts for a number of the instances of neglect we witness. And the selection of the largest insect present that the bird is hungry enough to eat – resulting, it may be, in the taking of an *Amauris miavius dominicanus* Trimen in preference to the far higher-grade but smaller *Precis cebrene* (Trimen) – is, as I have seen in both tame birds and wild, the rule where the bird *has* a choice and troubles to make it'.

Many of the different traits in feeding behaviour which Swynnerton noted also apply to our bird fauna, such as the difference between species in their reactions to moving prey. An extreme case is the 'blindness' of our young cuckoos which do not recognize a green caterpillar or a cryptic moth, even if fully exposed against a contrasting background, unless the insects move (pers.obs.). He also described birds which swallow prey whole on the wing; bash it against branches; batter it or rub it on the ground; hammer it or pick or tear it to pieces while holding it in their claws; pulp

---

* It is well known that individual birds of the same species may differ in how well they discriminate and also that their tastes change with time.

it by chattering it through their bills; or selectively discard or retain specific parts of the body (see also Calvert *et al.*, 1979).

There are also differences in the techniques employed by related species: thus the Asiatic bee-eaters (*Merops viridis* and *M. leschenaulti*) clip the wings off butterflies before swallowing them, while the bee-eater (*M. apiaster*) swallows them whole, wings and all (see also Fry, 1969). Different strategies are adopted against stings, glossy or slippery (torpedo-shaped) prey, or outsized butterflies.

Swynnerton was particularly impressed by the large numbers of relatively palatable butterflies his wild-captured birds would consume after refusing an aposematic species – something which has also impressed the writer. 'The fact remains that my various birds . . . would only eat *Amauris* when hungry but *Precis cebrene* nearly to repletion point. That the advantage may be a great one, sufficient to make the possessor worth mimicking, is shown by the immense meals that are sometimes eaten after the refusal of a low-grade butterfly; *e.g.* forty butterflies including fourteen large *Charaxes* by a roller after she had rejected a *Mylothris*, and thirty-seven including twelve large *Charaxes* after her rejection of a *Terias*.' However he found no evidence of visual instinctive recognition of unpalatable prey, but had doubts concerning rejection by *taste*, in which case he felt selection was often instinctive. What Swynnerton meant by 'instinctive' in this latter context was really 'instant' learning.

From observing caged bird predators' reaction to our aposematic Lepidoptera, I have come to believe that an unpleasant experience associated with the colour red is learned more rapidly than one linked to any other colour. The same observation has been made with fish learning a maze (Bisping *et al.*, 1974). Recently Schuler (1982), experimenting with naïve starlings and dummies with aposematic wasp-like yellow and black markings, has come to a somewhat similar conclusion: 'Hand-raised starlings (*Sturnus vulgaris*) at an age of six weeks initially rejected black and yellow-banded insect dummies at a higher rate than green, yellow or brown ones. The results indicate that in starlings the following principle holds: there is an initial tendency to avoid warningly coloured prey which for a permanent avoidance must be supplemented by unpleasant experience.* Individual differences in the tendency to avoid a pattern, such as those observed, provided the variation from which natural selection could mould – in cases of dangerous warningly coloured prey – innate reactions' (see also Davies & Green, 1976; Rothschild, in press). Similar innate recognition of coral snake pattern was shown by Smith (1975) to exist in motmots (*Eumomota superciliosa*). Indeed it would be strange if bird predators had not evolved a more fundamental response to the significance of warning colours.

Those of us who are interested in brightly coloured insects are apt to concentrate on the deterrents directed against bird predators, while the ant specialists focus their attention on the toxins and other features, like viscous entangling substances, directed towards predatory arthropods. I believe that the specific colours of aposematic insects today, and to a great extent the brilliant colours of berries and fruits, are

---

* Naïve chicks reject pyrazine-tainted water, but after a few days without any other unpleasant association, they will accept it (Guildford, Nicol & Rothschild, in prep.).

determined by avian colour vision, varied though it is believed to be in different species (Daw, 1973). The parallel evolution of the aposematic colour scheme in moths and butterflies in different continents today, would therefore be the result of selection based on avian vision. This, I think, accounts for the basic colour and pattern resemblance, for example of *Papilio antimachus* Drury and *Acraea* spp. in Africa, and of the mimetic ctenuchids and zygaenids which consort on flower-heads in Africa, Australia and Europe (Rothschild, 1979: pl.1).

One can expect the distribution of toxins and deterrents in bird prey to occur in the haemolymph more than in smaller arthropod prey. The manner in which the birds tackle prey is suggestive, for they often squeeze the bodies of their captives and the haemolymph squirts out of weak areas in the cuticle or exudes via exocrine gland apertures as in various arctiids. The blood cells in froths and sprays are indicators of toxin-bearing haemolymph. Probably pyrazines are thus present in both haemolymph and gland exudates. Blum and Nisho's study (see Blum, 1981) on the distribution of cardenolides in *Danaus plexippus* is most illuminating. The highest concentration was found in the haemolymph and then in the scales, which no doubt explains the initial observation by Parsons (1965) that the wings of *D. plexippus* contain three times as much cardiac glycosides as other parts of the body. And it is from the broken wing-veins and scales that a bird first comes into contact with cardenolides when it captures a butterfly. Pasteels (Pasteels & Daloze, 1977; Pasteels *et al.*, 1983) has also noted that the cardiac glycosides secreted by the chrysomelid beetles are oozed out of glands along the edge of the elytra and are the initial line of defence. This external distribution of deterrents contributes to the escape of prey with relatively insignificant injuries – a point in favour of individual selection of functioning in the evolution of warning coloration (Järvi *et al.*, 1981; Wiklund & Järvi, 1982).

I have pointed out on p. 12 that within the white butterfly mimicry complex there exists a striking toxicity spectrum. In Britain this is not due to lack of mustard oils in some of the foodplants but rather to their reduced storage by certain species within the complex. Among the aposematic tiger moths the pyrrolizidine alkaloids are likewise distributed most unevenly, but in their case this is the result of the polyphagous habits of the larvae which may or may not feed on plants containing them, or simply on those species of *Senecio* which contain less than others. Individual monarchs in a population (Brower *et al.*, 1972) – or in the case of its African relative *D. chrysippus* in whole local populations – may contain none, or very reduced amounts of cardiac glycosides. In Africa the paucity of these toxins reflects their absence in many species of asclepiad foodplants, and in the West African morph (*D. chrysippus alcippus*) a genetic element (Rothschild *et al.*, 1975a). This dotted type of distribution of the plant toxins within an aposematic species, or group of species, must be of considerable biological significance, for the same result is achieved in a variety of different ways. Similarly the resistance to beak injury can entail extra tough or extra pliant cuticle. The explanation offered by Rothschild *et al.*, (1979b) in regard to the tiger moths, namely that the phenomenon 'provides a stabilizing influence within the complex for it tends to distribute toxicity among a certain proportion of all the insects involved – thus minimizing the tendency to lose the specimens from both ends of the spectrum', is not entirely satisfactory for it tends to gloss over the question of bird behaviour. Although endoparasites and arthropods are in many cases the most important enemies of aposematic Lepidoptera in their early stages (Dempster, 1967; 1983) a deeper understanding of the significance of chemical defences and the *self-advertising life-style* should be sought first and foremost from the viewpoint of avian predators.

Mullerian Mimicry

Entomologists have developed the bad habit of thinking that Mullerian mimicry is merely another facet of 'common warning coloration' (see Poulton, 1908), although Müller (1878) first recognized the phenomenon in two genera of distasteful, unrelated butterflies (*Ituna* and *Thyridia*), not highly coloured, which were flying together. Mimicry rings can be formed by animals – especially ants and their associates – with no bright colours at all, but they are then far more difficult for us to recognize as such.

Mullerian mimics can be defined as two or more species, related or unrelated phylogenetically, which derive additional benefits and protection by resembling one another. When bird predation is involved, the resemblances almost always entail some similar bright colours, especially among the butterflies and moths. It is therefore probable that colour and pattern mutations played an important role in the evolution of this type of Mullerian mimicry. It is the *ongoing interaction* between the distasteful aposematic species and their specific predators, both in time and space, which distinguishes the aposematic Mullerian mimics from straightforward warning coloration, even if their attributes appear to be similar (Rothschild, 1979).

Among the British Lepidoptera we have an example of a rather special type of Mullerian mimicry, which has received less attention than it deserves. Thus there are families of warningly coloured butterflies such as pierids which do resemble each other for phylogenetic reasons yet, within the dual framework of family relationships and aposematic attributes, display all gradations of Mullerian mimics. Some species of these pierids, for instance those which are confined to the Orient or Europe and whose ranges therefore do not overlap, cannot qualify as Mullerian mimics but must be regarded as related species with similar defence mechanisms. No such reservations exist in areas like the Iberian peninsula where a number of pierid species occur together. These butterflies have the necessary prerequisites for the development of a successful Mullerian mimic; for from a certain distance they all look alike and their behaviour is also broadly similar. It should be noted that none of these species has converged in colour and pattern sufficiently to be virtually indistinguishable (although some of the clouded yellows are nearly so), presumably because the Mullerian mimicry is superimposed upon a long period of speciation within the aposematic life-style. A family like the Pieridae illustrate better than any other situation the sort of continuum formed by mimics within a single framework; for example the family includes at one end of the spectrum such rare types of warning colour as the inedible aposematic pupa (red, bright orange, black and white: *Delias* group) and at the other the hyper-cryptic palatable chrysalis such as that of the orange-tip.

A somewhat similar situation exists in the arctiid moths although there is not such a close-knit relationship with the chemistry of a particular family of host plants. The tiger moths seem able to tolerate a wider range of toxic hosts and in some parts of their range certain genera have, for example, specialized in apocynaceous plants although the family is basically a *Senecio*-feeder (Rothschild *et al.*, 1979b). In fact, the tiger moths qualify in some respects as poison-plant specialists – they are the bulbuls of the insect world. Furthermore their feeding habits range from polyphagy (including both 'apparent' and 'unapparent' plants) to strict monophagy, but by and large they have opted for self-advertisement, and like the pierids constitute a world-wide Mullerian mimicry complex.

VI. Problems and Speculations associated with Aposematic Lepidoptera and Toxic Plants

An unusually high proportion of butterflies and moths (and for that matter of other insects too) feeding on Asclepiadaceae are aposematic. It would seem that the cardiac glycosides in the foodplant are not so toxic for insects as for vertebrate herbivores, and relatively easy for them to accumulate and store. Those which are cryptic are non-storers and they metabolize or excrete the cardiac glycosides ingested with the foliage. Sometimes secondary plant toxins, for example nicotine, are too lethal for insects to incorporate (Rothschild, 1972b), and then the habitual feeders on tobacco are cryptic, not aposematic.

Ever since Fraenkel (1959; 1969) put forward a reasonable theory for the *raison d'être* of secondary plant substances and Reichstein *et al.* (1968; von Euw *et al.*, 1968) proved that monarchs and swallowtails really do sequester and store cardenolides and aristolochic acids, great emphasis has been placed on co-evolution (Ehrlich & Raven, 1965) and the ongoing battle between herbivore and host. Edgar (1984; Edgar & Culvenor, 1975) produced a delightful hypothesis based on the fact that certain primitive asclepiad plants contain both cardenolides and pyrrolizidine alkaloids, whereas most of to-day's species lack (or have lost) the latter substances. Many, if not all, danaid butterflies require pyrrolizidine alkaloids as precursors of their sex pheromones (Boppré, 1977; 1979) and he suggests that, to curb the breeding and to reduce the fitness of the rapacious danaid herbivores, the plants dropped these alkaloids and the butterflies now have to find their sex pheromone precursors and defensive substances elsewhere – not as larvae but as sequestering adults. Occasionally the two substances are still present in the same plant, as in *Parsonsia* spp., and then the larvae sequester and store both (Edgar, 1984). However, I fancy entomologists may be rather too enthusiastic about the 'ongoing conflict', and in reality we are not infrequently witnessing the evolution of peaceful coexistence and, perhaps even more often – though not always – the functioning of mutual benefit societies, rather than war. For instance, I doubt if it is a mere coincidence that monarch butterflies and white butterflies, which both sequester and store secondary plant toxins, are pollinators of the flowers on which their larvae feed. Are not the plants *protecting* their pollinators against bird predators and bacteria by providing them with mustard oils and cardenolides? The sacrifice of a portion of their foliage is a negligible price to pay in comparison with this enormous benefit.★ The loss of their pyrrolizidine alkaloids may have another explanation, for the cardenolides are also lacking in many of the Asclepiadaceae and the distribution of pyrrolizidine alkaloids in the Compositae is sporadic.

It is unlikely that all milkweed feeders become adapted to the plants' chemical attributes along similar evolutionary paths. The monarch, for instance, synthesizes its own heart irritant, which, although it is not emetic or a cardiac glycoside, may well have rendered it less sensitive to these substances. Probably most species which secrete their own toxins have a better chance than others of surviving on a toxic plant. There they would avoid destruction by large grazing herbivores (Trimen, 1887–89; Poulton, 1914; Rothschild, 1972b), which is a greatly under-rated aspect of the relationship. It is a short step from ingestion to sequestration and storage.

It is evident, however, that plant and butterfly have evolved together down the ages, the plants' gradually increasing ability to manufacture effective toxic secondary substances as *protection* against the grazing mammals, matched by the butterflies' increasing ability to adjust physiologically to the changes taking place. Possibly the latex protects both plant and butterfly from tissue damage from the cardiac glycosides. But this may be irrelevant for the monarch whose insensitivity to heart poisons due, according to Vaughan & Jungreis (1977) and Jungreis & Vaughan (1977) to the high levels of potassium in the blood which prevents these steroids from binding to ATPases, may have been a felicitous pre-adaptation rather than a facet of co-evolution.

A recent experiment (Rothschild *et al.*, 1979a) has shown on the other hand that a switch to storage of toxins can, apparently, be achieved with ease. The tobacco hornworm (*Manduca sexta*) feeds on tobacco and excretes the nicotine it ingests. When it was reared on another solanaceous plant, *Atropa belladonna*, it sequestered and stored atropine in relatively large amounts, and synthesized two metabolites, subsequently found in the frass. The resulting pupae proved lethal when fed to chickens.

In parts of Africa and the Canary Isles the death's-head hawk-moth, as we have noted (p. 26) not infrequently feeds on a species of *Datura*, but even if it stores hyoscine it probably does not do so often enough to have affected its appearance. But consider this moth's yellow- and brown-banded abdomen: it is already highly suggestive. It would have only to intensify the existing motif to assume warning coloration. The threat of that dangerous *modus vivendi* hangs over it.

Controversy rages round another aspect of the origins and evolution of aposematism. One school of thought insists that kin selection is a prerequisite for this life-style (Paxton & Harvey, 1983) and the gregarious habits of so many warningly coloured insects enhance the theory. But something analagous to kin selection, yet fundamentally different, though this has not been stressed up to now, very often occurs since many different aposematic species congregate on the same toxic foodplant. I originally pointed this out in connection with the insect aggregations on nettles

---

★ Nor should we forget that the silk moth caterpillar (*Samia cynthia ricini* (Jones)) distributes a plant fertilizer in the form of a fine crystalline cuticular powder (n-tricontanol and n-octacosanol) which improves the growth of the foliage on which it is feeding (Bowers & Thompson, 1965; Ries *et al.*, 1977).

(Rothschild, 1964b). Thus any highly coloured mutation would most likely be avoided by birds, reinforced by their experience of other aposematic species present in the immediate environment.

Another school of thought believes that individual selection takes place, which accounts for the evolution of unpalatability linked to warning colour. I have little doubt that all three types of selection occur but the latter is difficult to demonstrate although Wicklund & Järvi (1982) and Järvi *et al.* (1981) have made a brave attempt to do so with experiments in the laboratory. An amorphous mass of anecdotal evidence supports them. Thus, I myself have witnessed a bat capture, maul, bite and then release a female scarlet tiger and have subsequently reared the larvae from eggs she laid after her lucky escape. Similarly, the late Philip Sheppard witnessed a dragonfly capture and release a scarlet tiger, apparently unharmed (pers.comm.).

Biologists are like predators which have recently acquired a new searching image. Everything has to be fitted into the novel framework. It is most improbable that all forms of warning colour evolved along the identical pathway. For just as every case of mimicry is in reality a 'special case' so every case of aposematism is subtly different.

The drawers of the Rothschild collection bear witness to the fact that aposematic moths like the magpie moth and the garden tiger and the burnets provide entomologists with series of remarkably attractive varieties. It is quite likely that the cryptic forewings of, for example, the brindled beauty and the purple thorn (*Selenia tetralunaria* (Hufnagel)) vary just as much, but are less eye-catching, yet the tendency to produce colour and pattern aberrations may well be one of the prerequisites of the tiger moth type of aposematic life-style (see also Hutchinson, 1975). What we should not underrate are the number of *different* evolutionary pathways leading to self-advertisement. This danger is ever present. It has been pointed out elsewhere (Rothschild, 1979) that the relative numbers of cryptic and aposematic species tell their own tale – the road to warning coloration must be littered with evolutionary failures. Some plants may evolve a symbiotic relationship with specific caterpillars, but many Lepidoptera, compelled to share their host's defensive toxins, must have been destroyed by the unwanted attention which is consequently thrust upon them.

VII. A Different Aspect of Aposematic Lepidoptera

Warningly coloured insects are good indicators of chemically interesting plants, some of which contain substances of medical importance (Brown, 1980). Furthermore we can expect transformation of these chemicals within the herbivore. The cryptic species excrete the toxins but the aposematic species retain them, and in the course of development may produce new metabolites from the material ingested, for example, callimorphine synthesized from the pyrrolizidine alkaloids sequestered by arctiids from *Senecio* and *Symphytum* spp. (Edgar *et al.*, 1980).

Fed on *Euphorbia polychroma*, the spurge hawk-moth larva sequesters a diterpene from the plant and after passage through the caterpillar it can be extracted from the pupa and injected into the cancerous rat – a significant reduction in Walker's rat carcinoma follows (Marsh *et al.*, 1984). Other workers (Pettit *et al.*, 1968) have found tumour depressants in several species of Lepidoptera and this is clearly a field meriting further research.

Lepidoptera are prone to succumb to virus infections as every amateur entomologist knows to his cost. But at least one group of aposematic moths, the zygaenids, seem totally immune to their attack. I can find no record of zygaenids infected with virus disease. Possibly their apparent immunity could be exploited, and a vaccine produced to protect susceptible species.

The recent surge of interest in the influence of dietary beta-carotene on the reduction of human cancer rates (Peto *et al.*, 1981) and the concentration of these yellow pigments in the reproductive organs of many insects and other animals (Goodwin, 1950; Kayser, 1982), suggesting they may influence embryonic growth, is yet another facet of warning coloration requiring intensive research.

This is clearly only the edge of an immense field ripe for investigation, for next to nothing is known of the biochemistry of the aposematic Lepidoptera and one may well be able to exploit their incredible versatility and use them as tools for the production of new substances, ranging from antibiotic agents (McWhirter, in prep.) to tumour depressants (Marsh *et al.*, 1984) and aphrodisiacs.*

VIII. Conclusions

Vertebrates and a few species of large Hymenoptera with colour vision are the foremost predators of British aposematic Lepidoptera, both as adults and in the later larval stages. It is these animals which impose the warning life-style upon them. But their varied chemical defences (Eisner, 1970) are also directed selectively against a plethora of arthropod and other enemies of butterflies and moths, ranging from hornets, spiders, harvestmen or egg mites to bacteria and fungi and also perhaps virus. The exocrine and non-exocrine secretions of Lepidoptera, however, are purely defensive, not offensive like the venoms of many Hymenoptera and Arachnida, and therefore contain somewhat different categories of chemicals than those, for example, of spiders and hornets. These animals direct their stings and bites towards prey rather than predators – but they can serve for both purposes – although some substances like the biogenic amines may be shared. The relationship of the Lepidoptera with plants is close, and many of the chemicals they sequester, such as the mustard oils, cardiac glycosides and aristolochia acids, are known to be bactericidal and fungicidal as well as bird deterrent, and no doubt jointly protect their hosts and themselves. Sequestration and storage probably originated as a campaign directed against bacteria rather than vertebrates or dragonflies, but the emergence of the aposematic life-style reflects the arrival among the flowering plants of predators with colour vision.

Many of the alerting signals as well as the defensive substances synthesized or sequestered by Lepidoptera probably serve a dual purpose: protection from enemies and attraction for friends. Thus brilliant colours participate in both warning and sexual displays. Pyrazines which are also found in the mandibular glands of ants and in ant trails (Jones & Blum, 1983) no doubt assist in memorizing agreeable and significant, as well as unpleasant, events. The evocative scent of vanillin, which both

---

* It is known, for instance, that the pyrrolizidine alkaloids, despite their long-term deleterious effects, can stimulate the sex drive in man. There are similar suggestions for the cardiac glycosides stored by butterflies feeding on *Calatropis procera* (Schoental, unpublished results; Pobéguin, 1912; Kerharo & Bouquet, 1950).

butterflies and plants employ as attractants, can be jacked up to produce a most repellent stench (Rothschild, 1964b). The pyrrolizidine alkaloids which are powerful deterrents for various spiders (Brown, 1984) also function as sex stimulants or the precursors of pheromones for both arctiids and danaids (Boppré, 1977, 1979). The stridulation of some tiger moths can warn bats of their dangerous qualities (Roeder, 1967) and also serve to locate the opposite sex.

We are equipped to appreciate only the visual fraction of the aposematic life-style with its vivid colours and subtle patterns and mysterious glitter. We stand awed and envious trying to catch more than a few clicks and ripples of sound with our inferior ears, and sniffing ineffectually on the perimeter of the virtually unknown world of chemical messages.

Notes on the Literature

Owing to the importance of insect/plant relationship it is advisable to consult some more general books on ecological chemistry and the biochemistry of the plant hosts, in addition to my list of specific references. For examples, see below:

Harborne, J. B., 1977. *Introduction to ecological biochemistry,* (Edn 2). London & New York.

———, Boulter, D. & Turner, B. L., (Eds), 1971. *Chemotaxonomy of Leguminosae.* London.

Hawkes, J. G., Lester R. N. & Skelding A. D., (Eds), 1979. The biology and taxonomy of the Solanaceae. *Linn. Soc. Symp. Ser.* 7.

Hoch, J. H., 1961. *A survey of cardiac glycosides and genins.* Columbia. (For lists of plants containing cardiac glycosides.)

Manske, R. H. F. & Holmes, H. L., (Eds), 1950– . *The alkaloids,* 1– . London & New York.

Rosenthal, G. A. & Janzen, D. H., (Eds), 1979. *Herbivores, their interaction with secondary plant metabolites.* London & New York.

Vaughan, J. G., MacLeod, A. J. & Jones, B. M. G., 1976. *The biology and chemistry of the Cruciferae.* London.

There have been various Symposia during the past few years on the question of plant/insect interaction, for instance:

Gilbert, L. E. & Raven, P. H., (Eds), 1975. *Coevolution of animals and plants.* Austin, U.S.A. & London.

Harborne, J. B., (Ed.). 1978. *Biochemical aspects of plant and animal coevolution.* London & New York.

Van Emden, H. F., (Ed.), 1972. Insect/plant relationships. *Symp. R. ent. Soc. Lond.* **6**.

Wallace, J. W. & Mansell, R. L., (Eds), 1976. *Biochemical interaction between plants and insects.* New York & London.

Although published some time ago Swain's two volumes are invaluable:

Swain, T., 1963. *Chemical plant taxonomy.* London & New York.

———, 1966. *Comparative phytochemistry.* London & New York.

Papers and books written by the older naturalists (1850–1950) are the most fruitful source of information concerning the predators of aposematic species, their mimics and life-styles. Starting with Poulton and Hale Carpenter and working backwards, a wealth of relevant detail can be collected. Today the most lively brains in science are attracted to molecular biology and genetic engineering, but at the turn of the century their equivalent numbers were studying evolution and natural history. The quality of their contribution was of the highest order and is still invaluable for the student of warning colour and mimicry.

Acknowledgements

My grateful thanks are due to Keith Brown, Rachael Galun, Gadi Katzir, Allan Watson and Paul Whalley for reading the manuscript and for helpful criticisms and suggestions. I am also deeply indebted to Walter Blaney for all the trouble he has taken with the oscillograph tracings, and to D. C. Lees for providing the oscillograph tracings of the clearwing and hornet wing vibrations. Luciano Bullini kindly gave me permission to reproduce Paolo Ragazzini's marvellous photograph of the clearwing (*Pennisetia hylaeiformis*), and Lincoln Brower generously provided a photograph of the American hawk-moth caterpillar, *Eumorpha labruscae*. The late Howard Hinton made the drawings of the tarsal glands of the garden tiger (*Arctia caja*). He dissected and drew a very large number of legs of various families of moths, but unfortunately did not live to complete the joint piece of work we had planned.

I also wish to thank the Photographic Unit at the British Museum (Natural History) for the excellent U.V. photographs of the death's-head hawk-moth and the British Library (National Sound Archive) and British Beekeepers Association for the recordings of the queen bee piping.

All other photographs were taken by the author.

References

Ackery, P. R. & Vane-Wright, R. I., 1984. *Milkweed butterflies: their cladistics and biology*. London.

Allyn, A. C., Rothschild, M. & Smith, D. S., 1982. Microstructure of blue/green and yellow pigmented wing membranes in Lepidoptera with remarks concerning the function of pterobilins. I. Genus *Graphium*. *Bull. Allyn Mus.* No. 75: 1–20.

Aplin, R. T., D'Arcy Ward, T. & Rothschild, M., 1975. Examination of the large white and small white butterflies (*Pieris* spp.) for the presence of mustard oil glycosides. *J. Ent.* (A) **50**: 73–78.

———— & Rothschild, M., 1972. Poisonous alkaloids in the body tissues of the garden tiger moth (*Arctia caja* L.) and the cinnabar moth (*Tyria* (= *Callimorpha*) *jacobaeae* L.) (Lepidoptera). *In* Vries, A. de & Kochva, K., (eds), *Toxins of animal and plant origin*, **2**: 579–595. London.

Baker, R. R., 1970. Bird predation as a selective pressure on the immature stages of the cabbage butterflies *Pieris rapae* and *P. brassicae*. *J. Zool., Lond.* **162**: 43–59.

Bergstrom, G., 1979. Complexity of volatile signals in Hymenopteran insects. Some central problems regarding analytical techniques and biological interpretations in work with multicomponent secretions in bees, bumblebees and ants. *In* Ritter, R. J., (ed.), *Chemical Ecology: Odour communication in animals*: 187–200. New York.

———— & Lofqvist, J., 1971. *Camponotus ligniperda* Latr. a model for the composite volatile secretions of Dufour's gland in formicine ants. *In* Tahori, A. S., (ed.), *Chemical releasers in insects. Pesticide Chemistry.* **3**: 195–223. London.

Birch, M. C., 1979. Eversible structures. *In* Heath, J. & Emmet, A. M, (eds), *The moths and butterflies of Great Britain and Ireland*, **9**: 9–18. London.

Bisping, R., Benz, U., Boxer, P. & Longo, N., 1974. Chemical transfer of learned colour discrimination in goldfish. *Nature, Lond.* **249**: 771–773.

Bisset, G. W., Frazer, J. F. D., Rothschild, M. & Schachter, M., 1960. A choline ester and other substances in the Garden Tiger moth *Arctia caja* (L.). *J. Physiol., Lond.* **146**: 38–39.

Blest, A. D., 1963. Longevity, palatability and natural selection in five species of New World Saturniid moths. *Nature, Lond.* **197**: 1183–1187.

———, 1964. Protective display and sound production in some New World arctiid and ctenuchid moths. *Zoologica, Stuttgart* **49**: 161–181.

Blum, M. S., 1981. *Chemical defenses of arthropods.* New York & London.

Boppré, M., 1977. Pheromonbiologie am Beispiel der Monarchfalter (Danaidae). *Biol. Unserer Zeit.* **7**: 161–169.

———, 1979. *Untersuchungen zur Pheromonbiologie die Monarchfaltern (Danaidae).* Ph.D. Thesis: Ludwig-Maximilians-Universität, München.

Bourgogne, J., 1951. Ordre des Lépidoptères. *In* Grassé, P., (ed.), *Traité de Zoologie*, **10**(1): 174–448. Paris.

Bowers, M. D., 1980. Unpalatability as a defense strategy of *Euphydryas phaeton* (Lepidoptera: Nymphalidae). *Evolution* **34**: 586–600.

Bowers, W. S. & Thompson, M. J., 1965. Identification of the major constituents of the crystalline powder covering the larval cuticle of *Samia cynthia ricini* (Jones). *J. Insect Physiol.* **11**: 1003–1011.

Brower, J. van Z., 1958. Experimental studies of mimicry in some North American butterflies. Pt. III. *Danaus gilippus berenice* and *Limenitis archippus floridensis*. *Evolution* **12**: 273–285.

Brower, L. P., Brower, J. van Z. & Corvino, J. M., 1967. Plant poisons in a terrestrial food chain. *Proc. natn. Acad. Sci. U.S.A.* **57**: 893–898.

——— & Glazier, S., 1975. Localization of heart poisons in the Monarch butterfly. *Science* **188**: 19–25.

———, McEvoy, P. B., Williamson, K. L. & Flannery, M. A., 1972. Variation in cardiac glycoside content of Monarch butterflies from natural populations in Eastern North America. *Ibid.* **177**: 426–429.

———, Pough, F. H. & Meck, H. R., 1970. Theoretical investigations of automimicry: I. Single trial learning. *Proc. natn. Acad. Sci. U.S.A.* **66**: 1059–1066.

Brown, K. S., Jr., 1980. Insetos aposemáticos: indicadores naturais de planta medicinas. V. Simpósio de plantas medicinas do Brasil. *Ciên. Cult. S. Paulo* **32** (suppl.): 189–200.

———, 1984. Chemical ecology of dehydropyrrolizidine alkaloids in adult Ithomiinae (Lepidoptera: Nymphalidae). *Rev. bras. Biol.* **44**(3).

Bullini, L. & Sbordoni, V., 1970. Evoluzione del mimetismo in *Zygaena ephialtes* (L.). *Atti Ass. Genet. ital.* **15**: 207–209.

——— & ———, 1971. Ricerche sperimentali sul valore mimetico della forme efialtoidi rosse di *Zygaena ephialtes* (Lepidoptera, Zygaenidae). *Boll. Zool.* **38**: 502.

———, ——— & Ragazzini, P., 1969. Mimetismo Mülleriano in popolizoni Italiane di *Zygaena ephialtes* (L.) (Lepidoptera, Zygaenidae). *Archo zool. ital.* **54**: 181–214.

Busnel, R.-G., 1963. On certain aspects of animal acoustic signals. *In* Busnel, R.-G., (ed.), *Acoustic behaviour of animals*, Chap. 5, pp. 69–111. Amsterdam, London & New York.

Calvert, W. H., Hedrick, L. E. & Brower, L. P., 1979. Mortality of the Monarch Butterfly (*Danaus plexippus* L.): Avian predation at five overwintering sites in Mexico. *Science* **204**: 847–851.

Carpenter, G. D. Hale, 1941–42. Observations and experiments in Africa by the late C. F. M. Swynnerton on wild birds eating butterflies and the preference shown. *Proc. Linn. Soc. Lond.* **1941–42**: 10–46.

———, 1941. The relative frequency of beak-marks on butterflies of different edibility to birds. *Proc. zool. Soc. Lond.* (A) **111**: 223–231.

Chesher, G. B. & Collier, H. O. J., 1955. Identification of 5-hydroxytryptamine in nettle sting. *J. Physiol.* **130**: 41–42.

Collenette, C. L., 1935. Notes concerning attacks by British birds on butterflies. *Proc. zool. Soc. Lond.* **1935**: 201–217.

—— & Talbot, G., 1928. Observations on the bionomics of the Lepidoptera of Matto Grosso, Brazil. *Trans. ent. Soc. Lond.* **76**: 391–416.

Collins, C. T. & Watson, A., 1983. Field observations on bird predation on Neotropical moths. *Biotropica* **15**: 53–60.

Conner, W. E., Eisner, T., Van der Meer, R. K., Guerrero, A. & Meinwald, J., 1981. Precopulatory sexual interaction in an arctiid moth (*Utetheisa ornatrix*). Role of a pheromone derived from dietary alkaloids. *Behav. Ecol. Sociobiol.* **9**: 227–235.

Cott, H. B., 1940. *Adaptive coloration in animals.* London.

——, 1946. The Edibility of Birds: Illustrated by five years' experiments and observations (1941–1946) on the food preferences of the Hornet, Cat and Man, and considered with special reference to the theories of adaptive coloration. *Proc. zool. Soc. Lond.* **116**: 371–524.

Davies, N. B. & Green, R. E., 1976. The development and ecological significance of feeding techniques in the Reed Warbler (*Acrocephalus scirpaceus*). *Anim. Behav.* **24**: 213–229.

Davis, R. H. & Nahrstedt, A., 1979. Linamarin and Lotaustralin as the source of cyanide in *Zygaena filipendulae* L. (Lepidoptera). *Comp. Biochem. Physiol.* **64B**: 395–397.

Daw, N. W., 1973. Neurophysiology of colour vision. *Physiol. Rev.* **53**: 571–611.

Dazzini, M. V. & Finzi, P. V., 1974. Chemically known constituents of Arthropod defensive secretions. *Atti Accad. naz. Lincei Mem.* Ser. VIII **12**: 107–146.

de Jong, M. C. J. M. & Bleumink, K., 1977. Investigative studies of the dermatitis caused by the larva of the brown-tail moth *Euproctis chrysorrhoea* L. (Lepidoptera, Lymantridae). IV. Further characterization of skin reactive substances. *Arch. Dermatol. Res.* **259**: 263–281.

Dempster, J. P., 1967. The control of *Pieris rapae* with DDT. (1) The natural mortality of the young stages of *Pieris. J. Appl. Ecol.* **4**: 485–500.

——, 1983. The natural control of populations of butterflies and moths. *Biol. Rev.* **58**: 461–481.

de Ruiter, I., 1952. Some experiments on the camouflage of stick caterpillars. *Behaviour* **4**: 222–232.

Dixon, C. A., Erickson, J. M., Kellett, D. N. & Rothschild, M., 1978. Some adaptations between *Danaus plexippus* and its food plant, with notes on *Danaus chrysippus* and *Euploea core* (Insecta: Lepidoptera). *J. Zool., Lond.* **185**: 437–467.

Downey, J. C. & Allyn, A. C., 1973. Butterfly Ultrastructure. I. Sound production and associated abdominal structures in pupae of Lycaenidae and Riodinidae. *Bull. Allyn Mus.* No. 14: 1–47.

—— & ——, 1978. Sounds produced in pupae of Lycaenidae. *Ibid.* No. 48: 1–14.

Edgar, J. A., 1975. Danainae (Lep.) and 1, 2-dehydropyrrolizidine alkaloid-containing plants – with reference to observations made in the New Hebrides. *Phil. Trans. R. Soc. Lond.* **272**: 467–476.

——, 1984. Parsonsieae: ancestral larval food plants of the Danainae and Ithomiinae. *In* Vane-Wright, R. I. & Ackery, P. R. (eds), Biology of butterflies. *Symp. R. ent. Soc. Lond.* No. 11; 91–93.

——, Boppré, M. & Schneider, D., 1979. Pyrrolizidine alkaloid storage in African and Australian danaid butterflies. *Experentia* **35**: 1447–1448.

——, Cockrum, P. A. & Frahn, J. L., 1976. Pyrrolizidine alkaloids in *Danaus plexippus* and *Danaus chrysippus* L. *Ibid.* **32**: 1535–1537.

—— & Culvenor, C. C. J., 1975. Pyrrolizidine alkaloids in *Parsonsia* species (family Apocynaceae) which attract Danaid butterflies. *Ibid.* **31**: 393–394.

——, ——, Cockrum, P. A., Smith, L. W. & Rothschild, M., 1980. Callimorphine: identification and synthesis of the Cinnabar moth 'metabolite'. *Tetrahedron Letters* **21**: 1383–1384.

——, —— & Pliske, T. E., 1974. Coevolution of Danaid butterflies with their host plants. *Nature, Lond.* **250**: 646–648.

Edmunds, J., 1974. *Defence in animals*. Harlow & New York.

Ehrlich, P. R. & Raven, P. H., 1965. Butterflies and plants: a study in coevolution. *Evolution* **18**: 586–608.

Eisner, T., 1970. Chemical defense against predation in Arthropods. *In* Sondheimer, E. & Simeone, J. B., (eds), *Chemical Ecology*, Chap. 8, pp. 157–217. New York & London.

—— & Meinwald, Y. C., 1965. Defensive secretion of a caterpillar (*Papilio*). *Science* **150**: 1733–1735.

Eltringham, H., 1910. *African mimetic butterflies*. Oxford.

Emmelin, N. & Feltberg, W. 1947a. Pharmacologically active substances in the fluid of nettle hairs (*Urtica urens*). *J. Physiol.* **106**: 14–15.

—— & ——, 1947b. The mechanism of the sting of the common nettle *Urtica urens*. *Ibid.* **106**: 440–455.

Evans, C. S. & Bella, E. A., 1980. Neuroactive plant amino acids and amines. *Trends in Neurosciences.* **1980** (March): 70–72.

Everist, S. L., 1981. *Poisonous plants of Australia*. (Rev. edn). Sydney.

Feltwell, J. S. E. & Rothschild, M., 1974. Carotenoids in 38 species of Lepidoptera. *J. Zool., Lond.* **174**: 441–465.

—— & Valadon, G., 1970. Plant pigments identified in the Common Blue butterfly (*Polyommatus icarus* Rott., Lep. Lycaenidae). *Nature, Lond.* **225**: 969.

Fink, L. S. & Brower, L. P., 1981. Birds can overcome the cardenolide defence of monarch butterflies in Mexico. *Ibid.* **291**: 67–70.

Ford, E. B., 1941. Studies on the chemistry of pigments in the Lepidoptera with reference to their bearing on systematics. I. The anthoxanthins. *Proc. R. ent. Soc. Lond.* (A) **16**: 65–90.

Fraenkel, G. S., 1959. The raison d'être of secondary plant substances. *Science* **129**: 1466–1470.

——, 1969. Evaluation of our thoughts on secondary plant substances. *Entomologia exp. appl.* **12**: 473–486.

Frazer, J. F. D., 1965. The cause of urtication produced by larval hairs of *Arctia caja* (L.) (Lepidoptera: Arctiidae). *Proc. R. ent. Soc. Lond.* (A) **40**: 96–100.

—— & Rothschild, M., 1961. Defence mechanisms in warningly-coloured moths and other insects. *Proc. 11th Int. Congr. Ent. Vienna, 1960* **1**: 249–256.

Fry, C. H., 1969. The recognition and treatment of venomous and non-venomous insects by small bee-eaters. *Ibis* **111**: 23–29.

Goodman Gilman, A., Goodman, L. S. & Gilman, A., (eds)., 1980. *The pharmacological basis of therapeutics*, (Edn 6). New York.

Goodwin, T. W., 1950. Carotenoids and reproduction. *Biol. Rev.* **25**: 391–413.

Grant, V., 1978. Kin selection: a critique. *Biol. Zbl.* **97**: 385–392.

Guggenheim, M., 1951. *Die Biogenin Amine*: 157–158, 448. Basel & New York.

Harborne, J. B., 1977. Chemosystematics and coevolution. *Pure & appl. Chem.* **49**: 1403–1421.

Haskell, P. T., 1961. *Insect sounds*, 189 pp., 97 figs. London.

Heath, J. & Emmet, A. M., (eds), 1979. *The moths and butterflies of Great Britain and Ireland*, **9**, 288 pp, 13 col.pls, 19 figs, 203 maps. London.

Henry, T. A., 1949. *The plant alkaloids*. London.

Hoch, J. H., 1961. *A survey of cardiac glycosides and genins*. Columbia.

Hodges, R. W., *et al.* (Eds)., 1983. *Check list of the Lepidoptera of America North of Mexico*, xxiv, 284 pp. London.

Hsiao, T. H., Hsiao C. & Rothschild, M., 1980. Characterization of a protein toxin from dried specimens of the Garden Tiger moth (*Arctia caja* L.). *Toxicon* **18**: 291–299.

Hutchinson, G. E., 1975. Variations on a theme by Robert MacArthur. In *Ecology and evolution of communities*, pp. 492–521. Harvard.

Järvi, T., Sillen-Tullberg, B. & Wiklund, C., 1981. The cost of being aposematic. An experimental study of predation on larvae of *Papilio machaon* by the great tit *Parus major*. *Oikos* **36**: 267–272.

Jones, D. A., Parsons, J. & Rothschild, M., 1962. Release of hydrocyanic acid from crushed tissues of all stages in the life cycle of species of the Zygaeninae (Lepidoptera). *Nature, Lond.* **193**: 52–63.

Jones, F. M., 1932. Insect coloration and relative acceptability of insects to birds. *Trans. ent. Soc. Lond.* **80**: 345–385.

———, 1934. Further experiments on coloration and relative acceptability of insects to birds. *Trans. R. ent. Soc. Lond.* **82**: 443–453.

Jones, T. H. & Blum, M. S., 1983. Arthropod alkaloids: distribution, functions and chemistry. *In* Pelletier, S. W., (ed.), *Alkaloids: Chemical and biological perspectives* **1**: 33–84. New York.

Jordan, K., 1907–08. Familie: Zygaenidae, Widderchen. *In* Seitz, A., (ed.), *Gross-Schmetterlinge der Erde*, Abt.2, **10**: 5–56. Stuttgart.

Jungreis, A. M. & Vaughan, G. L., 1977. Insensitivity of Lepidopteran tissues to ouabain: absence of ouabain binding and Na$^+$–K$^+$ ATPases in larval and adult midgut. *J. Insect Physiol.* **23**: 503–509.

Kaiser, E. & Michl, H., 1958. *Die Biochemie der tierischen gifts.* Wien.

Kayser, H., 1982. Carotenoids in Insects. *In* Britton, G. & Goodwin, T. W., (eds), *Carotenoid chemistry & biochemistry*: 195–210. Oxford.

Kerharo, J. & Bouquet, A., 1950. *Sorciers, féticheurs et guérisseurs de la Côte d'Ivoire – Haute Volta.* Paris.

Kettlewell, H. B. D., 1973. *The evolution of melanism*, xv, 423 pp., 39 pls (3 col.). Oxford.

Krebs, J. R. & Davies, N. M., 1981. *An introduction to behavioural ecology.* Oxford.

Lane, C. D., 1957a. Preliminary note on insects eaten and rejected by a tame Shama (*Kittacincla malabarica* (Gm.)) with the suggestion that in certain species of butterflies and moths females are less palatable than males. *Entomologist's mon. Mag.* **93**: 172–179.

———, 1957b. Notes on the brush organs and cervical glands of the Ruby Tiger (*Phragmatobia fuliginosa* L.). *Entomologist* **90**: 148–151.

———, 1959. A very toxic moth: the five spot Burnet (*Zygaena trifolii* Esp.). *Entomologist's mon. Mag.* **95**: 93–94.

———, 1964. Round the blue lamp. *Ibid.* **99**: 189–195.

——— & Rothschild, M., 1965. A case of Mullerian mimicry of sound. *Proc. R. ent. Soc. Lond.* (A) **40**: 156–158.

Marsh, N. & Rothschild, M., 1974. Aposematic and cryptic Lepidoptera tested on the mouse. *J. Zool., Lond.* **174**: 89–122.

———, ——— & Evans, F., 1984. A new look at butterfly toxins. *In* Vane-Wright, R. I. & Ackery, P. R., (eds), Biology of butterflies. *Symp. R. ent. Soc. Lond.* No. 11: 135–139.

Marshall, G. A. K., 1902. Five years observations and experiments (1896–1901) on the bionomics of South African insects chiefly directed to the investigation of mimicry and warning colours. *Trans. ent. Soc. Lond.* **1902**: 287–584.

———, 1909. Birds as a factor in the production of mimetic resemblances among butterflies. *Ibid.* **1909**: 329–384.

Minnich, D. E., 1925. The reactions of the larvae of *Vanessa antiopa* Linn. to sounds. *J. exp. Zool.* **42**: 443–469.

Moore, B. P. & Brown, W. V., 1981. Identification of warning odour components, bitter principles and antifeedants in an aposematic beetle *Metriorrynchus rhipidius* (Coleoptera, Lycidae). *Insect Biochem.* **11**: 493–499.

Morley, J. & Schachter, M., 1963. Acetylcholine in non-nervous tissues of some Lepidoptera. *J. Physiol.* **168**: 706–715.

Mostler, G., 1935. Boebachtungen zur Frage der Wespenmimikry. *Z. Morph. Okol. Tiere* **29**: 381–454.

Müller, F., 1878. On the advantages of mimicry among butterflies. *Zool. Anz.* **1**: 54–55.

Nahrstedt, A. & Davis, R. H., 1983. Occurrence, variation and biosynthesis of the cyanogenic glucosides Linamarin and Lotustralin in species of the Heliconiini (Insecta: Lepidoptera). *Comp. Biochem. Physiol.* **75B**: 65–73.

Nicolaus, L. K., Cassel, J. F., Carlson, R. B. & Gustavson, C. R., 1983. Taste aversion conditioning of crows to control predation on eggs. *Science* **220**: 212–214.

Papageorgis, C. A., 1975. The adaptive significance of wing coloration of mimetic Neotropical butterflies. *Diss. Abstr. int. B*, **36**: 1571–1572.

Parsons, J. A., 1965. A digitalis-like toxin in the Monarch butterfly, *Danaus plexippus* L. *J. Physiol., Lond.* **178**: 290–304.

—— & Rothschild, M., 1964. Rhodanese in the larva and pupa of the Common Blue Butterfly (*Polyommatus icarus* (Rott.)) (Lepidoptera). *Entomologist's Gaz.* **15**: 58–59.

Pasteels, J., 1976. Evolutionary aspects in chemical ecology and chemical communication. *Proc. XV Int. congr. ent.*: 281–293.

—— & Daloze, D., 1977. Cardiac glycosides in the defensive secretion of Chrysomelid beetles: evidence for their production by the insects. *Science* **197**: 70–72.

——, Gregoire, J.-C. & Rowell-Rahter, M., 1983. The chemical ecology of defense in arthropods. *Ann. Rev. Ent.* **28**: 263–289.

Pavan, M. & Dazzini, M. V., 1976. Sostanze di difesa dei Lepidotteri. *Pubbl. Ist. Ent. agr. Univ. Pavia* No. 3: 3–23.

Paxton, R. J. & Harvey, P. H., 1983. On the evolution of Mullerian mimicry: hypotheses and tests. *Oikos* **41**: 146–148.

Peto, R., Doll, R., Buckley, J. D. & Sporn, M. B., 1981. Can dietary beta-carotene materially reduce human cancer rates? *Nature, Lond.* **290**: 201–208.

Pettit, G. R., Hartwell, J. L. & Wood, H. B., 1968. Arthropod antineoplastic agents. *Cancer Res.* **28**: 2168–2169.

Pliske, T. E. & Eisner, T., 1969. Sex Pheromone of the queen butterfly: biology. *Science* **164**: 1170–1172.

Pobéguin, H., 1912. *Plantes médicinales de la Guinée.* Paris.

Pocock, R. T., 1911. On the palatability of some British insects, with notes on the significance of mimetic resemblance. (With notes on the experiments by E. B. Poulton.) *Proc. zool. Soc. Lond.* **1911**: 809–868.

Poulton, E. B., 1890. *The colours of animals* (Edn 2), xiii, 360 pp. London.

——, 1908. *Essays on evolution.* Oxford.

——, 1914. Mimicry in North American Butterflies: a reply. *Proc. Acad. nat. Sci. Philad.* **1914**: 161–195.

Prell, H., 1920. Die Stimme des Totenkopfes (*A. atropos*). *Zool. Jb. Abt. Syst. Geog. Biol. Tiere* **42**: 235–272.

Quiroz, A. D., 1978. Venoms of lepidoptera. *In* Bettini, S., (ed.), *Arthropod venoms*, Chap. 20, pp. 555–612. Berlin.

Reichstein, T., von Euw, J., Parsons, J. A. & Rothschild, M., 1968. Heart poisons in the Monarch Butterfly. *Science* **161**: 861–866.

Rhoades, D. F., 1979. Evolution of plant chemical defense against herbivores. *In* Rosenthal, G. A. & Janzen, D. H., (eds), *Herbivores: their interaction with secondary plant metabolites*, pp. 4–54. New York & London.

Ries, S. K., Wert, V., Sweeley, C. C. & Leavitt, R. A., 1977. Triacontanol: a new naturally occurring plant growth regulator. *Science* **195**: 1339–1341.

Rocci, U., 1914. Stella resitenza degli Zigenini allacido cianidrico. *Z. allg. Physiol.* **16**: 42–64.

———, 1916. Sur une substance véneuse contenue dans les Zygènes. *Archs ital. Biol.* **66**: 73–96.

Roeder, K. D., 1967. *Nerve cells and insect behavior.* Cambridge, Mass.

Rothschild, M., 1961a. A female of the Crimson Speckled Footman (*Utetheisa pulchella* L.) captured at Ashton Wold. *Proc. R. ent. Soc. Lond.* (C) **26**: 35–36.

———, 1961b. Defensive odours and Mullerian mimicry among insects. *Trans. R. ent. Soc. Lond.* **113**: 101–121.

———, 1964a. An extension of Dr. Lincoln Brower's theory on bird predation and food specificity together with some observations on bird memory in relation to aposematic colour patterns. *Entomologist* **97**: 73–78.

———, 1964b. A note on the evolution of defensive and repellent odours of insects. *Ibid.* **97**: 276–280.

———, 1967. Mimicry, the deceptive way of life. *Nat. Hist. N.Y.* **76**: 44–51.

———, 1970. Crystals, fungi and poison glands. *Animals* **12**: 402–403.

———, 1972a. Secondary plant substances and warning colouration in insects. *In* van Emden, H. F., (ed.), Insect/plant relationships. *Symp. R. ent. Soc. Lond.* No. 6: 59–83.

———, 1972b. Some observations on the relationship between plants, toxic insects and birds. *In* Harborne, J. B., (ed.), *Phytochemical Ecology*, pp. 2–12. New York & London.

———, 1975. Remarks on carotenoids in the evolution of signals. *In* Gilbert, L. E. & Raven, P. H., (eds), *Coevolution of animals and plants*, pp. 20–51. Austin, U.S.A., & London.

———, 1979. Mimicry, Butterflies and Plants. *Symb. bot. Upsal.* **22**: 82–99.

———, 1981. The mimicrats must move with the times. *Biol. J. Linn. Soc.* **16**: 21–23.

———, 1984. Aide mémoire mimicry. *Ecol. Ent.* **9**: 311–319.

———, (in press). Inherited beak-wiping behaviour. *Ibis*

——— & Aplin, R. T., 1971. Toxins in tiger moths (Arctiidae: Lepidoptera). *In* Tahori, A. S., (ed.), *Pesticide Chemistry* **3**, Chemical Releasers in Insects, pp. 177–182. London.

———, Aplin, R., Baker, J. & Marsh, N., 1979a. Toxicity induced in the Tobacco Horn-worm (*Manduca sexta* L.) (Sphingidae: Lepidoptera). *Nature, Lond.* **280**: 487–488.

———, Aplin, R. T., Cockrum, P. A., Edgar, J. A., Fairweather, P. & Lees, R., 1979b. Pyrrolizidine alkaloids in Arctiid moths (Lep.) with a discussion on host plant relationships and the role of these secondary plant substances in the Arctiidae. *Biol. J. Linn. Soc.* **12**: 302–326.

——— & Edgar, J. A., 1978. Pyrrolizidine alkaloids from *Senecio vulgaris* sequestered and stored by *Danaus plexippus*. *J. Zool., Lond.* **186**: 347–349.

——— & Fairbairn, J. W., 1980. Ovipositing butterfly (*Pieris brassicae* L.) distinguishes between aqueous extracts of two strains of *Cannabis sativa* L. and THC and CBD. *Nature, Lond.* **286**: 56–59.

———, Gardiner, B., Valadon, G. & Mummery, R., 1975b. Lack of response to background colour in *Pieris brassicae* pupae reared on carotenoid free diet. *Nature, Lond.* **254**: 592–594.

——— & Haskell, P. T., 1966. Stridulation of the Garden Tiger moth *Arctia caja* L., audible to the human ear. *Proc. R. ent. Soc. Lond.* (A) **41**: 167–170.

——— & Kellett, D. N., 1972. Reactions of various predators to insects storing heart poisons (cardiac glycosides) in their tissues. *J. Ent.* (A) **46**: 103–110.

———, Keutmann, H., Lane, N. J., Parsons, J., Prince, W. & Swales, L. S., 1979c. A study of the mode of action and composition of a toxin from the female abdomen and eggs of *Arctia caja* L. (Lep. Arctiidae): an electrophysiological, ultrastructural and biochemical analysis. *Toxicon* **17**: 285–306.

——— & Marsh, N., 1978. Some peculiar aspects of Danaid/plant relationships. *Entomologia exp. appl.* **24**: 437–450.

———, Marsh, N. & Gardiner, B., 1978. Cardioactive substances in the Monarch butterfly and *Euploea core* reared on leaf-free artificial diet. *Nature, Lond.* **275**: 649–650.

———, Moore, B. P. & Brown, W. V., 1984. Pyrazines as warning odour components in the Monarch butterfly, *Danaus plexippus*, and in moths of the genera *Zygaena* and *Amata* (Lepidoptera). *Biol. J. Linn. Soc.* **23**: 375–380.

——— & Reichstein, T., 1976. Some problems associated with the storage of cardiac glycosides by insects. *Nova Acta Leopoldina*, Suppl. **7**: 507–550.

———, Reichstein, T., von Euw, J., Aplin, R. & Harman, R. R. M., 1970. Toxic Lepidoptera. *Toxicon* **8**: 293–299.

———, Rowan, M. G. & Fairbairn, J. W., 1977b. Storage of cannabinoids by *Arctia caja* and *Zonocerus elegans* fed on chemically distinct strains of *Cannabis sativa*. *Nature, Lond.* **266**: 650–651.

———, Valadon, G. & Mummery, R., 1977a. Carotenoids of the pupae of the Large White butterfly (*Pieris brassicae*) and the Small White butterfly (*Pieris rapae*). *J. Zool., Lond.* **181**: 323–339.

———, von Euw, J. & Reichstein, T., 1972. Some problems connected with warningly coloured insects and toxic defence mechanisms. *In* Schlettwein, C., (ed), *Impulse*, pp. 135–158. Basel.

———, von Euw, J., Reichstein, T., Smith, D. A. S. & Pierre, J., 1975a. Cardenolide storage in *Danaus chrysippus* (L.) with additional notes on *D. plexippus* (L.). *Proc. R. Soc.* (B) **190**: 1–31.

Rothschild, W. & Jordan, K., 1903. A revision of the Lepidopterous family Sphingidae. *Novit. zool.* **9**: Suppl. cxxxv, 972 pp., 67 pls.

Sato, R. & Nakajima, H., 1975. Foodplants of the Japanese Ennominae. *Japan Heterocerists' J.* Suppl. **2**: 46.

Sbordoni, V., Bullini, L., Scarpelli, L., Forestiero, S. & Rampini, M., 1979. Mimicry in the burnet moth, *Zygaena ephialtes*: population studies and evidence of a Mullerian-Batesian situation. *Ecol. Ent.* **4**: 83–93.

Schuler, W., 1982. Zur Funktion von Warnfarben: Die Reaktion junger Stare auf wespenähnlich schwarz-gelbe Attrappen. *Z. Tierpsychol.* **58**: 66–78.

Seifert, R. M., Buttery, R. G., Guadagni, D. G., Black, D. R. & Harris, J. G., 1970. Synthesis of some 2-methoxy-3-alkylpyrazines with strong bell pepper-like odors. *J. Agr. food chem.* **18**: 246–249.

Sheppard, P., 1967. *Natural selection and heredity*, (Edn 3). London.

Smith, S., 1975. Innate recognition of coral snake pattern by a possible avian predator. *Science* **187**: 759–760.

Smith, T. A., 1977a. Tryptamine and related compounds in plants. *Phytochemistry* **16**: 171–175.

———, 1977b. Phenethylamine and related compounds in plants. *Ibid.* **16**: 9–18.

South, R., 1961. *The Moths of the British Isles*, (Edn 4), **1**: 427 pp., 148 pls. London.

Swynnerton, C. F. M., 1915. Birds in relation to their prey: experiments on Wood-Hoopoes, Small Hornbills and a Babbler. *J. S. Afr. Ornith. Union* **11**: 32–108.

———, 1919. Experiments and observations bearing on the explanation of form and colouring, 1908–1913. *J. Linn. Soc.* **33**: 203–385.

Trave, R., Merlini, L. & Pavan, M., 1960a. Sulla natura chimica del secreto della larva del Lepidottero *Cossus ligniperda* Fabr. *Rc. Ist. lomb. Sci. Lett.* B. **94**: 151–155.

———, Garanti, L. & Pavan, M., 1960b. Sul secreto delle glandole mandibolari della larva di *Cossus cossus* L. (*C. ligniperda* Fabr.) (Lepidoptera). *XI Int. Kongr. Ent. Verh. Wien 1960* **3**: 73–76.

Trave, R., Garanti, L., Marchesini, A. & Pavan, M., 1966. Sulla natura chimica del secreto odorosa della larva del Lepidottero *Cossus cossus* L.   *Chim. Ind. (Milan)* **48**: 1167–1176.

Treat, A., 1963. Sound reception in Lepidoptera. *In* Busnel, R.-G., (ed.), *Acoustic behaviour of animals*, Chap. 16 & addendum, pp. 434–439, 800–801. Amsterdam, London & New York.

Trimen, R., 1887–89. *South African butterflies: a monograph of the extra tropical species*, **1–3**. London.

Turner, E. L., *in* Poulton, E. B., 1932. Observations on insects eaten or rejected by British birds. *Proc. ent. Soc. Lond.* **7**: 96–97.

Turner, J. R. G., 1971. Studies on Müllerian mimicry and its evolution in Burnet moths and Heliconid butterflies. *In* Creed, R., (ed.), *Ecological genetics and evolution*, pp. 224–260. Oxford.

Urquhart, F. A., 1960. *The Monarch butterfly.* Toronto.

———, 1976. Found at last: the Monarch's winter home.   *Natn. geogr. Mag.* **150**: 161–173.

Urzua, A., Salgedo, G., Cassels, B. K. & Eckhardt G., 1983. Aristolochic acids in *Aristolochia chilensis* and the *Aristolochia* feeder *Battus archidamas* (Lepidoptera).   *Colln. Czech. Chem. Commun.* **48**: 1513–1519.

Valadon, L. R. G. & Mummery, R. S., 1978. A comparative study of carotenoids in *Papilio* sp.   *Comp. Biochem. Physiol.* **61B**: 371–374.

Van Etten, C. H. & Tookey, H. L., 1979. Chemistry and Biological effects of glucosinolates. *In* Rosenthal, F. A. & Janzen, D. H., (eds), *Herbivores: their interaction with secondary plant metabolites*, pp. 471–501. New York & London.

Vaughan, G. L. & Jungreis, A. M., 1977. Insensitivity of Lepidopteran tissues to ouabain: physiological mechanisms for protection from cardiac glycosides.   *J. Insect Physiol.* **23**: 585–589.

von Euw, J., Reichstein, T. & Rothschild, M., 1968. Aristolochic acid-1 in the Swallowtail butterfly *Pachlioptera aristolochiae* (Fabr.) (Papilionidae).   *Israel J. Chem.* **8**: 659–670.

Waldbauer, G. P. & Sheldon, J. K., 1971. Phenological relationships of some aculeate Hymenoptera, their Dipteran mimics and insectivorous birds.   *Evolution* **25**: 371–382.

Walker, M., 1966. Some observations on the behaviour and life-history of the Jersey Tiger moth *Euplagia quadripunctaria* Poda (Lep: Arctiidae) in the 'Valley of the Butterflies', Rhodes.   *Entomologist* **99**: 1–24.

Walker, S., 1983. *Animal thought.* London.

Watson, R. W., 1975. New aberrations of *Tyria jacobaeae* L.   *Entomologist's Rec. J. Var.* **87**: 267, 1 col. pl.

Weatherstone, J., 1967. The Chemistry of Arthropod defensive secretions.   *Q. rev. chem. Soc.* **21**: 287–313.

Whalley, P., 1978. Derbyshire's darning needle.   *New Scientist*, 15 June, 1978: 740–741.

Wicklund, C. & Järvi, T., 1982. Survival of distasteful insects after being attacked by naïve birds: A reappraisal of the theory of aposematic coloration evolving through individual selection.   *Evolution* **36**: 998–1002.

Wilson, A., 1983. *Flavonoid pigments in butterflies.* Ph.D. Thesis, Reading University.

Windecker, W., 1939. *Euchelia (Hipocrita) jacobaeae* L. und das Schutztrachtenproblem.   *Z. Morph. Okol. Tiere* **35**: 84–138.

Witherby, H. F., Jourdain, F. C. R., Ticehurst, N. F. & Tucker, B. W., 1938. Crows to Flycatchers. *The Handbook of British birds.* **1**. London.

Wood, T. W., 1869. Insects in Disguise.   *Student Intellect. Obs.* **2**: 81–92.

Wurtman, R. J., 1982. Nutrients that modify brain function.   *Sc. Am.* **246**: 42–51.

Zlotkin, E., 1973. Chemistry of Animal Venoms.   *Experientia* **29**: 1453–1466.

———, Menache, M., Rochat, H., Miranda, F. & Lissitzky, S., 1975. Proteins toxic to arthropods in the venom of elapid snakes.   *J. Insect Physiol.* **21**: 1605–1611.

SYSTEMATIC SECTION

Scheme of Classification

The scheme of classification adopted throughout this work is that detailed in Kloet & Hincks (1972) but modified where necessary. The families will be treated in the ten volumes (volume number indicated below) according to the following plan:

ZEUGLOPTERA

Micropterigoidea
   Micropterigidae   1

DACNONYPHA

Eriocranioidea
   Eriocraniidae   1

EXOPORIA

Hepialoidea
   Hepialidae   1

MONOTRYSIA

Nepticuloidea
   Nepticulidae   1
   Opostegidae   1
   Tischeriidae   1

Incurvarioidea
   Incurvariidae   1
   Heliozelidae   1

DITRYSIA

Cossoidea
   Cossidae   2

Zygaenoidea
   Zygaenidae   2
   Limacodidae   2

Tineoidea
   Psychidae   2
   Tineidae   2
   Ochsenheimeriidae   2
   Lyonetiidae   2
   Hieroxestidae   2
   Gracillariidae   2
   Phyllocnistidae   2

Yponomeutoidea
   Sesiidae   2
   Choreutidae   2
   Glyphipterigidae   2
   Douglasiidae   2
   Heliodinidae   2
   Yponomeutidae   3
   Epermeniidae   3
   Schreckensteiniidae   3

Gelechioidea
   Coleophoridae   3
   Elachistidae   3
   Oecophoridae   4
   Ethmiidae   4
   Gelechiidae   4
   Blastobasidae   4

Stathmopodidae   4
Momphidae   4
Scythrididae   4

Tortricoidea
   Cochylidae   5
   Tortricidae   5

Alucitoidea
   Alucitidae   6

Pyraloidea
   Pyralidae   6

Pterophoroidea
   Pterophoridae   6

Hesperioidea
   Hesperiidae   7

Papilionoidea
   Papilionidae   7
   Pieridae   7
   Lycaenidae   7
   Nemeobiidae   7
   Nymphalidae   7
   Satyridae   7
   Danaidae   7

Bombycoidea
   Lasiocampidae   7
   Saturniidae   7
   Endromidae   7

Geometroidea
   Drepanidae   7
   Thyatiridae   7

Geometridae   8

Sphingoidea
   Sphingidae   9

Notodontoidea
   Notodontidae   9
   Thaumetopoeidae   9

Noctuoidea
   Lymantriidae   9
   Arctiidae   9
   Ctenuchidae   9
   Nolidae   9
   Noctuidae
     Noctuinae   9
     Hadeninae   9
     Cuculliinae to Hypeninae   10
     Agaristidae   10

NOTE. The dot distribution maps included in this volume must be regarded as provisional. Records from some localities mentioned in the text may not be shown on the maps when it has not been possible to localize them to a 10 km grid square, or when the records have been received too late for inclusion on the maps. Similarly, records may be shown on the maps from localities not mentioned specifically in the text, *e.g.* the Isle of Man, the Channel Islands. The vice-county maps have been compiled from records obtained from field observations, personal communications, collections and the literature.

Suborder DITRYSIA

Very small to very large; ocelli and chaetosemata present or absent; mandibles absent; galeae usually produced into haustellum; maxillary palpi one- to five-segmented or vestigial; labial palpi three-segmented, rarely two-segmented; wings rarely with a few aculeae, venation reduced in hindwing or in both wings; female with two genital openings, the ostium bursae on sternum 8 and the genital aperture on sternum 9–10. Pupa adecticous, obtect.

The suborder contains the bulk of the Lepidoptera, with a great diversity of forms having in common the complex genital system of the female. Wing-coupling is never of the jugate type, being either frenulate or amplexiform. In some of the more specialized superfamilies, such as the Bombycoidea, Papilionoidea, and Hesperioidea, the frenulum has often been lost. Aculeae have largely disappeared, persisting only in a few of the more primitive Tineoidea and then in restricted areas of the wing membrane. A heteroneurous venation distinguishes the Ditrysia from the Zeugloptera, Dacnonypha, and the Hepialoidea. The reduction of the venation in the hindwing, and sometimes also in the forewing, reaches extravagant limits in some of the narrow-winged species.

Although mandibles are generally absent (vestigial in a few families) the primitive five-segmented folded maxillary palpi are retained in some of the Tineidae and Lyonetiidae. Reduction, or even loss, of the maxillary palpi is characteristic of most superfamilies. The haustellum has become highly developed in some of the more specialized families which ingest sugary liquids, but in some the haustellum has become secondarily degenerate or even entirely lost (Common, 1970).

Fifty-five families in this suborder are represented in the British Isles.

Reference

Common, I. F. B., 1970. Ditrysia. *In* Mackerras, I. M. [Ed.], *The Insects of Australia*, 793–794. Melbourne.

Key to the families of Ditrysia

(Adapted from Sattler, K., *in* Brohmer, P., 1977. Lepidoptera. *Fauna von Deutschland* (Edn 13): 381–405.) Figures 13–23, 25–43, after Sattler

NOTE. This key applies only to British species.

1	Two pairs of functional wings	2
–	Hindwings much reduced or absent; flightless (♀)	77
2(1)	Hindwing deeply divided into three or more lobes	3
–	Hindwing not deeply divided, usually entire (a small basal dorsal lobe in some Geometridae)	4
3(2)	Each wing divided into six narrow lobes (figure 13) .... ALUCITIDAE (Volume 6)	
–	Forewing two-lobed (figure 14) or entire, hindwing three-lobed .... PTEROPHORIDAE (Volume 6)	
4(2)	Parts of either wing unscaled	5
–	Wings entirely scaled, at most a single hyaline spot on forewing	7
5(4)	Ocelli present (figure 15)	6
–	Ocelli absent ..... SPHINGIDAE (*Hemaris*) (Volume 9)	
6(5)	Hindwing with large clear hyaline area; forewing narrow; moths resembling Hymenoptera .... SESIIDAE (p. 369)	
–	All wings with small cloudy hyaline spots; forewing broad, much larger than hindwing (figure 16) .... CTENUCHIDAE (*Syntomis*) (Volume 9)	
7(4)	Antenna clubbed (figure 17)	8
–	Antenna not clubbed, may be gradually thickened towards apex	16
8(7)	Ocelli and frenulum present; forewing elongate, blackish with red spots .... ZYGAENIDAE: Zygaeninae (p.85)	
–	Ocelli and frenulum absent; wings usually held above body when the insect is at rest	9
9(8)	Antennae separated at base by width of an eye; a hair-pencil from base of antenna projecting over eye (figure 18); hindtibia usually with two pairs of spurs .... HESPERIIDAE (Volume 7)	
–	Antennae closer; no hair-pencil at base of antenna; hindtibia only with apical spurs	10
10(9)	Foretarsus without claws, usually brush-like (figure 19)	11
–	Foretarsus at least with one claw	13
11(10)	Hindwing cell open between 4 (M_3) and 5 (M_2) .... NYMPHALIDAE (Volume 7)	
–	Hindwing cell closed	12
12(11)	Antenna bare; forewing with veins not expanded at base .... DANAIDAE (Volume 7)	
–	Antenna scaled; forewing with veins much expanded at base (figure 20) .... SATYRIDAE (Volume 7)	

Figure 13 Alucitidae, forewing

Figure 14 Pterophoridae, forewing

Figure 15 Sesiidae: *Synanthedon*, head

Figure 16 Ctenuchidae: *Syntomis*, forewing

Figure 17 Clubbed antennae:
(**a**) Pieridae: *Pieris*
(**b**) Hesperiidae: *Hesperia*
(**c**) Zygaenidae: *Zygaena*

Figure 18 Hesperiidae, head

Figure 19 Nymphalidae, foretibia and tarsus

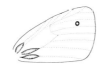

Figure 20 Satyridae: *Coenonympha*, forewing

Figure 21 Elachistidae: *Elachista*, hindwing

Figure 22 Tortricidae, hindwing

Figure 23 Phyllocnistidae: *Phyllocnistis*, hindtibia

Figure 24 Gelechiidae: *Aristotelia*, hindwing

13(10) Tarsal claws of all legs bifid ..... PIERIDAE (Volume 7)
–   Tarsal claws simple 14

14(13) Face bordering eye with dense band of white scales; up to 50mm wingspan 15
–   Face covered with erect hairs up to edge of eye; more than 70mm wingspan .... PAPILIONIDAE (Volume 7)

15(14) Hindwing with humeral vein; forewing tawny reticulated with brown .... NEMEOBIIDAE (Volume 7)
–   Hindwing without humeral vein; forewing coloured otherwise, most species blue, brown or copper .... LYCAENIDAE (Volume 7)

16(7)   Cilia of hindwing longer than breadth of hindwing (figure 21) 17
–   Cilia of hindwing at most shorter than breadth of hindwing (figure 22) 36

17(16) Head smooth-scaled 18
–   Head wholly or partly rough-scaled or haired 32

18(17) Hindtibia with long bristles 19
–   Hindtibia smooth-scaled or with long hairs; if bristles present, not longer than breadth of tibia 21

19(18) Hindtibia only with one row of evenly spaced dorsal bristles (figure 23); whitish moths up to 8mm wingspan.. .... PHYLLOCNISTIDAE (p. 363)
–   Hindtibia with bristles beneath or with tufts of bristles; moths not whitish, at least 8mm wingspan 20

20(19) Hindtibia with numerous bristles above and a few beneath .... EPERMENIIDAE (Volume 3)
–   Hindtibia with a few tufts of very long dorsal bristles .... STATHMOPODIDAE (Volume 4)

21(18) Head with scales broader than shaft of antenna (sometimes narrower in ♀) 22
–   Head with scales narrower than shaft of antenna 26

22(21) Moths mostly white, up to 9mm wingspan, forewing with black-margined metallic spot above tornus .... LYONETIIDAE (*Leucoptera*) (p. 214)
–   Wing markings otherwise 23

23(22) Hindwing very narrow with acute apex .... MOMPHIDAE (Volume 4)
–   Hindwing broader, oblong or oval 24

24(23) Hindwing strongly produced at apex (figure 24) .... GELECHIIDAE (Volume 4)
–   Hindwing not produced at apex 25

25(24) Hindwing narrow oblong, rather parallel-sided .... BLASTOBASIDAE (Volume 4)
–   Hindwing more or less oval .... OECOPHORIDAE (Volume 4)

26(21) Hindtibial spurs equal in length (figure 25) ....
.... SCHRECKENSTEINIIDAE (Volume 3)
– Inner hindtibial spurs longer than outer (figure 26) ... 27

27(26) Abdominal tergites each with a pair of scaleless patches
set with short spines (figure 27) ....
.... COLEOPHORIDAE (Volume 3)
– Abdomen without spined patches 28

28(27) Antennal scape with pecten (as figure 28) and medial
spurs of hindtibia well before middle (as figure 23) ... 29
– Without these characters in combination 30

29(28) Labial and maxillary palpi developed ....
.... GRACILLARIIDAE (p. 244)
– Only labial palpus developed ....
.... ELACHISTIDAE (Volume 3)

30(28) Ocelli present 31
– Ocelli absent .... SCYTHRIDIDAE (Volume 4)

31(30) Forewing with silver spots; hindtibia smooth-scaled
above .... HELIODINIDAE (p. 410)
– Forewing without silver spots; hindtibia long-haired ....
.... DOUGLASIIDAE (p. 408)

32(17) Antennal scape with eye-cap (as figure 29) ....
.... LYONETIIDAE (p. 212)
– Antennal scape with pecten, or simple 33

33(32) Head with a ridge of forwardly directed smooth scales
between antennae (figure 30) ....
.... HIEROXESTIDAE (p. 240)
– Head without this ridge 34

34(33) Hindtibia smooth-scaled ....
.... YPONOMEUTIDAE (Volume 3)
– Hindtibia long-haired 35

35(34) Labial palpus second segment with outstanding bristles
.... TINEIDAE (p. 152)
– Labial palpus second segment without bristles ....
.... GRACILLARIIDAE (p. 244)

36(16) Frenulum absent 37
– Frenulum present 39

37(36) Each wing with large eye-spot ....
.... SATURNIIDAE (Volume 7)
– Wings without obvious eye-spots 38

38(37) Eyes hairy (sometimes only sparsely); labial palpus with
basal segment only thinly haired ....
.... LASIOCAMPIDAE (Volume 7)
– Eyes bare; labial palpus with basal segment as thickly
haired as outer segments ....
.... ENDROMIDAE (Volume 7)

39(36) Haustellum scaled, fully developed 40
– Haustellum naked or reduced 45

Figure 25 Schreckensteiniidae: *Schreckensteinia*, hindtibia

Figure 26 Scythrididae: *Scythris*, hindtibia

Figure 27 Coleophoridae, abdominal tergite

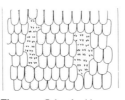

Figure 28 Momphidae, base of antenna, with pecten

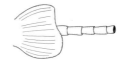

Figure 29 Nepticulidae, base of antenna, with eye-cap

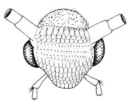

Figure 30 Hieroxestidae: *Oinophila*, head

Figure 31 Choreutidae: *Prochoreutis*, hindtibia

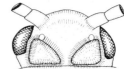

Figure 32 Zygaenidae: *Adscita*, head

Figure 33 Noctuidae, tarsal segment

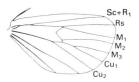

Figure 34 Noctuidae: *Pseudoips*, hindwing venation

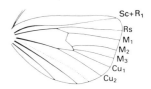

Figure 35 Thyatiridae: *Tethea*, hindwing venation

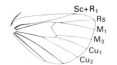

Figure 36 Notodontidae: *Clostera*, hindwing venation

Figure 37 Nolidae: *Nola*, base of antenna

40(39) Mid- and hindtibiae each with tufts of hair in middle and at end (figure 31) .... CHOREUTIDAE (p. 389)
– Mid- and hindtibiae evenly haired or smooth 41

41(40) Hindtibia smooth above; maxillary palpus present ....
.... PYRALIDAE (Volume 6)
– Hindtibia long haired above; maxillary palpus absent ....
.... 42

42(41) Hindwing of most species strongly produced at apex, otherwise not distinguished from Oecophoridae ....
.... GELECHIIDAE (Volume 4)
– Hindwing otherwise 43

43(42) Hindwing narrow oblong, rather parallel-sided in basal half .... BLASTOBASIDAE (Volume 4)
– Hindwing oval or broad oblong 44

44(43) Forewing plain white, grey, black and white or blackish with stigmata usually well marked ....
.... ETHMIIDAE (Volume 4)
– Forewing otherwise, mostly with complex pattern ....
.... OECOPHORIDAE (Volume 4)

45(39) Chaetosema occupying a large part of top of head (figure 32); hindtibia smooth without medial spurs; forewing green .... ZYGAENIDAE: Procridinae (p. 78)
– Chaetosema absent or very small; hindtibia mostly with two pairs of spurs 46

46(45) Ocelli present 47
– Ocelli absent 58

47(46) Hindtibia smooth-scaled or shortly haired 48
– Hindtibia long haired 54

48(47) Forewing dark with white or silver costal strigulae or plain brassy .... GLYPHIPTERIGIDAE (p. 400)
– Forewing marking otherwise 49

49(48) Hindtibia with medial spurs before middle; head with rough hair-scales bifurcate at apex, face broad-scaled ....
.... OCHSENHEIMERIIDAE (p. 208)
– Hindtibia with medial spurs in or beyond middle; head uniformly scaled or haired 50

50(49) Abdomen mostly white, red or yellow with a row of blackish mid-dorsal spots; wings, especially the hind-wings, usually brightly coloured ....
.... ARCTIIDAE: Arctiinae (Volume 9)
– Abdomen more uniformly coloured, without dorsal spots 51

51(50) Labial palpus upcurved; second segment scarcely thicker than third; wingspan under 16mm ....
.... YPONOMEUTIDAE: Acrolepiinae (Volume 3)
– Labial palpus second segment much thicker than third, or wingspan over 20mm 52

52(51) Tarsi strongly spined beneath (figure 33); wingspan of most species over 20mm ..... NOCTUIDAE (Volume 9)
– Tarsi not spined beneath; wingspan of most species under 20mm 53

53(52) Forewing with vein 2 (Cu$_2$) arising beyond three-quarters of cell; vein 1c (1A) obsolescent or obsolete .....
.... COCHYLIDAE (Volume 5)
– Forewing with vein 2 (Cu$_2$) arising before three-quarters of cell; vein 1c (1A) present at margin ....
.... TORTRICIDAE (Volume 5)

54(47) Ocelli separated from eye margins by width of an antenna .... PSYCHIDAE (p. 128)
– Ocelli touching eyes 55

55(54) Hindwing 8 (Sc+R$_1$) close to 7 (Rs) near end of cell ... 57
– Hindwing 8 (Sc+R$_1$) approximated or fused with cell near base then diverging from 7 (Rs) to wing margin (figure 34) 56

56(55) Antenna gradually thickened towards apex ....
.... AGARISTIDAE (Volume 10)
– Antenna not thickened towards apex ....
.... NOCTUIDAE (Volume 9)

57(55) Hindwing 6 (M$_1$) and 7 (Rs) well separated at base (figure 35) .... THYATIRIDAE (Volume 7)
– Hindwing 6 (M$_1$) and 7 (Rs) stalked or coincident (figure 36) .... NOTODONTIDAE (Volume 9)

58(46) First tarsal segment of mid- and hindleg about as long as tibia .... SPHINGIDAE (Volume 9)
– First tarsal segment of hindleg at most two-thirds length of tibia 59

59(58) Forewing black or brown with a few translucent spots; abdomen with yellow or blue and red rings, or, if completely yellow, hindtibia with only one medial spur ....
.... CTENUCHIDAE (Volume 9)
– Forewing marked otherwise; abdomen not with yellow or blue and red rings, sometimes completely yellow or with yellow apex; hindtibia without or with two middle spurs 60

60(59) Antennal scape with apicolateral projection (figure 37); forewing with scale-tufts .... NOLIDAE (Volume 9)
– Antennal scape without apicolateral projection; fore-wing without scale-tufts 61

61(60) Antennal scape with pecten 62
– Antennal scape without pecten 63

62(61) Hindtibia upperside long haired ....
.... TINEIDAE (p. 152)
– Hindtibia smooth-scaled ....
.... YPONOMEUTIDAE (Volume 3)

63(61) Hindwing with median vein forked in cell (figure 38) ....
.... COSSIDAE (p. 69)
– Hindwing with median vein not forked in cell 64

64(63) Wings unicolorous or with indefinite markings, blackish
or grey, thinly scaled or with scales replaced by hairs;
body usually long haired .... PSYCHIDAE (p. 128)
– Wings usually well scaled and with definite markings 65

65(64) Hindwing with 7 (Rs) and 8 (Sc+R₁) separate throughout their length but sometimes approximated or connected by a cross-vein 66
– Hindwing with 7 (Rs) and 8 (Sc+R₁) fused for some
distance 73

66(65) Hindwing with 7 (Rs) and 8 (Sc+R₁) well separate from
base ... YPONOMEUTIDAE (*Orthotaelia*) (Volume 3)
– Hindwing with 7 (Rs) and 8 (Sc+R₁) approximated or
connected by a cross-vein (figures 39,40) 67

67(66) Forewing with 4 (M₃) and 5 (M₂) parallel 68
– Forewing with 4 (M₃) and 5 (M₂) approximated at base ..
.... 70

68(67) Forewing with apex produced and hooked ....
.... DREPANIDAE (Volume 7)
– Forewing with apex rounded 69

69(68) Antenna simple; wingspan less than 32mm ....
.... LIMACODIDAE (p. 125)
– Antenna dentate (♀) or pectinate (♂); wingspan more
than 25mm .... LYMANTRIIDAE (Volume 9)

70(67) Hindwing 6 (M₁) and 7 (Rs) separate at base 72
– Hindwing 6 (M₁) and 7 (Rs) stalked or coincident 71

71(70) Wings light grey, thinly scaled; forewing with fine ante-
and postmedian lines, a disco-cellular mark and few
other markings; no dorsal scale-tooth; ♀ with large anal
tuft .... THAUMETOPOEIDAE (Volume 9)
– Wings usually coloured or marked otherwise; if with this
type of forewing marking then a scale tooth in middle of
dorsum .... NOTODONTIDAE (Volume 9)

72(70) Hindwing 8 (Sc+R₁) approximated or fused with cell
near wing base (figure 41) ....
.... GEOMETRIDAE (Volume 8)
– Hindwing 8 (Sc+R₁) separate from cell and not approxi-
mated to 7 (Rs) before end of cell (figure 35) ....
.... THYATIRIDAE (Volume 7)

73(65) Hindwing 8 (Sc+R₁) stalked with 7 (Rs) (figure 42) ......
.... DREPANIDAE (Volume 7)
– Hindwing 8 (Sc+R₁) separate from near base of cell .. 74

74(73) Hindwing 6 (M₁) and 7 (Rs) stalked 75
– Hindwing 6 (M₁) and 7 (Rs) connate or separate 76

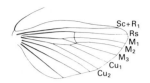

Figure 38 Cossidae:
Phragmataecia, hindwing venation

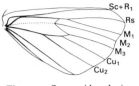

Figure 41 Geometridae: *Aspitates*,
hindwing venation

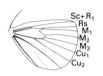

Figure 39 Limacodidae: *Apoda*,
hindwing venation

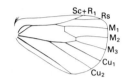

Figure 42 Drepanidae: *Cilix*,
hindwing venation

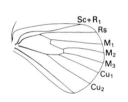

Figure 40 Drepanidae: *Drepana*,
hindwing venation

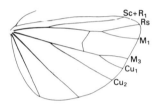

Figure 43 Arctiidae: *Eilema*,
hindwing venation

75(74) Hindwing with 8 (Sc+R$_1$) and 7 (Rs) forming a basal loop (figure 41); forewing usually triangular, at least as broad as hindwing .... GEOMETRIDAE (Volume 8)

– Hindwing 8 (Sc+R$_1$) fused with 7 (Rs) at base, not forming a loop (figure 43); forewing usually elongate, narrower than hindwing ....
.... ARCTIIDAE: Lithosiinae (Volume 9)

76(74) Hindwing 5 (M$_2$) parallel to 4 (M$_3$) (figure 41) ....
.... GEOMETRIDAE (Volume 8)

– Hindwing 5 (M$_2$) approximated to 4 (M$_3$) at base (figure 34) .... NOCTUIDAE (Volume 9)

77(1) Legs normally fully developed 78

– Legs, antennae and mouthparts strongly reduced; wings absent; imago usually remaining in larval case ....
.... PSYCHIDAE (p. 128)

78(77) Wings completely absent, or if rudimentary wings present, all tibiae without spurs ....
.... PYRALIDAE (*Acentria*) (Volume 6)

– At least a vestige of forewing present; at least apical spurs on mid- and hindtibia 79

79(78) Hindtibia without medial spurs 80
– Hindtibia with medial spurs 81

80(79) Wings and body only with hairs; antenna dentate; legs very short ..... LYMANTRIIDAE (*Orgyia*) (Volume 9)

– Wings and body with hairs and scales; antenna long, simple; legs more than half body length ....
.... GEOMETRIDAE (*Lycia*) (Volume 8)

81(79) Ocelli present 82
– Ocelli absent 83

82(81) Antenna shorter than forewing; moth in autumn ....
.... TORTRICIDAE (*Exapate*) (Volume 5)

– Antenna longer than forewing; moth in spring ....
.... OECOPHORIDAE (*Dasystoma*) (Volume 4)

83(81) Labial palpus ascending above head; wings longer than body .... OECOPHORIDAE (*Diurnea*) (Volume 4)

– Labial palpus short; wings much shorter than body, sometimes absent .... GEOMETRIDAE (Volume 8)

COSSIDAE
B. Skinner

A family of world-wide distribution, especially well represented in southern Africa and Australia. There are 85 Palaearctic species placed in 11 genera, and three species in three genera occur in the British Isles.

This family of large moths is usually included among the Macrolepidoptera. The moths are strictly nocturnal in habit.

Imago. Antenna usually pectinate in male, simple in female; haustellum very short or absent; maxillary palpus minute, one- or two-segmented; labial palpus short or moderate. Wings narrow; venation, forewing with median vein strong and forked in discal cell, 1c (1A) present; hindwing with median vein usually forked, 1c (1A) sometimes reduced.

Larva. Thorax with large sclerotized prothoracic plate. Feeds internally in living wood or pith.

Pupa. Long, cylindrical, abdomen spined, segments 3–7 movable in male. Protruded from tunnel at ecdysis.

Key to species (imagines) of the Cossidae

1 Large species, wingspan at least 75mm; wings greyish brown .... *Cossus cossus* (p. 72)

– Smaller species, wingspan not exceeding 70mm; wings not greyish brown; 2

2(1) Wings white, thinly scaled, with numerous blue-black spots .... *Zeuzera pyrina* (p. 71)

– Wings pale brownish ochreous ....
.... *Phragmataecia castaneae* (p. 70)

Zeuzerinae

Imago. Antenna in male pectinate only in basal half; tibial spurs absent on all legs.

PHRAGMATAECIA Newman

Phragmataecia Newman, 1850, *Zoologist* 8: 2931.

Represented in the Palaearctic region by five species and in South Africa by one species. One species occurs in England.

Imago. Labial palpus very short, setose. Hindtibia setose. Hindwing with vein 8 (Sc+R$_1$) free (figure 44).

PHRAGMATAECIA CASTANEAE (Hübner)
The Reed Leopard

Phalaena (Bombyx) castaneae Hübner, 1790, *Beitr.Gesch. Schmett.* **2**: 9.

Zeuzera arundinis Hübner, [1822], *Verz.bekannt.Schmett.*: 196.

Type locality: not stated.

Description of imago (Pl.3, figs 4,5)

Wingspan of male 30–42mm, of female 38–56mm. Antenna short in both sexes; in male strongly bipectinate except towards tip, in female weakly bipectinate. Head and thorax greyish brown. Forewing elongate with rounded apex; pale greyish brown; costa and dorsum darker, tinged with yellow; veins pale brown, sometimes with small blackish brown spots especially towards termen. Hindwing white, lightly dusted with pale brown. Abdomen greyish brown; narrower and much longer in female (male 11–14mm; female 16–18mm, in dried specimens).

Variation, apart from size, limited to the extent and intensity of the blackish brown spots on the forewing.

Life history

Ovum. *c.*1.5mm long by *c.*0.8mm diameter. Elliptical, rounded at both ends. Surface smooth; whitish or almost white with a pearly gloss (Buckler, 1887). In captivity (and probably in the wild) the eggs are inserted singly or in pairs into a leaf-sheath of common reed (*Phragmites australis*) and hatch in 12–14 days.

Larva. Full-fed 40–50mm long. Head small, shiny and yellowish brown. Prothoracic plate glossy brownish. Body slender and cylindrical, and nearly uniform in girth

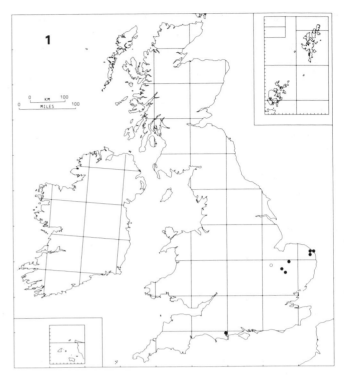

Phragmataecia castaneae

throughout its length. Skin wrinkled with a shiny appearance. Ochreous white with a pale reddish brown subdorsal stripe. Thoracic legs short, prolegs reduced to small point-like structures, anal claspers attenuated and almost absent.

Feeds in a stem of common reed for about 21 months, becoming full-fed in April or May of the third year.

Pupa. Very slender and cylindrical; head with a small beak-like projection. Brown, paler on the wings and darker on the abdomen (Buckler, *loc.cit.*). Pupation takes place head upwards within a section of the reed-stem, in which it moves freely up or down.

Imago. Univoltine, occurring from mid-June through most of July. The male comes readily to light but the female is rarely seen except occasionally on the wing, or at rest on a reed-stem by the light of a hand-lamp.

Distribution (Map 1)

A denizen of reed-beds first discovered in Britain about 1841 at Holme Fen, Cambridgeshire. Subsequently found in the same county at Whittlesea Mere, and at Wicken and Chippenham Fens. In 1873 it was introduced in the egg stage at Ranworth, Norfolk, from which the existing Broadland populations may have originated. At present it is well established at Wicken and Chippenham Fens and locally in the Norfolk Broads between Horning and

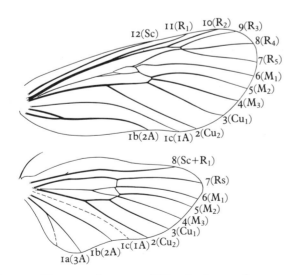

Figure 44 *Phragmataecia castaneae* (Hübner), wing venation

Horsey. It also occurs in a small pond near Wareham, Dorset, where it was first noted in 1928 (Andrewes, 1928). Eurasiatic. In western Europe locally distributed from Spain through France and Belgium to Denmark, Sweden and the U.S.S.R.; China and Japan.

ZEUZERA Latreille

Zeuzera Latreille, 1804, *Nouv.Dict.Hist.nat.* **24** (Tab.): 186.

A genus of world-wide distribution represented by five species in the Palaearctic region, of which one occurs in England and Wales.

Imago. Labial palpus very short with appressed scales. Hindtibia rough-scaled. Hindwing with vein 8 (Sc+R$_1$) connected to 7 (Rs) just beyond cell (figure 45).

ZEUZERA PYRINA (Linnaeus)
The Leopard Moth

Phalaena (*Noctua*) *pyrina* Linnaeus, 1761, *Fauna Suecica* (Edn 2): 306.

Phalaena (*Noctua*) *aesculi* Linnaeus, 1767, *Syst.Nat.* (Edn 12) **1** (2): 833.

Type locality: Sweden.

Description of imago (Pl.3, figs 1–3)
Wingspan of male 46–58mm, of female 60–78mm. Antenna of male bipectinate from base to middle and thence simple to tip; that of female simple and thread-like. Head white, frons black. Thorax downy and white, with three pairs of oval subdorsal black spots, approximated anteriorly. Forewing of both sexes semitransparent, white with a dense pattern of blue-black spots in rows between veins; costa and basal area tinged yellowish brown. Hindwing semitransparent, white and covered, except the anal area, with numerous blue-black dots between veins. All cilia white with series of black dots at ends of veins. Abdomen blackish grey, lightly covered with short white hairs, the anal segment of the male with a flat white tuft and that of the female with a well-developed and conspicuous ovipositor.

Variation usually limited to the intensity of the black spotting but in ab. *confluens* Cockayne (fig.3) the central and dorsal spots of the forewing of the female are more or less confluent. This very uncommon aberration has been recorded from Warwickshire, Hertfordshire, Surrey, Sussex and Kent.

Life history
*Ovum. c.*1.3mm long by *c.*0.7mm diameter at centre, oval, tapering slightly towards the micropyle. Surface irregular,

shining, sharply ribbed towards the micropyle; yellowish salmon (Freeman, 1905). The eggs are laid singly or in small batches on the surface or in a crevice of the stem or branch of the foodplant.

Larva. Full-fed *c.*43–55mm long. Head rather small, rounded, blackish brown. Body dull white tinged with yellow or brown; first thoracic and anal segments covered dorsally with blackish brown sclerotized plates; remainder of body covered with numerous blackish brown pinacula each bearing a short fine bristle.

It feeds for two or possibly three years in a stem or branch of a wide variety of trees and shrubs including ash (*Fraxinus excelsior*), elm (*Ulmus* spp.), oak (*Quercus* spp.), horse-chestnut (*Aesculus hippocastanum*), hawthorn (*Crataegus* spp.), sallow (*Salix* spp.), sycamore (*Acer pseudoplatanus*), birch (*Betula* spp.), lilac (*Syringa vulgaris*), privet (*Ligustrum* spp.), hornbeam (*Carpinus betulus*), beech (*Fagus sylvatica*), wayfaring-tree (*Viburnum lantana*), honeysuckle (*Lonicera* spp.), blackcurrant (*Ribes nigrum*), apple (*Malus* spp.), plum (*Prunus* spp.) and pear (*Pyrus* spp.). Occasionally infestations of larvae cause damage of economic importance to fruit and other commercially grown trees. Stems and branches up to 10cm in diameter are attacked but never those over that thickness (Haggett, 1950). In captivity it will accept potato and other root vegetables. When full-fed, usually from mid-May to mid-June, it pupates head downwards or horizontally in the larval gallery without constructing any form of cocoon.

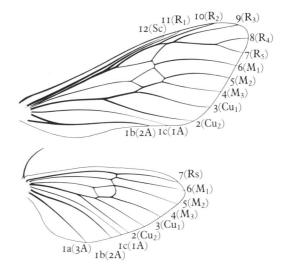

Figure 45 *Zeuzera pyrina* (Linnaeus), wing venation

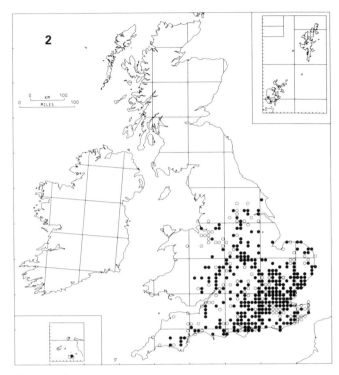

Zeuzera pyrina

*Pupa. c.*30mm long. Flattened ventrally; head with a beak-like projection, abdominal segments with rows of strong spines; reddish brown. Emergence takes place through a carefully prepared exit hole leaving the empty pupa case partly protruding.

Imago. Univoltine, appearing from late June to early August. During the day it may occasionally be found at rest on tree-trunks and fences; the females have been noted resting on lawns and short grass where they are frequently attacked by ants (Goater, 1974). After dusk both sexes, but mostly males, are attracted to light in small numbers. Predation by birds is well known in this species and has been discussed by Barrett (1895) and Haggett (*loc.cit.*).

Distribution (Map 2)

An inhabitant of woodland, commons, parkland and gardens, and found in moderate numbers throughout southern England except the extreme south-west where it is very scarce; absent from the Isles of Scilly. Elsewhere it is found especially in East Anglia, the Midlands, east Wales, ranging northwards to Yorkshire and Lincolnshire. Although mentioned by Donovan (1936), the only confirmed record from Ireland is a single specimen taken in Co. Mayo on 12 July 1978 (Myers, 1979). Eurasiatic.

Widely distributed from Spain to southern Fennoscandia and the U.S.S.R. eastwards to Japan. Introduced into North America where it is occasionally a pest.

Cossinae

Imago. Antenna in male pectinate to apex; hindtibia with middle spurs present.

COSSUS Fabricius

Cossus Fabricius, 1794, *Ent.syst.* **3** (1): [1]; **3** (2): [3].
Trypanus Rambur, 1866, *Cat.syst.Lépid.Andalousie* (2): 326.

Occurs in the Palaearctic region where it is represented by 15 species, south-east Asia and South Africa. One species occurs in Great Britain and Ireland.

Imago. Haustellum absent; labial palpus moderate, ascending, with dense appressed scales. Hindtibia rough-scaled with two pairs of spurs. Hindwing with vein 8 (Sc+R$_1$) free (figure 46).

COSSUS COSSUS (Linnaeus)
The Goat Moth
Phalaena (*Bombyx*) *cossus* Linnaeus, 1758, *Syst.Nat.* (Edn 10) **1**: 504.
Cossus ligniperda Fabricius, 1794, *Ent.syst.* **3** (2): 3.
Type locality: [Sweden].

Description of imago (Pl.3, figs 6,7)

Wingspan of male 68–84mm, of female 88–96mm. Antenna strongly pectinate in both sexes, in female shorter. Patagium ochreous grey; thorax greyish brown with posterior transverse black bar. Forewing greyish brown, frequently clouded with whitish grey, transversely intersected by numerous wavy blackish brown striae and two well-defined dark lines – one extending from two-thirds of costa to tornus, and another, much shorter, from the costa near apex. Hindwing smoke-grey, paler terminally, cilia slightly reticulate. Abdomen stout, dark grey with each segment edged with a pale grey fringe.

Variation usually limited to the amount of whitish clouding of the forewing, which in the male may be extensive.

Life history

*Ovum. c.*1.7mm long by *c.*1mm diameter. Slightly elongate, striate both longitudinally and transversely; dull pale brown. The eggs are laid either singly or in small batches in a crevice or damaged part of the host tree.

Larva. Full-fed *c*.85mm long. Head black and glossy with strongly developed black mandibles. Body smooth, fleshy, somewhat flattened, deeply indented between segments, covered with fine short hairs; pinkish yellow ventrally, reddish brown dorsally; prothoracic plate large with narrow yellowish median sulcus.

It feeds from three to four years in the living wood of a wide variety of deciduous trees; most frequently willow (*Salix* spp.), birch (*Betula* spp.), elm (*Ulmus* spp.) and ash (*Fraxinus excelsior*) but sometimes oak (*Quercus* spp.), alder (*Alnus glutinosa*), poplar (*Populus* spp.), apple (*Malus* spp.) and other fruit trees. Heavy infestation often causes serious damage to the host tree. In captivity it will accept uncooked beetroot, other root vegetables and apples. It should be reared in a metal or plastic, but not wooden, cage or container. The vernacular name of this species is derived from the strong goat-like smell emitted by the larva. When full-grown in the autumn it may wander away from the host tree in search of more suitable quarters in which to pupate and is most frequently observed at such times. The oval silken cocoon composed of particles of wood or earth is constructed either in the host tree or just below the surface of the ground. In captivity rough sawdust provides a suitable pupating medium.

Pupa. *c*.45mm long. Very stout, flattened ventrally, dor-

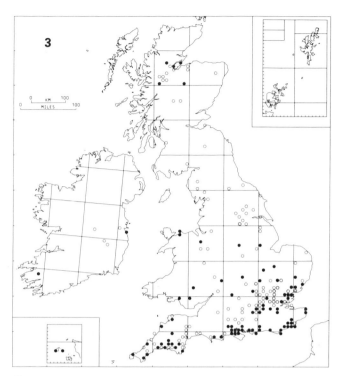

Cossus cossus

sal region rounded; head with sharp projecting beak; wings rather prominent; intersegmental divisions of abdomen well marked; the segments with rows of stiff short spines; anal segment blunt. Dark red-brown, abdominal segments lighter brown with paler incisions and dark brown spiracles (Barrett, 1895).

Imago. Univoltine, appearing in June and July. Found occasionally at rest on fences or tree-trunks, and in very small numbers at light. It has also been reported on numerous occasions to attend the sugar patch, although why this is, is inexplicable as the moth is unable to feed.

Distribution (Map 3)

May be found in a wide variety of habitats, from dense woodland to isolated trees growing along river banks and roadsides. A local species, occurring in widely scattered localities throughout much of England, Wales and central Scotland. Formerly more widespread, especially in north-east England and Scotland, and now considered to be generally uncommon, although small colonies are easily overlooked. In Ireland it is stated to be sporadically distributed over the southern half of the country (Baynes, 1964), and the paucity of very recent records more likely reflects the lack of recorders than the true distribution. Eurasiatic. Generally distributed throughout Europe to the U.S.S.R. and Japan.

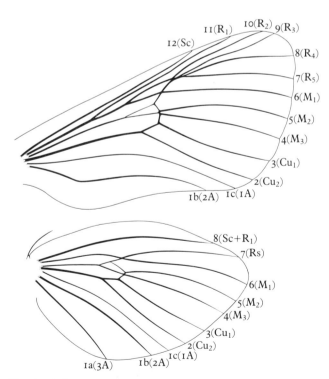

Figure 46 *Cossus cossus* (Linnaeus), wing venation

References

Andrewes, H. L, 1928. *Phragmatoecia castaneae* in Dorset, etc. *Entomologist* **61**: 259–260.

Barrett, C. G., 1895. *The Lepidoptera of the British Islands*, **2**. London.

Baynes, E. S. A., 1964. *A revised catalogue of Irish Macrolepidoptera*, 110 pp. Hampton.

Buckler, W., 1887. *The larvae of the British butterflies and moths*, **2**, 169 pp., 35 col.pls. London.

Donovan, C., 1936. *A catalogue of the Macrolepidoptera of Ireland*, 100 pp. Cheltenham & London.

Freeman, R., 1905. Eggs of Lepidoptera. *Entomologist's Rec. J. Var.* **17**: 76–78.

Goater, B., 1974. *The butterflies & moths of Hampshire and the Isle of Wight*, 439 pp. Faringdon.

Haggett, G., 1950. The life history and habits of *Zeuzera pyrina* Linn. (*aesculi* Linn.) in Britain. *Entomologist* **83**: 73–81, pl. 2.

Myers, A. A., 1979. Confirmation of *Zeuzera pyrina* (Linnaeus) (Lepidoptera: Cossidae) as an Irish species. *Entomologist's Gaz.* **30**: 30.

ZYGAENIDAE
W. G. Tremewan

Species of Zygaenidae occur in all zoogeographical regions except New Zealand, although the group is poorly represented in the Malagasy region and the Pacific islands. In the British Isles the family is represented by the genera *Adscita* Retzius and *Zygaena* Fabricius which belong respectively to the subfamilies Procridinae and Zygaeninae; the species of the former genus are known in the vernacular as Foresters, those of the latter genus as Burnets. The adults and larvae are aposematic, that is to say, they are brightly or contrastingly coloured, and toxins have been found in all phases of their life history.

Imago. Ocelli and chaetosemata present; haustellum well developed, naked; antenna clavate, weakly dentate, or bipectinate; labial palpus moderate, ascending, terminal segment short, pointed; maxillary palpi present. Tibia of foreleg with epiphysis; tibia of hindleg with medial spurs present, vestigial or absent. Hindwing with vein 8 $(Sc+R_1)$ connected to upper margin of cell (R) by a bar, 7 (Rs) and 6 (M_1) sometimes connate; frenulum and retinaculum present.

Ovum. Ovoid, flattened; micropyle situated at one extremity of the horizontal axis (Tutt, 1899; Döring, 1955); chorion finely reticulate, transparent, colour of egg resulting from the yolk. Deposited in rows, in a single layer, or in several layers to form an irregularly shaped batch.

Larva. Head retractile. Thoracic legs and five pairs of abdominal prolegs present. Thorax and abdomen with weakly plumose setae arranged in groups or verrucae. With one, two or several periods of diapause.

Pupa. Obtect; adecticous. Abdominal segments 1 and 2 (male, female) and 8–10 (male) or 7–10 (female) fused, each tergite with a transverse row of spines anteriorly. Enclosed in a cocoon which it vacates or from which it protrudes before the moth emerges.

NOTE. In the present work the host plant records and observations on ecology and ethology are my own, unless stated otherwise, and are the result of extensive field-work conducted throughout Britain during the last 30 years. The nomenclature follows, with slight amendments, the revised check list of the British Zygaenidae (Tremewan, 1976).

Derivation of the vernacular names

(1) 'Forester'
The earliest use of the word 'forester' as a name for an *Adscita* species appears to be that of Harris (1766: 5, 112,

[147], pl. 34) who referred to *Adscita statices* (Linnaeus) as the 'Forrester' or 'Forester'. According to *The Shorter Oxford English Dictionary* (1952), forester is a Middle English word derived from the Old French *forestier*. Three of the meanings that are cited are: 'One who lives in a forest 1513', 'A bird or beast of the forest 1630' and 'A name of some moths of the family *Zygaenidae* 1819'. The date of origin (1819) in the last citation is probably attributable to Samouelle (1819: 245, 397) who referred to *Adscita statices* as '*Ino statices* (forester)' and 'The Forester'. Two possible explanations of forester in relation to *Adscita* species are: (a) that it alludes to *A. statices* being an inhabitant of the forest (it does occur occasionally in woodland clearings and in rides); (b) that there may be some allusion to the green coloration of the moths (certain foresters of Sherwood Forest allegedly wore a bright green cloth known as Lincoln green).

The earliest reference to an *Adscita* species that I have been able to trace in the English literature is by Petiver (1699: 35; 1767: 5), who referred to *A. statices* as 'Our green Meadow Butterfly' and commented that he had observed it '*in several* Meadows *when the Grass is high*'.

(2) 'Burnet'

The earliest use of the word 'burnet' as a name for a *Zygaena* species appears to be that of Wilkes ([1749]: 7, [19], 46, pl. [91]), who referred to *Zygaena filipendulae* (Linnaeus) as 'The Burnet-Moth' or 'The Burnet Moth'. In a paragraph describing the differences between butterflies and moths, there is a further (and rather delightful) reference by Wilkes (p. [11]): '*N.B.* There is a Species of Fly betwixt the *Moth* and the *Butterfly*, whose Horns are more flat and hollow; this is describ'd in the second Book of this Work, by the Name of the *Burnet*.' According to *The Shorter Oxford English Dictionary* (1952), burnet, as an adjective, is an obsolete English word meaning dark brown, and is derived from the Old French *burnete*, *brunette*, a diminutive of *brun*. As a noun, burnet is a Middle English word meaning 'Any plant belonging to the genera *Sanguisorba* and *Poterium* . . ., as the Great or Common Burnet, the Lesser or Salad Burnet, etc. . . ., from the colour of their flowers. Burnet-fly and burnet-moth are referred to *Anthrocera* or *Zygaena filipendulae*. Although the derivation of burnet as a name for *Zygaena* species remains uncertain, it may not be without significance that Wilkes associated *Z. filipendulae* with 'The Burnet Rose' (*Rosa pimpinellifolia*), which he also illustrated on pl. [91]. Another explanation might be that 'Burnet-moth' is an allusion to its habit of sometimes resting on the flower heads of salad burnet (*Sanguisorba minor*).

The earliest reference to a *Zygaena* species that I have been able to trace in the English literature is by Moffet (1634: 97), who described and illustrated *Z. filipendulae*

with five spots, the basal two (1, 2) being represented by a single spot in the figure. Petiver (1699: 36; 1767: 5) also described it as 'maculis quinque' and commented that it was '. . . *more common than the last* [*A. statices*], *and found in the same Places*'; the description is followed by the statement 'Moffet's greenish Leopard with 5 scarlet Spots'. Albin (1720: pl. 82), with reference to *Z. filipendulae*, stated 'this is commonly called the *Wood Leopard*'. Presumably the spotted forewings of *Z. filipendulae* are reflected in 'Leopard' and 'Wood Leopard', but these names very quickly became obsolete.

Conservation

At present, *Z. viciae* ([Denis & Schiffermüller]) is the only species of Zygaenidae which is endangered in the British Isles and, consequently, is protected under the Wildlife and Countryside Act, 1981; its protection is necessary because it is known only from one small colony in Argyll. The Welsh subspecies of *Z. purpuralis* (Brünnich) has not been recorded since 1962 and therefore should not be collected if rediscovered; moreover, restraint should be used when collecting *Z. exulans* (Hohenwarth), *Z. loti* ([Denis & Schiffermüller]) and *Adscita globulariae* (Hübner) as these species are local and occur in small colonies.

The pros and cons of insect conservation in Britain have been dealt with recently in an excellent paper by Stubbs (1982). However, in discussing here the ecological requirements of the Zygaenidae relative to conservation, it is also necessary to comment briefly on some of the more general aspects.

The protection of the habitat must be considered the most important and fundamental principle in conservation. In spite of this, emphasis has been placed on the restriction of or even total ban on collecting, the paucity of some species of insects having been attributed to 'collectors'. The majority of collectors are responsible field entomologists who, by providing data on biology, ecology and distribution, have contributed much to the cause of conservation; ironically they are often blamed by those who may have been responsible for the destruction of habitats. Happily this situation is changing, albeit slowly, and the contributions that field entomologists have made and continue to make are being recognized (Stubbs, 1982: 62).

The Zygaenidae, like many species of butterflies, are useful as 'indicator species' in conservation work because they are conspicuous and fly by day. They are also particularly vulnerable as they sometimes occur in very small colonies, and their biotopes include wet meadows, marshland, sandhills, coastal cliffs, chalk downland and roadside verges, *viz.* habitats which are constantly threatened.

Wet meadows have been subjected to drainage and subsequent overgrazing, while the marshland habitats of

Zygaena trifolii (Esper) in Cornwall are seriously threatened by opencast mining of alluvial tin. Many coastal habitats are under great pressure from the recreational activities of man. Such activities have also affected biotopes on chalk downland where the vegetation is regularly mown in order to create amenity areas, and many thousands of hectares have been permanently destroyed by agriculture and forestry.

Much of the chalk downland that does remain has changed drastically. The use of selective herbicides in some areas, while producing improved grazing, has simultaneously exterminated the natural flora, in fact the very host plants on which some of the endangered species live. Moreover, the advent of myxomatosis and the consequent reduction in rabbit populations was followed by an accelerated growth of hawthorn and other scrub, which in some areas reached a climax and completely obliterated the chalkland flora. However, in those areas which did not acquire a climax vegetation, the problem has been alleviated to some extent by a gradual recovery in the rabbit populations and an increase in the number of deer; the deliberate intervention of man, *e.g.* the introduction of management policies such as scrub clearance, has also been helpful.

Although the protection of the habitat is of paramount importance, there is also a need for habitat management, as pointed out by Stubbs (1982: 65), and the ecological conditions favourable to the typical chalkland species must be maintained. The recent clearance and eradication of scrub followed by the introduction of sheep grazing on chalk downland has undoubtedly improved the quality and quantity of the herbaceous flora, but the extent of the grazing and the consequent results need to be monitored very carefully. While wishing to stress how beneficial grazing has been, it is nevertheless necessary to draw attention to the detrimental effects of overgrazing; undoubtedly this has occurred in parts of the North Downs in Surrey and on chalk downland in Hampshire, in spite of criticism from experienced field entomologists. Unfortunately their opinions appear to have gone unheeded, partly because too much emphasis has been placed on the theory, as yet not entirely proven, that one of the optimum ecological requirements of a single species, the butterfly *Lysandra bellargus* (Rottemburg) (Lycaenidae), is a turf of 1–4cm high (Thomas, 1983).

The overgrazing of some downland sites near Dorking, Surrey, during the winter of 1982–83 adversely affected the colonies of *Zygaena filipendulae* and *Z. lonicerae* (Scheven) that occur there – the lack of grass-stems on which the larvae spin their cocoons resulted in abnormally heavy predation of the pupae by birds, while shortage of suitable roosting sites for the adults, a factor that is probably of even greater importance to blue butterflies, resulted in greater losses due to heavier predation from crab spiders. The situation might have been more disastrous but for the wet spring of 1983. The general values of rainfall in April and May 1983 for England and Wales were 191 per cent and 175 per cent of average, respectively – April was the wettest since 1920, and May the wettest since 1979, while the three months' total including March had been exceeded only three times this century (Meteorological Office, pers.comm.). At Dorking, 118.5mm of rainfall were recorded for April (against an average of 50mm) and 96.9mm for May (average 58mm). There is no doubt that irreversible damage would have been inflicted on the flora and fauna had the overgrazing been followed by a dry spring. Even with the wet spring, some of the turf during the following summer was reminiscent of agricultural pasture rather than chalk grassland; in these particular areas *Zygaena* species (adults and cocoons) and butterflies were conspicuously scarce, suggesting that a high mortality rate of the larvae had occurred during the spring (in the overgrazed areas, larvae that had broken diapause would have had little protection from a sudden cold spell).

Overgrazing and the problems that arise from it could be prevented by considering the number of sheep per hectare, the time of year when the grazing takes place, and the length of the grazing period. The first consideration depends on the density of the herbage and is inextricably linked to the extent of the grazing period; grazing should only be done between late October and the end of January, and domestic animals should not be introduced during the summer months when natural grazing by wild herbivores, such as rabbits and deer, provides sufficient control. It is also recommended that a rotational system is employed whereby a particular slope or field is grazed by sheep every other winter. This would be beneficial to a site where the winter grazing is followed by an abnormally dry spring and summer.

The decline of *Z. trifolii* on the chalk downs of southeast England is discussed elsewhere (Tremewan, 1980a: 144; 1982a: 11); while climatic factors may be involved, the species also appears to be intolerant of habitat disturbance. In two sites allocated as amenity areas the grass is mown once a year, and *Z. trifolii* has completely disappeared from both – in those sites subjected to overgrazing it has declined to near extinction.

Extensive and sometimes genetically interesting colonies of *Z. filipendulae* and *Z. lonicerae* can be found along roadside verges and embankments. Such man-made but nevertheless important habitats are often colonized very quickly after their construction; this applies especially to *Z. lonicerae* which moves in after the rapid colonization by red clover (*Trifolium pratense*), a host plant that thrives in freshly seeded grassland devoid of high vegetation. The detrimental effects caused to such habitats by the use of herbicides are obvious, whilst annual mowing, although essential to protect the host plants from competing rank vegetation, should not be done while unhatched cocoons still remain on the stems of grasses and other plants. One detrimental factor on newly constructed embankments is the deliberate planting of hawthorn scrub and trees, presumably to prevent soil erosion, after the herbaceous plants have become established, but such a policy seems to be unnecessary if the embankment is small or consists of a gentle slope. In the Federal Republic of Germany, common bird's-foot trefoil (*Lotus corniculatus*) is deliberately included with the grasses when newly constructed verges are seeded, a policy that is strongly advocated and one that could easily be adopted in Britain.

Collecting and preparation techniques

The collection and preparation of Zygaenidae can present problems – the species are resistant to cyanide while some killing agents, such as ammonia, cause structural damage to the scales and permanently change the colour of their wings. In my experience ethyl acetate is the best killing agent as it will leave the specimens sufficiently relaxed for setting almost immediately, provided they are not kept in the medium for longer than 20 minutes. A method widely used on the Continent (Leinfest, 1952; Wiegel, 1958), and one that is recommended for expeditions abroad, is to pin the specimens alive in the field, using pins that have been previously 'nicotined'. The pins are prepared by pressing them against cotton wool soaked in nicotine extracted from a plug of tobacco, then placed in a pocket-sized, cork-lined or preferably plastazote-lined collecting-box and allowed to dry. When pins prepared in this manner are used, the specimens are killed almost immediately and remain in good condition, with no loss of colour or damage to their scales; however, freshly emerged specimens usually 'bleed' freely and it is recommended that any haemolymph arising from the thorax be removed by using absorbent paper, otherwise it will discolour the wings should they come into contact. At the end of the day the specimens are transferred to a storebox in which they are allowed to dry by occasionally leaving the lid open (which will also prevent the formation of mould); they can then be relaxed and set at a later stage. Pinned specimens are easier to prepare than those which have been stored in paper envelopes; the latter method results in broken antennae, and almost invariably the wings spring upwards after the specimens are removed from the setting boards. Unlike many Lepidoptera, *Zygaena* and *Adscita* will relax easily. An efficient relaxing-box can be made by using an airtight plastic container in the bottom of which some water-absorbent material is placed (the substance used by florists and known by the trade-name 'Oasis' is ideal for this purpose); mould is prevented by adding a few drops of household disinfectant, such as Dettol, to the water. The specimens are pinned into the box and will relax within 24 hours, after which they can be set in the conventional manner. It should be noted that prolonged exposure to moisture can permanently change the wing coloration of *Zygaena* and *Adscita*, species of the latter genus being especially affected. After allowing them to dry naturally for two or three weeks, they should be placed (still *in situ* on the setting boards) for 10–12 hours in an oven set at a fairly constant temperature of 45–50°C; this method will usually prevent any subsequent springing of the wings. On removal from the oven, the specimens are naturally very brittle and should remain on the boards for a day or so at normal room temperatures. When drying in an oven the temperature should be checked regularly to prevent damage to or destruction of the specimens.

Key to subfamilies and genera (imagines) of the Zygaenidae

1   Forewing with veins 9 (R_3) and 8 (R_4) stalked (figure 47). Antenna of both sexes simple, clavate, *i.e.* dilated towards apex to form a pointed club. Forewing ground colour black with a moderate, blue, blue-green or green sheen, spotted or streaked with red; hindwing red with a black border from apex to tornal area ....
.... Zygaeninae: *Zygaena* (p. 85)

–   Forewing with veins 9 (R_3) and 8 (R_4) separate (figure 48). Antenna of both sexes weakly clavate, that of male bipectinate, that of female weakly dentate. Forewing uniformly yellowish green, green or blue-green, with a strong submetallic sheen; hindwing fuscous ....
.... Procridinae: *Adscita* (p. 78)

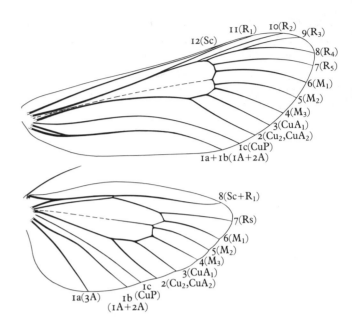

Figure 47 *Zygaena (Zygaena) filipendulae* (Linnaeus), wing venation

Procridinae

ADSCITA Retzius

Adscita Retzius, 1783, *in* Degeer, *Gen.et Spec.Ins.*: 8, 35.

A Palaearctic genus which is represented in the British Isles by three species; all are local and occur in very small colonies. The most widely distributed is *A. statices* (Linnaeus) which occurs in southern England and Wales northwards to the Western Isles of Scotland, and in Ireland. *A. geryon* (Hübner) is also widely distributed but restricted to calcareous districts in England and Wales. The rarest is *A. globulariae* (Hübner), which is very local and restricted to a few localities on the chalk downs of southern England.

Imago. Head, thorax and abdomen smooth. Antenna of male bipectinate, that of female weakly dentate. Tibia of hindleg with one pair of spurs (apical). Forewing with veins 9 (R_3) and 8 (R_4) separate (figure 48). Hindwing with upper margin of cell (R) comparatively straight or weakly angulate at connection with vein 8 ($Sc+R_1$) (figure 48). Submetallic yellowish green, green or blue-green, mainly diurnal species, but occasionally also nocturnal.

Larva. Feeds on Compositae, Cistaceae, Polygonaceae and Globulariaceae.

Pupa. Enclosed in a flimsy cocoon constructed of silk.

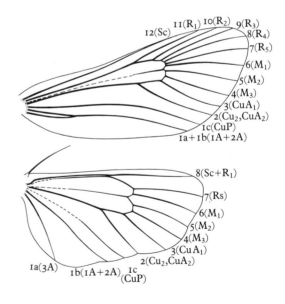

Figure 48 *Adscita (Adscita) statices* (Linnaeus), wing venation

Key to species (imagines) of the genus *Adscita*

NOTE. Species of *Adscita* can be separated most easily by characters of the genitalia.

1   Antenna of male tapering apically, pointed, with only three terminal segments lacking pectinations (figure 49a); antenna of female weakly dentate, almost of same thickness from base to apex. Male genitalia (figure 50, p. 80) with a long sclerotized spine arising from sacculus, vesica of aedeagus without a cornutus; female genitalia (figure 51, p. 81) with ductus bursae weakly sclerotized, ostium circular. (S. England) .... *globulariae* (p. 82)

–   Antenna of male thickened apically, obtuse, with eight or more terminal segments lacking pectinations; antenna of female weakly dentate, gradually thickening apically. Male genitalia with valva lacking a sclerotized spine, vesica of aedeagus with one or two cornuti; female genitalia with ductus bursae heavily sclerotized, or at least partly so, ostium not circular 2

2(1) Antenna of male with 9–11 terminal segments lacking pectinations (figure 49c). Male genitalia (figure 50, p. 80) with aedeagus stout, vesica with a small cornutus and a large, curved cornutus; female genitalia (figure 51, p. 81) with ductus bursae heavily sclerotized only below ostium. Wingspan of male usually more than 25mm, that of female usually more than 22mm. (Britain, Ireland) .... *statices* (p. 84)

–   Antenna of male with 8–10 terminal segments lacking pectinations (figure 49b). Male genitalia (figure 50, p. 80) with aedeagus slender, vesica with a small, straight cornutus; female genitalia (figure 51, p. 81) with ductus bursae heavily sclerotized for more than half its length. Wingspan of male usually less than 25mm, that of female usually less than 22mm. (England, Wales) .... *geryon* (p. 83)

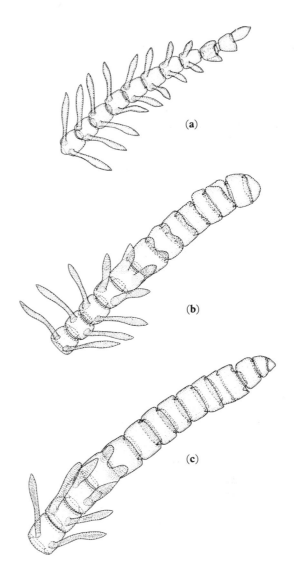

Figure 49 Terminal segments of male antennae
(**a**) *Adscita* (*Jordanita*) *globulariae* (Hübner)
(**b**) *Adscita* (*Adscita*) *geryon* (Hübner)
(**c**) *Adscita* (*Adscita*) *statices* (Linnaeus)

Figure 50

Adscita (Jordanita) globulariae (Hübner)
(**a**) male genitalia

Adscita (Adscita) geryon (Hübner)
(**b**) male genitalia

Adscita (Adscita) statices (Linnaeus)
(**c**) male genitalia

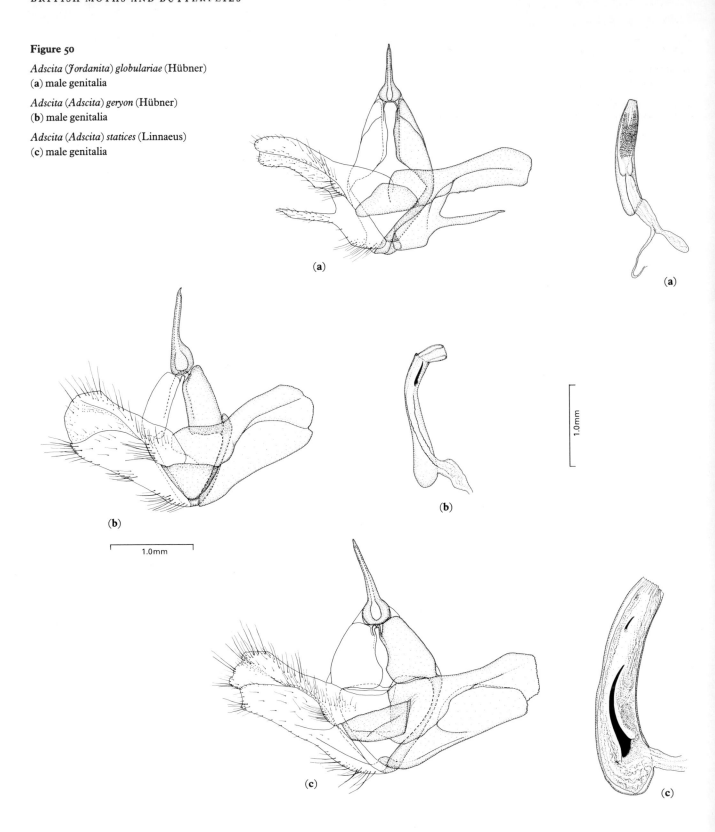

(a)

(a)

(b)

1.0mm

(b)

1.0mm

(c)

(c)

Figure 51

Adscita (Jordanita) globulariae (Hübner)
(**a**) female genitalia

Adscita (Adscita) geryon (Hübner)
(**b**) female genitalia

Adscita (Adscita) statices (Linnaeus)
(**c**) female genitalia

(**a**)

1.0mm

1.0mm

(**b**)

(**c**)

Subgenus JORDANITA Verity

Jordanita Agenjo, 1940, *Eos, Madr.* **13**: 46, 47 [unavailable name; without designation of type-species].

Jordanita Verity, 1946, *Redia* **31**: 134 [name made available by designation of type-species].

Imago. Antenna of male tapering apically, pointed, that of female weakly dentate, of almost even thickness throughout. Male genitalia with a variably developed, sclerotized spine arising from sacculus; female genitalia with ductus bursae lacking strong sclerotization.

Larva. Feeds on Compositae and Globulariaceae.

ADSCITA (JORDANITA) GLOBULARIAE (Hübner)
The Scarce Forester

Sphinx globulariae Hübner, 1793, *Samml.Vögel und Schmett.*: 12, pl.67.

Procris (Rhagades) acanthophora Agenjo, 1937, *Eos, Madr.* **12**: 302.

Type locality: Germany; [Jena].

Description of imago (Pl.4, figs 1,2)

Wingspan: male 23–30mm, female 19–23mm. Antenna of male bipectinate, tapering towards and pointed at apex, with the three terminal segments lacking pectinations (figure 49a, p.79); antenna of female weakly dentate, relatively long and slender and almost of the same thickness from base to apex. Forewing with a strong submetallic sheen, varying from yellowish green or green to blue-green. Hindwing light fuscous. Antenna, head, thorax, abdomen and legs concolorous with forewing.

Genitalia. *Male* (figure 50, p. 80): valva with a long sclerotized spine arising from sacculus; aedeagus relatively small, stout, vesica lacking cornuti. *Female* (figure 51, p. 81): ostium circular, heavily sclerotized; ductus bursae short, weakly sclerotized.

Similar species. Both sexes of *A. globulariae* can be distinguished from *A. geryon* (Hübner) and *A. statices* (Linnaeus) by the structure of the antennae. In the male of *A. globulariae* the antenna tapers towards and is pointed at the apex, with only the three terminal segments lacking pectinations; in the female the antenna is longer than in the other two species, and is more slender and almost of the same thickness from the base to the apex. The three species are also distinguished readily by characters of the genitalia.

Life history

Ovum. Yellow. June; deposited in rows on the leaves of the host plant, hatching after about three weeks (Jackson, 1959: 112).

Larva. Full-fed *c.*12mm long. Head blackish brown.

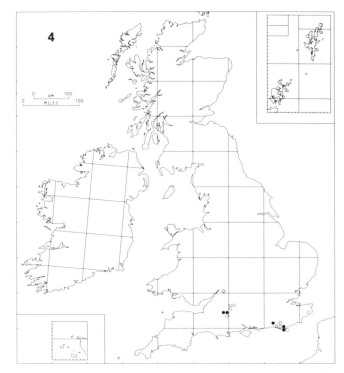

Adscita (Jordanita) globulariae

Prothoracic and anal plates dark brown; thoracic legs black; thorax and abdomen pale greyish brown, darker dorsally, with a fairly broad, whitish, cream or pale yellow dorsal line curving towards middle of dorsum intersegmentally and sometimes edged ventrally with red; verrucae pink; setae brown or white; spiracles small, oval, black. A detailed description of the larva (and pupa) is given by Cockayne & Hawkins (1932: 19). July to May; on common knapweed or hardhead (*Centaurea nigra*) and greater knapweed (*C. scabiosa*), blotch-mining the leaves, usually from the underside, feeding until early September, then overwintering and recommencing to feed in the following spring (Bramwell, 1919).

Pupa and cocoon. Pupa *c.*10–12mm long. Pale brown or yellowish brown. May and early June; in a cocoon spun just below the surface of the soil or on the ground amongst leaf-litter. Cocoon constructed of soft but fairly tough, pale brownish grey silk and enclosed in an elongate-oval, fragile outer spinning of silk and earth particles (Cockayne & Hawkins, 1932: 21). After eclosion of the moth the pupal exuviae sometimes remain protruding from the cocoon.

Imago. Univoltine. June and early July; inhabiting grassy banks and slopes on chalk downs and limestone hills. The males can be seen flying actively in sunshine, but the females are somewhat lethargic and usually rest on flower-heads or grass-stems. According to Jackson (1959: 114) there is a nocturnal flight of the males between 23.00 hrs and midnight, when they range over quite considerable distances; occasionally specimens of this sex have been taken at light.

Distribution (Map 4)

In the British Isles known only from England where it is very local and scarce, being restricted to a few localities in calcareous districts in Kent, Sussex, Wiltshire and Gloucestershire. Widely distributed from Europe to the Caucasus Mountains.

Subgenus ADSCITA Retzius

Adscita Retzius, 1783, *in* Degeer, *Gen.et Spec.Ins.*: 8, 35.
Procris Fabricius, 1807, *in* Illiger, *Magazin Insektenk.* **6**: 289.
Ino Leach, [1815], *in* Brewster, *Edinburgh Encycl.* **9**: 131.

Imago. Antenna of male thickened apically, obtuse, with eight or more terminal segments lacking pectinations, that of female weakly dentate, gradually thickening apically. Male genitalia lacking sclerotized spine on sacculus; female genitalia with ductus bursae sclerotized.
Larva. Feeds on Polygonaceae and Cistaceae.

ADSCITA (ADSCITA) GERYON (Hübner)
The Cistus Forester

Sphinx geryon Hübncr, [1813], *Samml.eur.Schmett.* **2**: pl.28, figs 130,131.
Type locality: Europe.

Description of imago (Pl.4, figs 3,4)

Wingspan: male 19–25mm, female 18–22mm. Antenna of male bipectinate, thickened at apex, with 8–10 terminal segments lacking pectinations (figure 49b, p. 79); antenna of female weakly dentate, slender, gradually thickening apically. Forewing with a strong submetallic sheen, varying from yellowish green to green. Hindwing light fuscous. Antenna, head, thorax, abdomen and legs concolorous with forewing.

Genitalia. *Male* (figure 50, p. 80): valva simple; aedeagus relatively small, slender, vesica with a small, straight cornutus. *Female* (figure 51, p. 81): ductus bursae heavily sclerotized.

Similar species. Differs from *A. statices* (Linnaeus) by its usually smaller size, and readily distinguished by characters of the genitalia.

Life history

Ovum. Flattened-ovoid; yellow. June; deposited singly, or in two rows which are generally placed side by side.
Larva. Full-fed *c.*12mm long. Head black. Prothoracic plate dark brown or blackish brown, edged with yellowish anteriorly; thoracic legs black; thorax and abdomen purplish or reddish brown, a narrow, whitish mediodorsal line; dorsal verrucae purplish red, edged with cream. July to May, overwintering. On common rock-rose (*Helianthemum nummularium*), at first blotch-mining the leaves, later feeding on the underside of a leaf and leaving the upper epidermis intact, and in the spring eating the entire leaf (Hellins *in* Buckler, 1887: 91–94).
Pupa and cocoon. Pupa *c.*9mm long. Shining, dark olive, abdomen lighter. May to early June; in a cocoon low down and attached to the stems of the host plant, or spun

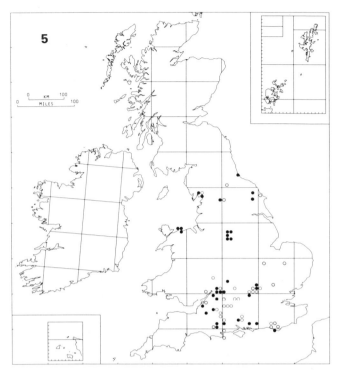

Adscita (Adscita) geryon

amongst moss near the roots. Cocoon somewhat fusiform in shape, with base flattened; constructed of loosely spun, white or greyish white silk, sometimes tinged with yellow. *Imago.* Univoltine. Late May to early July; inhabiting grassland on chalk downs and limestone hills. The males are active in sunshine, but the females are more frequently found at rest on grass-stems and other herbage.

Distribution (Map 5)

A local species which, in Britain, is restricted to calcareous districts in England and Wales. In England its range extends from Kent to Dorset and Somerset, northwards to Cumberland (Cumbria) and Durham. In Wales known only from Denbighshire (Clwyd) and Caernarvonshire (Gwynedd), the Great Ormes Head being a noted locality. Unknown from Ireland. Widely distributed in Europe, its range extending to north-western Turkey.

ADSCITA (ADSCITA) STATICES (Linnaeus)
The Forester

Sphinx statices Linnaeus, 1758, *Syst.Nat.* (Edn 10) **1**: 495.
Procris lutrinensis Heuser, 1960, *Z.pfälz.Ges.Förd.Wiss.* no. 1 [not seen].
Procris heuseri Reichl, 1964, *NachrBl.bayer.Ent.* **13**: 100.
Type locality: Europe [Sweden; Uppsala].

Description of imago (Pl.4, figs 5,6)

Wingspan: male 25–31mm; female 22–25mm. Antenna of male bipectinate, thickened at apex, with 9–11 terminal segments lacking pectinations (figure 49c, p. 79); antenna of female weakly dentate, slender, gradually thickening apically. Forewing with a strong submetallic sheen, varying from yellowish green or green to blue-green. Hindwing light fuscous. Antenna, head, thorax, abdomen and legs concolorous with forewing.

Genitalia. *Male* (figure 50, p. 80): valva simple; aedeagus relatively large, stout, vesica with a small cornutus and a large, curved cornutus. *Female* (figure 51, p. 81): ostium broad; corpus bursae double; ductus bursae S-shaped, heavily sclerotized only below ostium.

Similar species. Differs from *A. geryon* (Hübner) *q.v.* by its generally larger size, and readily distinguished by characters of the genitalia.

Life history

Ovum. Flattened-ovoid; light yellow. June and early July; deposited in small batches.

Larva. Full-fed *c.*12mm long. Head and thoracic legs black; prothoracic plate dark brown; thorax and abdomen varying from pale green or greenish yellow to pinkish or dirty white, with a dark mediodorsal line; verrucae pink, pinkish brown or brown, dorsal verrucae edged with cream. July to early May; on common sorrel (*Rumex acetosa*), at first mining the leaves, usually on the underside, then feeding exposed on the lower leaves of the plant (Hellins *in* Buckler, 1887: 87–90; G. M. Tarmann, pers. comm.).

Pupa and cocoon. Pupa *c.*9mm long. Light brown. May to June; in a cocoon spun near the ground and concealed amongst herbage. Cocoon bluntly fusiform, with a flattened base; rather flimsy, constructed of loosely spun, white silk; pupa visible within.

Imago. Univoltine. Late May to early July; inhabiting damp meadows, rides and clearings in woods, sandhills, chalk downs and limestone hills. The moth flies in sunshine and during dull weather can be found resting on herbage; on warm evenings a flight of the males occasionally occurs an hour or so before sunset. Both sexes are attracted to and feed at flowers, especially those of ragged robin (*Lychnis flos-cuculi*).

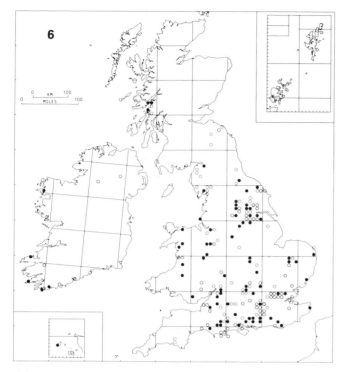

Adscita (Adscita) statices

Distribution (Map 6)

Although local, *A. statices* is the commonest and most widely distributed species of the genus in the British Isles, its range extending from southern England into Wales and northwards to Argyll and Inverness-shire in Scotland. Widely distributed in Ireland. The Channel Islands. Throughout Europe to Turkey and the Caucasus Mountains.

Zygaeninae

ZYGAENA Fabricius

Zygaena Fabricius, 1775, *Syst.Ent.*: 550.

A Palaearctic genus which is represented in the British Isles by seven species all of which are absent from the Orkneys and Shetlands. The commonest and most widely distributed species is *Z. filipendulae* (Linnaeus) which also occurs on many of the islands around Britain and Ireland. *Z. viciae* ([Denis & Schiffermüller]) is now known only from one colony in Argyll, having become extinct in the New Forest, Hampshire, in about 1927 (Tremewan, 1966). Two other species with disjunct distributions are *Z. lonicerae* (Scheven) and *Z. purpuralis* (Brünnich); the former occurs from southern England to Cumberland (Cumbria) and Northumberland, in the Isle of Skye, and in Ireland, the latter is represented by three subspecies in North Wales, western Scotland and western Ireland. *Z. trifolii* (Esper) is widely distributed in southern England and in Wales, and also occurs on the Isle of Man, the most northerly point of its entire range. *Z. loti* ([Denis & Schiffermüller]) and the boreo-alpine species *Z. exulans* (Hohenwarth) are restricted to Scotland.

It is perhaps curious that six of the seven British species are to be found in the high rainfall areas of Scotland, with five occurring in the west Highlands – the seventh, *Z. trifolii*, is a west Mediterranean species and therefore would not be expected to occur so far north. The disjunct distributions of *Z. viciae*, *Z. lonicerae* and *Z. purpuralis* probably reflect different waves of immigration from mainland Europe although, contrary to the theories of Beirne (1943; 1947; 1952), Harrison (1947: 142; 1955: 177) and Ford (1955), the arrival and subsequent establishment of all seven species must have been postglacial. Dennis (1977) contends that the establishment of all the British species of butterflies was postglacial, and it is my opinion that this hypothesis can be applied equally to the British *Zygaena* species (and, for that matter, the *Adscita* species).

Imago. Head, thorax and abdomen rough-haired. Antenna clavate, *i.e.*, dilating apically to form a pointed club. Tibia of hindleg usually with two pairs of spurs (medial and apical), but medial pair often vestigial or absent. Forewing with veins 9 (R_3) and 8 (R_4) stalked (figure 47, p. 78). Hindwing with upper margin of cell (R) distinctly angulate at connection with 8 (Sc+R_1) (figure 47, p. 78). Brightly coloured, predominantly black and red, mainly diurnal species, although some are also nocturnal in the Middle East.

Larva. Feeds on Leguminosae, Empetraceae, Umbelliferae, Compositae and Labiatae.

Pupa. Enclosed in a parchment-like cocoon.

Rearing

Provided that the utmost care is taken with regard to hygiene, and fresh food is supplied daily to the larvae, most species can be reared successfully from the egg phase; the most difficult part is bringing the larvae safely through diapause, which normally takes place in the fourth instar (p. 91). As the larvae have an obligate diapause, no attempt should be made to force them through as invariably this will be fatal.

The larvae are reared best in transparent plastic boxes of two sizes: approximately 5 × 3 × 2cm and 8 × 4 × 2cm. The smaller size is used for newly hatched or young larvae which should be split into groups and transferred to the larger boxes at a later stage. Eventually the final instar larvae are segregated, one per box; while this involves additional time and work, the results are more satisfactory and the risk of spreading disease, should it ever occur, is reduced to a minimum.

When the larvae have entered diapause, which can be recognized by loss of the normal coloration immediately after a moult (p. 92), and by the cessation of feeding, they should be kept in a cool room until the autumn; by this time some frass may have accumulated in the boxes which should be cleaned before the larvae are put into a refrigerator for the winter. In late October or early November the boxes containing the larvae are placed in plastic bags in which a little water has been sprinkled – not more than a quarter of the bag should be filled with boxes, the remainder of the bag being wrapped loosely around them. The plastic bags containing the larvae are then placed in the bottom of the refrigerator and kept at 5–7°C; to prevent desiccation of the larvae it is advisable to continue sprinkling a little water into the bags, should they become dry, and inspection is recommended every two or three weeks. In the spring the larvae should be removed from the refrigerator and placed in a warm room; feeding will not recommence until they have moulted, and moulting can be helped considerably by keeping the larvae in a reasonably humid atmosphere (this can be achieved by sprinkling a little water daily on to the underside of the lids of the boxes).

When the final instar larvae are fully fed, they should be placed in large paper-lined boxes in which they will spin their cocoons, either on the paper or on the dry stems of grass which have been provided. The cocoons should be kept in large plastic boxes on the bottom of which some absorbent paper has been placed; every night a little water should be sprinkled on the underside of the lid of each box to prevent desiccation of the pupae.

In order to have a constant supply of fresh food for the larvae, it is recommended that cultures of the host plants are grown in the garden or in a greenhouse. If it is necessary to supplement this food supply by using plants growing in the wild, freshly picked sprigs will remain fresh for up to a week if they are sprinkled with water, placed in a plastic box and kept in the refrigerator.

If a pairing of newly emerged adults is required, it can usually be obtained by putting the moths in a gauze-covered cage which is placed in a sunny position in the garden or, on cool or sunless days, in the greenhouse. After copulation, which usually lasts for 24 hours, the female should oviposit readily in a pill-box if kept in a warm place.

Life history
Ovum

Eggs of *Zygaena* species are classified as ova jacentia. Ovoid, flattened, horizontally resting, the micropyle situated at one end of the horizontal axis of the egg (Tutt, 1899; Döring, 1955). Chorion finely reticulate, transparent, the colour of the egg resulting from the yolk. When first laid, yellow with one pole transparent, the colour changing to orange-yellow and finally lead-grey on the day prior to hatching.

The eggs are coated with a glutinous substance secreted by the colleterial or sebaceous glands of the female and adhere to the surface on which they are deposited in groups, each group, according to the species, consisting of a single layer, or several layers forming an irregularly shaped batch. *Z. lonicerae* normally deposits its eggs in a single layer, but they are placed several layers high if the available surface, such as a narrow blade of grass, is not sufficiently wide and flat (Tremewan, pers.obs.). Oviposition takes place during the day and generally within a few hours after copulation is completed; light is not essential for this function as females captured during or immediately after copulation will oviposit readily in pill-boxes in captivity, provided that they are kept in a warm place. An added inducement for them to lay in captivity is to enclose a sprig of the foodplant in the box. In the British Isles it is usual for *Zygaena* to oviposit on plants other than their host plants, the prerequisite being a flat surface large enough to accommodate the batch of eggs (Pl. 6, fig.10); on hatching the larvae then have to search for their host plant but this apparently causes no great hardship as they normally do not commence feeding until 24 hours after hatching, at least in captivity. In addition to common bird's-foot trefoil (*Lotus corniculatus*), I have found the eggs deposited on the underside of leaves of knapweeds (*Centaurea* spp.), plantains (*Plantago* spp.), wild strawberry (*Fragaria vesca*) and grasses (Gramineae) which were growing in the immediate vicinity of the host plant, and females of *Z. lonicerae* have even been observed ovipositing on the underside of the leaves of a sallow (*Salix* sp.)

about a metre or so from the ground (Tremewan, 1982a: 9). The egg phase lasts for an average of 11 days under normal temperatures but I have recorded as little as eight days, while the period may be prolonged to as much as 17 days or more during inclement weather. The female oviposits more than once, and it seems that copulation takes place several times; dissection of old worn females has shown that sometimes two or three spermatophores are present in the bursa copulatrix (Tremewan, pers.obs.). Parthenogenesis has not been recorded.

Larva

Head retractile, prognathous; epicranium with well-developed, inverted Y-shaped suture, the inverted V-portion enclosing the frons; medial arm of suture almost half length of head; adfrontal sclerites narrow; labrum distinct, emarginate on middle of distal margin. Ocelli normally six in number, situated posterolateral to antenna. Antenna three-segmented, arising between head capsule and mandible. Spinneret located on hypopharynx of labium, preceded on both sides by labial palpi, the latter flanked by the maxillary palpi.

Integument of thorax and abdomen shagreened. Thorax three-segmented, each segment bearing a pair of well-developed, curved, segmented legs; a pair of well-developed, circular spiracles on prothorax. Abdomen ten-segmented; segments 3–6 and 10 each with a pair of prolegs, those on 3–6 being the ventral prolegs, those on 10 the anal prolegs; crochets uniordinal, uniserial, arranged in a mesoseries; segments 1–8 each with a pair of circular spiracles.

Head with primary setae consisting of simple hairs; thorax and abdomen with primary and subprimary, and secondary, weakly plumose setae arranged in groups known as verrucae. Only primary setae are present in the first instar, and these are supplemented with subprimary and secondary setae in the second instar, and with secondary setae in subsequent instars.

The chaetotaxy of *Z. trifolii* is described below and illustrated in setal maps of the epicranium (figure 52), pro- and mesothoracic segments and abdominal segment 6 (figure 53, p. 89) and abdominal segment 10 (figure 54, p. 90); the terminology follows Hinton (1946a).

(i) Cranial setae and punctures

C1 and C2 long, of approximately equal length, C2 above and mesad from C1; F1 shorter than C1 and C2, Fa close to and mesad from F1; AF1 very short, AFa caudad from AF1, AF2 absent; A1 long, A2 shorter, caudad and laterad from A1, A3 very short, caudad from A2, Aa close to and mesad from A3; P1 as long as A2, near adfrontal sclerite, P2 very short, caudad and mesad from P1, Pa closer to L1 than to P2, Pb close to and caudad from P2; L1 near to and anterolaterad from Pa, La absent; V1 and V2 very short, caudad and laterad from Pb, Va caudad and laterad from V2, V3 absent; O1 very short, near fourth and fifth ocellus, O2 longer than A2, caudad and laterad from O1, O3 short, above and caudad from fifth ocellus, Oa caudad from SO2; SO1 as short as or shorter than A2, SO2 longer than SO1, near first ocellus, SO3 longer than SO2, caudad from SO1, SOa below SO2; G2, G1 and Ga caudad from SO3, forming a straight line.

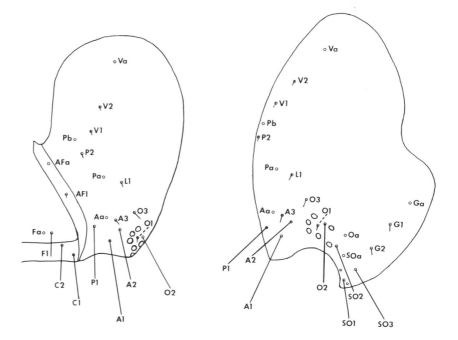

Figure 52

Zygaena (Zygaena) trifolii (Esper), chaetotaxy of epicranium

(ii) Thoracic and abdominal setae

Tactile setae. XD1 and XD2 primary, of approximately equal length, present only on prothoracic segment; XD1 dorsal to XD2, forming a vertical line.

D1 and D2 primary on thoracic and abdominal segments, of approximately equal length; D1 posterodorsad from XD1 on prothoracic segment and anterodorsad from D2 on pro-, meso- and metathoracic segments and abdominal segment 9, forming an oblique line (less oblique on metathoracic segment and abdominal segment 9); D1 dorsal to D2 on abdominal segments 1–8, forming an almost vertical line; D1 anterior to D2 on segment 10, forming a horizontal line.

SD1 primary on thoracic and abdominal segments, SD2 primary on thoracic segments and abdominal segments 1–8; SD2 approximately equal in length to SD1 on thoracic segments, microscopic on abdominal segments 1–8; SD2 anterodorsad from SD1 on prothoracic segment, forming an oblique line; SD2 dorsal to SD1 on meso- and metathoracic segments, forming an almost vertical line; SD1 dorsal to spiracle on abdominal segments 1–7, anterodorsad from spiracle on segment 8, anteroventrad from D2 on segment 9 and L1 on segment 10; SD2 microscopic on abdominal segments 1–8, those on 1–7 anterodorsad from spiracle, that on abdominal segment 8 anterior to spiracle.

L1 and L2 of approximately equal length, L1 primary on thoracic and abdominal segments, L2 primary on prothoracic segment and abdominal segments 1–9, subprimary on meso- and metathoracic segments. L1 anteroventrad from SD1 and anterodorsad from spiracle on prothoracic segment, L2 anterodorsad from L1, forming an oblique line; L1 ventral to SD1 on meso- and metathoracic segments, forming an almost vertical line; L1 posteroventrad from spiracle, L2 anteroventrad from L1, forming a weakly oblique line, on abdominal segments 1–8; L1 posteroventrad from SD1, forming an oblique line, L2 ventral to L1, forming a vertical line, on abdominal segment 9; L1 anteroventrad from D2, forming an oblique line, on abdominal segment 10.

Before considering the L3 setae, which are subprimary, it should be noted that, in the second instar, secondary setae are present in the XD1/XD2, D1/D2, SD1/SD2 groups of the thorax and abdomen, the L1/L2/L3 groups of the thorax, the L1 group of the abdomen, and the SV1/SV2 groups of the thorax – because of this, and the fact that the primary and subprimary setae are indistinguishable in character from the secondary setae, it is not always possible to recognize L3 in the second instar; the following is therefore tentative (with the exception of abdominal segments 3–6, where L3 can be positively identified).

L3 situated with L1+L2 on prothoracic segment; L3 with L2 on meso- and metathoracic segments; L3 adjacent to and on the same pinacula as L2 on abdominal segments 1, 2 and 7; L3 immediately dorsad from ventral proleg on abdominal segments 3–6. As L3 is considered to be situated with L2 on the meso- and metathoracic segments, then by analogy L3 is also situated with L2 on abdominal segments 1, 2 and 7.

SV1 longer than SV2; SV1 primary on thoracic and abdominal segments; SV2 primary on thoracic segments and abdominal segments 2–7 and 10, absent on abdominal segments 1, 8 and 9. SV2 anterodorsad from SV1 on thoracic segments, forming an oblique line (less oblique on meso- and metathoracic segments); SV2 ventral to SV1 and forming a vertical line on abdominal segments 2 and 7; SV1 and SV2 situated on ventral proleg on abdominal segments 3–6, SV1 laterally on outside of proleg, SV2 anteromesad from SV1. SV3 absent (except on abdominal segment 10 – see below).

V1 primary on thoracic and abdominal segments; V1 situated mesal to and near base of leg on pro-, meso- and metathoracic segments; on segments 1, 2 and 7–9 of the abdomen, V1 is nearer to the medioventral line than MV3; on abdominal segments 3–6 and 10, V1 is situated on the mesal side of the proleg.

Proprioceptors. The proprioceptors are primary, occurring in all instars.

MXD1 present on prothoracic segment, situated near posterodorsal margin of segment and posterior to D2.

MD1 present on meso- and metathoracic segments, situated near anterior margin of segment and anterior to SD2; MD1 present on abdominal segments 1–9, situated near anterior margin of segment and anterodorsad from SD1.

MSD1 and MSD2 present on meso- and metathoracic segments, situated anterior to L1.

MV1, MV2 and MV3 present on meso- and metathoracic segments; MV1 and MV3 situated on anterior margins of these segments and forming a vertical line, MV3 situated anterior to leg, MV2 nearer to MV1 than to MV3, situated posteroventrad from MV1 and posterodorsad from MV3 to form a triangle; MV3 present on prothoracic segment, situated anterior to leg; MV3 present on abdominal segments 1–10, situated near anterior margins of segments 1–9, that on segment 1 anterior to and midway between L3 and V1 and posterior to leg of metathoracic segment, that on segment 2 nearer to V1 than to L3, that on segments 3–6 anteromesad from ventral proleg, that on segments 7–9 anterodorsad from V1, that on segment 10 situated anteroventrally on base of anal proleg.

Abdominal segment 10. The setae of abdominal segment 10 are enigmatic; however, the fact that this segment has more than the usual complement of setae may be ex-

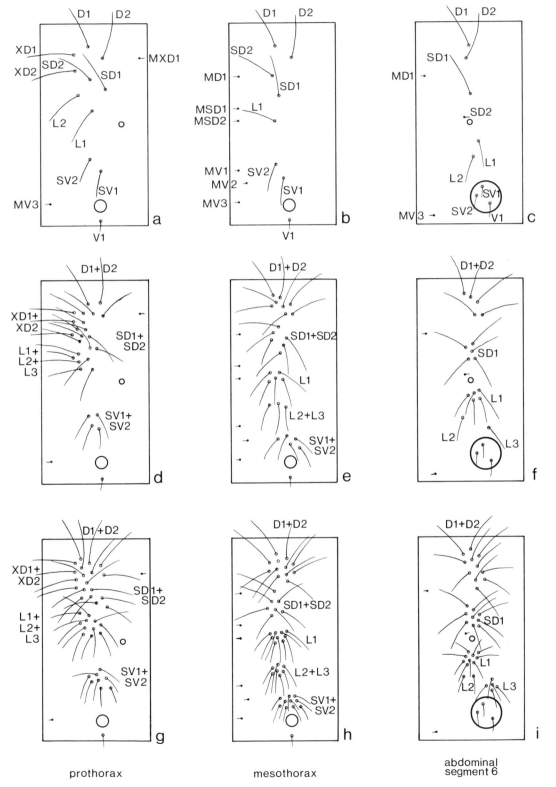

Figure 53

Zygaena (Zygaena) trifolii (Esper), chaetotaxy of pro- and mesothoracic segments and abdominal segment 6
(**a–c**) first instar
(**d–f**) second instar
(**g–i**) third instar

prothorax

mesothorax

abdominal
segment 6

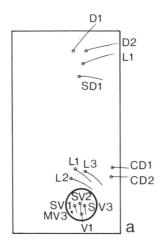
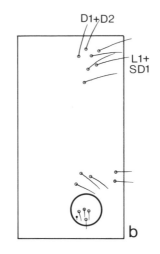
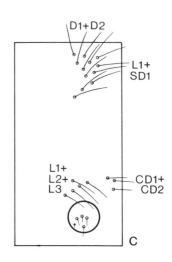

plained by the generally accepted theory that, in some groups of Lepidoptera, the eleventh and twelfth segments have fused with the tenth during embryonic development (Matsuda, 1976: 410–411).

On abdominal segment 10 of the first instar larva, there are four primary setae situated dorsally and dorsolaterally. Immediately above the anal proleg are three primary setae, and posterior to these are two further primary setae situated on the paraproct; five primary setae are situated on the anterior/mesal area of the anal proleg. Secondary setae occur in the dorsal/subdorsal area from the second instar onward, and also in the area above the anal proleg and on the paraproct in the third and subsequent instars, but they are absent on the anterior/mesal area of the proleg in all instars.

The four primary setae in the dorsal/subdorsal area are considered to be D1, D2, SD1 and L1, and are situated on what is presumably a vestige of the true tenth segment; D2 is caudad from D1, forming a horizontal line; L1 is antero-ventrad from D2, SD1 is anteroventrad from L1, the three setae forming an oblique line. According to Matsuda (1976: 57), the terminal appendages on abdominal segment 10 probably represent the true eleventh segment; it follows that the setae situated on the paraproct, above the anal proleg and on the anterior/mesal area of the anal proleg, belong to what is now a vestige of the eleventh segment. Those on the paraproct are considered to be D1 and D2, and in the Geometridae have been named CD1 and CD2 by Singh (1951: 68). Those above the anal proleg are considered to be L1, L2 and L3, the most dorsal being L1, with L2 anteroventrad and L3 posteroventrad from L1. The setae on the anterior and mesal sides of the anal proleg are considered to belong to the SV, V and MV groups; Mutuura (1956: 101), commenting briefly on

abdominal segment 10 of lepidopterous larvae, also considered such setae to be the V and SV setae of Hinton (1946a). There can be little doubt that the seta which is nearest the medioventral line, and situated on the mesal side of the anal proleg near its base, is V1, and that the proprioceptor anterodorsad from it is MV3, these being the homotypes of V1 and MV3 on abdominal segment 9. Of the three remaining setae, one is on the anterior side of the anal proleg and ventrad from MV3, the other two are comparatively close together on the mesal side and ventrad from V1; the seta on the anterior side of the anal proleg is considered to be SV1, and the two on the mesal side and ventrad from V1 could conceivably be SV2 and SV3.

Pattern and coloration

Beck (1982) has described larvae of *Orthosia* Ochsenheimer (Noctuidae) by correlating the pattern with the setae of the first abdominal segment. This appears to be a most satisfactory method and is adapted here for *Zygaena* larvae, the black markings being correlated with the verrucae or setal groups of the abdominal segments; the thoracic segments are disregarded for this purpose as they are fewer in number and, because of the specialized nature of these segments, the verrucae are somewhat displaced compared with most of those on the abdomen. A larva in dorsolateral aspect is therefore described as having dorsal, subdorsal and lateral rows of black spots (figure 55, p. 91). The dorsal row consists of one or two spots per segment, if two then the larger spot is anterior. The subdorsal spots (if present) are immediately above the level of the abdominal spiracles, and the lateral spots (if present) are below and associated with setal groups L1, L2 and L3. In the dorsal series, the larger, anterior spot on each segment is always the first to appear in the early instars, and it is also retained during diapause when the remaining spots are temporarily lost or lose their normal coloration. A bright yellow spot is situated on the posterior part of each segment, and immediately below the posterior black dorsal spot if this is present; these yellow spots usually occur on the second and third thoracic segments and on the first to eighth abdominal segments. The yellow spots are not readily discernible in larvae with a yellowish ground colour, but they are very conspicuous in those which have a dark ground colour, such as *Z. exulans*.

Increase or decrease in the size of the black spots can produce striking aberrations. Two very rare forms of *Z. filipendulae* (Pl.5, figs 11,12) and *Z. lonicerae* lack the posterior dorsal spot on each segment, and the subdorsal and lateral series of spots are also absent. Of regular occurrence in *Z. filipendulae* are those forms which have

the black spots enlarged to such an extent that the two dorsal rows often coalesce. Larvae of *Z. filipendulae* having asymmetrical markings occur occasionally.

So-called spirally segmented larvae occur infrequently. A larva of *Z. trifolii* with spiral segmentation was recorded by Harris (1946: 248; 1947: 13), and I have also noted such abnormality in the same species. Occasionally, damage from a predator will produce abnormal markings, but this should not be confused with spiral segmentation.

Ethology

An obligate diapause takes place in the fourth or occasionally the fifth instar of the larval phase of all the British species; sometimes the brood, or a small proportion, will go back into diapause for a second or even several times. When this occurs the larvae moult and re-enter diapause in which they remain until the spring of the second year (literally aestivation through to hibernation). The larvae of *Z. exulans* apparently overwinter several times (Cockayne, 1932: 100; Grosvenor, 1932a: 62), and this has been observed also in montane species of *Zygaena* from Iran (Tremewan, 1975: 237, 244) and elsewhere. Dryja (1959) maintained that double or multiple diapause in the larvae of the continental species *Z. ephialtes* (Linnaeus) is hereditary and may be dependent on polygenes that determine the number and length of the periods of diapause. However, recent research has shown that the number and sequence of instars in diapause is also controlled by exogenous factors (C. M. Naumann, pers.comm.).

Multiple diapause could be advantageous for the survival of populations. At night and during inclement weather by day the adults of many *Zygaena* species rest fully exposed on grass-stems and other herbage. If prolonged torrential rain and thunderstorms occur during the peak emergence, almost the whole population of adults may be exterminated very quickly. The survivors (including the moths yet to emerge) might not necessarily be sufficient for the colony to remain viable; its survival would then rely mainly on the larvae in diapause at that time. Multiple diapause might also reduce direct inbreeding, thereby producing a genetically more viable population.

It has also been suggested (C. M. Naumann, pers. comm.) that a second diapause could be advantageous to a population that is subjected to a cold spring – the larvae that enter diapause for the second time would not be affected by unfavourable weather conditions and thus would not develop into adults which would produce larvae too late in the season to effectively feed for three instars and enter the obligate diapause, which cannot take place until the fourth instar. Lane (1961: 80) considers that when diapause takes place a second time it might be important, if not essential, to the survival of a colony. It is well known that the moths in a colony build up to very large numbers and that it can be reduced quite suddenly to

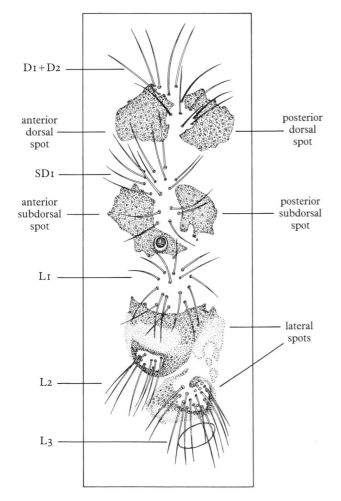

Figure 55 *Zygaena (Zygaena) trifolii* (Esper), abdominal segment 3 of final instar larva showing correlation of the pattern with the setal groups

D1+D2

anterior dorsal spot

SD1

anterior subdorsal spot

L1

L2

L3

posterior dorsal spot

posterior subdorsal spot

lateral spots

a few individuals or even none. The cause of such fluctuations has been attributed to parasites. Lane contends that the *Zygaena* larvae which re-enter diapause in the spring might not be attacked by parasites; if the latter build up to such an extent that the *Zygaena* population is drastically reduced, those *Zygaena* larvae in diapause for the second time would then ensure the survival of the colony by carrying over to the following year.

It appears that when in diapause the larvae are not affected by heavy rain or floods. In Co. Tyrone, *Z. lonicerae* occurred in low-lying meadows which were partially flooded in winter (Greer, 1918: 188), and at one period one of the best localities was covered with at least 30cm of water. In west Cornwall many of the habitats of *Z. trifolii* are also subjected to regular flooding in winter.

During diapause, *Zygaena* larvae, unlike those of other species of Lepidoptera, have one or more special instars during which no feeding takes place. When the larva is about to enter diapause it spins a silken pad on which it remains until it moults before actually entering diapause. It then loses its usual coloration, the ground colour becoming pale yellow, brownish yellow or brown (in the larva of *Z. exulans* the coloration becomes dull); all the black spots are lost or become pale brown with the exception of the dorsal spots in the anterior part of each segment – the small but conspicuous yellow spots also remain. The usual coloration returns a few days prior to moulting at the end of the diapause and can then be seen through the old skin of the larva. Almost invariably, moulting takes place before the larva recommences to feed. Tutt (1899: 417)

stated that the larva feeds in the spring before moulting, and once I observed frass from overwintering larvae of *Z. purpuralis* and *Z. loti* which had nibbled leaves of their host plants prior to ecdysis, but such behaviour is abnormal.

If the larva overwinters once, then the larval phase consists of seven instars, diapause taking place in the fourth or occasionally the fifth instar (figure 56a). However, if diapause takes place for the second time then the larval phase consists of eight instars (figure 56b), and when diapause occurs three times (figure 56c) or more then the number of instars is increased accordingly.

It follows that the length of the larval phase is also affected by the number of times the larva enters diapause. If this occurs only once then the larval phase is approximately 11 months, if twice then it is prolonged to approximately 22 months, if more than twice then it is extended accordingly (figures 56a–c).

Ecdysis of a *Zygaena* larva differs from that of other lepidopterous larvae: it is effected by the larva crawling out of its old skin which has split mediodorsally from the thoracic to the caudal region.

Larvae of *Zygaena* species are diurnal in habits and normally feed during the day, but in captivity they will feed readily also at night providing the temperature is sufficiently high. Five of the British species feed on herbaceous Leguminosae, one is confined to a species of Labiatae, another to a species of Empetraceae. Cannibalism has not been recorded in British populations of *Zygaena*, but it has occurred in two Moroccan subspecies of *Z. trifolii* (Tremewan, pers.obs.), and in *Zygaena* larvae

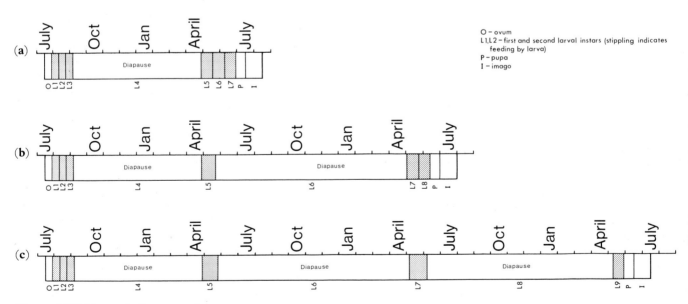

O – ovum
L1,L2 – first and second larval instars (stippling indicates feeding by larva)
P – pupa
I – imago

Figure 56 Life history of a *Zygaena* species

which were fed on a semi-artificial diet (C. M. Naumann, pers.comm.).

Predation

In spite of the toxic properties of *Zygaena* larvae (Povolný & Weyda, 1981; Franzl & Naumann, 1984; Naumann & Reimann, in prep.), there are authentic records of predation by birds. Button (1865: 290) noted that the stomach contents of a hoopoe (*Upupa epops*) included larval skins of *Z. filipendulae*, and Newman (1872: 77) and Curtis (1917: 149) recorded cuckoos (*Cuculus canorus*) feeding on the larvae of this species. Larval predation by starlings (*Sturnus vulgaris*) was recorded by Newman (1872: 77), and Carpenter (1943: 158) observed great tits (*Parus major*) extracting larvae from newly spun cocoons. Carpenter (1945: 283) considered that 'distastefulness to birds is of less protective value during the breeding season when great stress is caused by fledglings driving their parents to desperation by their insistent clamour for food'. The feeding by the bug *Picromerus bidens* on a dead larva in a damaged (by birds?) cocoon of *Z. filipendulae* is probably fortuitous (Tremewan, pers.obs.).

Pupa

Zygaena pupae belong to the group known as pupae incompletae (Chapman, 1893a: 100, 119; Kuznetsov, 1967: 237) or incomplete, obtect, adecticous pupae (Hinton, 1946a: 4; 1946b: 290, 292); the abdominal segments 3–7 of the male and 3–6 of the female are freely movable, the appendages are loosely fused to the body, and articulated mandibles are absent. The following description is generalized and is based on the pupae of the British species.

Head with occiput, labrum, paraclypei and labial palpi; maxillae reaching abdominal segments 5–6; antennae reaching abdominal segment 4; antennae, labial palpi and maxillae passively movable.

Thorax consisting of pro-, meso- and metathorax; prothorax with pronotum and propleura, the pronotum divided into two lateral halves by a longitudinal mediodorsal suture; mesonotum of metathorax divided by a distinct longitudinal mediodorsal suture; metathorax with well-developed metanotum; prothoracic spiracles concealed, situated on propleura; appendages (wings, pro-, meso- and metathoracic legs) passively movable; metathoracic legs reaching abdominal segments 5–7.

Abdomen with ten segments. Segments 1 and 2 fused (male, female), segments 3–7 (male), 3–6 (female) free, segments 8–10 (male), 7–10 (female) fused. Pedes spurii (remnants of the larval prolegs) usually clearly visible on segments 4–6 and 10. Abdominal spiracles present on segments 1–8; those on 1 and 2 concealed by hindwings, those on 8 reduced. Tergites 3–8 (male) or 3–7 (female) each with a transverse row of admincula or hooked spines placed anteriorly; admincula terminating well before or

hardly reaching spiracles on tergite 3, those on tergites 4–8 of male and 4–7 of female usually reaching spiracles or extending just beyond; shafts of admincula projecting cephalad, their hooks projecting caudad. A transverse row of caudally projecting, short, stout spinulae on anterior part of tergite 8 (female only), terminating before spiracles; a shorter transverse row of spinulae on anterior and posterior parts of tergite 9 (male, female), those on posterior part less in number or sometimes absent; apex of segment 10 with cremaster consisting of a cluster of short, stout spinulae (without hamuli).

In the male the genital opening is situated in the middle of sternite 9 and consists of a longitudinal slit with two lateral tubercles. In the female the ostium bursae forms a longitudinal slit on sternite 8; the ostium oviductus is slit-like and situated on sternite 10.

Ethology

Under normal climatic conditions the pupal phase of the British *Zygaena* species lasts for 14 days, the moth emerging 17 days after completion of the cocoon. The length of the pupal phase is rarely affected by high temperatures, but frequently it can be prolonged during adverse weather conditions, and I have recorded *Z. lonicerae* and *Z. trifolii* emerging 22 days after completion of the cocoon. Moore (1892: 37) stated that a cocoon of *Z. filipendulae* contained a living pupa after a period of 17 months, but it is not known whether the moth eventually emerged, while Webb (1896: 255) recorded that he reared *Z. trifolii* and *Z. filipendulae* from pupae which went into a second year; as the obligate diapause is in the larval phase the authenticity of these records is highly suspect.

Just before the moth emerges, the pupa ruptures the anterior part of the cocoon, using the vertex of the head to break the silken threads and the tergal admincula to force its way through until the head, thorax and fused first and second abdominal segments are exposed. After eclosion of the moth the pupal exuviae remain protruding from the cocoon in *Z. viciae*, *Z. filipendulae*, *Z. trifolii* and *Z. lonicerae*, but are loosely retained in *Z. purpuralis* and *Z. exulans*, or fall to the ground in *Z. purpuralis* and *Z. loti*. Grosvenor (1923a: 71) suggested that in some species the retention of the pupal exuviae by the cocoon is necessary for the successful emergence of the moth, but *Z. lonicerae*, *Z. trifolii* and *Z. filipendulae* will sometimes emerge successfully from pupae resulting from larvae that failed to construct a cocoon (Tremewan, pers.obs.; Beavis, 1973: 267).

Pupae of *Zygaena* sometimes produce quite a distinct rasping sound; this is caused by the pupa rubbing its abdominal spines against the walls of the cocoon. Beavis (1973: 267) has likened the sound to that of a grasshopper. The function of sound production is unknown in *Zygaena*, although Hinton (1948: 268; 1955: 81) suggested that it is defensive when it is caused by other lepidopterous pupae. Some *Zygaena* pupae will produce sound without any apparent stimulus, while others will sometimes respond immediately if the cocoon is touched or tapped. Sound production by *Zygaena* pupae is probably incidental, as the spines which produce the sound function when the pupa breaks through the cocoon before eclosion of the moth.

Predation

Predation of the pupae is well known (*e.g.* South, 1897: 181; Colthrup, 1918: 52) but there has been much speculation over the identity of the predators. Darlow (1947: 134, 136) suggested that the predation is attributable to birds, but Grosvenor (1912: 216) considered that neither birds nor mice are responsible. Usually the cocoons are ruptured or torn open at the anterior end although there are records of damage at or near the centre (Nash, 1918: 215; Hodgson, 1945: 176), and Grosvenor (1912: 216) recorded that the lower (posterior) end is roughly torn off; in my opinion such damage can be attributed more readily to birds than to small rodents as the latter are more likely to make a comparatively neat hole. Mention has already been made of birds extracting larvae from newly spun cocoons (Carpenter, 1945: 283), and there is no apparent reason why pupae should not be taken in a similar manner. Greer (1920: 155; 1941: 55) recorded reed buntings (*Emberiza schoeniclus*) extracting pupae from cocoons of *Z. lonicerae* in Co. Tyrone. Carpenter (1937: 176) originally suggested that ants were responsible because of staining from pupal fluid around the edge of the hole, but subsequently (Carpenter, 1945: 284) considered the damage to be characteristic of bird predation as no ants were seen attacking undamaged cocoons. However, as such staining is lacking in many damaged cocoons, it is not necessarily characteristic of an attack by a bird. The feeding by ants, the bug *Picromerus bidens* (Tremewan, pers.obs.), an unidentified pentatomid bug (Carpenter, 1937: 176) and the scorpion fly *Panorpa communis* (Tremewan, 1980b: 274) on pupae of *Z. filipendulae* and *Z. lonicerae* must be fortuitous.

Cocoons spun on wire fences or on twigs of low-growing bushes are often prone to attack (Nash, 1918: 215; Carpenter, 1944: 239), while predation is usually less frequent on those spun on grass-stems. This may be due to the cocoons being more conspicuous on bushes and fences, while those on grass-stems are possibly afforded some degree of protection as, depending on their colour, they may be seen less easily in this situation. However, at Ranmore Common, Surrey, I have observed heavy predation of cocoons of *Z. filipendulae* that were spun on grass-stems near a wire fence which appeared to have provided a convenient perch or observation post for avian predators; as one proceeded up the slope and away from the fence it was noted that the predation became progressively less.

Nash (1918: 215) and Carpenter (1943: 158) suggested that it would be difficult for a bird to attack a cocoon which is spun high on a grass-stem, the latter not affording a foothold for the predator; however, grass-stems bearing damaged or empty cocoons are often inclined towards the ground (Tremewan, pers.obs.), and it is suggested that this has been caused by the weight of a bird attacking the cocoon.

Cocoon

The larvae of all *Zygaena* species construct a cocoon before pupating. The cocoons of *Z. viciae*, *Z. filipendulae*, *Z. lonicerae* and *Z. trifolii* are fusiform or spindle-shaped (figures 57e,f, p. 95), those of *Z. exulans* and *Z. purpuralis* are bluntly fusiform and somewhat rounded at each end (figures 57a,b, p. 95), that of *Z. loti* is ovoid or elliptical (figures 57c,d, p. 95). They are attached to stems of grasses and other herbage, and occasionally they are spun on leaves, rocks and stones, on twigs of low-growing bushes, and even on wire fences. The fusiform cocoons which are spun on the stems of grasses are placed in a vertical position with the head of the pupa uppermost.

After completing the silken construction, the larva smears the inside of the whole structure with excretory products which harden and combine with the silk to form the familiar parchment-like cocoon (Beavis, 1973: 267). According to Wigglesworth (1972: 576, 584) the smeared coating on the cocoons of some lepidopterous species consists of urates from the Malpighian tubules; in the continental *Z. fausta* (Linnaeus), and in *Z. purpuralis*, it has been identified as calcium oxalate monohydrate (Naumann, 1977: 34; Naumann *et al.*, 1983: 31), which is a poisonous substance insoluble in water.

The cocoons of *Z. filipendulae* (Pl.6, fig.4), *Z. trifolii* (marsh form) (Pl.6, fig.6) and *Z. lonicerae* (Pl.6, fig.8) are usually spun exposed on the stems of grasses, rushes and other herbage. They vary considerably in colour and show every gradation from white to beige to straw-yellow – although it has been suggested that procrypsis might be involved there is no substantial evidence to support this hypothesis. Pickett (1903: 268) observed that cocoons of *Z. filipendulae* spun on iron railings were darker than usual and dirty brown in colour, and he suggested that adaptation to the surroundings was apparently attempted. A larva of *Z. filipendulae* which he placed in a willow chip-box spun a cocoon which was the same colour as the box; another larva which was placed in a white-lined glass-topped box produced an almost satin-white cocoon. However, personal observations and experiments suggest that procryptic coloration is not involved, and Dryja (1959: 394) contends that in the continental *Z. ephialtes* (Linnaeus) the colour of the cocoon is controlled by hereditary factors.

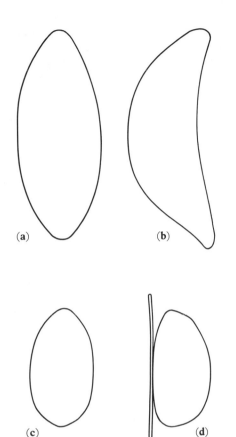

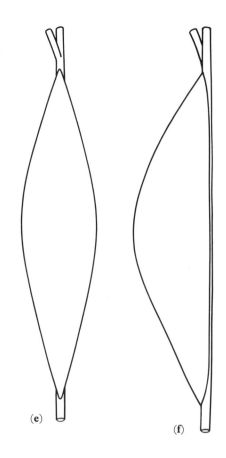

Figure 57
Dorsal and lateral views of
cocoons (×3)

(**a,b**) *Zygaena (Zygaena)
exulans* (Hohenwarth),
bluntly fusiform

(**c,d**) *Zygaena (Zygaena) loti*
([Denis & Schiffermüller]),
ovoid

(**e,f**) *Zygaena (Zygaena)
filipendulae* (Linnaeus),
fusiform

(a)

(b)

(c)

(d)

(e)

(f)

The cocoons of the chalk-down form of *Z. trifolii* are placed low and concealed amongst herbage (Pl.6, fig.7), but occasionally a cocoon is spun high on a grass stem and then the pupa is almost invariably parasitized. Kaye (1903: 306) reported that almost every cocoon of *Z. filipendulae* that he found spun low down in tufts of grass produced dipterous parasites. Such observations suggest that the behaviour of the larva might be affected when it is parasitized.

Imago
Zygaena species inhabit downland, coastal cliffs, sand hills, woodland clearings, marshes and roadside verges, and usually occur at low elevations (except the boreo-alpine species *Z. exulans*). However, there are records of *Z. filipendulae* and *Z. purpuralis* occurring up to 300m (975 ft) and 400m (1300 ft) respectively on the Isle of Rhum (Harrison, 1940: 135, 137), and Adkin (1895: 91) recorded the former species at 600m (1950 ft) in Sutherland, but records from elevations as high as this are unusual in the British Isles.

All the British species are diurnal and are most active from mid-morning to late afternoon, their flight being dependent on high temperatures and, to a lesser extent, sunshine (during mild weather in west Cornwall I have observed *Z. filipendulae* flying actively in heavy drizzle). Their flight is normally slow and direct but *Z. loti* flies rapidly and *Z. viciae* will dash wildly away if disturbed. They are attracted to and feed at flowers, particularly those of field scabious (*Knautia arvensis*), small scabious (*Scabiosa columbaria*), thyme (*Thymus* spp.), marjoram (*Origanum vulgare*), knapweeds (*Centaurea* spp.), especially greater knapweed (*C. scabiosa*), hawkweeds (*Hieracium* spp.), brambles (*Rubus* spp.), vetches (*Vicia* spp.), especially tufted vetch (*V. cracca*), trefoils (*Lotus* spp.), horse-shoe vetch (*Hippocrepis comosa*) and orchids (Orchidaceae), and undoubtedly they play an important role in pollination. The pollinia of orchids are frequently seen adhering to their haustella (Buxton, 1912: 245; 1915: 183; 1916a: 16; Harrison, 1953: 66), and the pollination of the pyramidal orchid (*Anacamptis pyramidalis*) by *Z. trifolii* has been described and illustrated by Proctor & Yeo

(1973: 241, pl.44, figs 1a,b). During inclement weather moths can be found at rest on flower-heads or on stems of grass and other herbage, and if disturbed in dull weather or during the evening they fall to the ground and feign death; *Z. filipendulae*, *Z. trifolii*, *Z. lonicerae* and *Z. purpuralis* rest exposed but *Z. exulans*, *Z. loti* and *Z. viciae* are well down in the herbage and difficult to find. From a distance those species which rest exposed resemble the flower-heads of plantains (*Plantago* spp.), salad burnet (*Sanguisorba minor*) and glaucous sedge (*Carex flacca*). Rothschild (1964: 76) has suggested that when at rest the moths are easily mistaken for the ripe, black or brown seed pods of various field vetches, and she considered that the observer has to approach closely before the true nature of the 'pod' is suddenly revealed so that the aposematic coloration strikes the eye with the added impact of surprise.

Eclosion normally takes place by day, with a peak emergence between mid-morning and noon (Harker, 1891: 91; Burrows, 1916: 149; Tremewan, pers.obs.). Buxton (1916b: 38) recorded *Z. filipendulae* emerging in great numbers from 10.00 to 11.30 hrs when the sun was shining on the cocoons, but stated that on sunless days the emergence continued irregularly from 08.00 to 21.00 hrs; in captivity it is also erratic and can occur from 06.00 to 23.00 hrs (Tremewan, pers.obs.).

In the British Isles *Zygaena* species are univoltine although Fletcher reared 11 examples of *Z. trifolii* from eggs deposited in the same year (Tutt, 1899: 418); I have also reared the same species, *Z. lonicerae* and *Z. filipendulae* from the egg in one year, and without a period of diapause. There are records of freshly emerged moths of *Z. lonicerae* and *Z. filipendulae* occurring in September and October in the wild (Mathews, 1859: 6789; Towndrow, 1871: 443; Lang, 1878: 69; Edwards & Towndrow, 1899: 6; Tutt, 1899: 417). At the end of August, over a period of several years, I have taken occasional specimens of freshly emerged *Z. filipendulae* at Gwithian, Cornwall, where the normal flight period is July to mid-August; in 1972, a very late season, a fresh example of the same species was seen at Scorrier, Cornwall, on 19 September (Tremewan, 1973: 54). Although it has been considered that these and other examples (Dobson, 1955: 277) represent a second generation there is no evidence to support this; as suggested by Tutt (1899: 418), it is possible that they came from larvae which had hatched the previous year and became full-fed after breaking a second period of diapause. If this is correct then it must be assumed that some abnormal conditions stimulated the larvae to break diapause which, in such a case, could not have lasted for more than a few weeks. In captivity larvae will sometimes come out of diapause and commence to feed in the autumn; they then develop to produce moths during the early winter, or

re-enter diapause after feeding for one instar. Although Dryja (1959) maintains that diapause is genetically controlled in *Z. ephialtes* (Linnaeus), it appears that in many species it may also be affected by exogenous factors (Naumann & Tremewan, 1980: 116).

Dispersal

Zygaena species occur in colonies which vary in size and density and fluctuate considerably from one season to another. Migration is unknown although there are records of local movement and it is surprising how quickly a habitat can be colonized when conditions become suitable.

In west Cornwall, Hodge (1915: 268) recorded that *Z. filipendulae* was common on the Lelant side of the Hayle estuary but rare on the Hayle side. Ten years later (Hodge, 1925: 272) he stated that the species was common on the Hayle side of the estuary and suggested that it was spreading gradually eastwards by crossing the water; an increase in numbers occurred on the east side in the two following years (Hodge, 1926: 320; 1927: 282). Personal observations have shown that *Z. filipendulae* still occurs on the east side of the Hayle estuary, although fluctuating from one year to another. Hodge's suggestion of a positive easterly movement by *Z. filipendulae* is based on assumption as moths were not actually seen crossing the water, his observations having been made on empty cocoons in the autumn. *Z. filipendulae* is widespread on the sand-hills from Gwithian to Hayle, so that dispersal could have occurred from the east, or the Hayle colony could have arisen from wind-borne gravid females from Lelant, as the prevailing winds are westerly. It would appear that Hodge merely observed normal fluctuation of the species.

In Surrey, however, Buckstone (1926: 6) recorded that during a walk along the Downs from Newlands Corner to Dorking on 24 June 1916 he saw numerous individuals of *Z. trifolii* flying in a westerly direction; specimens of both sexes were encountered every few yards and over half of those netted were males.

Grosvenor (1933a: 63) considered that *Zygaena* migrated annually and that the movement, although not spectacular, was nevertheless methodical. He contended that the reason for such migration was due to food supply and that individuals would move to the greatest source. In 1907 Grosvenor (1927: 92) discovered a colony of *Z. trifolii* in a marshy field which was neglected and became overgrown from 1914–1918, so that the host plant, greater bird's-foot trefoil (*Lotus uliginosus*), was eradicated; on a nearby railway embankment *Lotus* occurred in profusion although *Z. trifolii* was absent. The area was not visited between 1915 and 1919 but in 1920 *Z. trifolii* was absent from the overgrown marshy field but plentiful on the embankment, although worn and going over when it should have been just emerging.

In west Cornwall, greater bird's-foot trefoil is not restricted to marshes and damp meadows but also grows abundantly in hedgerows and along roadside verges. It will also colonize quite dry agricultural grassland very quickly if this becomes neglected. The colonization of such meadows by *Z. trifolii* follows rapidly (Tremewan, 1954: 233; 1980a: 144), the colonies almost certainly arising from wind-borne gravid females, as the possible source is often several kilometres away. Dispersal could also take place along the hedgerows and roadside verges where I have found freshly emerged moths on different occasions.

It is probable that wind is also responsible for the dispersal of *Zygaena* species to areas which are isolated from their normal habitat and which are unsuitable for them. Owen (1954: 53) recorded a worn female, which he stated was probably *Z. lonicerae*, from a bombed site at Cripplegate in the City of London. Bradley & Mere (1964: 61, 66, 73) reported the occurrence of single specimens of *Z. trifolii* and *Z. lonicerae* in Buckingham Palace garden, but here accidental importation cannot be ruled out.

The increase and expansion of *Z. lonicerae* in southern England during the last 30 years or so and the decline to apparent extinction of *Z. trifolii* in marshland habitats in south-east England have already been commented on (Tremewan, 1961b: 113; 1980a: 143; 1982a: 9; Chalmers-Hunt, 1978: (187)). What is more remarkable is the fact that in some boggy or marshland habitats in Surrey and Hampshire the larvae of *Z. lonicerae* feed solely on greater bird's-foot trefoil (Tremewan, 1982a: 10), the host plant of *Z. trifolii decreta* Verity; in Staffordshire similar observations have been made by Warren (1983: 3) who also comments on the recent expansion of *Z. lonicerae* in that county. I have suggested (Tremewan, 1982a: 11) that in some of these localities *Z. trifolii* has been replaced by *Z. lonicerae*, but the expansion of *Z. lonicerae* in southern England, and the apparent extinction of *Z. trifolii* in its marshland habitats in south-east England and its general decline on chalk downs in this region, may be due also to climatic factors. Heath (1981: 5) has commented briefly on the effects of short-term climatic changes on butterfly populations in Europe and suggested that long-term climatic changes have contributed to major expansions and contractions of some species. As *Z. trifolii* is a west Mediterranean species which in Britain is on the northern edge of its range, the succession of late springs and cold wet summers during the last two or three decades may have been responsible for its decline in south-east England. On the other hand, the more widely distributed Euro-Siberian species *Z. lonicerae* may be better adapted to colder conditions and in fact might even have benefited from them.

Mating and scent organs

Pairing takes place immediately after eclosion of the female (Pl.6, fig.11), and sometimes even before its wings are fully expanded; moreover, it is not unusual to see a number of males fluttering or settling around an intact cocoon from which a female is about to emerge. The male is attracted to the female by a directive scent or pheromone. The paired scent-producing glands of the female are situated dorsally between segments 8 and 9, anterior to the ovipositor lobes (papillae anales); when the female is in the characteristic 'calling position' the ovipositor lobes are extended, and the scent-producing glands can then be seen as pulsating brownish yellow swellings. These odiferous (pheromone) glands have been referred to incorrectly by Hewer (1932: 36, fig.1) as sebaceous glands. Hinton (1981: 199) stated that the glands open near the end of the common oviduct and that they are incorrectly shown opening into the rectum by Hewer (1932: fig. 1). On the Continent, Renou & Decamps (1982) have studied the electrophysiological responses of male *Z. filipendulae* and *Z. hippocrepidis* (Hübner) to various synthetic pheromonal compounds. The male, as in many species of Lepidoptera (Wigglesworth, 1972: 711), produces a scent which is considered to be an aphrodisiac which incites the female to copulate. In some species of *Zygaena*, the scent of the male is apparently distributed by the paired erectile hair-tufts situated dorsolaterally on the intersegmental region of the eighth and ninth segments; of the British species only *Z. loti* lacks such hair-tufts. A morphological and histological study of these structures in some of the European species of *Zygaena* has been made by Kames (1980). When a male approaches a receptive female, it vibrates its wings rapidly and flutters around and over the female with its abdomen extended and the valvae held apart, pairing usually taking place shortly afterwards. Buckstone (1909a: 96; 1909b: 73) recorded two males of *Z. filipendulae* in copula with one female, but the mechanism of this phenomenon was not explained!

It appears that the directive scent produced by the females of different *Zygaena* species is very similar, as interspecific pairings occur regularly (see p. 98). There are also records of male *Z. filipendulae* being attracted to the females of species belonging to other families of Lepidoptera. Sutton (1922: 280), Ford (1926: 20), Smith (1947: 68) and Wheeler (1970: 26) recorded that numbers of *Z. filipendulae* were attracted to females of *Lasiocampa quercus* (Linnaeus) (Lasiocampidae), although the latter species failed to attract *Zygaena trifolii* on the same occasion (Ford, 1926: 20). Williams (1914: 185) recorded a male *Z. filipendulae* paired with a female *Tyria jacobaeae* (Linnaeus) (Arctiidae), and Fassnidge (1922: 190) captured a male *Adscita statices* in copula with a female *Zygaena trifo-*

lii. Decamps *et al.* (1981) have found that the males of some continental *Zygaena* species are attracted to the pheromones produced by female tortricids (Tortricidae).

The mechanism of copulation in *Zygaena* species has been described by Hewer (1934). The duration of copulation is usually as much as 24–48 hours, the aposematic coloration of the moths presumably affording some degree of protection from predators. The male and probably the female pair more than once: I have observed a male to pair with several females in captivity, and a worn female captured in the wild will sometimes have as many as three spermatophores in the bursa copulatrix (Tremewan, pers. obs.).

Unlike most diurnal Lepidoptera, *Zygaena* species appear to be unable to fly while in copula, and if a pair is thrown into the air it drops quickly to the ground, either sex fluttering its wings to guide the descent (Wheeler, 1918: 153). In Iran, however, a pair of *Z. haematina* Kollar has been recorded at light (Wiltshire, 1968: 50), but it is not stated whether the specimens were actually seen flying or were merely at rest on herbage.

Hybridization

Cross-pairings between different species occur frequently in the wild, especially between *Z. filipendulae*, *Z. lonicerae* and *Z. trifolii*, all of which, incidentally, form the same species-group. The hybrids resulting from male *Z. filipendulae* crossed with female *Z. lonicerae* (or the reciprocal cross) are sterile, but those resulting from male *Z. lonicerae* crossed with female *Z. trifolii* (or the reciprocal cross) are fertile and will breed *inter se* (Fletcher, 1891: 115; 1893: 53; Tutt, 1899: 419; 1906: 36). The cross-pairing of different species and the rearing of hybrids in captivity are discussed by Fletcher (1893), South (1907) and Onslow (1918). Although records of cross-pairing in the wild are numerous (Hamm, 1899: 269; South, 1904a: 15; 1904b: 64; 1905; Greer, 1918: 188; Hayward, 1923: 43; Chalmers-Hunt, 1957: 199), authentic records of wild hybrids are few but include those of Cockayne & Darlow (1941: 113, pl.6, fig.2 – *Z: lonicerae* × *Z. filipendulae*), Darlow (1947: 135 – ditto) and Tremewan (1961b: 111; 1961c: 7, pl.C3, fig.5 – *Z. filipendulae* × *Z. trifolii*).

As mentioned above, cross-pairings of *Z. lonicerae* and *Z. trifolii* can produce fertile offspring yet the two species are partially sympatric but still maintain their specific distinctness. On the chalk downs of southern England they are temporally separated (Tremewan, 1980a: 144) because the flight period of the ecological subspecies *Z. trifolii palustrella* Verity normally ends before the first males of *Z. lonicerae* appear in late June. I have already commented (p. 97) on the apparent extinction of *Z. trifolii decreta* in the marshland habitats in south-east England, and suggested that one of the causes may have been the expansion of *Z. lonicerae* and its subsequent invasion of such biotopes. However, in the limestone habitats of the Cotswolds, where *Z. trifolii palustrella* occurs later than in south-east England, they actually fly together but there is no evidence of hybridization (Tremewan, pers.obs., 1983).

Hybrids inherit characters from both parent species, and those resulting from *Z. filipendulae* crossed with either *Z. lonicerae* or *Z. trifolii* usually have only a vestige of spot 6. Such hybrids can be easily confused with aberrant examples of *Z. filipendulae* in which spot 6 is reduced; any specimen caught in the wild and suspected to be a hybrid should be subjected to an examination of the genitalia which in genuine hybrids have intermediate characters (Cockayne & Darlow, 1941: 113, pl. 6, fig. 2). As might be expected, characters of both parent species, such as coloration, morphology and host plant specificity, are also inherited by hybrid larvae. Compared with *Z. filipendulae*, the larvae of *Z. lonicerae* have very long setae, but hybrid larvae from these two species have the setae intermediate in length (Pl.5, figs 17–20); they will also feed indiscriminately on the host plants of either parent species.

It is well known that eggs from interspecific pairings are not always fertile, and infertility, due to incompatibility of species, is discussed by Hewer (1934: 526).

Variation and genetics

Zygaena species are extremely variable in wing pattern and coloration. The normal crimson or scarlet of the forewing spots and hindwings can be replaced by orange-red, orange, yellow, brownish red, brown, blackish brown or sooty black, but such colour-forms are usually either uncommon or very rare. In *Z. filipendulae* the yellow form occurs regularly and has been recorded widely throughout its range but the orange form is local and the orange-red form is rare. Although the sooty black form occurred for some years along the Lancashire coast from St. Annes-on-Sea to Fleetwood, it has always been very rare and is otherwise known almost entirely from single specimens captured in widely separated localities. Yellow, orange and 'pseudo-orange' forms of *Z. lonicerae* are rare, but the yellow form has occurred regularly on the Cotswold Hills, Gloucestershire. In *Z. trifolii*, orange, yellow, brown and sooty black forms are known, the last two being rare, although the sooty black aberration occurred regularly and in some numbers in Tilgate Forest and at Chailey, Sussex, and there are single records from elsewhere. Yellow, orange and brownish red forms of *Z. purpuralis* are known from Scotland (Tremewan & Manley, 1964: 149–151; Harper & Langmaid, 1975: 139), and the yellow form is also recorded from Ireland (Allen, 1894: 217); the Welsh populations of *Z. purpuralis* were noted for the black form which occurred over a number of years.

In spite of the regular occurrence of these colour-forms in the British Isles, much research still needs to be done on their genetics. The yellow and orange forms of *Z. filipendulae* are recessive to red (Grosvenor, 1932b: 68; 1933b: 88, 109; G. Reiss, pers.comm.; Tremewan, pers.obs.), and these two forms produce an intermediate form when paired together (Newman, 1916a: 20; 1916b: 127; G. Reiss, pers.comm.; Tremewan, pers.obs.), which is to be expected when two recessive forms are crossed. The yellow form of *Z. trifolii* is recessive to red (G. Reiss, pers. comm.; Tremewan, pers.obs.), and this applies also to the yellow form of *Z. lonicerae* (Tremewan, pers.obs.). I have coined the term 'pseudo-orange' to denote those forms which appear to be orange to the naked eye but whose colour can be seen under magnification to be derived from a mixture of red and yellow scales; if the percentage of red scales is greater than that of the yellow scales the form then appears dark orange or reddish orange, while those forms which appear pale orange have a greater percentage of yellow scales in the mixture. The pseudo-orange form is known in *Z. filipendulae*, *Z. trifolii* and *Z. lonicerae*, and in the last species it is recessive to red but dominant to yellow (Tremewan, pers.obs.). The sooty black form of *Z. trifolii* is recessive to red (Grosvenor, 1926: 52; 1927: 95).

In *Z. filipendulae* a rare aberration occurs in which spot 6 is vestigial or even absent. A very rare form of *Z. trifolii* has spot 6 present and appears to be dominant to the normal five-spotted form (Tremewan, pers.obs.); the heterozygotes have the sixth spot present but reduced, while the homozygous form has spot 6 larger and attached to spot 5.

There are two basic forms of confluence in *Zygaena*: one has the forewing spots coalesced to form well-defined longitudinal streaks which extend from the base to the distal area of the forewing; the other, often referred to as a temperature form, has between the forewing spots variable amounts of suffused red scaling which invades almost the whole area of the wing in extreme examples. Although both forms of confluence are rare in *Z. filipendulae*, in breeding experiments they almost invariably occur in a 50 per cent ratio in the F_1 generation (Tremewan, pers.obs.). Confluence of the forewing spots in *Z. trifolii* is common in many colonies but it occurs only rarely in *Z. lonicerae*. Suffused confluent forms have been produced in controlled temperature experiments (Burgeff, 1956), and by injecting wolfram acid solution (H_2WO_4) into freshly formed pupae (Dabrowski, 1963; 1966); in addition, these experiments produced forms in which the forewing spots were reduced.

A comprehensive paper listing the aberrations of the British species, with coloured illustrations of the more striking forms, was published by Tremewan (1961c), and the original and other references to such forms were catalogued by Reiss & Tremewan (1967).

Gynandromorphism

Gynandromorphs are apparently unknown from the British Isles; the record by Cockayne (1935: 514) is referable to an asymmetrically marked specimen of *Z. trifolii* which appears to have originated from the Continent (Tutt, 1899: 422).

Homoeosis and teratology

As Cockayne (1922; 1926; 1927) has collated the records of homoeosis and teratology in Lepidoptera, those of *Zygaena* species which he recorded will not be repeated here; however, the references to the British records which I have abstracted from the literature have been included in the bibliography at the end of this family.

Homoeosis is very rare but in Surrey I have taken two examples of *Z. lonicerae* showing forewing pattern and coloration on the hindwings, and in Cornwall I captured a *Z. filipendulae* which has splashes of red hindwing coloration in the dark ground colour of the left forewing.

It appears that duplication or triplication of wings has occurred most frequently in *Z. filipendulae*, although there are also records of such abnormalities in *Z. trifolii* and *Z. lonicerae*. Examples of *Z. filipendulae* and *Z. lonicerae* with extra antennae are known, and moths which have retained the larval head capsule have been recorded in these species.

Predation

Although the adults have aposematic coloration and secrete a defensive fluid they are attacked and sometimes eaten by birds; other predators include spiders (Araneae) and bugs (Heteroptera). On Garinish Island, Co. Kerry, Lawless (1872: 77) observed numbers of birds, chiefly starlings (*Sturnus vulgaris*) and buntings (*Emberiza* spp.), catching *Z. filipendulae*, removing the wings and flying away with the bodies; sometimes two and in one case three were taken at a time by a single bird. In Dorset, Curtis (1917: 148) observed a number of cuckoos (*Cuculus canorus*) which appeared to take *Z. filipendulae* which were resting on grass-stems, and Bolam (1926: 551) recorded that he occasionally observed this species being eaten by house sparrows (*Passer domesticus*), sky larks (*Alauda arvensis*), meadow pipits (*Anthus pratensis*) and a whinchat (*Saxicola rubetra*). On two separate occasions in west Cornwall I observed a female blackbird (*Turdus merula*) taking and eating *Z. filipendulae* which were at rest on herbage. However, when Sheppard (1964) investigated the effectiveness of warning coloration in moths, he noted that *Z. filipendulae* was not taken by two sparrows and a blackbird, and there is evidence of rejection by birds in the wild. For example, on the North Downs, Surrey, I have frequently observed examples of *Z. filipendulae*, *Z. trifolii* and *Z. lonicerae* with characteristic marks on the forewings which appear to have been made by birds.

Bristowe (1941; 1958) recorded that various species of spiders attacked *Z. filipendulae* and *Z. trifolii* but, with few exceptions, these were rejected and not eaten. However, two major predators of *Z. trifolii* on the North Downs are the spiders *Xysticus bifasciatus* and *Tibellus oblongus* (Tremewan, pers.obs.).

Hamilton & Heath (1976: 337) recorded *Picromerus bidens* (misidentified as *Pentatoma rufipes*) feeding on *Z. filipendulae* in Co. Kerry, and I have seen this species of bug and also *Zicrona caerulea* attacking and feeding on *Zygaena trifolii* in Surrey and Cornwall; *Zicrona caerulea* will also attack *Zygaena filipendulae* (Tremewan, pers. obs.).

Near Lands End, Cornwall, Andrewes (1937: 155) observed the bush-cricket *Platycleis albopunctata* eating an adult of *Z. filipendulae* and the condition of the prey suggested that it had been captured alive.

The scorpion fly *Panorpa communis* has been recorded feeding on a dead specimen of *Z. lonicerae* in Hampshire (Tremewan, 1980b: 274).

There is no evidence to suggest that reptiles, such as lizards, are predatory on *Zygaena* species. Butler (1869: 27) recorded that green lizards (*Lacerta viridis*) rejected adults of *Z. filipendulae*, and Ford (1963: 77) observed a lizard stalking a *Zygaena* species on the Downs in Sussex, but no attack was made.

Parasites

Species of parasitic Hymenoptera (Braconidae, Ichneumonidae and Pteromalidae) and Diptera (Tachinidae and possibly Phoridae) are associated with *Zygaena* which act as primary or secondary hosts; these parasitoids, as they are also known, are listed in Tables 1–3. Only authentic British records have been included in the present work, and these are based either on specimens reared by myself and determined by the appropriate specialists, or on information supplied by specialists. Authenticated records of parasites from one or more species of *Zygaena* whose specific identity has not been determined or confirmed are entered under *Zygaena* spp. Most literature records have been ignored because of the possibility of incorrect determination of the hosts and parasites but, for those who are interested, references to such records have been included in the bibliography.

Oviposition by hymenopterous parasites occurs during the larval and pupal phase of the host. The larvae of the primary parasite *Apanteles zygaenarum* vacate the host-larva before the latter spins its cocoon and the gregarious, yellowish silken cocoons of the *Apanteles* spun on grass-stems and other herbage are a familiar sight. However, the larvae of most species of hymenopterous parasites, such as the endoparasitic *Casinaria orbitalis* and the ectoparasitic *Mesostenidea obnoxius*, pupate within the *Zygaena* cocoon from which the adult parasite subsequently emerges.

Mesochorus temporalis is an example of a secondary parasite which parasitizes *Apanteles zygaenarum*. The primary parasites of *Zygaena* are host-specific, but the secondary parasites are polyphagous and also interact with other parasite complexes of Lepidoptera and even those of other orders.

Oviposition by dipterous parasites occurs during the larval phase of the host. The larvae of the tachinids normally vacate the cocooned host-larva or pupa before pupating themselves. *Phryxe magnicornis* usually pupates among litter on the ground, while *Exorista fasciata* pupates inside the host-cocoon.

For general information on hymenopterous and dipterous parasites of Lepidoptera, including *Zygaena*, the reader is referred to the account by Shaw & Askew (Vol. 1: 24).

Although red mites (Acarina) are common parasites of adults and occasionally larvae of *Zygaena*, there appear to be few literature records. Green & Wilkinson (1951: 144) recorded red mites on larvae of *Z. trifolii* from Skokholm Island, Pembrokeshire (Dyfed), and I have often observed them on adults of *Z. filipendulae*, *Z. trifolii* and *Z. lonicerae*; one that I found on *Z. filipendulae* was determined as *Metathrombium poriceps*, which is thought to be the larval stage of *Trombidium holosericeum*.

The occurrence of pathogens in *Zygaena* appears to be unrecorded from the British Isles. Fungal infections have occurred in the larvae of *Z. viciae*, *Z. filipendulae* and *Z. trifolii*, and on the Continent Weiser (1951) has described a fungus which infects the epithelium of the intestine in the larva of *Z. carniolica* (Scopoli).

Buxton (1915: 183) recorded a triungulin larva on the head of an adult *Z. filipendulae*, but such an occurrence must be incidental as the larvae of oil beetles (*Meloe* spp.) are parasitic and host-specific on solitary bees (*Anthophora* spp.).

Toxicity

Toxins have been found in all phases of the life history of *Zygaena* species. Seitz (1907: 18) and Rocci (1914) first drew attention to the fact that the moths are strongly resistant to cyanide; subsequently, Rocci (1915; 1916) revealed the poisonous nature of the haemolymph and proposed (1915: 103; 1916: 95) the names 'zigenina' or 'zygénine' for the toxic principle that it contained. More recently, Jones *et al.* (1962: 53) have shown that the greenish yellow haemolymph of the moths, when drawn off from the haemocoele or collected from specialized 'bleeding areas', is strongly positive for cyanide; they also showed that the crushed tissues of ova, larvae, pupae and adults of *Z. filipendulae* and *Z. lonicerae*, and ova and adults of *Z. trifolii*, release hydrocyanic acid, although old or worn specimens were only slightly positive, the amount varying according to age. Moreover, the abdomen of *Z. lonicerae* contains histamine in concentrations generally

	Z. exulans	Z. loti	Z. viciae	Z. filipendulae	Z. trifolii	Z. lonicerae	Z. purpuralis	Zygaena spp.
BRACONIDAE	•			•	•	•		
Aleiodes bicolor (Spinola)				•				
Apanteles zygaenarum Marshall	•			•	•	•		
Meteorus unicolor (Wesmael)				•				
ICHNEUMONIDAE				•	•	•		•
Agrothereutes hospes (Tschek)								•
Gambrus ornatulus (Thomson)				•				
Mesostenidea obnoxius (Gravenhorst)				•		•		•
Casinaria orbitalis (Gravenhorst)				•	•	•		
Charops cantator (Degeer)				•	•			•
Pimpla instigator (Fabricius)				•				

Table 1. Hymenopterous primary parasites of British *Zygaena* species

	Z. exulans	Z. loti	Z. viciae	Z. filipendulae	Z. trifolii	Z. lonicerae	Z. purpuralis	Zygaena spp.
ICHNEUMONIDAE				•				•
Acrolyta distincta (Bridgmann)				•				
Acrolyta submarginata (Bridgmann)				•				
Encrateola laevigata (Ratzeburg)								•
Lysibia nana (Gravenhorst)								•
Gelis corruptor (Förster)								•
Gelis instabilis (Förster)								•
Itoplectis alternans (Gravenhorst)				•				
Itoplectis maculator (Fabricius)				•				
Mesochorus temporalis Thomson				•				•
PTEROMALIDAE				•		•		
Pteromalus vibulenus (Walker)				•		•		
Pteromalus semotus (Walker)				•				

Table 2. Hymenopterous secondary parasites of British *Zygaena* species

	Z. exulans	Z. loti	Z. viciae	Z. filipendulae	Z. trifolii	Z. lonicerae	Z. purpuralis	Zygaena spp.
TACHINIDAE				•	•	•		
Exorista fasciata (Fallén)				•	•			
Exorista larvarum (Linnaeus) sensu auctt.						•		•
Pales pavida (Meigen)				•	•	•		
Platymya fimbriata (Meigen)				•				
Phryxe magnicornis (Zetterstedt)				•	•	•		
Phryxe nemea (Meigen)				•				
Phryxe vulgaris (Fallén)						•		•

Table 3. Dipterous parasites of British *Zygaena* species

associated with venomous organs (Bisset *et al.*, 1960: 261). Intraperitoneal injections of extracts from eggs and gravid females of *Z. trifolii* were lethal to the mouse, but extracts from worn females of this species and males of *Z. lonicerae* had little or no apparent effect (Marsh & Rothschild, 1974: 97).

The source of cyanide in all phases of *Z. filipendulae* has been identified by Davis & Nahrstedt (1979: 395; 1982: 329) as two cyanogenic glucosides, linamarin and lotaustralin; initially these were found to be present in amounts equivalent to as much as two per cent of the body weight of freshly emerged gravid females, but subsequently they were found to be present also in the males. Davis & Nahrstedt (1979: 396) considered that most of the cyanide in the female could be located in the eggs, and suggested that it was unlikely to have a defensive role in the adult.

The larvae of *Z. filipendulae* and *Z. lonicerae* will feed readily on the strain of common bird's-foot trefoil (*Lotus corniculatus*) which contains cyanogenic glucosides, although the preferred host plants of *Z. lonicerae* are meadow-vetchling (*Lathyrus pratensis*) and red clover (*Trifolium pratense*) in neither of which cyanogenic glucosides have been detected. Nevertheless, all phases of *Z. lonicerae* released hydrocyanic acid when the larvae had been reared on meadow-vetchling (Jones *et al.*, 1962: 53).

Davis & Nahrstedt (1982: 329) found that the proportions of the glucosides in the eggs of *Z. filipendulae* were very similar to those found in female adults, and that the cyanide in newly hatched larvae seems to be derived from the eggs, the content of the cyanide increasing during the growth of the larva after diapause. This increase in cyanide during the growth of the larva may reflect direct acquisition or sequestering from cyanogenic strains of the host plant, but in those larvae which feed on acyanogenic strains of the host plant the increase must be due to biosynthesis, the cyanide being derived from precursors in the host plant (Davis & Nahrstedt, 1982: 332).

The adults of *Zygaena* species have a distinctive odour and when disturbed or attacked they secrete a colourless, transparent, volatile defensive liquid which bubbles forth from around the palpi and along the proximal portion of the galeae (Lane, 1959: 93); spontaneous emissions of this strong-smelling fluid were tested and found not to release hydrocyanic acid (Jones *et al.*, 1962: 53). If an adult *Zygaena* is compressed laterally, haemolymph is released from the cervical region and from the legs, at the base of each coxa and at the end of each tarsus, and this has been considered to be a further defensive mechanism (Lane, 1959: 93). Rothschild (1961: 102) considered the defensive odour in *Zygaena* species to be similar to that of the yellow defensive secretions of ladybirds (Coccinellidae); however, I consider the odour of *Zygaena* to be less offensive. Gas chromatography revealed that *Z. lonicerae* and

Callimorpha dominula (Linnaeus) (Arctiidae), which have a similar aposematic colour scheme, have remarkably similar chromatograms, suggesting convergent evolution of scent patterns (Rothschild, 1961: 105, fig.7).

Jones *et al.* (1962: 53) have suggested that hymenopterous and dipterous parasites must be equipped to deal with the specialized medium in which they live, and those that were tested were positive for rhodanese, an enzyme which is present in their body tissues and which detoxifies hydrocyanic acid to thiocyanate.

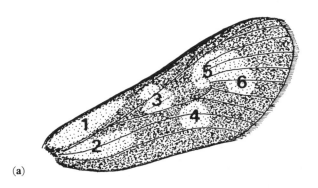

(a)

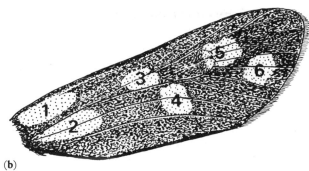

(b)

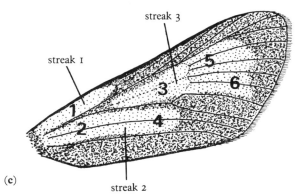

(c)

Figure 58 Forewing patterns and spot terminology
(a) *Zygaena (Zygaena) loti* ([Denis & Schiffermüller]) 1, 2, 3, 4, 5+6
(b) *Zygaena (Zygaena) filipendulae* (Linnaeus) 1, 2, 3, 4, 5, 6
(c) *Zygaena (Mesembrynus) purpuralis* (Brünnich) 1, 2+4, 3+5+6

Key to species (imagines) of the genus *Zygaena*

NOTE. Due to the wide range of variation in species of *Zygaena*, it is impractical to construct a key to include every aberration; the following is based on typical specimens only.

To avoid confusion, specialists in *Zygaena* have adopted a system whereby the forewing spots and streaks are numbered (see figures 58a–c).

1   Forewing with five or six spots (figures 58a,b) 2
–   Forewing with three longitudinal streaks (figure 58c). (N. Wales, W. Scotland, W. Ireland) .... *purpuralis* (p. 114)

2(1)   Forewing with five spots 3
–   Forewing with six spots. (Britain, Ireland) .... *filipendulae* (p. 108)

3(2)   Forewing with spot 1 approximately equal in length to spot 2; coloration opaque 4
–   Forewing with spot 1 distinctly longer than spot 2 and extending along costa as far as spot 3; coloration translucent. (Aberdeenshire: montane habitat) .... *exulans* (p. 104)

4(3)   Distal spot on forewing rounded, consisting of spot 5 only 5
–   Distal spot on forewing a kidney-shaped blotch, consisting of spots 5+6 (figure 58a). (Argyll: Morvern, Mull, Ulva) .... *loti* (p. 106)

5(4)   Spot 3 on forewing relatively large, rounded, or confluent with spot 4 6
–   Spot 3 on forewing small, elongate. (Argyll) .... *viciae* (p. 106)

6(5)   Forewing with apex attenuate, spots 3 and 4 separate; hindwing with apex pointed. (England, Wales, Scotland: Isle of Skye, Ireland) .... *lonicerae* (p. 112)
–   Forewing with apex distinctly rounded, spots 3 and 4 confluent; hindwing with apex rounded. (S. England, Wales, Isle of Man) .... *trifolii* (p. 110)

Key to species (larvae) of the genus *Zygaena*

NOTE. This key is based on typical forms of the final instar, unless stated otherwise. The basic pattern of the larva is described in dorsolateral aspect (p. 90) and illustrated in figure 55, p. 91.

1   Ground colour greenish yellow or yellow, whitish green, or green 2
–   Ground colour olive-brown or olive-green, greenish grey, or blackish grey 9

2(1)   Dorsal, subdorsal and lateral series of black spots present 3

–   Only dorsal series of black spots present, the subdorsal and lateral series absent 6

3(2)   Ground colour greenish yellow or yellow 4

–   Ground colour whitish green 5

4(3)   Black dorsal spots separate; ground colour greenish yellow or yellow. (Britain, Ireland; on *Lotus corniculatus*) ... . *filipendulae* (Pl.5, figs 7,8) (p. 108)

–   Black dorsal spots enlarged and coalescing, especially in thoracic region; ground colour greenish yellow or yellow. (Britain, Ireland; on *Lotus corniculatus*) .... *filipendulae* (aberration) (Pl.5, figs 9,10) (p. 108)

5(3)   Black dorsal spots separate; setae short, *c.*0.8mm long. (S. England, Wales, Isle of Man; on *Lotus corniculatus*, *L. uliginosus*) .... *trifolii* (Pl.5, figs 13,14) (p. 110)

–   Black dorsal spots larger, usually separate but often just coalescing; setae longer, *c.*2.4mm long. (Britain, Ireland; on *Trifolium pratense*, *Lathyrus pratensis*, *Lotus uliginosus*) .... *lonicerae* (Pl.5, figs 15,16) (p. 112)

6(2)   Ground colour emerald-green, or green 7

–   Ground colour greenish yellow or yellow 8

7(6)   Ground colour emerald-green, minutely speckled with black; a black dorsal spot present only in anterior part of each segment. (Argyll; on *Lathyrus pratensis*, *Lotus corniculatus*) .... *viciae* (Pl.5, figs 5,6) (p. 106)

–   Ground colour green, without black specks; a black dorsal spot present in anterior and posterior part of each segment. (Argyll; on *Lotus corniculatus*) .... *loti* (early to penultimate instars) (p. 106)

8(6)   Setae short, *c.*0.8mm long. (Britain, Ireland; on *Lotus corniculatus*) .... *filipendulae* (aberration) (Pl.5, figs 11,12) (p. 108)

–   Setae longer, *c.* 2.4mm long. (Britain, Ireland; on *Trifolium pratense*, *Lathyrus pratensis*) .... *lonicerae* (aberration) (p. 112)

9(1)   Ground colour olive-brown or olive-green, or very dark green or blackish green; a black dorsal spot present only in anterior part of each segment 10

–   Ground colour dark greenish grey; a black dorsal spot present in anterior and posterior part of each segment. (Argyll; on *Lotus corniculatus*) .... *loti* (Pl.5, figs 3,4) (p. 106)

10(9)   Ground colour olive-brown or olive-green; black dorsal spots distinct. (N. Wales, W. Scotland, W. Ireland; on *Thymus praecox*) .... *purpuralis* (Pl. 5, figs 21,22) (p. 114)

–   Ground colour very dark green or blackish green; black dorsal spots indistinct. (Aberdeenshire: montane habitat; on *Empetrum nigrum*) .... *exulans* (Pl.5, figs 1,2) (p. 104)

Key to species (cocoons) of the genus *Zygaena*

NOTE. The cocoons of *Z. filipendulae* and *Z. trifolii decreta* are very similar and are not easily distinguished. If a cocoon is spun on a flat surface, such as that of a rock, fence or leaf, the base, *i.e.*, the part attached to the surface, is broadened considerably, especially in cocoons of the fusiform type; this should be taken into account when determining the shape of a cocoon. Although the ranges of length/breadth ratios are based on measurements of cocoons selected at random, care was taken to avoid the selection of abnormally large or small examples; the same applies to lengths, when these are given.

1   Shape distinctly fusiform (figures 57e,f, p. 95), or bluntly fusiform (figures 57a,b, p. 95). If texture smooth, then cocoon irregularly and longitudinally ribbed, or wrinkled 2

–   Shape distinctly ovoid (figures 57c,d, p. 95). Length/breadth ratio (3 measured) 1.69–1.84:1; colour dirty white, texture smooth, shining. Concealed amongst herbage on or near the ground. (Argyll) .... *loti* (Pl.6, fig. 2) (p. 106)

2(1)   Distinctly fusiform (figures 57e,f, p. 95) 3

–   Bluntly fusiform (figures 57a,b, p. 95) 7

3(2)   Length 21mm or more 4

–   Length less than 21mm. Length/breadth ratio (6 measured) 2.90–5.00:1. Pale yellow; usually strongly ribbed, shining. Concealed low amongst herbage; usually spun on underside of a leaf, or on a grass-stem. (Argyll) .... *viciae* (Pl.6, fig. 3) (p. 106)

4(3)   Translucent, texture rough, irregularly ribbed; pupa indistinctly visible within. Broadly fusiform; length/breadth ratio (10 measured) 3.00–3.69:1. Pale whitish yellow or greenish yellow to white. Spun exposed on grass-stems, etc. (Britain, Ireland) .... *lonicerae* (Pl.6, fig. 8) (p. 112)

–   Opaque, texture smooth; pupa not easily visible within 5

5(4)   Cocoon spun exposed on stems of herbage etc. Length more than 24mm 6

–   Cocoon spun low down and concealed amongst herbage; usually attached to grass-stems but occasionally on a leaf. Length usually less than 24mm. Length/breadth ratio (10 measured) 3.00–4.00:1. Pale whitish yellow to yellow. (S. England; calcareous habitats) .... *trifolii palustrella* (Pl.6, fig.7) (p. 110)

6(5)   Cocoon spun on stems of *Juncus* (occasionally on other herbage). Length/breadth ratio (10 measured) 3.33–4.16:1. Dirty white or pale yellow to bright dark yellow. (SW. England, Wales, Isle of Man: marshes and damp meadows) .... *trifolii decreta* (Pl.6, fig.6) (p. 110)

– Cocoon spun on grass-stems and other herbage. Length/breadth ratio (10 measured) 3.53–4.71:1. Dirty white to straw-yellow. (Britain, Ireland: dry habitats) ....
............................*filipendulae*(Pl.6, figs 4,5)(p. 108)

7(2) Translucent, texture rough, not wrinkled; pupa indistinctly visible within. Length/breadth ratio (6 measured) 2.26–2.80:1. Yellowish white. Spun low down and partially concealed on heather and crowberry.(Aberdeenshire: montane habitat) ...... *exulans*(Pl.6, fig.1)(p. 104)

– Opaque, texture smooth, surface irregular, shining; pupa not visible within. Length/breadth ratio (10 measured) 2.12–2.61:1. Dirty white or whitish brown. Concealed low down amongst herbage; sometimes spun exposed on rocks or stones. (N. Wales, W. Scotland, W. Ireland) .... *purpuralis*(Pl.6, fig.9)(p. 114)

Subgenus ZYGAENA Fabricius

Zygaena Fabricius, 1775, *Syst.Ent.*: 550.
Anthrocera Scopoli, 1777, *Intr.Hist.nat.*: 414.
Thermophila Hübner, [1819], *Verz.bekannt.Schmett.*: 117.

Imago. Forewing pattern five- or six-spotted, spots often confluent in pairs or, more rarely, forming streaks; spot 6, when developed, rounded, or confluent with spot 5 to form a kidney-shaped blotch (*loti*-group); forewing ground colour black, forewing spots and hindwing normally red, rarely orange or yellow. Hindwing usually with a well-developed, black border from apex to tornal area. Male genitalia with processes of uncus usually long, well developed; female with ductus bursae usually strongly sclerotized.

Larva. Feeds on Leguminosae and Empetraceae.

ZYGAENA (ZYGAENA) EXULANS (Hohenwarth)
The Scotch or Mountain Burnet

Sphinx exulans Hohenwarth, 1792, *in* Reiner & Hohenwarth, *Botanische Reisen*: 265, pl.6, fig.2.

Type locality: Austria; Gross-Glockner region, 2400m [7870 ft].

The nominate subspecies does not occur in the British Isles where the species is represented by subsp. *subochracea* White.

Subsp. *subochracea* White

Zygaena exulans var. *subochracea* White, 1872, *Scott.Nat.* 1: 175.

Type locality: Scotland; Aberdeenshire, Braemar, 2400–2600ft [730–790m].

Description of imago (Pl.4, figs 7,8)

Wingspan 22–33mm. Head, thorax and abdomen strongly haired, especially in the male, black; patagium of female whitish yellow; legs of male black, inwardly tinged yellowish brown, those of female distinctly yellowish brown. Forewing five-spotted (1, 2, 3, 4, 5), thinly scaled, coloration translucent; ground colour blue-black or green-black with a weak sheen, in female dusted with whitish or golden yellow, especially along veins; spots crimson, in female indistinctly edged with whitish yellow, spot 1 extending along costa as far as spot 3, the latter elongate. Hindwing more translucent than forewing, with a small but distinct hyaline area at base; a relatively broad, black border, broadest in the male, around apex and along termen, terminating abruptly well before tornus (at vein 1A+2A). Cilia dark grey (male) or grey (female), with basal third black.

Variation occurs in the size of the forewing spots, which

are often reduced, and the breadth of the hindwing border.

This species is distinguished by the strongly haired thorax and abdomen, the thinly scaled wings and translucent coloration, the rounded apex of the forewing, and spot I characteristically extended along the costa as far as spot 3; the montane habitat is also a useful guide to its identification.

Life history

Ovum. Yellow, paler medially, with one pole transparent. July; deposited in several layers to form an irregularly shaped batch.

Larva (Pl.5, figs 1,2). Full-fed *c.*17mm long. Head and thoracic legs black; thorax and abdomen velvety grey-black, greenish laterally; black dorsal spots inconspicuous, one per segment, in anterior part of each segment from third thoracic to ninth abdominal; yellow spots conspicuous, from second thoracic to eighth abdominal segment; a small, shining, black intersegmental spot immediately below level of yellow spots; peritreme of spiracles black. Earlier instars are usually lighter and more generally flushed with green. August to May, with several periods of diapause; feeding principally on crowberry (*Empetrum nigrum*) and eating the terminal shoots and unripe berries (Tremewan, pers.obs.), although Cockayne (1932: 100), who gives an account of collecting the larvae, recorded it also on cowberry (*Vaccinium vitis-idaea*), bilberry (*V. myrtillus*) and heather (*Calluna vulgaris*). In captivity the larvae can be reared on common bird's-foot trefoil (*Lotus corniculatus*) which is not a natural host plant although it is common in the Scottish biotopes; when fully grown the larvae should be placed singly in suitable containers and provided with sprigs of heather on which they will spin their cocoons – it is essential that they are segregated at this stage, otherwise some of the larvae yet to spin up are liable to damage the cocoons that are already constructed.

Pupa and cocoon (Pl.6, fig.1). Pupa *c.*12–14mm long. Shining, black. June; in a cocoon spun low down and usually partially concealed on heather, crowberry or other vegetation. Cocoon bluntly fusiform, yellowish white, thin and translucent, the pupa indistinctly visible within.

Imago. Univoltine. Mid-June to mid-July; inhabiting the flat tops of mountains where the mainly prostrate vegetation consists predominantly of heather and crowberry, with lichens and scattered plants of bilberry, mountain-everlasting (*Antennaria dioica*) and trailing azalea (*Loiseleuria procumbens*). The moth flies strongly in sunshine and feeds at the flowers of common bird's-foot trefoil, or it can be found resting on these flowers or those of mountain-everlasting; during inclement weather, however, it conceals itself amongst the low-growing vegetation of

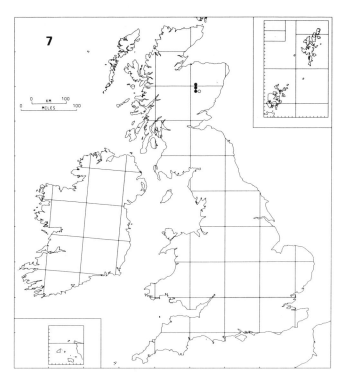

Zygaena (Zygaena) exulans

its wind-swept habitat. Tugwell (1886), James (1912) and Showler (1955) give interesting accounts of collecting this species at Braemar where it was first discovered on 17 July 1871 (White, 1871: 68).

Distribution (Map 7)

In Britain known only from Scotland where it occurs in a few localities near Braemar, Aberdeenshire, at elevations of 700–850m (2295–2790ft); records from Loch Etive, Argyll (Sheldon, 1901: 136), and Lochnagar, Aberdeenshire (Cowie, 1903: 22), are unconfirmed (Tremewan & Carter, 1967: 3; Tremewan & Classey, 1970: 67). Unknown from Ireland. A boreo-alpine species, occurring in Fennoscandia and on the major mountain ranges of western, southern and south-eastern Europe, its distribution extending eastwards to Siberia and Mongolia; in the northern part of its range, *i.e.* in Lapland, it occurs at elevations as low as 150m (490ft), but proceeding southwards the altitude at which it occurs gradually increases.

ZYGAENA (ZYGAENA) LOTI ([Denis & Schiffer-müller)]
The Slender Scotch Burnet

Sphinx loti [Denis & Schiffermüller], 1775, *Schmett. Wien.*: 45.

Type locality: [Austria]; Vienna district.

The nominate subspecies does not occur in the British Isles where the species is represented by subsp. *scotica* (Rowland-Brown).

Subsp. *scotica* (Rowland-Brown)

Anthrocera achilleae scotica Rowland-Brown, 1919, *Entomologist* **52**: 225.

Zygaena fulvia exerge *caledoniae* Verity, 1930, *Memorie Soc.ent.ital.* **9**: 21.

Zygaena achilleae subsp. *caledonica* Reiss, 1931, *Int.ent.Z.* **25**: 341.

Type locality: Scotland; Argyll [Morvern].

Description of imago (Pl.4, figs 9,10)

Wingspan 25–30mm. Head, thorax and abdomen of male strongly haired, dull black, patagium, tegula and thorax mixed with whitish grey; female less strongly haired, with patagium, tegula and thorax more strongly mixed with whitish grey, abdomen with a greenish sheen; legs of male black, inwardly yellowish brown, those of female almost entirely yellowish brown. Forewing six-spotted (1, 2, 3, 4, 5+6); ground colour black with a moderate, green sheen, more pronounced in female; spots crimson, in female sometimes indistinctly edged with yellowish white, spot 1 sometimes extended along costa as far as spot 3, spots 2 and 4 often weakly connected, spots 5 and 6 confluent, forming a reniform blotch. Hindwing crimson, male usually with an indistinct, narrow black border. Cilia of male black, in female whitish grey with basal half black.

Variation occurs in the size of the spots; these can be reduced, especially spot 6, or enlarged, and spots 2 and 4 are frequently connected with red scaling.

Similar species. Differs from *Z. filipendulae* (Linnaeus) by its yellowish brown legs, the thinner scaling and consequently comparatively translucent coloration of the wings, and the confluence of spots 5 and 6 which form a characteristic reniform blotch in the terminal part of the forewing. In the field, *Z. loti* can be distinguished from *Z. filipendulae* by its faster flight and comparatively dull coloration (James, 1936: 271; Tremewan, pers.obs.).

Life history

Ovum. Pale yellow, with one pole transparent. Late June and early July; deposited in a batch consisting of a single layer.

Larva (Pl.5, figs 3,4). Full-fed *c.*16mm long. Head and

thoracic legs black; thorax and abdomen dark grey (bright green in early to penultimate instars), darker dorsally, paler and tinged with green laterally; dorsal spots black, two per segment, from second thoracic to ninth abdominal, that in posterior part of each segment smaller with a conspicuous bright yellow spot beneath; subdorsal and lateral spots absent; peritreme of spiracles black. July to late May, with one or more periods of diapause; feeding on common bird's-foot trefoil (*Lotus corniculatus*).

Pupa and cocoon (Pl.6, fig.2). Pupa *c.*12mm long. Head, wings and appendages shining, yellowish brown, thoracic region and abdomen greenish brown; a suffused blackish brown dorsal spot in anterior part of each abdominal segment to the ninth, a yellow spot below and in posterior part of each segment to the seventh. Late May and early June; in a cocoon spun on the ground and concealed amongst grass and other herbage. Cocoon ovoid, shining, dirty white.

Imago. Univoltine. Mid-June to early July. The moth flies swiftly in sunshine but during inclement weather rests low down amongst bracken and is then difficult to find; it inhabits low cliffs and grassy banks near the sea, and apparently only where its host plant, common bird's-foot trefoil, grows in association with bell-heather (*Erica cinerea*), heather (*Calluna vulgaris*) and bracken (*Pteridium aquilinum*) (Woodbridge, 1934: 238; Tremewan, 1968a: 3).

Distribution (Map 8)

In the British Isles known only from Argyll (Morvern, Mull, Ulva) in Scotland; records from Raasay, Eilean nan Each and Rhum, Inverness-shire, are doubtful. An account of its distribution and recent history in Scotland is given by Tremewan (1968b). From Europe its range extends to Turkey, the Caucasus and Siberia.

ZYGAENA (ZYGAENA) VICIAE ([Denis & Schiffer-müller])
The New Forest Burnet

Sphinx viciae [Denis & Schiffermüller], 1775, *Schmett. Wien.*: 45.

Type locality: [Austria]; Vienna district.

The nominate subspecies does not occur in the British Isles where the species is represented by subspp. *ytenensis* Briggs and *argyllensis* Tremewan.

Description of imago (Pl.4, figs 11–14)

Wingspan 21–31mm. Head, thorax and abdomen weakly haired, black with a greenish sheen, or strongly haired, black; antenna slender, weakly clubbed, especially in female; legs black, inwardly yellowish brown, especially in male. Forewing five-spotted (1, 2, 3, 4, 5); ground

Zygaena (Zygaena) loti

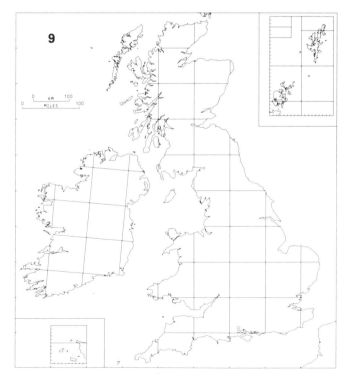

Zygaena (Zygaena) viciae

colour black with a moderate, green or blue-green sheen; spots crimson, spot 3 elongate, spot 4 larger, subquadrate. Hindwing crimson, a moderately broad, black or blue-black border extending from apex to near tornus. Cilia dark grey (male) or grey (female), with basal third black.

Subsp. *ytenensis* Briggs

Zygaena trifolii var. *ytenensis* Briggs, 1888, *Young Nat.* **9**: 82.

Zygaena meliloti var. *anglica* Reiss, 1931, *Int.ent.Z.* **25**: 344.

Type locality: England; New Forest.

Description of imago (Pl.4, figs 11,12)

Head, thorax and abdomen weakly haired; forewing comparatively narrow; hindwing border narrower than in subsp. *argyllensis* Tremewan, especially in female.

Forms with variable confluence of the forewing spots (f. *confluens* Tutt) were not infrequent, while the six-spotted form (f. *sexpunctata* Tutt) occurred more frequently in females; examples with a red abdominal cingulum were very rare.

Subsp. *argyllensis* Tremewan

Zygaena viciae argyllensis Tremewan, 1967, *Entomologist's Gaz.* **18**: 159, pl.7, figs 1–18.

Type locality: Scotland; Argyll.

Description of imago (Pl.4, figs 13,14)

Head, thorax and abdomen strongly haired; legs darker, almost entirely black; red coloration colder; forewing broader, with apex more rounded; spots larger, tending to coalesce; hindwing border broader than in subsp. *ytenensis* Briggs.

Forms occur with spot 5 extended basad and connected with spot 3 by crimson scales; rarely spot 1 is extended half-way along the costa, or spots 2 and 4 are weakly confluent.

Similar species. Distinguished from *Z. trifolii* (Esper) by its slender, weakly clubbed antennae, the more rounded apex of the forewing, and the characteristically small, elongate spot 3 contrasting with the relatively large, subquadrate spot 4.

Life history

Ovum. Pale yellow, with one pole transparent. July; deposited in a batch consisting of a single layer.

Larva (Pl.5, figs 5,6). Full-fed *c.*14mm long. Head and thoracic legs black; thorax and abdomen dark emerald-green, minutely speckled with black, tinged with grey anteriorly and posteriorly, with a narrow, whitish mediodorsal line; a small black dorsal spot in anterior part of each segment from second thoracic to ninth abdominal, a yellow spot beneath and in posterior part of each segment from second thoracic to seventh abdominal; peritreme of spiracles black. Late July to early June, with one or more periods of diapause; feeding on meadow-vetchling (*Lathyrus pratensis*) and common bird's-foot trefoil (*Lotus corniculatus*) (Tremewan, pers.obs., 1970).

Pupa and cocoon (Pl.6, fig.3). Pupa *c*.11–12mm long. Head, wings and appendages shining, black; thorax bright emerald-green, darker laterally; abdomen yellow tinged with emerald-green, with sooty black, transverse bands laterally and ventrally, caudal region blackish, a minute black dorsal spot in anterior part of each segment to the ninth, a cream spot beneath in posterior part of each segment to the seventh; peritreme of spiracles black. Mid-June to early July; in a cocoon attached to the underside of a leaf or bracken frond, or spun on a grass-stem or dead blade of grass near the ground and concealed amongst the herbage (Lyle, 1912: 129; Tremewan, 1965b: 120). Cocoon fusiform, longitudinally ribbed, shining, pale yellow.

Imago. Univoltine. In the New Forest, subsp. *ytenensis* formerly occurred in rides and clearings from late June to late July. In Scotland, subsp. *argyllensis* begins to emerge in early July and inhabits a steep slope on dry grassy cliffs where it flies actively in sunshine; during dull or wet weather, however, the moth rests low amongst the herbage and is then difficult to find.

Distribution (Map 9)

Z. viciae ytenensis occurred in small colonies in the New Forest, Hampshire, from where it was last recorded in 1927; a detailed account of its history and former distribution there is given by Tremewan (1966). The Scottish subspecies *argyllensis* was first discovered by F. C. Best (1963:149) who captured three examples in a remote part of Argyll on 10 July 1963; further specimens were taken by Best in 1964. In spite of intensive searching in other parts of Argyll during the last two decades, it is still known only from the original locality. Its habitat and aspects of its ecology in Scotland are described by Tremewan (1965b). Its range abroad extends from Spain through central and southern Europe to Siberia and Mongolia.

Conservation

This species is listed as endangered and is protected under the Wildlife and Countryside Act, 1981, which prohibits its collection, consequently the Argyll record is excluded from the distribution map.

ZYGAENA (ZYGAENA) FILIPENDULAE (Linnaeus)
The Six-spot Burnet

Sphinx filipendulae Linnaeus, 1758, *Syst.Nat.* (Edn 10) **1**: 494.

Type locality: [Sweden].

The nominate subspecies does not occur in the British Isles where the species is represented by subsp. *stephensi* Dupont.

Subsp. *stephensi* Dupont

[*Anthrocera hippocrepidis* (Hübner) sensu Stephens, 1828, *Ill.Br.Ent.* (Haust.) **1**: 109. Misidentification.]

Zygaena stephensi Dupont, 1900, *Bull.Soc.Etude Sci. nat.Elbeuf* **18**: 77.

Zygaena filipendulae var. (?ab.) *tutti* Rebel, 1901, *in* Staudinger & Rebel, *Cat.Lepid.Palaearct.Faunengeb.*: 384.

Zygaena filipendulae var. (?ab.) *lismorica* Reiss, 1931, *Int. ent.Z.* **25**: 345.

Zygaena filipendulae f. loc. *degenerata* Tremewan, 1958, *Entomologist's Gaz.* **9**: 190.

Zygaena filipendulae ssp. *anglicola* Tremewan, 1960, *Entomologist's Gaz.* **11**: 189.

Type localities: England; Darenth Wood, Kent, and Coombe Wood, Surrey.

Description of imago (Pl.4, figs 15–24)

Wingspan 25–39mm. Head, thorax and abdomen weakly to strongly haired, black, with a weak to strong, blue, blue-green or green sheen; legs black, inwardly light brown or yellowish brown, especially in the male. Forewing six-spotted (1, 2, 3, 4, 5, 6); ground colour black, with a strong, blue, blue-green or green sheen; spots red, varying from crimson through vermilion to scarlet. Hindwing red (as in forewing spots), with a narrow, black border extending along termen from apex and terminating before tornus. Cilia black or blue-black, basal third more intense; in female, cilia of forewing lighter than those of hindwing. In Scotland, specimens from the Western Isles have the thorax and abdomen strongly haired and the forewing spots, especially 5 and 6, are frequently confluent in pairs (Harrison, 1940: 135; 1945: 25; Tremewan & Manley, 1964: 153); although they are distinct from other populations in Britain it is preferable to consider them as part of a cline.

The forewing spots vary in size and shape. A common form has the spots confluent in pairs (1+2, 3+4, 5+6) and another, almost equally as common, has the distal pair of spots confluent (5+6) and sometimes enlarged and suffused. Forms in which the spots are variably confluent longitudinally to form an irregularly shaped blotch (f. *conjuncta* Tutt, fig.16) are very rare. Occasionally specimens occur with the spots of the forewing suffused and confluent (fig.17); such forms are genetical and are also caused by the effects of extreme temperatures on the pupa. Rarely, one or more of the spots are absent as, for example, in f. *spoliata* Cockayne which lacks spot 4; less rare are those specimens which have spot 6 reduced or vestigial, and in extreme examples spot 6 is absent. The hindwing border varies considerably and in f. *nigrolimbata* Cockayne (fig.23) it is very broad, especially in the apical and tornal areas, and extends along the inner margin to the base; in this form the forewing spots are often reduced. The f. *grisescens* Oberthür (fig.18) has the forewing ground colour, the hindwing border and the cilia translucent bluish or purplish grey, the normal red coloration being replaced by pink; the head, antennae, thorax, legs and abdomen are also purplish grey or whitish grey. A number of recurrent colour-forms are known, all of which are rare and are simple recessives. Perhaps the commonest and most widespread of these is f. *flava* Robson (fig.19) in which the red coloration of the forewing spots and hindwings is replaced by yellow – the yellow coloration itself varies from lemon-yellow to deep chrome-yellow. Less common are f. *aurantia* Tutt (fig.20) with the red coloration replaced by orange, and f. *intermedia* Tutt (fig.21), which has the spots and hindwings orange-red. The pseudo-orange form (see p. 99) is uncommon. Undoubtedly the rarest of the colour forms is that in which the spots and hindwings are brownish or sooty black (fig.24).

Life history

Ovum. Yellow, with one pole transparent. July and August; deposited in an irregularly shaped batch consisting of several layers.

Larva (Pl.5, figs 7–12). Full-fed *c*.19mm long. Head and thoracic legs black; thorax and abdomen greenish yellow or yellow; dorsal spots black, two per segment from second thoracic to ninth abdominal, that in anterior part of each segment large, subquadrate, that in posterior part smaller, elongate; an inconspicuous, yellow spot beneath each posterior dorsal spot from second thoracic to seventh abdominal segment; subdorsal and lateral spots black, the subdorsal spots similar in shape to the dorsal spots; peritreme of spiracles black. August to early June, with one or more periods of diapause; feeding on common bird's-foot trefoil (*Lotus corniculatus*), and recorded (Tremewan, 1982b) also on greater bird's-foot trefoil (*Lotus uliginosus*).

Pupa and cocoon (Pl.6, figs 4,5). Pupa *c*.13–17mm long. Shining, black, or occasionally dark brown. June to early August; in a cocoon spun exposed on the stems of grass and other herbage and occasionally on twigs of bushes and on wire fences, rocks and stones; in Scotland the cocoon is usually placed much lower on the herbage or on heather

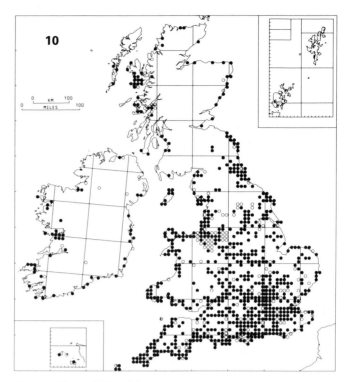

Zygaena (Zygaena) filipendulae

(James, 1936: 271; Tremewan & Manley, 1964: 150; Tremewan & Carter, 1967: 6, pl.2, figs 3,4). Cocoon fusiform, irregularly ribbed, varying from bright yellow to dirty white.

Imago. Univoltine. Mid-June to August, the emergence period varying considerably from one locality to another; inhabiting chalk downs, cliffs, sand-hills, rough meadows and roadside verges where its host plant, common bird's-foot trefoil, grows abundantly.

Distribution (Map 10)

Z. filipendulae is the commonest and most widely distributed species of the genus in Britain, ranging from southern England (including the Isles of Scilly) northwards to Sutherland and the Outer Hebrides; in Scotland it is usually absent from the higher ground and is restricted mainly to coastal areas. Widely distributed and locally common in Ireland. The Channel Islands. Throughout Europe (except Portugal and south-west Spain) north to Lapland, its range extending through Turkey to northwest Iran and the Caucasus Mountains.

ZYGAENA (ZYGAENA) TRIFOLII (Esper)
The Five-spot Burnet

Sphinx trifolii Esper, 1783, *Schmett.* **2**: 223, pl.34, figs 4,5.

Type locality: Germany; Frankfurt am Main.

The nominate subspecies does not occur in the British Isles where the species is represented by subspp. *decreta* Verity and *palustrella* Verity.

Description of imago (Pl.4, figs 25–38)

Wingspan 24–37mm. Head, thorax and abdomen moderately (male) or weakly haired (female), black, with a blue-green or green sheen, especially in the female; legs black, inwardly yellowish brown, especially foretibia. Forewing five-spotted (1, 2, 3+4, 5); ground colour black, with a strong, blue-green or green sheen; spots red, varying from crimson to vermilion, 3 and 4 confluent. Hindwing red (as in forewing spots), a moderately broad, black border extending from apex, narrowing before and terminating at tornus. Cilia blue-black, basal third black.

Subsp. *decreta* Verity

Zygaena trifolii major Tutt, 1897, *Entomologist's Rec.J. Var.* **9**: 88 [primary homonym of *Zygaena lonicerae major* Frey, 1888].

Zygaena trifolii race *decreta* Verity, 1926, *Entomologist's Rec.J.Var.* **38**: 57.

Type locality: England; Sussex [Tilgate Forest].

Description of imago (Pl.4, figs 25–30)

Larger than subsp. *palustrella* Verity, scaling denser, coloration brighter and more intense.

Subsp. *palustrella* Verity

Anthrocera trifolii minor Tutt, 1899, *Nat.Hist.Br.Lepid.* **1**: 480 [secondary homonym of *Zygaena cocandica minor* Erschoff, 1874].

Zygaena trifolii race *palustrella* Verity, 1926, *Entomologist's Rec.J.Var.* **38**: 11.

Type locality: England; Surrey [North Downs].

Description of imago (Pl.4, figs 31–38)

Smaller than subsp. *decreta* Verity, scaling thinner, coloration duller, weakly translucent.

An extremely variable species with the following forms usually occurring in either subspecies. Almost as common as the typical form is that in which spots 3 and 4 are separate (fig.26); in a rare but recurring form (f. *obsoleta* Tutt, fig.32) spot 4 is absent and the remaining spots are reduced. Specimens with variable confluence of the spots are frequent, especially in some colonies, and extreme forms (figs 27,36) have spots 1–5 broadly confluent to form an irregularly shaped, longitudinal blotch. A form (f.

extrema Tutt, fig.37) which is probably genetical, but can also be caused by extreme temperatures, has the forewing spots confluent and suffused, leaving a narrow area of ground colour along the costa, termen and dorsum. Reddish orange, pale orange (f. *carnea* Cockayne) and yellow (f. *lutescens* Cockerell, fig.34) forms are uncommon, and brown (fig.38) and blackish brown or sooty black forms (f. *daimon* Porritt, fig.28) are very rare; occasionally such colour forms also have confluent spots, resulting in striking aberrations. Another rare form (fig.30) has spot 6 present and often attached to spot 5, and the very rare, yellow six-spotted form has been reared in captivity (Tremewan, pers.obs.). Specimens with a very broad hindwing border, leaving only a small amount of red coloration in the discal area of the hindwing (fig.33), are rare.

Similar species. Z. lonicerae (Scheven), from which it differs by its usually smaller size, relatively shorter forewing, more rounded apices of both wings, and the usually broader hindwing border.

Life history

Ovum. Pale yellow, one pole transparent. July to early August (subsp. *decreta*), or late May to mid-June (subsp. *palustrella*); deposited in several layers to form an irregularly shaped batch.

Larva (Pl.5, figs 13,14). Full-fed *c*.12–18mm long. Head and thoracic legs black; thorax and abdomen pale yellowish or whitish green; dorsal spots black, two per segment from second thoracic to ninth abdominal, that in anterior part of segment large, subquadrate, that in posterior part of segment smaller, elongate; an inconspicuous yellow spot beneath each posterior dorsal spot from second thoracic to ninth abdominal segment; black subdorsal spots present, two per segment, that in anterior part of each segment generally larger and constricted or broken posteriorly; lateral spots black; peritreme of spiracles black. August to June (subsp. *decreta*), feeding on greater bird's-foot trefoil (*Lotus uliginosus*), or late June to early May (subsp. *palustrella*), feeding on common bird's-foot trefoil (*Lotus corniculatus*); the larvae of both subspecies have one or more periods of diapause.

Pupa and cocoon (Pl.6, figs 6,7). Pupa *c*.10–18mm long. Shining, black, or blackish brown. June to July, in a cocoon which is usually spun high on the stems of soft-rush (*Juncus effusus*) and other marshland vegetation and is not concealed (subsp. *decreta*, fig.6); late April to May, the cocoon usually spun low near the ground and well concealed amongst grass and other herbage (subsp. *palustrella*, fig.7). Cocoon fusiform, irregularly ribbed; varying from bright yellow to cream-white or dirty white.

Imago. Univoltine. The British populations of *Z. trifolii* consist of two ecotypes or ecological subspecies; although they do not differ so strongly in superficial characters, their ecology and ethology are well defined and warrant their separation into two taxa. The subspecies *decreta* occurs from late June to early August and inhabits marshes, wet moorland and damp meadows, being found occasionally also on comparatively dry coastal cliffs; its larva feeds on greater bird's-foot trefoil and the cocoon is usually spun high on the stems of soft-rush and other vegetation. The subspecies *palustrella* occurs from mid-May to mid-June and is restricted to chalk downs and limestone hills; the larva feeds on common bird's-foot trefoil and the cocoon is concealed amongst the downland herbage where it is difficult to find.

Distribution (Map 11; see also map 12)

Locally common in England, Wales and the Isle of Man, but absent from Scotland and Ireland. The subspecies *decreta* was formerly widespread throughout southern England, from Norfolk and Kent westwards to Cornwall; however, it appears to have become extinct in the southeast counties during the last two decades or so (Tremewan, 1980a; 1982a), but it is still widespread and often common in south-west England, its range extending eastwards to the New Forest, Hampshire, and northwards through Wales to Anglesey; the Isle of Man. The subspecies *palustrella* is restricted to chalk downland south of the River Thames, *i.e.* the North Downs, South Downs and Salisbury Plain, and to limestone hills in Gloucestershire. On the chalk it is locally common but has declined in recent years through overgrazing or destruction of its habitat (Tremewan, 1980a: 144; 1982a: 11); on limestone it is very local and known only from one or two localities on the Cotswold Hills. It occurred at Tintern, Monmouthshire (Gwent), and was very common at Cotswold Park, Gloucestershire, but it is now extinct in these limestone habitats which have been destroyed by afforestation. The Channel Islands. Throughout western and central Europe; Sicily; North Africa (Morocco, Algeria, Tunisia).

It has not been possible to confirm every record of *Z. trifolii* and *Z. lonicerae* (Scheven) from those areas where their ranges overlap in England and Wales, consequently all the records of these two species are included in an aggregate map (Map 12). However, a separate map for each species is also provided, based on authentic records only, but the distribution patterns in these maps are, of necessity, incomplete.

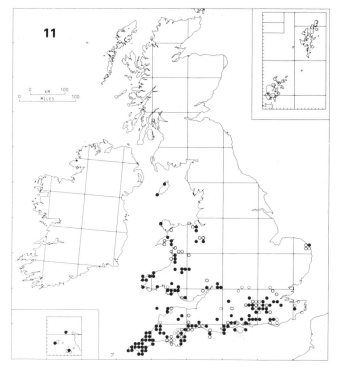

Zygaena (Zygaena) trifolii

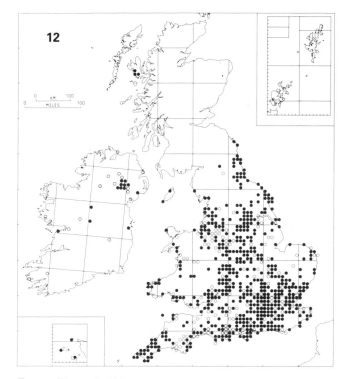

Zygaena (Zygaena) trifolii and *Zygaena (Zygaena) lonicerae*

ZYGAENA (ZYGAENA) LONICERAE (Scheven)
The Narrow-bordered Five-spot Burnet
Sphinx lonicerae Scheven, 1777, *Naturforscher, Halle* **10**: 97.

Type locality: Germany; Regensburg.

The nominate subspecies does not occur in the British Isles where the species is represented by subspp. *latomarginata* (Tutt), *jocelynae* Tremewan and *insularis* Tremewan.

Description of imago (Pl.4, figs 39–53)
Wingspan 22–40mm. Head, thorax and abdomen weakly to strongly haired, black, often with a blue, blue-green or green sheen, especially in the female; legs black, inwardly yellowish brown, especially in the male. Forewing with apex attenuate; five-spotted (1, 2, 3, 4, 5); ground colour black, with a strong, blue, blue-green or green sheen, spots red, varying from crimson to vermilion. Hindwing red (as in forewing spots), a moderately narrow, black border extending from apex and terminating before tornus. Cilia dark grey with basal third black or blue-black.

Subsp. latomarginata (Tutt)
Anthrocera lonicerae var. *latomarginata* Tutt, 1899, *Nat.Hist.Br.Lepid.* **1**: 468.
Zygaena lonicerae race *transferens* Verity, 1926, *Entomologist's Rec.J.Var.* **38**: 59.
Zygaena lonicerae race *britanniae* Verity, 1926, *Ibid.* **38**: 61.
Zygaena lonicerae race *misera* Verity, 1926, *Ibid.* **38**: 73.
Type locality: England; Filey, Yorkshire.

Description of imago (Pl.4, figs 39–48)
Head, thorax and abdomen moderately haired, sheen weak; red coloration crimson; forewing spots smaller than in other subspecies, giving the insect an overall darker appearance; hindwing border broader. The foregoing description applies to specimens from Filey; those from southern England (formerly separated as *transferens* Verity) have the head, thorax and abdomen weakly haired, with a pronounced sheen, especially in the female, the forewing spots often larger, and the red coloration warmer and brighter. In spite of these differences, the populations occurring from southern England northwards to Northumberland are considered to represent a cline, therefore their separation into two subspecies is unwarranted.

In contrast to *Z. trifolii* (Esper), variation in *Z. lonicerae* occurs only rarely, and some populations are so stable that the specimens hardly ever deviate from the typical form, which has all the forewing spots separate. An infrequent form (f. *centripuncta* Tutt, fig.41) has spots 3 and 4 confluent; specimens showing variable confluence leading to an extreme form which has spots 1–5 united and forming an elongate, roughly triangular blotch (fig.42) are also rare. Suffused confluent specimens are rather more frequent and are probably genetical, although sometimes they can be produced by subjecting the pupae to extreme temperatures. In f. *eboracae* Prest the forewing ground colour and hindwing border are translucent grey and the normal red coloration is translucent pink; a similar but more extreme form (f. *grisescens* Cockayne, fig.43) has the forewing spots and hindwing of the normal red coloration, but the forewing ground colour is purplish grey, with the hindwing border and the cilia of both wings golden yellow. Yellow (fig.44), orange (fig.45) and pseudo-orange (fig.46) forms are very rare although they have occurred regularly in one locality on the Cotswolds, Gloucestershire; the yellow and orange forms have occurred singly elsewhere.

Subsp. jocelynae Tremewan
Zygaena lonicerae jocelynae Tremewan, 1962, *Entomologist's Gaz.* **13**: 10.
Type locality: Scotland; Isle of Skye, Inverness-shire.

Description of imago (Pl.4, figs 49–51)
Larger than the other subspecies; head, thorax and abdomen strongly haired, sheen weak or absent; red coloration cold crimson; forewing spots larger, especially spots 4 and 5, spots 3 and 4 narrowly separated but often confluent; weakly suffused confluence of spots not uncommon (fig.51).

Subsp. insularis Tremewan
Zygaena lonicerae ssp. *insularis* Tremewan, 1960, *Entomologist's Gaz.* **11**: 191.
Type locality: Ireland; Mullinures, Co. Armagh.

Description of imago (Pl.4, figs 52,53)
Head, thorax and abdomen moderately haired, sheen moderate; forewing spots larger than in subsp. *latomarginata* (Tutt), inclined to confluence.

Similar species. *Z. trifolii* (Esper), from which it differs by its generally larger size, relatively longer forewing with the apex attenuate, the more strongly pointed apices of both wings and the usually narrower hindwing border.

Life history
Ovum. Pale yellow to dark yellow, with one pole transparent. July; deposited in a batch consisting of a single layer.
Larva (Pl.5, figs 15,16). Full-fed *c.*19mm long. Head and thoracic legs black; thorax and abdomen pale whitish green; dorsal spots black, two per segment, from first thoracic to ninth abdominal, that in anterior part of each

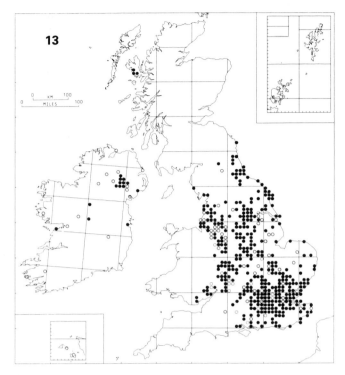

Zygaena (Zygaena) lonicerae

downs, cliffs, rough meadows, roadside verges, embankments, rides and woodland clearings and marshes.

Distribution (Map 13; see also map 12)

The subspecies *latomarginata* is widely distributed and locally common in southern England from Kent to Somerset, its range extending into Wales (Monmouthshire (Gwent) and Glamorgan) and northwards to Yorkshire, Durham, Cumberland (Cumbria) (Hancock & Kydd, 1980) and Northumberland; an account of collecting this subspecies at the type locality (Filey) is given by Tremewan & Classey (1970: 67). In Scotland, *Z. lonicerae* is represented by subsp. *jocelynae*, which is known only from the Isle of Skye where it is locally common in several localities on the west coast; for an account of collecting and rearing this distinctive subspecies see Tremewan & Manley (1964: 149–152), Tremewan & Carter (1967: 4–8) and Tremewan (1965a: 87–88). In Ireland, where it is represented by subsp. *insularis*, it is widely distributed, occurring more frequently in the northern part of the country. The Channel Islands. Europe (except Portugal and western Spain) through Turkey and the Caucasus to Siberia.

segment large, subquadrate, that in posterior part narrow; a conspicuous, bright yellow spot beneath posterior dorsal spot from second thoracic to eighth abdominal segment; black subdorsal spots similar to but smaller than dorsal spots; lateral series of black spots present; readily distinguished by the long setae which are three times the length of those of the other British species and give the larva a comparatively hairy appearance. Late July to June, with one or more periods of diapause. The two main host plants are meadow-vetchling (*Lathyrus pratensis*) and red clover (*Trifolium pratense*); it is also found occasionally on common bird's-foot trefoil (*Lotus corniculatus*), white clover (*Trifolium repens*), bitter vetch (*Lathyrus montanus*) and sainfoin (*Onobrychis viciifolia*), and in some of its marshland biotopes in southern England it feeds solely on greater bird's-foot trefoil (*Lotus uliginosus*) (Tremewan, 1982a: 10; Warren, 1983: 3).

Pupa and cocoon (Pl.6, fig.8). Pupa *c.*16–18mm long. Shining, black. June to early July; in a cocoon spun exposed on the stems of grass and other herbage. Cocoon fusiform, relatively broader than that of *Z. filipendulae* (Linnaeus) and *Z. trifolii*, comparatively translucent, irregularly ribbed, varying from pale whitish yellow or greenish yellow to white.

Imago. Univoltine. Late June to July; inhabiting chalk

Subgenus MESEMBRYNUS Hübner

Mesembrynus Hübner, [1819], *Verz.bekannt.Schmett.*: 119.

Imago. Forewing pattern five- or six-spotted, spots frequently confluent and developed into three longitudinal streaks (1, 2+4, 3+5+6), or enlarged and confluent in pairs; coloration of forewing spots and hindwing red or pink, sometimes orange or yellow. Hindwing usually with black border reduced, rarely broadened and extending into discal area. Scaling of both wings often thin and coloration translucent. Male genitalia with processes of uncus of medium length or reduced to short lobes; in female, ductus bursae usually lacking sclerotization or only weakly sclerotized.

Larva. Feeds on Umbelliferae, Compositae and Labiatae.

ZYGAENA (MESEMBRYNUS) PURPURALIS (Brünnich)

The Transparent Burnet

Sphinx purpuralis Brünnich, 1763, *in* Pontoppidan, *Danske Atlas* 1: 686, pl.30, fig.

Type locality: Denmark [Sjælland].

The nominate subspecies does not occur in the British Isles where the species is represented by subspp. *segontii* Tremewan, *caledonensis* Reiss and *sabulosa* Tremewan.

Description of imago (Pl.4, figs 54–64)
Wingspan 25–34mm. Head, thorax and abdomen strongly haired, black; legs black, inwardly yellowish brown, especially in the female. Wings thinly scaled, coloration translucent. Forewing with three longitudinal streaks (1, 2+4, 3+5+6), streak 3 hatchet-shaped; ground colour black with a weak, green or blue-green sheen, streaks vermilion. Hindwing vermilion, a narrow black border usually present at apex, sometimes extending to near tornus. Cilia of forewing light grey, basal third darker, cilia of hindwing blackish grey.

Subsp. *segontii* Tremewan

Zygaena purpuralis ssp. *segontii* Tremewan, 1958, *Entomologist's Gaz.* 9: 188.

Type locality: Wales; Abersoch, Caernarvonshire (Gwynedd).

Description of imago (Pl.4, figs 54–56)
Generally smaller than the other subspecies; red coloration colder, more translucent; forewing streaks narrower, well separated, especially in the male; hindwing border present only at apex.

A very rare form (f. *obscura* Tutt, fig.55) has the red coloration of the forewing streaks and hindwings suffused with black.

Subsp. *caledonensis* Reiss

Zygaena purpuralis var. *caledonensis* Reiss, 1931, *Int.ent.Z.* 25: 341.

Type locality: Scotland; Oban, Argyll.

Description of imago (Pl.4, figs 57–62)
Larger than subsp. *segontii* Tremewan; forewing streaks broader; hindwing border broader, generally extending from apex to near tornus.

Brownish red (fig.58), orange (fig.59), yellow (fig.60) and suffused confluent (fig.61) specimens occur not infrequently (Tremewan & Manley, 1964: 149–152; Harper & Langmaid, 1975: 139).

Subsp. *sabulosa* Tremewan

Zygaena purpuralis hibernica Reiss sensu auctt.

Zygaena purpuralis ssp. *hibernica* f. loc. *sabulosa* Tremewan, 1960, *Entomologist's Gaz.* 11: 186 [unavailable name].

Zygaena (Mesembrynus) purpuralis sabulosa Tremewan, 1976, *Entomologist's Gaz.* 27: 150.

Type locality: Ireland; Ballyvaughan, Co. Clare.

Description of imago (Pl.4, figs 63,64)
Generally larger than the other subspecies; red coloration warmer; forewing streaks broader and often confluent, especially in the female; hindwing border usually present only at apex.

A yellow specimen (f. *lutescens* Tutt) was taken in Co. Galway by J. E. R. Allen (1894: 217), but this form is apparently rare in Ireland.

Distinguished from confluent forms of other *Zygaena* species by the thinly scaled wings, translucent coloration and the formation of the forewing streaks (1, 2+4, 3+5+6), the hatchet-shaped streak (3+5+6) being characteristic.

Life history

Ovum. Pale yellow, one pole transparent. June and July; deposited in an irregularly shaped batch consisting of two or more layers.

Larva (Pl.5, figs 21,22). Full-fed *c.*16mm long. Head and thoracic legs black; thorax and abdomen olive-green or yellowish green with an indistinct, very narrow, yellowish mediodorsal line; dorsal spots black, one per segment, situated in anterior part of each segment from second thoracic to tenth abdominal, those on ninth and tenth abdominal segments smaller; an inconspicuous, dull, pale yellow spot beneath and in posterior part of each segment

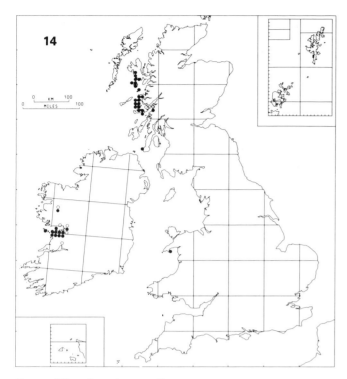

Zygaena (Mesembrynus) purpuralis

it may now be extinct in Caernarvonshire (Gwynedd) where it occurred in small colonies along the cliffs near Abersoch; it appears that it was last observed in this locality in 1962 (Tremewan, pers.obs.). In Scotland, however, subsp. *caledonensis* is widely distributed and locally common, and in some years even abundant, in the Inner Hebrides (Skye, Canna, Rhum, Mull, Ulva) and coastal areas of Argyll from Ardnamurchan and the Oban district to the Mull of Kintyre, its most southerly location there (Tremewan, 1969; Elliott, 1980). In Ireland subsp. *sabulosa* is widely distributed and often abundant in the Burren district of Cos Clare and Galway, its range extending northwards through Co. Galway to Ballinrobe and Partry (Chalmers-Hunt, 1977a) in Co. Mayo; it also occurs on the Aran Islands (Inishmore) (Emmet, 1968: 53). Recently, Pelham-Clinton (1981) recorded three specimens from south-east Ireland which were taken in 1950 by R. C. Faris, allegedly from Curracloe and Kilmore Quay, Co. Wexford. Abroad it is distributed from the Pyrenees through central and southern Europe to Turkey and Central Asia.

from third thoracic to eighth abdominal, that on eighth abdominal segment smaller and very indistinct; peritreme of spiracles black; setae whitish, those of dorsal and subdorsal series banded with black at three-quarters. Early instar larvae are darker and have the thorax and abdomen olive or pinkish olive with conspicuous yellow spots. July to May, with one or more periods of diapause; feeding on wild thyme (*Thymus praecox*).

Pupa and cocoon (Pl.6, fig.9). Pupa *c*.10–13mm long. Head, thorax, wings and appendages shining, yellowish brown, head and thorax marked with brown; abdomen brownish yellow, a dorsal spot clearly visible in anterior part of each segment. Late May to June; in a cocoon spun low near the ground and usually concealed amongst herbage, or occasionally attached to a rock or stone. Cocoon bluntly fusiform, surface irregular, shining, dirty white or whitish brown.

Imago. Univoltine. Early June to July; inhabiting steep, grassy, south-facing slopes and undercliffs on or near the coast, dunes, and limestone areas inland. The moth flies in sunshine, but during dull or wet weather it can be found resting fully exposed on grass-stems and flower-heads.

Distribution (Map 14)

There is no recent record of subsp. *segontii* from Wales and

References

Titles not referred to in the text are marked with an asterisk *

Adkin, R., 1895. [*Zygaena filipendulae* at 2000ft in Sutherland.] *Entomologist* **28**: 91.

Albin, E., 1720. *A natural history of English insects.* [12] pp., 100 pls. London.

Allen, J. E. R., 1894. [A yellow specimen of *Zygaena pilosellae.*] *Entomologist's Rec. J. Var.* **5**: 217.

Andrewes, C. H., 1937. A note on *Platycleis grisea* (F.) (Tettig.) devouring *Zygaena filipendulae* at Sennen, Cornwall. *Proc. R. ent. Soc. Lond.* (A)**12**: 155.

*Arnold, G., 1873. Vitality of life in larva of *Z. filipendulae.* *Entomologist* **6**: 434.

*Audcent, H., 1942. A preliminary list of the hosts of some British Tachinidae (Dipt.). *Trans. Soc. Br. Ent.* **8**: 1–42.

*Barrett, C. G., 1893–95a. *The Lepidoptera of the British Islands*, **2**: 372 pp., pls 41–86. London. [Zygaenidae: pp. 108–138, pls 58–60; 1894].

*———, 1895b. Extraordinary aberrations in Lepidoptera. *Entomologist's mon. Mag.* **31**: 219.

Beavis, I. C., 1973. Notes on the pupation and emergence of *Zygaena lonicerae* Scheven. *Entomologist's Rec. J. Var.* **85**: 267.

Beck, H., 1982. Projekt: Neuartiges Bestimmungsbuch für Lepidopterenlarven. *Neue ent. Nachr.* no. 1: 3–14, figs 1–8, 24, 25, figs.

Beirne, B. P., 1943. The distribution and origin of the British Lepidoptera. *Proc. R. Ir. Acad.* **49**(B): 27–59, figs 1–27 (distr. maps).

———, 1947. The origin and history of the British Macro-Lepidoptera. *Trans. R. ent. Soc. Lond.* **98**: 273–372, figs 1–45.

———, 1952. *The origin and history of the British fauna*, xii, 164 pp., 60 text figs. London.

Best, F. C., 1963. *Zygaena viciae* Schiff. (*meliloti* Esp.) (Lep.) in Scotland. *Entomologist's Gaz.* **14**: 149.

*Bignell, G. C., 1880. *Zygaena filipendulae* and its parasites. *Entomologist* **13**: 16–17.

*Billups, T. R., 1886. [*Hemiteles fulvipes* Gr. bred from *Zygaena filipendulae.*] *Abstr. Proc. S. Lond. ent. nat. Hist. Soc.* **1885**: 35.

Bisset, G. W., Frazer, J. F. D., Rothschild, M. & Schachter, M., 1960. A pharmacologically active choline ester and other substances in the garden tiger moth, *Arctia caja* (L.). *Proc. R. Soc.* (B)**152**: 255–262, figs 1–4, pl. 18, figs 5, 6.

*Blair, K. G., 1951. Some records of *Gelis* Thunberg (= *Pezomachus* Gravenhorst) (Hym., Ichneumonidae). *Entomologist's mon. Mag.* **87**: 194–195.

Bolam, G., 1926. The Lepidoptera of Northumberland and the Eastern Borders. *Hist. Berwicksh. Nat. Club* **25**(1925): 515–573.

Bradley, J. D. & Mere, R. M., 1964. Natural history of the garden of Buckingham Palace. *Proc. Trans. S. Lond. ent. nat. Hist. Soc.* **1963**(2): 55–74.

Bramwell, F. G. S., 1919. The larva of *Ino globulariae.* *Entomologist* **52**: 214–215.

*Bridgman, J. B., 1883. Further additions to Mr. Marshall's Catalogue of British Ichneumonidae. *Trans. ent. Soc. Lond.* **1883**: 139–171.

*———, 1884. Ichneumons and their hosts. *Entomologist* **17**: 69–71.

*Briggs, C. A., 1888. The New Forest *Zygaena meliloti.* *Young Nat.* **9**: 82–83.

*Briggs, T. H., 1871. On the forms of *Zygaena trifolii*, with some remarks on the question of specific difference, as opposed to local or phytophagic variation, in that genus. *Trans. ent. Soc. Lond.* **1871**: 417–440.

*———, 1873. Note on the larva of *Zygaena meliloti.* *Entomologist's mon. Mag.* **10**: 116–117.

*Bright, P., 1910. [*Anthrocera filipendulae* with five wings.] *Entomologist* **44**: 44.

Bristowe, W. S., 1941. *The comity of spiders*, **2**: vii–xiv, 229–560 pp., text figs 16–96, pls 20–22. London.

———, 1958. *The world of spiders.* xiii, 304 pp., 36 pls, 116 text figs. London.

Buckler, W., 1887. *The larvae of the British butterflies and moths*, **2**: xi, 172 pp., pls 18–35. London.

Buckstone, A. W. W., 1909a. [Two males of *Anthrocera filipendulae* in copula with one female.] *Proc. S. Lond. ent. nat. Hist. Soc.* **1908–09**: 96.

———, 1909b. [Two males of *Anthrocera filipendulae* in copula with one female.] *Entomologist* **42**: 73.

———, 1926. Migration of insects. *Ibid.* **59**: 5–8.

Burgeff, H., 1956. Über die Modifizierbarkeit von Arten und geographischen Rassen der Gattung *Zygaena* (Lep.). *Nova Acta Leopoldina* (N.F.) **18**(127): 1–59, pls 1–6.

Burrows, C. R. N., 1916. The time of emergence of lepidopterous imagines. *Entomologist's Rec. J. Var.* **28**: 148–151.

Butler, A. G., 1869. Remarks upon certain caterpillars, &c., which are unpalatable to their enemies. *Trans. ent. Soc. Lond.* **1869**: 27–29.

Button, D. T., 1865. The larvae of *Zygaena filipendulae* a favourite food of the hoopoe. *Entomologist* **2**: 290.

Buxton, P. A., 1912. Various bionomical notes. *Entomologist's Rec. J. Var.* **24**: 244–246.

———, 1915. Fauna of Calday Island, Pembrokeshire. *Ibid.* **27**: 182–184.

———, 1916a. Pollination of orchids by insects. *Ibid.* **28**: 16.

———, 1916b. The hour of emergence of lepidopterous imagines. *Ibid.* **28**: 38.

Carpenter, G. D. H., 1937. A note on some parasites of *Zygaena* (Lep.). *J. Soc. Br. Ent.* **1**: 176–178.

*———, 1938. A note on some parasites of *Zygaena* (Lep.): addendum. *Ibid.* **1**: 216.

———, 1943. Birds as enemies of the larvae of *Zygaena filipendulae* L. (Lep.). *Entomologist's mon. Mag.* **79**: 157–159.

———, 1944. Natural selection in the Six-spot Burnet moth. *Nature, Lond.* **154**: 239–240.

———, 1945. Bionomic notes on a colony of *Zygaena filipendulae* L. (Lep.). *J. Soc. Br. Ent.* **2**: 280–284.

Chalmers-Hunt, J. M., 1957. Feral mating of *Zygaena lonicera* [*sic*] Esp. and *Z. filipendulae* L. *Entomologist's Rec. J. Var.* **69**: 199.

———, 1977a. Lepidoptera in Co. Mayo. *Ibid.* **89**: 347.

———, 1977b–78. *Lepidoptera of Kent.* Zygaenidae, pp. (182)–(188) (as supplement to *Entomologist's Rec. J. Var.*).

Chapman, T. A., 1893a. On some neglected points in the structure of the pupae of heterocerous Lepidoptera, and their probable value in classification; with some associated observations on larval prolegs. *Trans. ent. Soc. Lond.* **1893**: 97–119.

*———, 1893b. Second broods of *Vanessa io* and *atalanta*. *Entomologist's Rec. J. Var.* **4**: 242–243.

*———, 1911. On insect teratology (remarks to introduce a discussion on 'teratological specimens'). *Proc. S. Lond. ent. nat. Hist. Soc.* **1910–11**: 39–53, pls 1, 2.

*Christy, W. M., 1894. [*Zygaena trifolii* with the left hindwing replaced by another wing exactly similar to the forewing and the right hindwing absent.] *Entomologist's Rec. J. Var.* **5**: 217.

*———, 1895. Notes on the yellow and other varieties of *Zygaena trifolii*. *Entomologist* **28**: 214–215.

*Cockayne, E. A., 1908. *Anthrocera achilleae* Esp. added to the British list. *Entomologist's Rec. J. Var.* **20**: 73.

———, 1922. Structural abnormalities in Lepidoptera. *Lond. Nat.* **1921**: 10–69, pl. 1.

———, 1926. Homoeosis and heteromorphosis in insects. *Trans. ent. Soc. Lond.* **74**: 203–230, pls 61–64.

———, 1927. Extra wings in Lepidoptera. *Ibid.* **75**: 163–176, pls 17–19.

———, 1932. A holiday at Braemar. *Entomologist's Rec. J. Var.* **44**: 99–102.

———, 1935. The origin of gynandromorphs in the Lepidoptera from binucleate ova. *Trans. R. ent. Soc. Lond.* **83**: 509–521.

*———, 1951. *Zygaena achilleae* Esper ssp. *scotica* Rowland-Brown. *Entomologist's Rec. J. Var.* **63**: 143.

*———, 1954, Aberrations of British Macrolepidoptera. *Ibid.* **66**: 65–68, pl. 2, figs 1–11.

——— & Darlow, H. M., 1941. Wild hybrids of *Zygaena filipendulae* L. × *Z. lonicerae* Esp. *Ibid.* **53**: 113–114, pl. 6, figs 1–3.

——— & Hawkins, C. N., 1932. The early stages of *Procris globulariae* Hb., and of *P. cognata* H.-S. *Ibid.* **44**: 17–23.

Colthrup, C. W., 1918. Some field notes for 1916–17. *Ibid.* **30**: 50–54.

*Couldwell, C., 1919. Aberrations in Lepidoptera. *Naturalist, Hull* **1919**: 182.

Cowie, W., 1903. Macro-lepidoptera of Aberdeen and neighbourhood. *Trans. Aberd. wkg Men's nat. Hist. scient. Soc.* **1**: 20–35.

*Crocker, W., 1912a. [*Anthrocera filipendulae* with an under wing on left side in place of usual upper wing.] *Entomologist* **45**: 106.

*———, 1912b. *L. favicolor* and teratological *A. filipendulae*. *Entomologist's Rec. J. Var.* **24**: 132.

Curtis, W. P., 1917. The coloration problem. II. *Ibid.* **29**: 5–11, 33–38, 54–57, 76–82, 121–126, 145–150.

Dabrowski, J. S., 1963. Changes of design of the butterflies of the genus *Zygaena* Fabr. (Lepidoptera: Zygaenidae) obtained by intrachrysalid injections. *Folia biol. Kraków* **11**: 339–346, text figs 1–3, pl. 1.

Dabrowski, J. S., 1966. Changes of the wing pattern in the moths of the genus *Zygaena* Fabr. (Lepidoptera, Zygaenidae) obtained by intrapupal injections. *Acta ent. bohemoslovaca* **63**: 411–419, text figs 1–7, pls 1–3.

Darlow, H. M., 1947. Observations on variation and hybridisation in *Zygaena lonicerae* Esp. and *Zygaena filipendulae* L. (Lep.). *Entomologist's Rec. J. Var.* **59**: 133–136.

Davis, R. H. & Nahrstedt, A., 1979. Linamarin and lotaustralin as the source of cyanide in *Zygaena filipendulae* L. (Lepidoptera). *Comp. Biochem. Physiol.* **64**(B): 395–397.

—— & ——, 1982. Occurrence and variation of the cyanogenic glucosides linamarin and lotaustralin in species of the Zygaenidae (Insecta: Lepidoptera). *Ibid.* **71**(B): 329–332.

Decamps, C., du Merle, P., Gourio, C. & Luquet, G., 1981. Attraction d'espèces du genre *Zygaena* F. par des substances phéromonales de tordeuses (Lepidoptera, Zygaenidae et Tortricidae). *Annls Soc. ent. Fr.* (N.S.) **17**: 441–447, fig. 1, tables I–IV.

Dennis, R. L. H., 1977. *The British butterflies. Their origin and establishment*, xviii, 318 pp., 20 text figs, 15 tables. Faringdon, Oxon.

Dobson, A. H., 1955. A note from south Devon. *Entomologist's Rec. J. Var.* **67**: 276–277.

*Dodson, M., 1937. Development of the female genital ducts in *Zygaena* (Lepidoptera). *Proc. R. ent. Soc. Lond.* (A) **12**: 61–68, figs 1–5.

Döring, E., 1955. *Zur Morphologie der Schmetterlingseier.* 154 pp., 61 pls. Berlin.

*Druitt, A., 1933. Dipterous and hymenopterous parasites of *Zygaena trifolii* Esper (Lep.). *J. ent. Soc. S. Engl.* **1**: 65.

Dryja, A., 1959. *Badania nad polimorfizmem Kraśnika Zmiennego (Zygaena ephialtes L.)*, 403 pp., 7 pls. Warszawa.

*Duffield, C. A. W., 1961. The Burnet complex. *Entomologist's Rec. J. Var.* **73**: 25–28.

*Edelsten, H. M., 1933. A tachinid emerging from an adult moth. *Proc. R. ent. Soc. Lond.* **8**: 131.

*——, 1934. [*Phryxe vulgaris* emerging from an adult of *Zygaena lonicerae*.] *Proc. Trans. S. Lond. ent. nat. Hist. Soc.* **1933–34**: 48.

*——, 1947. *Zygaena achilleae* bred. *Entomologist* **80**: 48.

Edwards, W. & Towndrow, R. F., 1899. *The butterflies and moths of Malvern*, viii, 42 pp. Malvern.

Elliott, B., 1980. A further record of *Zygaena purpuralis caledonensis* Reiss (Lepidoptera: Zygaenidae) from south Argyll. *Entomologist's Gaz.* **31**: 1.

Emmet, A. M., 1968. Lepidoptera in west Galway. *Ibid.* **19**: 45–58.

*Esson, L. G., 1919. *Zygaena achilleae* in Argyllshire. *Entomologist* **52**: 189.

*Fairclough, R., 1981. A late date for *Adscita statices* (L.) (Lepidoptera: Zygaenidae). *Entomologist's Gaz.* **32**: 4.

Fassnidge, W., 1922. Hybridisation in nature. *Entomologist* **55**: 190.

*Fitch, E. A., 1880a. *Mesostenus obnoxius* Gr. *Ibid.* **13**: 17–18.

*——, 1880b. Hymenopterous parasites of Lepidoptera. *Ibid.* **13**: 67–69.

*——, 1883. Hymenopterous parasites of Lepidoptera. *Ibid.* **16**: 64–69.

*Fletcher, T. B., 1895. Moths with undeveloped wings. *Naturalist's J.* **4**: 163.

Fletcher, W. H. B., 1891. [Hybrids of *Zygaena lonicerae* × *Z. filipendulae*.] *Entomologist's mon. Mag.* **27**: 115.

——, 1893. Notes on some experiments in hybridising burnet moths (Zygaenae). *Ibid.* **29**: 53–54.

Ford, E. B., 1926. Zygaenidae attracted by the female of *Lasiocampa quercus* L. *Proc. ent. Soc. Lond.* **1**: 20–21.

——, 1955. *Moths*, xix, 266 pp., 32 col., 24 half-tone pls, 7 text figs, 12 distr. maps. London.

Ford, R. L. E., 1963. *British reptiles and amphibians*, 88 pp., 20 pls, text figs. London.

Franzl, S. & Naumann, C. M., 1984. Morphologie und Histologie der Wehrsekretbehälter erwachsener Raupen von *Zygaena trifolii* (Lepidoptera, Zygaenidae). *Ent. Abh. Mus. Tierk. Dresden* **48**: 1–12, figs 1–10.

*Frazer, J. F. D. & Rothschild, M., 1960. Defence mechanisms in warningly-coloured moths and other insects. *XI Int. Congr. Ent.* **3**: 249–256.

*Graham, M. W. R. de V., 1969. The Pteromalidae of northwestern Europe (Hymenoptera: Chalcidoidea). *Bull. Br. Mus. nat. Hist.* (Ent.) Suppl. **16**: 908 pp., 686 text figs.

Green, J. & Wilkinson, W., 1951. Mites on insects of Skokholm Island. *Entomologist's mon. Mag.* **87**: 143–146.

Greer, T., 1918. *Zygaena filipendulae* and *Z. lonicerae* hybrids. *Entomologists' Rec. J. Var.* **30**: 187–188.

——, 1920. Some Lepidoptera from east Tyrone in 1919. *Ibid.* **32**: 154–156.

*——, 1921. The Macro-lepidoptera of County Tyrone. *Entomologist* **54**: 282–285.

——, 1941. Random notes from east Tyrone, 1940. *Entomologist's Rec. J. Var.* **53**: 54–55.

Grosvenor, T. H. L., 1912. The season 1912. *Ibid.* **24**: 213–217.

*——, 1922. Zygaenidae hybrids. *Ibid.* **34**: 55.

———, 1923a. Notes on the genus *Zygaena*. *Proc. S. Lond. ent. nat. Hist. Soc.* **1922–23**: 64–72.

*———, 1923b. [*Zygaena filipendulae* with the right hindwing an almost exact duplication of the forewing.] *Entomologist* **56**: 265.

*———, 1923c. [*Zygaena filipendulae* with the right hindwing an almost exact duplication of the forewing.] *Entomologist's Rec. J. Var.* **35**: 184, 187.

———, 1926. Annual address to the members of the South London Entomological and Natural History Society. *Proc. S. Lond. ent. nat. Hist. Soc.* **1925–26**: 35–53.

———, 1927. Annual address to the members of the South London Entomological and Natural History Society. *Ibid.* **1926–27**: 88–97.

———, 1932a. [*Zygaena exulans* overwintering three times.] *Ibid.* **1931–32**: 62.

———, 1932b. [Genetics of the yellow form of *Zygaena filipendulae*.] *Ibid.* **1931–32**: 68.

———, 1933a. Annual address to the members of the South London Entomological and Natural History Society. *Trans. Proc. S. Lond. ent. nat. Hist. Soc.* **1932–33**: 61–69.

———, 1933b. [Genetics of the yellow form of *Zygaena filipendulae*.] *Ibid.* **1932–33**: 88–89, 109–110.

Hamilton, I. & Heath, J., 1976. Predation of *Pentatoma rufipes* (L.) (Hemiptera: Pentatomidae) upon *Zygaena filipendulae* (L.) (Lepidoptera: Zygaenidae). *Ir. Nat. J.* **18**: 337.

Hamm, A. H., 1899. Cross-pairing of *Anthrocera lonicerae* and *A. filipendulae*. *Entomologist's Rec. J. Var.* **11**: 269–270.

*———, 1942. Records of bred Tachinidae (Dipt.) chiefly from the Oxford district. *Entomologist's mon. Mag.* **78**: 191–192.

*Hammond, H. E. & Smith, K. G. V., 1955. On some parasitic Diptera and Hymenoptera bred from lepidopterous hosts. Part II. *Entomologist's Gaz.* **6**: 168–174, fig. 1.

*——— & ———, 1960. On some parasitic Diptera and Hymenoptera bred from lepidopterous hosts. Part IV. *Ibid.* **11**: 50–54.

Hancock, E. F. & Kydd, D. W., 1980. *Zygaena lonicerae* (Scheven) (Lepidoptera: Zygaenidae) in Cumbria. *Ibid.* **31**: 92.

Harker, G. A., 1891. Times of emergence. *Entomologist's Rec. J. Var.* **2**: 91.

Harper, M. W. & Langmaid, J. R., 1975. Lepidoptera in Perthshire and Inverness-shire, June 1974, including the rediscovery of *Ancylis tineana* Hübner (Lep.: Tortricidae). *Ibid.* **87**: 137–142.

Harris, M., 1766. *The Aurelian, a natural history of English moths and butterflies . . .*, 150 pp., 44 pls. London.

Harris, W. H. A., 1946. [Spirally segmented larvae of *Zygaena trifolii*.] *Entomologist* **79**: 248.

———, 1947. [Spirally segmented larvae of *Zygaena trifolii*.] *Proc. Trans. S. Lond. ent. nat. Hist. Soc.* **1946–47**: 13.

Harrison, J. W. H., 1940. The genus *Zygaena* in the Western Isles of Scotland. *Entomologist's Rec. J. Var.* **52**: 134–137.

———, 1945. Further observations on the genus *Zygaena* in the Inner and Outer Hebrides. *Ibid.* **57**: 25–27.

———, 1947. The Pleistocene races of certain British insects and distributional overlapping. *Ibid.* **59**: 141–145.

*———, 1949. The recurrence of *Zygaena lonicerae* Esp. in Durham. *Entomologist* **82**: 139–140.

———, 1953. Insects at the flowers of the pyramidal orchid (*Anacamptis pyramidalis*) and the fragrant orchid (*Gymnadenia conopsea*). *Ibid.* **86**: 66.

———, 1955. The Lepidoptera of the Lesser Skye Isles. *Entomologist's Rec. J. Var.* **67**: 141–147, 169–177.

Hayward, K. J., 1923. Hybridisation in nature. *Entomologist* **56**: 43.

Heath, J., 1981. Threatened Rhopalocera (butterflies) in Europe. *Nature Environ. Ser.* no. 23: vi, 157 pp.

Hewer, H. R., 1932. Studies in *Zygaena* (Lepidoptera). Part I. (A) the female genitalia; (B) the male genitalia. *Proc. zool. Soc. Lond.* **1932**: 33–75, figs 1–33.

———, 1934. Studies in *Zygaena* (Lepidoptera). Part II. The mechanism of copulation and the passage of the sperm in the female. *Ibid.* **1934**: 513–527, text figs 1–10, pls 1, 2.

*Hewett, W., 1889. *Zygaena lonicerae* var. *Entomologist* **22**: 73–74.

*———, 1890. Variation in *Zygaena lonicerae*. *Entomologist's Rec. J. Var.* **1**: 59–60.

Hinton, H. E., 1946a. On the homology and nomenclature of the setae of lepidopterous larvae, with some notes on the phylogeny of the Lepidoptera. *Trans. R. ent. Soc. Lond.* **97**: 1–37, figs 1–24, tables 1–6.

———, 1946b. A new classification of insect pupae. *Proc. zool. Soc. Lond.* **116**: 282–328, figs 1–64.

———, 1948. Sound production in lepidopterous pupae. *Entomologist* **81**: 254–269, tables 1–4, figs 1–8.

———, 1955. Protective devices of endopterygote pupae. *Trans. Soc. Br. Ent.* **12**: 49–92, figs 1–23.

———, 1981. *Biology of insect eggs*, **1**: 473 pp., 135 text figs, 155 pls. Oxford.

Hodge, H., 1915. *Zygaena filipendulae* and *Macrothylacia rubi* on the Hayle estuary. *Entomologist* **48**: 268.

———, 1925. Local distribution of *Zygaena filipendulae* in west Cornwall. *Ibid.* **58**: 272.

Hodge, H., 1926. *Zygaena filipendulae* at the Hayle estuary, Cornwall. *Ibid.* **59**: 320.

———, 1927. *Zygaena filipendulae* on the Hayle estuary. *Ibid.* **60**: 282.

Hodgson, S. B., 1945. *Zygaena lonicerae* pupae eaten by birds. *Ibid.* **78**: 176.

*Hyde, G. E., 1951. [*Zygaena filipendulae* with the left hindwing replaced by a wing of the forewing pattern.] *Proc. Trans. S. Lond. ent. nat. Hist. Soc.* **1949–50**: 31, 34, pl. 3, fig. F.

Jackson, R. A., 1959. Some observations on *Procris globulariae* (Hb.) the Scarce Forester: Lepidoptera (Zygaenidae). *Entomologist* **92**: 111–115.

*Jacobs, S. N. A., 1925. [*Zygaena filipendulae* with three fore-wings on the right side.] *Ibid.* **58**: 256.

*———, 1926. [A remarkable teratological example of *Zygaena filipendulae*.] *Proc. S. Lond. ent. nat. Hist. Soc.* **1925–26**: 73, pl. 4, fig.

James, R. E., 1912. Supplementary notes from Braemar. *Entomologist's Rec. J. Var.* **24**: 253–259.

*———, 1932. *Zygaena achilleae* in Argyllshire. *Entomologist* **65**: 224–226.

———, 1936. Further notes on *Zygaena achilleae* and *Z. filipendulae* in the western Highlands. *Ibid.* **69**: 271–273.

Jones, D. A., Parsons, J. & Rothschild, M., 1962. Release of hydrocyanic acid from crushed tissues of all stages in the life-cycle of species of the Zygaeninae (Lepidoptera). *Nature, Lond.* **193**: 52–53.

Kames, P., 1980. Das abdominale Duftorgan der Zygaenen-Männchen (Lepidoptera, Zygaenidae). Teil I: Freiland-beobachtungen, morphologische und histologische Untersuchungen an einigen europäischen Arten der Gattung *Zygaena* Fabricius, 1775. *Ent. Abh. Mus. Tierk. Dresden* **43**: 1–28, figs 1–33.

Kaye, W. J., 1903. Some notes on collecting Lepidoptera at Wye and Boxhill. *Entomologist's Rec. J. Var.* **15**: 306–308.

*Kirby, W. F., 1861. *Zygaena minos* in Scotland. *Zoologist* **19**: 7716.

*Klots, A. B., 1970. Lepidoptera, pp. 115–130, figs 143–154. *In* Tuxen, S. L. (Ed.), *Taxonomist's glossary of genitalia in insects*, 359 pp., 248 figs. Copenhagen.

Kuznetsov, N. Ya., 1967. *Fauna of Russia and adjacent countries*, Lepidoptera **1**: iv, 305 pp., 212 figs, 7 maps. Jerusalem [English translation].

Lane, C., 1959. A very toxic moth: The Five-spot Burnet (*Zygaena trifolii* Esp.). *Entomologist's mon. Mag.* **95**: 93–94.

———, 1961. Observations on colonies of the Narrow-bordered Five-spot Burnet (*Zygaena lonicerae* von Schev.) near Bicester. *Entomologist* **94**: 79–81.

*———, 1962. Differences in the egg-laying habits of the Five-spot Burnet (*Zygaena lonicerae* von Schev.) and the Six-spot Burnet (*Z. filipendulae* L.) (Lep., Zygaenidae). *Entomologist's Gaz.* **13**: 11–12, figs A, B.

Lang, H. C., 1878. *Zygaena filipendulae* double-brooded. *Entomologist* **11**: 69–70.

Lawless, E., 1872. Irish captures in 1870 and 1871. *Ibid.* **6**: 74–78, 97–100.

Leinfest, J., 1952. Über das Töten von Zygaenen. *Ent. Z., Frankf.a.M.* **62**: 131–132.

Lyle, G. T., 1912. New Forest notes, 1911. *Entomologist* **45**: 126–130.

*———, 1916. Contributions to our knowledge of the British Braconidae. *Ibid.* **49**: 228–232.

Marsh, N. & Rothschild, M., 1974. Aposematic and cryptic Lepidoptera tested on the mouse. *J. Zool., Lond.* **174**: 89–122, pl. 1.

Mathews (*sic*) [Mathew], G. F., 1859. Second brood of *Zygaena lonicerae*. *Zoologist* **17**: 6789.

Matsuda, R., 1976. *Morphology and evolution of the insect abdomen*, viii, 534 pp., 155 figs. Oxford.

Moffet, T., 1634. *Insectorum sive minimorum animalium theatrum . . .*, 18, 326 pp. Londini.

Moore, H., 1892. Retarded development. *Entomologist's Rec. J. Var.* **3**: 37.

*Morley, C. & Rait-Smith, W., 1933. The hymenopterous parasites of the British Lepidoptera. *Trans. R. ent. Soc. Lond.* **81**: 133–183.

*Mosley, S. L., 1896–97. An illustrated catalogue of varieties of British Lepidoptera. *Naturalist's J.* **6**, Supplement: 1–28, pls 3–16.

Mutuura, A., 1956. On the homology of the body areas in the thorax and abdomen and new system of the setae on the lepidopterous larvae. *Bull. Univ. Osaka Prefect.* (B)**6**: 93–122, pls 1–11.

Nash, W. G., 1918. Burnet pupae attacked by birds. *Entomologist* **51**: 215.

Naumann, C. M., 1977. Rasterelektronenoptische Untersuchungen zur Feinstruktur von Lepidopteren-Gespinsten. *Mitt. münch. ent. Ges.* **67**: 27–37, figs 1–13.

———, Richter, G. & Weber, U., 1983. Spezifität und Variabilität im *Zygaena-purpuralis*-Komplex (Lepidoptera, Zygaenidae). *Thes. zool.* **2**: 263 pp., 137 figs, 15 text figs, 4 distr. maps.

——— & Tremewan, W. G., 1980. On the biology of *Zygaena* (*Mesembrynus*) *tamara* Christoph, 1889 (Lepidoptera: Zygaenidae). *Entomologist's Gaz.* **31**: 113–121, pls 3–5.

Newman, E., 1872. [Cuckoos feeding on larvae of *Zygaena*.] *Entomologist* 6: 77 (footnote).

Newman, L. W., 1916a. [Genetics of orange and yellow forms of *Anthrocera filipendulae*.] *Entomologist's Rec. J. Var.* 28: 19–20.

———, 1916b. [Genetics of orange and yellow forms of *Zygaena filipendulae*.] *Proc. S. Lond. ent. nat. Hist. Soc.* 1915–16: 127–128.

Onslow, H., 1918. Hybrids of *Zygaena filipendulae* and of *Z. lonicerae*. *Entomologist's Rec. J. Var.* 30: 148–149.

Owen, D. F., 1954. A further analysis of the insect records from the London bombed sites. *Entomologist's Gaz.* 5: 51–60.

*Parsons, J. & Rothschild, M., 1964. Rhodanese in the larva and pupa of the Common Blue butterfly (*Polyommatus icarus* (Rott.)) (Lepidoptera). *Ibid.* 15: 58–59.

Pelham-Clinton, E. C., 1981. New Irish localities for *Zygaena purpuralis* (Brünnich) (Lepidoptera: Zygaenidae). *Ibid.* 32: 258.

*Perkins, V. R., 1880. *Zygaena filipendulae* and its parasites. *Entomologist* 13: 69.

Petiver, J., 1695–1703. *Musei Petiveriani*, 96 pp. Londini.

———, 1767. *Musei Petiveriani*, edn 2, 30 pp. London.

Pickett, C. P., 1903. Pupation of *Anthrocera filipendulae*. *Entomologist's Rec. J. Var.* 15: 268.

Povolný, D. & Weyda, F., 1981. On the glandular character of larval integument in the genus *Zygaena* (Lepidoptera: Zygaenidae). *Acta ent. bohemoslovaca* 78: 273–279, pls 1–4.

Proctor, M. & Yeo, P., 1973. *The pollination of flowers*, 418 pp., 4 col., 56 half-tone pls, 132 text figs. London.

*Reid, P. C., 1919. A hunt for *Zygaena achilleae*. *Entomologist* 52: 188–189.

Reiss, H. & Tremewan, W. G., 1967. A systematic catalogue of the genus *Zygaena* Fabricius (Lepidoptera: Zygaenidae). *Series ent.* 2: xvi, 329 pp.

Renou, M. & Decamps, C., 1982. Les phéromones sexuelles des Zygènes (Lepidoptera, Zygaenidae): données d'électroantennographie. *C.r. hebd. Séanc. Acad. Sci., Paris* (Sér. III, Sci. Vie) 295: 623–626.

*Richardson, N. M., 1889. Substitution of a wing for a leg in *Zygaena filipendulae*, and notes on the yellow variety of that species. *Entomologist's mon. Mag.* 25: 289–290.

*———, 1890. On a case of apparent substitution of a wing for a leg in a moth (*Zygaena filipendulae*). *Proc. Dorset nat. Hist. antiq. Fld Club* 11: 64–73, pl. [1], fig. 1.

*———, 1891a. [*Zygaena filipendulae* with five wings.] *Entomologist's Rec. J. Var.* 2: 20.

*———, 1891b. [*Zygaena filipendulae* with five wings.] *Proc. ent. Soc. Lond.* 1891: x.

Rocci, U., 1914. Sulla resistenza degli Zigenini all'acido cianidrico. *Z. allg. Physiol.* 16: 42–64.

———, 1915. Di una sostanza velenosa contenuta nelle Zigene. *Atti Soc. ligust. Sci. nat. geogr.* 26: 71–107.

———, 1916. Sur une substance vénéneuse contenue dans les Zygènes. *Archs ital. Biol.* 66: 73–96.

Rothschild, M., 1961. Defensive odours and Müllerian mimicry among insects. *Trans. R. ent. Soc. Lond.* 113: 101–121, figs 1–11, pls 1, 2.

———, 1964. An extension of Dr. Lincoln Brower's theory on bird predation and food specificity, together with some observations on bird memory in relation to aposematic colour patterns. *Entomologist* 97: 73–78.

*——— & Lane, C., 1964. [Larvae of *Zygaena* 2 years 9 months old.] *Proc. R. ent. Soc. Lond.* (C)29: 10–11.

*Rowland-Brown, H., 1919. *Anthrocera achilleae* Esper in Scotland. Notes on its distribution and variation. *Entomologist* 52: 217–226.

Samouelle, G., 1819. *The entomologist's useful compendium . . .*, 496 pp., 12 pls. London.

Seitz, A., 1907–13. Gattung: *Zygaena* F., pp. 18–34, pls 4–8. *In* Seitz, A., *Die Gross-Schmetterlinge der Erde*, 2: vii, 479 pp., 56 pls. Stuttgart.

Shaw, M. R. & Askew, R. R., 1976. Parasites, pp. 24–54, figs 8–10. *In* Heath, J. (Ed.), *The moths and butterflies of Great Britain and Ireland* 1: 343 pp., 13 pls, 85 text figs, 152 distr. maps. Colchester.

Sheldon, W. G., 1901. On a probable new locality for *Anthrocera exulans*. *Entomologist's Rec. J. Var.* 13: 136–137.

*———, 1908. The Scotch *Anthrocera achilleae*. *Ibid.* 20: 185.

*———, 1919. The re-discovery of *Anthrocera achilleae* in Scotland. *Entomologist* 52: 213–214.

Sheppard, P. M., 1964. Protective coloration in some British moths. *Ibid.* 97: 209–216, pl. 9.

Showler, A. J., 1955. A week in Scotland, and notes on *Zygaena exulans* Hoch. [*sic*]. *Entomologist's Rec. J. Var.* 67: 316–317.

Singh, B., 1951. Immature stages of Indian Lepidoptera No. 8 – Geometridae. *Indian Forest Rec.* (N.S., Ent.) 8: 67–158, pls 1–10.

Smith, S. G., 1947. Notes on assembling of *Lasiocampa quercus* Linn. race *callunae*. *Rep. Proc. Chester Soc. nat. Sci.* 1947: 66–68.

*South, R., 1887. [An apparently apterous specimen of *Zygaena filipendulae* L.] *Abstr. Proc. S. Lond. ent. nat. Hist. Soc.* 1887: 79.

*———, 1894. Abnormal example of *Zygaena trifolii*. *Entomologist* 27: 253, fig.

South, R., 1897. *Zygaena filipendulae* var. *hippocrepidis*. *Ibid.* **30**: 181–183.

———, 1904a. Cross-pairing of *Zygaena trifolii* and *Z. filipendulae*. *Ibid.* **37**: 15–16.

———, 1904b. [Cross-pairing of *Zygaena trifolii* and *Z. filipendulae*.] *Proc. S. Lond. ent. nat. Hist. Soc.* **1903**: 64–66.

———, 1905. [Cross-pairing of *Zygaena trifolii* and *Z. filipendulae*.] *Entomologist* **38**: 118.

———, 1907. [Hybrid from *Anthrocera filipendulae* ♀ × *A. trifolii* ♂.] *Proc. S. Lond. ent. nat. Hist. Soc.* **1905–06**: 70–71.

Stubbs, A. E., 1982. Conservation and the future for the field entomologist. *Proc. Trans. Br. ent. nat. Hist. Soc.* **15**: 55–66, figs 1–3.

Sutton, G. P., 1922. Zygaenidae attracted by *Lasiocampa quercus* ♀, etc. *Entomologist* **55**: 280.

*Symes, H., 1958. The life history of *Procris globulariae* Hb. *Entomologist's Rec. J. Var.* **70**: 279–281.

*Talbot, G., 1921. [*Z. filipendulae* with five wings.] *Ibid.* **33**: 139.

Thomas, J. A., 1983. The ecology and conservation of *Lysandra bellargus* (Lepidoptera: Lycaenidae) in Britain. *J. appl. Ecol.* **20**: 59–83, figs 1–14.

Towndrow, R. F., 1871. *Zygaena filipendulae* in October. *Entomologist* **5**: 443.

*Treat, A. E., 1975. *Mites of moths and butterflies*, 362 pp., frontispiece, 150 text figs. Ithaca & London.

Tremewan, W. G., 1954. Notes on variation in *Zygaena trifolii* Esper, *Z. trifolii* ssp. *palustris* Oberthür, and *Z. filipendulae* Linnaeus. *Entomologist's Rec. J. Var.* **66**: 233–234.

*———, 1958. Notes on the British species of the genus *Zygaena* Fabricius. *Entomologist's Gaz.* **9**: 187–196.

*———, 1959. *Procris globulariae* Hübner: an historical note and the provision of a neotype. *Entomologist* **92**: 116–119, pl. 6, figs 1–5.

*———, 1960. Additional notes on the British species of the genus *Zygaena* Fabricius (Lep., Zygaenidae). *Entomologist's Gaz.* **11**: 185–194.

*———, 1961a. The British species of the genus *Procris* Fabricius (Lep., Zygaenidae). *Ibid.* **12**: 19–23, text figs 1–12.

———, 1961b. The Burnet complex – a reply. *Entomologist's Rec. J. Var.* **73**: 110–113.

———, 1961c. The aberrations of the British species of the genus *Zygaena* Fabricius (Lepidoptera: Zygaenidae). *Coridon* (A) no. 1: 1–10, pls C1–C3.

*———, 1961d. A catalogue of the types and other specimens in the British Museum (Natural History) of the genus *Zygaena* Fabricius, Lepidoptera: Zygaenidae. *Bull. Br. Mus. nat. Hist.* (Ent.) **10**: 239–313, pls 50–64.

*———, 1962. A new subspecies of *Zygaena lonicerae* Scheven (Lep., Zygaenidae) from Scotland. *Entomologist's Gaz.* **13**: 10–11.

———, 1965a. A note on rearing *Zygaena lonicerae jocelynae* Tremewan (Lep., Zygaenidae). *Ibid.* **16**: 87–88.

———, 1965b. Collecting *Zygaena* Fabricius (Lep., Zygaenidae) in Scotland in 1965. *Ibid.* **16**: 119–124.

———, 1966. The history of *Zygaena viciae anglica* Reiss (Lep., Zygaenidae) in the New Forest. *Ibid.* **17**: 187–211, text fig. 1, pl. 3, figs 1, 2.

*———, 1967. *Zygaena viciae* Denis & Schiffermüller (Lep., Zygaenidae) in west Scotland, with the description of a new subspecies. *Ibid.* **18**: 159–160, pl. 7, figs 1–24.

———, 1968a. Collecting *Zygaena* Fabricius (Lep., Zygaenidae) in Scotland in 1967. *Ibid.* **19**: 3–8, pl. 1, figs 1, 2, pl. 2, figs 1–4.

———, 1968b. The history of *Zygaena loti scotica* (Rowland-Brown) (Lep., Zygaenidae) in west Scotland. *Ibid.* **19**: 203–218, text fig. 1.

———, 1969. A new record of *Zygaena purpuralis caledonensis* Reiss (Lep., Zygaenidae) from Argyll. *Ibid.* **20**: 2.

*———, 1970. On *Adscita statices* (Linnaeus) (Lep., Zygaenidae, Procridinae). *Ibid.* **21**: 156–157.

*———, 1972. Late Lepidoptera in Cornwall in 1972. *Ibid.* **23**: 226.

———, 1973. *Zygaena filipendulae* (Linnaeus) (Lep., Zygaenidae) in September. *Ibid.* **24**: 54.

———, 1975. On *Zygaena* Fabricius (Lep., Zygaenidae) from Iran. *Ibid.* **26**: 229–248, pl. 6, figs 1–12.

———, 1976. Nomenclatural notes on British Zygaenidae (Lepidoptera). *Ibid.* **27**: 149–154.

*———, 1977. Protracted emergence of *Zygaena trifolii palustrella* Verity (Lep., Zygaenidae). *Ibid.* **28**: 214.

———, 1980a. On the status of *Zygaena* (*Zygaena*) *trifolii decreta* Verity (Lepidoptera: Zygaenidae) in south-east England. *Ibid.* **31**: 143–145.

———, 1980b. *Panorpa communis* L. (Mecoptera: Panorpidae) feeding on *Zygaena lonicerae* (Scheven) (Lepidoptera: Zygaenidae). *Ibid.* **31**: 274.

———, 1982a. *Zygaena* (*Zygaena*) *trifolii decreta* Verity in southeast England, with records of a new host-plant of *Z.* (*Z.*) *lonicerae* (Scheven) (Lepidoptera: Zygaenidae). *Ibid.* **33**: 9–11.

———, 1982b. An unusual host-plant of *Zygaena* (*Zygaena*) *filipendulae* (L.) (Lepidoptera: Zygaenidae). *Ibid.* **33**: 12.

——— & Carter, D. J., 1967. Collecting *Zygaena* Fabricius (Lep., Zygaenidae) in Scotland in 1966. *Ibid.* **18**: 3–9, text fig. 1, pl. 1, figs 1, 2, pl. 2, figs 1–4.

———— & Classey, E. W., 1970. Collecting *Zygaena* Fabricius (Lep., Zygaenidae) in Scotland and Yorkshire in 1968. *Ibid.* **21**: 65–72, text fig. 1, pl. 1, figs 1–9.

———— & Manley, W. B. L., 1964. Collecting *Zygaena* Fabricius (Lep., Zygaenidae) in Scotland in 1963. *Entomologist's Rec. J. Var.* **76**: 149–153.

Tugwell, W. H., 1886. In search of *Zygaena exulans*. *Entomologist* **19**: 217–223.

*————, 1895. On *Zygaena exulans* and var. *subochracea* White. *Ibid.* **28**: 8–11.

*Tutt, J. W., 1890. *Zygaena lonicerae* imago with head of larva. *Entomologist's Rec. J. Var.* **1**: 174.

————, 1899. *A natural history of the British Lepidoptera*, **1**: 560 pp. London.

————, 1906. *Ibid.*, **5**: xiii, 558 pp. London.

*————, 1908. *Anthrocera achilleae* Esp. as a British species. *Entomologist's Rec. J. Var.* **20**: 73–74.

*Van Emden, F. I., 1954. Diptera: Cyclorrhapha. Calyptrata (I). Section (a). Tachinidae and Calliphoridae. *Handbk Ident. Br. Insects* **10**(4a): 133 pp.

*Wainwright, C. J., 1940. The British Tachinidae (Diptera): second supplement. *Trans. R. ent. Soc. Lond.* **90**: 411–448, figs 1–11.

*Walker, J. J., 1907. Some notes on the Lepidoptera of the 'Dale Collection' of British insects, now in the Oxford University Museum. *Entomologist's mon. Mag.* **43**: 154–158.

Warren, R. G., 1983. *Zygaena lonicerae transferens* Verity (Lepidoptera: Zygaenidae) in Staffordshire. *Entomologist's Gaz.* **34**: 3.

*Watkins, C. J., 1902. Collecting in the Cotteswolds in 1902. *Entomologist's Rec. J. Var.* **14**: 349–350.

Webb, S., 1896. Protracted pupal period of *Papilio machaon*, *Zygaena trifolii* and *Z. filipendulae*. *Ibid.* **7**: 255.

Weiser, J., 1951. Příspěvek k poznání plísní cizopasících v hmyzu. *Ent. Listy* **14**: 130–135, figs 1–5.

Wheeler, A. S., 1970. *Zygaena filipendulae* (Linnaeus) (Lep., Zygaenidae) attracted to *Lasiocampa quercus* (Linnaeus) (Lep., Lasiocampidae). *Entomologist's Gaz.* **21**: 26.

Wheeler, G., 1918. Paired Lepidoptera in flight. *Entomologist's Rec. J. Var.* **30**: 152–153.

White, F. B., 1871. Capture of a *Zygaena* new to the British lists. *Entomologist's mon. Mag.* **8**: 68.

*————, 1872. The Scottish form of *Zygaena exulans* Hochenwarth [*sic*]. *Scott. Nat.* **1**: 174–175.

Wiegel, K.-H., 1958. Die Nikotintötungsmethode und die Behandlung von Lepidopteren, insbesondere Zygaenen, beim Sammeln. *NachrBl. bayer. Ent.* **7**: 35–38, 45–47.

Wigglesworth, V. B., 1972. *The principles of insect physiology*, edn 7. viii, 827 pp., 412 text figs. London.

Wilkes, B., [1749]. *The English moths and butterflies* . . ., 8, [22], 63, [4] pp., 120 pls. London.

Williams, B. S., 1914. An irregular pairing in nature. *Entomologist's Rec. J. Var.* **26**: 185.

Wiltshire, E. P., 1968. Studies in the geography of Lepidoptera, VIII. Notes on the ecology and distribution of Zygaenidae in the Middle East. *Proc. Trans. Br. ent. nat. Hist. Soc.* **1**: 47–54, distr. map.

Woodbridge, F. C., 1934. *Zygaena achilleae*. *Entomologist* **67**: 238–239.

*Wormell, P., 1983. Lepidoptera in the Inner Hebrides. *Proc. R. Soc. Edinb.* (B)**83**: 531–546.

LIMACODIDAE
B. Skinner

A family of world-wide distribution, formerly also known as Heterogeneidae or Cochlididae, with 15 species in the Palaearctic region placed in six genera of which two occur in England. The family, in spite of its systematic position, is included among the macrolepidoptera.

Moths of this family may be recognized by their stout and rather short hairy bodies and broad wings, superficially resembling Bombycoidea. The British species *Heterogenea asella* ([Denis & Schiffermüller]) is one of the smallest in the family and is an exception in its general appearance, and indeed the British species are hardly typical of the family, a great variety of form and colour being found in the tropics. Their colouring is cryptic, some sombre, but many species are bright green and brown. The size extends to about 80mm wingspan. They are mainly forest insects and largely nocturnal.

Imago. Antenna bipectinate in male; haustellum very small or absent; maxillary palpus one- to three-segmented or absent; labial palpus short, two- or three-segmented. Wings broad; venation (figure 59) of forewing with median vein present in discal cell, veins 9 (R_3), 8 (R_4) and 7 (R_5) stalked, 1a present; hindwing with median vein present in discal cell, vein 8 $(Sc+R_1)$ fused with 7 (R_5) near base or connected to 7 (R_5) by 8 (R_1), 1c $(1A)$ present.

Larva. Head retractile; antenna long; thoracic legs reduced; prolegs absent; abdominal segments with ventral suckers or lateral ridges. The larvae superficially resemble those of Lycaenidae. They feed on the upper surface of leaves of trees and adhere, either by suction or by adhesive silk, so tightly to the leaf that they can rarely be dislodged by beating. They feed exposed but are cryptically coloured.

Pupa. In an oval or pyriform papery cocoon, with a hinged lid at one end which opens on eclosion of the moth.

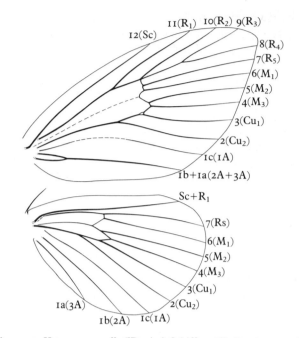

Figure 59 *Heterogenea asella* ([Denis & Schiffermüller]), wing venation

Key to species (imagines) of the Limacodidae

1   Forewing dark blackish brown .... *Heterogenea asella* ♂ (p. 126)

–   Forewing ochreous 2

2(1)   Forewing dark ochreous brown without lines .... *H. asella* ♀ (p. 126)

–   Forewing yellow-ochreous, with darker lines at one-half and three-quarters .... *Apoda limacodes* (p. 125)

APODA Haworth

Apoda Haworth, 1809, *Lepid.Br.*: 137.
Cochlidium Hübner, 1822, *Syst.-alph.Verz.*: 58.

Occurs in the Palaearctic region where it is represented by three species; West Africa and North America.

Imago. Labial palpus porrect. Hindtibia with two pairs of spurs. Thorax, abdomen and femora setose. Forewing with vein 10 (R_2) separate; hindwing with vein 4 (M_3) and 5 (M_2) approximated, 6 (M_1) and 7 (Rs) connate, 8 ($Sc+R_1$) anastomosing with cell from near base to middle (figure 59).

APODA LIMACODES (Hufnagel)
The Festoon

Phalaena limacodes Hufnagel, 1766, *Berlin.Mag.* **3**: 402.
Apoda avellana (Linnaeus) sensu auctt.
Bombyx testudo [Denis & Schiffermüller], 1775, *Schmett. Wien.*: 65.
Phalaena funalis Donovan, 1794, *Br.Ins.* **3**: 76.
Type locality: Germany; Berlin.

Description of imago (Pl.3, figs 11–14)
Wingspan of male 24–28mm, of female 27–32mm. Antenna of male weakly pectinate, that of female simple. Head, thorax and abdomen ochreous brown. Forewing broad, in male ochreous brown shaded with darker brown and with blackish brown cilia; in female pale ochreous brown with darker brown cilia. Both sexes have two dark brown transverse lines, one extending inwards from middle of costa to dorsum, the other outwards from costa to termen just above the tornus. Hindwing of male purplish brown with blackish brown cilia; of female pale ochreous brown with brownish cilia tipped with white.

Ab. *ochracea* Seitz is a rare form in which the male has the paler coloration of the female; it has been recorded from the New Forest, Hampshire and Chattenden, Kent. In ab. *suffusa* Seitz (fig.12) the forewing of the male is more or less suffused with russet-black; it has been noted in the New Forest and the Ashford district of Kent. Ab. *assella* Esper (fig.13) is a rare and more extreme melanic form of the male and has occurred at Liphook, Hampshire and Chattenden, Kent. Ab. *maculata* Seitz is another scarce form of the male in which the dark ochreous brown colour of the forewing is irregularly spotted with yellow; it has been noted from Pamber Forest and the New Forest, Hampshire, and north Kent.

Life history
Ovum. Ovoid flattened, with lozenge-shaped reticula-

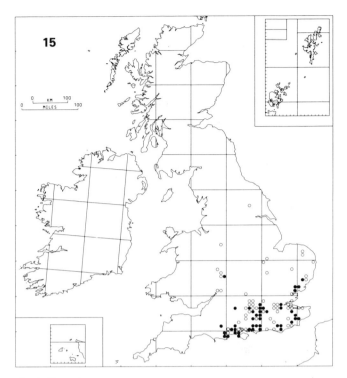

Apoda limacodes

tions; pale yellow, almost transparent when first laid, becoming grey prior to hatching. Laid at the angle of the veins on the underside of a leaf of the foodplant (Tutt, 1899). Duration of this stage is about 14 days.

Larva. Full-fed *c*.14mm long. Head small, retractile, whitish green. Body woodlouse-shaped, bright green, covered with small yellowish spots dorsally and laterally; subdorsal line pale yellow edged with purplish red spots. Thoracic legs white; prolegs absent, being replaced by minute sucker-like structures.

It feeds from late July to early October on oak (*Quercus* spp.) and beech (*Fagus sylvatica*), and overwinters in a silken cocoon in which it pupates the following June.

Pupa. *c*.6mm long; thick and stumpy, broadest about middle of abdomen; creamy white with head and thorax tinged with light brown; dorsal surface of abdominal segments 2–8 each with a broad, transverse, yellowish band; and the whole surface covered with minute yellowish brown spines (Tutt, *loc.cit.*). The cocoon is reddish brown, oval, and surrounded by flossy silk. It is spun on top of a leaf which falls to the ground in autumn. The pupal stage lasts approximately 15 days.

Imago. Univoltine, occurring from mid-June to late July. Both sexes have been noted flying in hot sunshine, usually high up among beech and oak trees. In dull weather it may occasionally be dislodged from the lower branches of its foodplant, but in recent years the species is more usually encountered after dark when both sexes, especially the male, come to light.

Distribution (Map 15)
An inhabitant of mature beech and oak woodland, found locally in the southern English counties from Kent to Dorset, and extremely locally in Worcestershire, Gloucestershire, Nottinghamshire, Buckinghamshire, Oxfordshire, Essex and Suffolk. Eurasiatic. In Europe widely distributed from Spain to southern Fennoscandia; Asia Minor.

HETEROGENEA Knoch
Heterogenea Knoch, 1783, *Beitr.Insektengesch.* **3**: 60.

Occurs in the Palaearctic region where it is represented by three species; North America.

Imago. Labial palpus ascending. Hindtibia with middle spurs absent, apical spurs long. Thorax, abdomen and femora not setose. Forewing with vein 10 (R_2) from 9 (R_3) near base. Hindwing vein 5 (M_2) from transverse vein, more or less parallel to 4 (M_3), veins 6 (M_1) and 7 (Rs) remote at base, 8 ($Sc+R_1$) connected to middle of cell (figure 59).

HETEROGENEA ASELLA ([Denis & Schiffermüller])
The Triangle
Bombyx asella [Denis & Schiffermüller], 1775, *Schmett. Wien.*: 65.
Phalaena (Heterogenea) cruciata Knoch, 1783, *Beitr.Insektengesch.* **3**: 60.
Type locality: [Austria]; Vienna district.

Description of imago (Pl.3, figs 8–10)
Wingspan of male 14–19mm, of female 16–22mm. Antenna of both sexes short, simple. Head, thorax and abdomen of male dark purplish brown; of female light ochreous brown. Forewing triangular-shaped; male glossy unicolorous blackish brown; female unicolorous ochreous brown. Hindwing of both sexes blackish brown with ochreous brown cilia.
  Variation is confined to the ground colour which in the male may be almost black tinged with purple (ab. *nigra* Tutt (fig.9)); or in the female pale ochreous yellow (ab. *flavescens* Tutt).
Similar species. Owing to its small size this species may be overlooked by macrolepidopterists. On the infrequent occasions when it comes to light it may be recognized on the sheet by its broad triangular forewings, quite unlike those of a tortricid, and by its rather head-down attitude with somewhat deflexed wings.

Life history
Ovum. Ovoid, flattened, covered with lozenge-shaped reticulations. Colourless and almost transparent when first laid, becoming pale yellowish brown prior to hatching. In captivity the eggs are laid in small agglomerate batches.
Larva. Full-fed *c.*12mm long. Head very small, smooth, glossy, retractile, pale yellowish green. Body woodlouse-shaped, ground colour yellowish green; ventral surface pale flesh-pink tinged with green; dorsal surface with a broad olive-brown diamond-shaped mark extending towards both head and anal segment. Thoracic legs minute

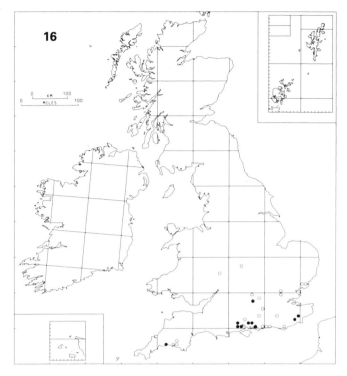

Heterogenea asella

(Hufnagel), have been at light, to which both sexes come sparingly, especially on very warm nights.

Distribution (Map 16)

A very local and rather scarce woodland species occurring in south-east Kent, Sussex and south Hampshire, the New Forest being a well-known locality. Elsewhere single colonies exist in south Oxfordshire, Wiltshire and Buckinghamshire, and in the past it has occurred in Essex, Surrey, Warwickshire, Worcestershire and south Devon. Single specimens taken near Looe, Cornwall, on 26 June 1960 and 4 July 1969 constitute the only West Country records this century. Eurasiatic. In Europe widely distributed from France to southern Fennoscandia and the U.S.S.R.

References

Barrett, C. G., 1895. *The Lepidoptera of the British Islands*, **2**. London.

Forster, W. & Wohlfahrt, T. A., 1960. *Die Schmetterlinge Mitteleuropas* **3**. *Spinner und Schwärmer (Bombyces und Sphinges)*, vii, 239 pp., 28 col.pls. Stuttgart.

Lanktree, P. A. D., 1960. Lepidoptera: some unusual larval foodplants. *Entomologist's Rec. J. Var.* **72**: 187–190.

Tutt, J. W., 1899. *A natural history of the British Lepidoptera*, **1**, 560 pp. London.

and hardly perceptible; on margin of venter is a soft projecting ridge of flexible skin which appears to serve the purpose of legs and propels the larva forward with an undulatory motion (Buckler *in* Barrett, 1895).

It feeds from August to October on the leaves of oak (*Quercus* spp.) or beech (*Fagus sylvatica*), overwintering in a silken cocoon in which it pupates the following June. There is an account of the larva being found, but not reared, on poplar in West Sussex (Lanktree, 1960). Abroad it is reputed to feed also on hornbeam (*Carpinus betulus*), birch (*Betula* spp.), hazel (*Corylus avellana*) or lime (*Tilia* spp.).

Pupa. *c*.5mm long. Short, stumpy, with abdomen bending under, giving it a rather rounded form; wings and appendages not attached to abdominal segments beyond the second. Surface of pupa smooth, polished, transparent whitish brown (Tutt, 1899). The cocoon is described by Tutt (*loc.cit.*) as a quarter of an inch long, dark brown marbled with pale grey. It is either on a leaf or in the fork of a twig, and Forster & Wohlfahrt (1960) state that on the lower branches of beech, before the leaves fall, cocoons in the forks of twigs are not difficult to find.

Imago. Univoltine, appearing from mid-June to late July. Although it is reputed to fly in sunshine, especially in the afternoon, most recent captures, as in *Apoda limacodes*

PSYCHIDAE
P. Hättenschwiler

A large family of highly specialized moths. Their distribution is world-wide, extending from northern Finland to the southern tip of South America; the majority, however, occur in the warmer regions. In Britain there are 20 species representing three subfamilies; two of these species were each recorded only once in doubtful circumstances.

The family comprises primitive to highly developed species, this being especially pronounced and visible in the females. Only two British species have fully winged females capable of flight, and in the remainder the females are apterous, some also lacking antennae and legs. Three species are parthenogenetic in Britain, though males of two of these are known in central Europe and the third may be a parthenogenetic race of a bisexual species found in southern England and the Channel Islands.

Imago. Neither sex can take any food. The maxillary palpi and haustellum are drastically reduced to a point of invisibility or nearly so and the lifetime of this stage is therefore brief, lasting from a few hours to rarely more than a day for males though up to two weeks for females. Mating normally takes place in or on the case of the female at a time of day which is constant for each species. Many of the males fly by day, often in sunshine.

The development of a new generation of females from unfertilized eggs is not uncommon in this family. There are three parthenogenetic species known in the British Isles, *Dahlica triquetrella* (Hübner), *D. lichenella* (Linnaeus) and *Luffia ferchaultella* (Stephens). Although males of these *Dahlica* species have never been found in this country, they are known from central Europe. It is not known for certain whether *Luffia ferchaultella* is a parthenogenetic form of *L. lapidella* (Goeze) or a distinct species.

Ovum. The stage usually lasts from three to five weeks. Except in the species with winged females, the eggs are laid in the larval case which was also used for pupation; in some species they are even laid within the pupal shell.

Larva. Immediately on hatching, the young caterpillar begins the construction of the case or bag in which it will remain until the adult stage. Some species begin the work even inside the parent female's case; others do this later, but they all construct a miniature case prior to taking any food. During growth the larva repeatedly works on the case; much time is spent on repairs and even more on enlargement. The larval case has a silken, cocoon-like interior which is wide enough for the larva to turn round completely. The exterior is decorated with various materials such as sand, plant matter, often from the foodplant,

or particles of dead insects; dry material seems to be preferred. The shape of the case, the material used and the manner in which it is attached are characteristic for each species. The cases offer very good camouflage and protection against enemies and weather conditions. The case of a psychid can generally be distinguished from that of a coleophorid by its exterior decoration. A coleophorid case is constructed in a section of mined leaf, in a seed-head or directly of silk and has no other plant material attached to it except in a few species where there is a thin sprinkling of seeds from the heads in which they are feeding. Some species of the Tineidae also construct cases to which they attach hair, feathers, textile fragments or sand. Whereas the cases of the Psychidae and Coleophoridae have the fore and rear ends differently constructed, tineid cases are more or less identical at each end and the larvae often change the direction in which they operate.

The larval period generally lasts almost one year but may extend to two or three years in a minority of species. The case has an opening at either end. The larva protrudes through the opening at the fore end for feeding and crawling, which it does with its three pairs of thoracic legs, pulling the case behind it (figure 60). It also fixes the fore end of the case to a tree-trunk, rock or other surface; it does this for moults, normally five or more in number, for overwintering and pupation. The opening at the rear end serves for the purposes of excretion and emergence of the adult. The larva continuously closes the openings not in use with silken threads.

Figure 60 Psychid larva, protruding from larval case

Pupa. Pupation takes place within the larval case with the head towards the rear opening. Behaviour at the time of emergence differs between groups and is described in the introductions to subfamilies.

NOTE. There are certain deviations in this work from the classification given by Kloet & Hincks (1972). The Solenobiinae and Taleporiinae are here treated as a single subfamily under the latter name. The genera *Proutia* Tutt and *Whittleia* Tutt are reinstated and *Dahlica* Enderlein is introduced for the species formerly listed under *Solenobia* Duponchel; there are also two changes in the order of genera. Supposed British specimens of *Bankesia conspurcatella* (Zeller) were misidentified and are referred to *douglasii* Stainton which is now placed in the genus *Bankesia* Tutt.

Leraut (1984) proposes a number of systematic and nomenclatural changes, some of which are controversial. These have not been adopted here.

Key to species (male and winged female imagines) of the Psychidae

1	Antenna simple or ciliate	2
–	Antenna bipectinate	6
2(1)	Ocelli absent; foretibia without spur .... *Dahlica inconspicuella* (p. 134)	
–	Ocelli present and/or foretibia with spur (spur or epiphysis may be very small and may be obscured by scales)	3
3(2)	Antenna fasciculate (figure 66a, p. 137)	4
–	Antenna ciliate in rings round the base of each segment (figure 66b, p. 137) .... *Bankesia douglasii* (p. 137)	
4(3)	Antenna with uniform brownish scales	5
–	Antenna with light and dark scales forming annulations .... *Narycia monilifera* (p. 132)	
5(4)	Hindwing with scales narrower or equal in width to those on forewing .... *Taleporia tubulosa* (p. 136)	
–	Hindwing with scales one and one-half times to twice as wide as those on forewing .... *Diplodoma herminata* (p. 131)	
6(1)	Forewing with 1a (3A) and 1b (2A + 1A) anastomosing and then separating again to reach margin individually (figure 61a)	7
–	Forewing with 1a (3A) and 1b (2A + 1A) anastomosing and remaining united (figure 61b)	9
7(6)	Hindwing with vein 6 (M₁) absent	8
–	Hindwing with vein 6 (M₁) present .... *Sterrhopterix fusca* (p. 150)	
8(7)	Body clothed with black hair; wingspan 16–22mm .... *Acanthopsyche atra* (p. 147)	
–	Body clothed with long grey hair; wingspan 22–28mm .... *Pachythelia villosella* (p. 148)	
9(6)	Wings covered with fine, hair-like scales; moths small, wingspan 9–16mm	10
–	Wings covered with broad scales	11

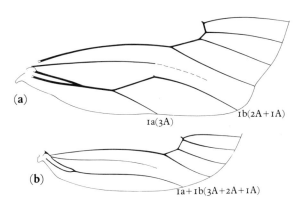

Figure 61 Anal area of forewings
(**a**) *Pachythelia villosella* (Ochsenheimer) (detail of figure 75)
(**b**) *Psyche casta* (Pallas) (detail of figure 71)

10(9)	Wings and body black ... *Epichnopterix plumella* (p. 145)	
–	Wings white with grey, undulating striations .... *Whittleia retiella* (p. 146)	
11(9)	Wings and body irrorate pale grey; wingspan 11–17mm	12
–	Wings and body brown, unpatterned; wingspan 7–12mm .... *Luffia lapidella* (p. 139)	
12(11)	Pectinations of antenna lacking scales (figures 68a,b, p. 141)	13
–	Pectinations of antenna scaled (figure 68c, p. 141) 14	
13(12)	Antenna with pectinations short, approximately one to one and one-half times length of one antennal segment (figure 68a, p. 141) .... *Bacotia sepium* (p. 141)	
–	Antenna with pectinations long, approximately two to four times length of one antennal segment (figure 68b, p. 141) .... *Proutia betulina* (p. 142)	
14(12)	Male antenna with 18–21 segments; foretibia with spur index 0.76–0.82 (figure 69c, p. 142); all wings uniform brown .... *Psyche casta* (p. 143)	
–	Male antenna with 21–25 segments; foretibia with spur index 0.65–0.72 (figure 69c, p. 142); forewing brown, hindwing grey-brown .... *P. crassiorella* (p. 144)	

Key to the cases of full-grown larvae of the Psychidae

Cases vary individually to some extent; therefore a comparison with the respective figures should be made. Special attention is necessary for the cases with plant matter attached irregularly (see couplets 9–11); here individual examples may occur with the plant matter arranged longitudinally. Cases of earlier instars may appear quite different from those of full-grown larvae and may lead to erroneous determination.

1   Case covered with sand or lichen 2
–   Case covered with plant matter other than lichen 9

2(1)   Case with cross-section round or almost round 3
–   Case with cross-section triangular 4

3(2)   Case almost cylindrical with rear end rounded; occasionally decorated with small fragments of bark or moss; length 6–8mm (Pl.7, figs 40,41)   *Bacotia sepium* (p. 141)
–   Case tapered (conical); length 5–7mm (Pl.7, figs 42–44) .... *Luffia lapidella* (p. 139)/*L. ferchaultella* (p. 140)

4(2)   Case tapered at both ends; length 5–9mm 5
–   Case not tapered; length 14–20mm (Pl. 7, fig.39) .... *Taleporia tubulosa* (p. 136)

5(4)   Case light in colour, covered with sand; fore end decorated with parts of dead insects; length 6–8mm (Pl. 7, fig.34) .... *Dahlica triquetrella* (p. 134)
–   Case dark in colour, covered with sand and lichen; length 5–9mm 6

The cases of the following four species cannot be separated with certainty:

6(5)   Case almost completely covered with lichen, often green; length 5–6mm (Pl.7, fig.33). Larva with head black and thoracic plates dark brown .... *Narycia monilifera* (p. 132)
–   Case covered mainly with sand 7

7(6)   Case with lichen but sand dominant, dark grey to dark green; length 5–7mm (Pl.7, fig.37). Larva with head black and body dark grey ..... *Dahlica lichenella* (p. 135)
–   Case with practically no lichen 8

8(7)   Case dark, almost completely covered with sand and very fine particles of bark; length 5–6mm (Pl.7, fig.35). Larva with head black and body yellowish .... *D. inconspicuella* (p. 134)
–   Case covered with sand, often also decorated with fragments of leaves and bark; length 6–8mm (Pl.7, fig.36). Larva with head brown and body whitish .... *Bankesia douglasii* (p. 137)

9(1)   Case with plant matter attached irregularly 10
–   Case with plant matter attached longitudinally 12

10(9)   Case with cross-section round or nearly so 11
–   Case with cross-section triangular, consisting of a soft outer and hard inner case; length 8–12mm (Pl.7, fig.38) .... *Diplodoma herminata* (p. 131)

11(10)   Case length 8–10mm, diameter 2.5–3.0mm (Pl.7, figs 49,50) .... *Proutia betulina* (p. 142)
–   Case length 13–16mm, diameter 5–7mm (Pl.7, figs 61,62) .... *Sterrhopterix fusca* (p. 150)

12(9)   Plant matter same length as case or slightly longer; total length usually under 15mm 13
–   Plant matter shorter than total length of case; length of case 20mm or more 16

13(12)   Case with flat grass attached over its entire length 14
–   Case with grass stems attached at fore end only, remainder hanging free, splayed outwards at rear end 15

14(13)   Length of case 8–12mm, diameter 2–3mm (Pl.7, figs 53,54) .... *Epichnopterix plumella* (p. 145)
–   Length of case 6–8mm, diameter 1.5–2.0mm (Pl.7, figs 51,52) .... *Whittleia retiella* (p. 146)

The cases of the two following species cannot be separated with certainty, though on average they differ in length:

15(13)   Case length 8–12mm (Pl.7, figs 45,46) .... *Psyche casta* (p. 143)
–   Case length 10–15mm (Pl.7, figs 47,48) .... *P. crassiorella* (p. 144)

16(12)   Case covered mainly with grass-stems; length 18–20mm or less (Pl.7, figs 55,56) ...... *Acanthopsyche atra* (p. 147)
–   Case covered with pieces of heather, leaves, grass-stems or small twigs; female case covered for entire length, male in fore half only, leaving silken tube visible in rear half; length 35–50mm (Pl.7, figs 57,58) .... *Pachythelia villosella* (p. 148)

Taleporiinae

The two subfamilies Solenobiinae and Taleporiinae of Kloet & Hincks (1972) are treated here as a single subfamily under the name Taleporiinae, which is divided into three tribes with characteristics as follows.

NARYCIINI (*Diplodoma* Zeller and *Narycia* Stephens)

The females are fully developed with wings, legs and antennae. The larva pupates in its case, the period spent in this stage varying considerably. At the time of emergence, the pupa manoeuvres itself to the rear opening and protrudes approximately half-way; eclosion takes place in this position. The ova are laid on bark and not in the case.

SOLENOBIINI (*Dahlica* Enderlein)

Wings in female reduced to almost invisible stumps; legs and antenna fully developed. The larva pupates in its case, the duration of the pupal stage being one to three weeks. Eclosion as in the Naryciini. Copulation takes place on the outside of the case. The eggs are then laid inside the case, covered with abdominal hairs from the female.

TALEPORIINI (*Taleporia* Hübner and *Bankesia* Tutt)

In this tribe, as with Solenobiini, females are almost wingless but have functional legs and antennae. The behaviour, pupation, emergence and mating are as described under Solenobiini. The eggs are laid within the case. *Bankesia* species attach the case for pupation in the autumn and emerge in the spring, *Taleporia* species pupate in spring and emerge two to three weeks later.

NARYCIINI

DIPLODOMA Zeller

Diplodoma Zeller, 1852, *Linn.ent.* **7**: 332, 359–362.

A small genus with two species in the Palaearctic region; one of these occurs in the British Isles.

Imago. Head rough-scaled; antenna less than half length of forewing, ciliate; ocelli present; labial palpus with three segments and hair-tuft. Foretibia with one spur; midtibia with apical spurs and hindtibia with medial and apical spurs in pairs. Wings fully developed in both sexes, with broad scales; ground colour dark fuscous, spotted with yellowish markings.

Larva. Feeds on lichen and decayed leaves from a double case consisting of a hard inner element with a loose, soft outer cover.

DIPLODOMA HERMINATA (Geoffroy)

Tinea herminata Geoffroy, 1785, *in* Fourcroy, *Ent.Paris* **2**: 322.
Lampronia marginepunctella Stephens, 1835, *Ill.Br.Ent.* (Haust.) **4**: 358.

NOMENCLATURE. Leraut (1984) has proposed introducing the name *laichartingella* Goeze, 1783 as a replacement name for *herminata* Geoffroy, 1785 (see also *Antenna* **9**: 49), but it is not adopted here as in our opinion the identity of the species named by Goeze has not been established.
Type locality: France; Paris.

Description of imago (Pl.7, figs 9,10)

Wingspan 10–15mm. Head pale yellowish ochreous; antenna ciliate. All wings broad-scaled. Forewing brownish fuscous with scattered pale yellow spots and strigulae; a subquadrate antemedian whitish yellow spot on dorsum. Hindwing dark greyish brown.

Life history

Ovum. Laid singly in cracks of bark.

Larva (Pl.7, fig.38). Head pale brown. Body dull whitish; prothoracic plate brown; thoracic segments brownish-tinged laterally. Solitary, feeding on lichen growing on wood, fences or rocks, also on decaying plant matter and dead insects, particularly in spiders' webs under bark and in hollow trees. Case triangular in cross-section, having a hard inner case and a soft outer case covered with fragments of plants, insects, small stones and a variety of other material; length 10–13mm, tapered at both ends. September to May.

Pupa. In the case attached to a solid surface at a height up to 1m above ground level. May.

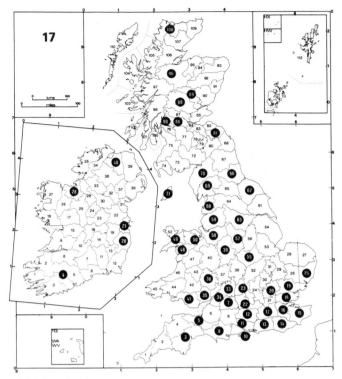

Diplodoma herminata

Imago. Univoltine, emerging in May and June. In colder climates a generation may take two or three years.

Distribution (Map 17)

Locally not uncommon in wooded areas, throughout most of the British Isles. Europe, extending eastwards to the U.S.S.R. and the Middle East.

NARYCIA Stephens

Narycia Stephens, 1836, *Nom.Br.Insects* (2): 118.

A genus containing two European species, one of which occurs in Britain.

Imago. Antenna ciliate; ocelli absent; labial palpus with three segments. Foretibia without spurs but with a long tuft of dark scales, mistakenly described as a spur by some authors; midtibia with apical spurs, hindtibia with medial and apical spurs. Wings developed in both sexes, fully scaled.

NARYCIA MONILIFERA (Geoffroy)

Tinea monilifera Geoffroy, 1785, *in* Fourcroy, *Ent.Paris* 2: 325.

Tinea melanella Haworth, 1828, *Lepid.Br.*: 566.

NOMENCLATURE. Leraut (1984) has proposed introducing the name *duplicella* Goeze, 1783 as a replacement name for *monilifera* Geoffroy, 1785 but it is not adopted here as in our opinion the identity of the species named by Goeze has not been established.

Type locality: France; Paris.

Description of imago (Pl.7, figs 1,2)

Wingspan in male 9–12mm, in female 7–10mm. Head blackish; antenna half length of forewing, with 26–30 segments, ciliate, white with black annulations. All wings broad-scaled. Forewing blackish fuscous with scattered pale yellowish spots, the more prominent of which form an ill-defined antemedian fascia and small costal and dorsal postmedian spots, more clearly defined in male. Hindwing dark grey. Abdomen in female with dark anal tuft.

Life history

Ovum. Laid in cracks in bark.

Larva (Pl.7, fig.33). Head black. Body yellowish white; pro- and mesothoracic plates dark brown. Solitary, feeding on algae growing on trees, fences and other surfaces; case lying along the feeding surface, triangular in cross-section, tapered at both ends and covered with sand and algae, varying in colour according to the materials used; length 5–6mm, width 1.5–2.0mm. August to May, overwintering half-fed.

Pupa. In the case attached to a tree-trunk, from ground level up to 2m. May to June.

Imago. Univoltine, flying in June and July. Often rests on tree-trunks.

Distribution (Map 18)

Common in southern England and Wales, becoming more local northwards to southern Scotland. Europe, occurring more plentifully in southern countries.

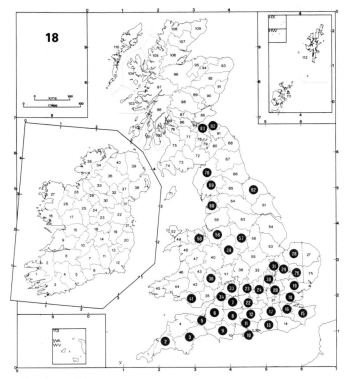

Narycia monilifera

SOLENOBIINI

DAHLICA Enderlein

Dahlica Enderlein, 1912, *Zool.Anz.* **40**: 264.

Brevantennia Sieder, 1953, *Z.wien.ent.Ges.* **38**: 120.

Solenobia sensu auctt.

A genus containing approximately 25 European species, three of which are found in Britain. All have apterous females and some species are parthenogenetic.

Imago.   Male. Antenna shortly ciliate (figure 62); ocelli absent; labial palpus with only two segments. Foretibia without spurs, midtibia with apical spurs, hindtibia with medial and apical spurs. Wings long and slender, fully scaled, dark brown or grey with paler pattern which varies considerably. Venation somewhat variable. Forewing with nine veins emanating from cell, accessory cell usually present. Hindwing mostly with five veins from cell (figure 63).   *Female.* Apterous. Antenna simple, number of segments varying from 12 to 26; labial palpus reduced to one minute segment. Legs short; number of spurs and tarsal segments variable, normally one or two very small spurs on mid- and/or hindtibia. Ovipositor extensile.

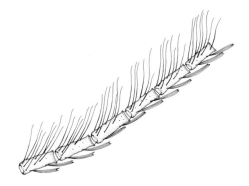

Figure 62 *Dahlica* spp., part of antenna

British authors have held confusing opinions over the parthenogenetic species. Ford (1946) refers to *Dahlica lichenella* (Linnaeus) as having a case measuring 8–9mm in length and occurring locally in southern England; this description, however, applies to *D. triquetrella* (Hübner). Ford then writes with reference to *D. inconspicuella* (Stainton), 'A parthenogenetic form of this species occurs and sometimes the cases of this form can be found in large numbers'. Meyrick (1928) was also of the opinion that *D. inconspicuella* could include a parthenogenetic form. A comparison made by Sauter (1956) of the specimens collected by Ford, however, showed that these belonged to *D. lichenella* and have no connection with *D. inconspicuella*. Based on these observations, the following three species are known to occur in Great Britain and Ireland: *D. inconspicuella*, which is bisexual; *D. triquetrella* and *D. lichenella*, both in their parthenogenetic form only (Sauter, *loc.cit.*). The distribution of these species is imperfectly known because of misidentifications.

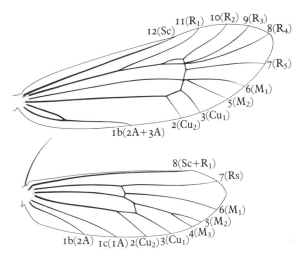

Figure 63 *Dahlica inconspicuella* (Stainton), wing venation

DAHLICA TRIQUETRELLA (Hübner)

Tinea triquetrella Hübner, [1813], *Samml.eur.Schmett.* **5**: pl.55, fig.373.

Type locality: [Europe].

Description of imago (Pl.7, fig.7)

In Great Britain and Ireland only the parthenogenetic form of this species has been recorded; therefore the description of the male is omitted. *Female.* Antenna with 15–24 segments, simple; eyes small, ocelli absent. All legs with five tarsal segments. Body yellow; abdomen with brown dorsal plates and smaller divided ventral plates; abdominal segment 8 with a ventral field of spines (figure 64a), their shape differing from those of *D. lichenella* (Linnaeus); a ventral tuft of hairs which become detached when the eggs are laid within the case, serving to protect and secure them; these long, wavy hairs each have what appears to be a small knob at the end, visible only under high magnification, the similar hairs of *D. lichenella* being without this knob.

Life history

Ovum. Laid in the larval case and covered with hairs from the ventral tuft.

Larva (Pl.7, fig.34). Head brown. Body pale yellow; prothoracic plate brown. Feeds on lichens, dead insects and decaying plants, being found on tree-trunks in woods, on rocks and in other places where lichen or moss is available. Case prostrate, triangular in cross-section and tapered at both ends, decorated with sand, frass and parts of dead insects; length 7–9mm, width 3mm. April or May to February, overwintering full-fed.

Pupa. In the larval case. On emergence a section of the exuviae covering the head and legs is detached. This section readily distinguishes the females of the parthenogenetic species; in *D. triquetrella* the neck section is not indented (figure 65).

Imago. During March or April the adult, still inside the pupa, manoeuvres itself to the posterior opening of the larval case and protrudes half-way, in which position emergence takes place during the early morning. After a short while the adult commences laying its eggs inside the case.

Distribution (Map 19)

The distribution in Britain is imperfectly known owing to confusion with related species (see p. 133); reliably recorded from Kent and Westmorland (Cumbria) but records from south Essex, Cheshire, south Lancashire and Co. Durham lack confirmation; possibly widespread at low density. Found in the greater part of Europe as far north as Finland. Males occur only in those areas which were not glaciated during the last Ice Age.

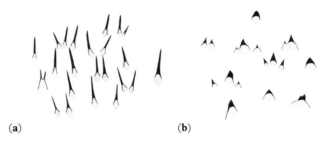

Figure 64 Anal spines of imagines
(a) *Dahlica triquetrella* (Hübner)
(b) *Dahlica lichenella* (Linnaeus)

DAHLICA INCONSPICUELLA (Stainton)

Lesser Lichen Case-bearer

Taleporia inconspicuella Stainton, 1849, *Syst.Cat.Br. Tineidae and Pterophoridae*: 6.

Type locality: England.

Description of imago (Pl.7, figs 3,4)

Male. Wingspan 9–13mm. Antenna with 26–31 segments, ciliate (figure 62, p. 133); eyes small, distance apart nearly twice diameter of eye. Wings long and slender, venation as in figure 63, p. 133. Forewing broad-scaled, whitish grey with scattered darker spots. Hindwing more narrowly scaled, grey. *Female.* Apterous. Antenna with 12–18 segments, simple; eyes small. Legs with four tarsal segments. Body yellowish with brown plates on head and all segments dorsally; anal hair-tuft whitish, composed of simple hairs.

Life history

Ovum. Laid in the case.

Larva (Pl.7, fig.35). Head blackish. Body yellowish grey; prothoracic plate and pairs of spots on meso- and metathorax blackish. Feeds on lichens, from a prostrate case which is triangular in cross-section, tapered at both ends and coated with lichen, very small particles of wood and fine sand; length 5–6mm, width 1.5–2.0mm. June to March, overwintering full-fed.

Pupa. In the case affixed to the bark of trees, rocks or fences. March to April.

Imago. The adults emerge in March and April, the males nocturnally, the females in the early morning.

Distribution (Map 20)

Local throughout England as far north as Yorkshire. Unknown elsewhere; supposed specimens from Europe were found upon examination to belong to species of the *D. nickerlii* group.

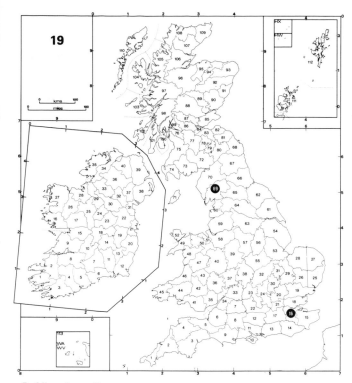

Dahlica triquetrella

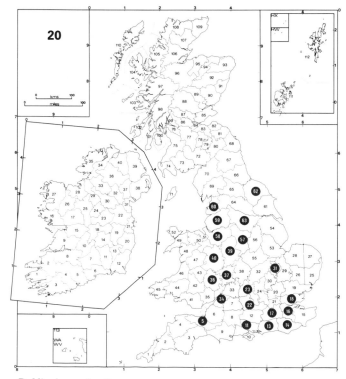

Dahlica inconspicuella

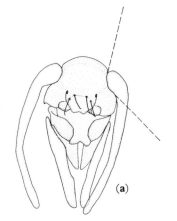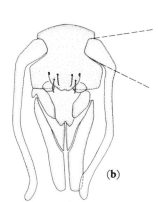

Figure 65 Headplates of exuviae
(**a**) *Dahlica triquetrella* (Hübner)
(**b**) *Dahlica lichenella* (Linnaeus)

DAHLICA LICHENELLA (Linnaeus)
Lichen Case-bearer
Phalaena (Tinea) lichenella Linnaeus, 1761, *Fauna Suecica* (Edn 2): 370.
Type locality: Sweden.

Description of imago (Pl.7, fig.8)
In Great Britain only the parthenogenetic form of this species has been recorded; so the description of the male is omitted. *Female.* Antenna with 14–19 segments. Legs with four, less often three, tarsal segments. Spines of ventral field of abdominal segment 8 less pointed (figure 64b, p. 134) than those of *D. triquetrella* (Hübner); ventral tuft with simple hairs. Otherwise similar to *D. triquetrella*.

Life history
Ovum. Laid in the case and covered with hairs from the ventral tuft.

Larva (Pl.7, fig.37). Head black. Body dark grey; thoracic plates black. Feeds on lichen, moss and decayed plant matter. Case prostrate, tapered at both ends, normally covered with lichen, minute fragments of wood and frass, the colour varying considerably depending on the material used; length 5–7mm, width 1.5–2.5mm. June to March, overwintering full-fed.

Pupa. In the case attached to tree-trunks, fences or rocks up to 2m above ground level. Frontal section of pupal exuviae (*cf. D. triquetrella*) with neck section indented (figure 65b,). March to April.

Imago. Emergence takes place during the early morning in March or April.

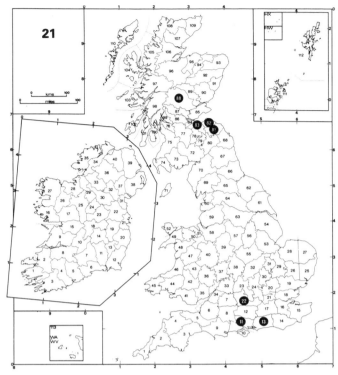

Dahlica lichenella

Distribution (Map 21)

Known in Britain only from southern England, southern Scotland and Perthshire; probably overlooked elsewhere. The parthenogenetic form is found throughout Europe.

TALEPORIINI

TALEPORIA Hübner

Taleporia Hübner, [1825], *Verz.bekannt.Schmett.*: 400.

Solenobia Duponchel, 1843, *in* Godart & Duponchel, *Hist.nat.Lépid.Fr.* Suppl. **4**: 197.

There are probably six species in Europe, one of which occurs in Britain.

Imago. Male. Wings fully developed and fully scaled; all veins present, 7 (R_5) and 8 (R_4) stalked. Antenna shortly ciliate; ocelli present; labial palpus with three segments. Foretibia with one spur, midtibia with apical spurs and hindtibia with medial and apical spurs. *Female.* Apterous. Antenna simple; eyes developed; labial palpus reduced to one segment. Legs fully developed but without spurs. Mating takes place on the outside of the case.

TALEPORIA TUBULOSA (Retzius)

Tinea tubulosa Retzius, 1783, *in* DeGeer, *Gen.et Spec.Ins.*: 44.

Type locality: not stated.

Description of imago (Pl.7, figs 11,12)

Male. Wingspan 15–20mm. Antenna with 40–45 segments, shortly ciliate (figure 66a); eyes large. Forewing rather long and slender with accessory cell and often intercalary cell present; dark brown with more or less pronounced yellowish white spots giving a reticulate appearance. Hindwing dark grey. *Female.* Apterous. Antenna with 27–32 segments; eyes large, ocelli absent. Legs short. Body pale brown with a tuft of greyish white scales on the ventral surface of abdominal segment 7. Ovipositor extensile.

Life history

Ovum. Laid within the case after mating has taken place on its external surface. Hatches in 4–5 weeks.

Larva (Pl.7, fig.39). Head blackish brown. Body pale yellowish brown; thoracic segments each with a dark brown dorsal plate. Lichen and decaying plants are taken for food but other insects are not disdained. Case triangular in cross-section, constructed of silk coated with lichen, fine sand and bark and, at the fore end, particles of dead insects; length 14–20mm, width approximately 2mm. August to April but sometimes living two years, depending on climatic conditions.

Pupa. In the case affixed to tree-trunks, fences, stones or other firm surfaces, up to 2m above ground level. April to May.

Imago. The adults emerge in late May and June. Mating takes place in the early morning on the outside of the case.

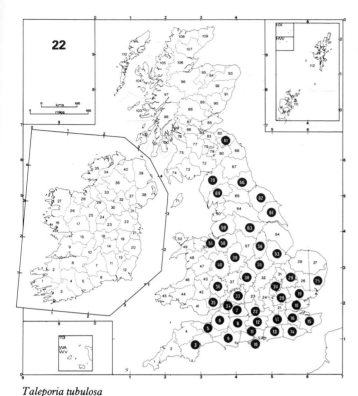

Taleporia tubulosa

Distribution (Map 22)

Inhabits woodland where there is little undergrowth, rocks, old walls and, less often, meadows. Locally common in Britain as far north as southern Scotland. Throughout Europe from Spain northwards to Finland.

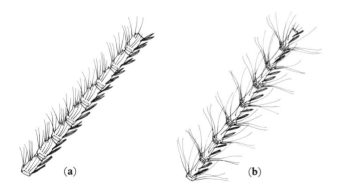

Figure 66 Part of antennae
(**a**) *Taleporia tubulosa* (Retzius)
(**b**) *Bankesia* spp.

BANKESIA Tutt

Bankesia Tutt, 1899, *Entomologist's Rec.J.Var.* **11**: 191.

A genus which includes several species in Europe and one in Britain.

Imago. *Male.* Antenna with cilia arranged in rings round the base of each segment (figure 66b); eyes large, ocelli present; labial palpus very long with three segments. Foretibia with minute spur, midtibia with apical spurs and hindtibia with apical and medial spurs. Wings long and slender. Forewing with ten veins emanating from cell and with accessory cell (figure 67). *Female.* Apterous. Antenna short with only 5–6 segments; eyes small, ocelli absent; labial palpus reduced to a single segment. Legs without spurs, tarsal segments reduced.

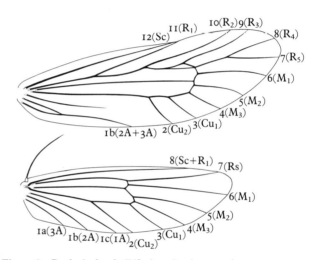

Figure 67 *Bankesia douglasii* (Stainton), wing venation

BANKESIA DOUGLASII (Stainton)

Solenobia douglasii Stainton, 1854, *Ins.Br.Lepid.*: 19.
Talaeporia (Bankesia) staintoni Walsingham, 1899, *Entomologist's Rec.J.Var.* **11**: 258.
Bankesia conspurcatella sensu auctt.

Type locality: England; Birch Wood, Kent.

NOTE ON SYNONYMY. *Bankesia douglasii* was named from a single male taken at the type locality and long supposed to be unique. The specimen is in BMNH. Kloet & Hincks (1972) listed this species in *Solenobia* Duponchel.

A psychid occurring beside Southampton Water, Hampshire, has hitherto been referred by authors to *Bankesia conspurcatella* (Zeller). The latter is a species found in Italy, southern France and Spain but not in Britain. Walsingham (1899) described '*Bankesia staintoni* sp.n. (= *conspurcatella*, Stn., nec Z.)' and below on p. 258

writes 'Mr. Tutt now informs me that the Brussels Speci-mens are most certainly the same species as our British *conspurcatella*'. Bradley (1966) disagreed that this was a separate species from *conspurcatella* and returned *staintoni* to the synonymy of *conspurcatella*, but his opinion is not accepted in this work.

Description of imago (Pl.7, fig.6)

Male. Wingspan 11–15mm. Antenna more than half length of forewing with 30–32 segments, each segment with a ring of cilia at its base (figure 66b, p. 137); eyes large, distance apart twice diameter of eye. Forewing long and slender with very broad scales; veins 7 (R_5) and 8 (R_4) stalked (figure 67, p. 137); brownish yellow with dark reddish brown spots in middle of dorsum, at end of cell and on posterior half of costa. Hindwing with six veins emanating from cell, pale grey, scales narrow. *Female*. Apterous. Eyes much smaller than in male. Abdominal segment 7 with a ventral tuft of greyish scales.

Life history

Ovum. Laid in the case, hatching in 4–5 weeks.

Larva (Pl.7, fig.36). Head brown. Body whitish; thoracic plates dark fuscous; first abdominal segment pale fuscous. Case triangular in cross-section, tapered at both ends, covered with sand, soil and lichen; length 6–8mm, width 2.5–3.0mm. April to September.

Pupa. In the case attached to cracks of bark, rocks or fences. October to March.

Imago. Univoltine. The males emerge nocturnally, the females in the early morning. March to April.

Distribution (Map 23)

Very local in southern England where it has been found in Hampshire and Kent – where it was rediscovered in 1984 (N. F. Heal, pers.comm.), and Worcestershire. The Channel Islands. Western Europe in Belgium, Holland and northern France.

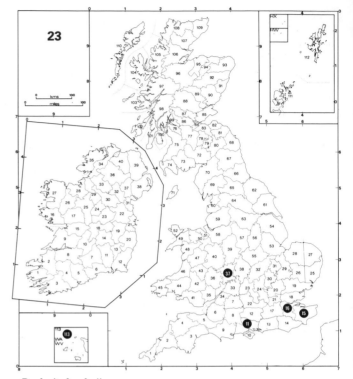

Bankesia douglasii

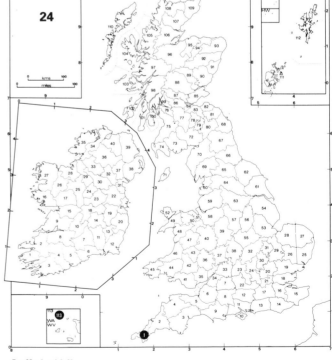

Luffia lapidella

Psychinae

The genera are arranged somewhat differently from those in Kloet & Hincks (1972). *Luffia* Tutt is placed before *Bacotia* Tutt; *Proutia* Tutt and *Whittleia* Tutt are reinstated as distinct genera.

The subfamily is divided into two tribes with characteristics as follows.

PSYCHINI (*Luffia* Tutt, *Bacotia* Tutt, *Proutia* Tutt and *Psyche* Schrank)

The females have legs and antennae but no wings. Emergence takes place within the case and the pupal exuviae do not protrude. The mature female moves through the rear opening to the outside of the case to await copulation. The ova are then laid within the case.

EPICHNOPTERIGINI (*Epichnopterix* Hübner and *Whittleia* Tutt)

The females have no wings, legs or antennae. The development of the adult female takes place within the case. When ready for copulation, the female is positioned still partly within the pupal shell with the head protruding through the rear opening of the case. The male is attracted to the receptive female by scent and settles at the rear end of the case. Rupture of the female pupal shell occurs at the anterior end. The extensile abdomen of the male must pass between the pupal case and the female's body and then extend the entire length of the body before coming into contact with the genitalia. The ova are then laid in the pupa shell.

PSYCHINI

LUFFIA Tutt

Luffia Tutt, 1899, *Entomologist's Rec.J.Var.* **11**: 191.

A genus with three or four species in Europe, two of which occur in the British Isles. The species are either bisexual with apterous females or parthenogenetic.

Imago. *Male.* Wingspan 7–12mm. Antenna with 18–24 segments, shortly bipectinate; eyes relatively large, ocelli absent; labial palpus reduced, with a tuft of hair. Legs with one spur on foretibia, an apical pair on midtibia and medial and apical pairs on hindtibia. *Female.* Apterous. Antenna with 6–10 segments, simple; ocelli absent. Legs with tarsi reduced to 1–4 segments, the number differing between species.

Early stages. The cases are round in cross-section and tapered. Pupation and emergence take place within the case, the exuviae remaining within. The female then moves through the rear opening to the outside of the case for mating but lays the eggs within the case.

LUFFIA LAPIDELLA (Goeze)

Tinea lapidella Goeze, 1783, *Ent.Beyträge* **3**(4): 168.
Type locality: Europe.

Description of imago (Pl.7, fig.15)

Male. Wingspan 7–12mm. Antenna with 18–24 segments, shortly bipectinate; eyes large, distance apart approximately diameter of eye. Forewing broad-scaled, whitish with darker grey pattern; cilia nearly white. Hindwing with scales approximately half width of those of forewing; light grey. *Female.* Apterous. Antenna with 7–9 segments, simple; eyes small. Legs with 3–4 tarsal segments. Body yellow; head and all segments with brownish dorsal plates divided on abdominal segments; each abdominal segment with a ring of scales; anal hair-tuft yellowish white.

Life history

Ovum. Laid in the case.

Larva (Pl.7, figs 42,43). Head black. Body greyish yellow; thoracic segment 1 black, 2 and 3 each with a dark plate divided by a yellow spot; abdominal segment 1 with a shining yellow dorsal plate. Case round in cross-section, conical, covered with lichen and sand; length 5–7mm, width 2–3mm. Lichen is taken for food, mainly when growing on rocks but also on wood, preferably in dry, sunny places. August to May, overwintering as a larva.

Pupa. In the case attached to the rock or trunk where the larva was feeding. May to June.

Imago. Univoltine. The moths emerge nocturnally in June or July and mate in the early morning.

Distribution (Map 24)

Cornwall (Smith, 1983); the Channel Islands; otherwise not recorded in Britain. In Europe occurs in France, Spain, Italy, southern Switzerland and Yugoslavia, its range overlapping with that of the following species (Narbel-Hofstetter, 1964).

LUFFIA FERCHAULTELLA (Stephens)

Talaeporia ferchaultella Stephens, 1850, *Zoologist* **8** (Appx): 109.

Type locality: England; Camberwell, Surrey (now Greater London).

NOTE ON STATUS. Of this species only the parthenogenetic wingless female is known. It is not certain what status should be given to *L. ferchaultella*; here it is treated as a distinct species but possibly it is a form of *L. lapidella* (Goeze).

Description of imago

Female. Antenna with 6–8 segments. Legs with 1–3 tarsal segments. Otherwise resembles *L. lapidella*.

Life history

Ovum. Laid in the case.

Larva (Pl.7, fig.44). As described for *L. lapidella*. Case also similar, its colour varying from green to black, depending on the lichen used; often there are rings of different colours. More often found on trees or wood, but also found on rocks; prefers shady places with high humidity. Lichen is taken for food. Sometimes found in vast numbers on tree-trunks and fences. August to May.

Pupa. In the case attached to the trunk, fence or rock where the larva fed. May to June.

Imago. The adult emerges in the early morning and immediately starts egg-laying.

Distribution (Map 25)

Widespread and locally abundant in the southern counties of England; Wales; Isle of Man; eastern Ireland. In Europe it occurs in Belgium, the Netherlands, and north-western Germany, and in France along the Channel and Mediterranean coasts.

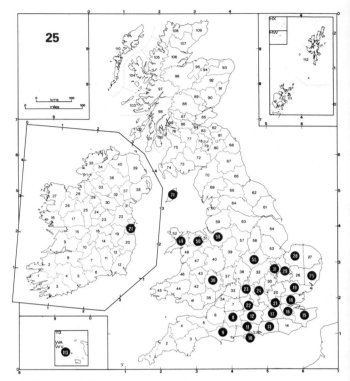

Luffia ferchaultella

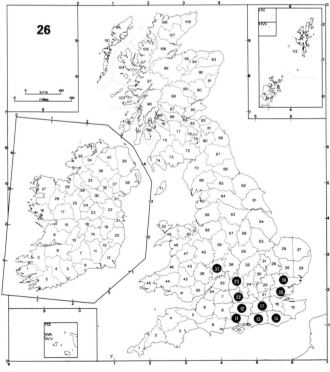

Bacotia sepium

BACOTIA Tutt

Bacotia Tutt, 1899, *Entomologist's Rec.J.Var.* **11**: 207.

A genus with only one species in Europe occurring also in England; two more are known from Nepal and Japan.

Imago.   Male. Wings fully developed. Antenna bipectinate, the pectinations devoid of scales (figure 68a); ocelli absent; labial palpus reduced. Foretibia with one long spur (figure 69a, p. 142), midtibia with apical spurs, hindtibia with apical and medial spurs. Forewing slender with acute apex, accessory cell present; scales broad. Hindwing with narrower scales.   *Female.* Apterous. Antenna simple. Legs with number of tarsal segments reduced.

BACOTIA SEPIUM (Speyer)

Psyche sepium Speyer, 1846, *Isis, Leipzig* **1846**: 31.

NOMENCLATURE. Leraut (1984) has proposed introducing the name *claustrella* Bruand, 1845 as a replacement name for *sepium* Speyer, 1846 (see also *Antenna* **9**: 49), but it is not adopted here as in our opinion the identity of the species named by Bruand has not been established.

Type locality: not stated.

Description of imago (Pl.7, figs 13,14)
Male. Wingspan 13–15mm. Antenna with 26–28 segments, shortly pectinate, the pectinations devoid of scales (figure 68a); eyes large, distance apart approximately 1.0–1.5 times diameter of eye. Forewing relatively slender, apex acute; scales broad; brown, a darker spot at end of cell. Hindwing with narrower scales; greyish, paler than forewing.   *Female.* Apterous. Antenna with 6–8 segments, simple; eyes small. Legs with only 3–5 tarsal segments. Body yellow with dark head and brown plates dorsally; a ventral tuft of greyish brown anal hairs; ovipositor long.

Life history
Ovum. Laid in the case, hatching in 4–5 weeks.
Larva (Pl.7, figs 40,41). Head black. Body blackish; plates on thoracic segments and abdominal segment 1 black. Case round in cross-section and with rear end rounded; covered with lichen and occasionally larger fragments of bark and leaves; length 6–7mm, width 2.5–3.5mm. The case is carried projecting from the bark at a right angle, presenting a fine example of protective resemblance, simulating exactly a bud and thereby making detection very difficult. The larva feeds on lichen growing on tree-trunks. August to May, completing growth after overwintering.
Pupa. In the case projecting from the trunk at a right angle. May to June.

Imago. Univoltine. The moths emerge in early morning in June and July.

Distribution (Map 26)
Very local and not common in the southern counties of England as far north as Worcestershire. Throughout Europe, eastwards to the U.S.S.R. and northwards to Finland.

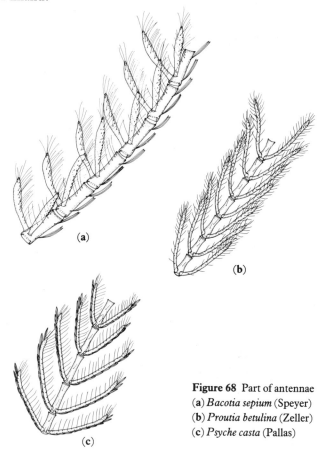

Figure 68 Part of antennae
(**a**) *Bacotia sepium* (Speyer)
(**b**) *Proutia betulina* (Zeller)
(**c**) *Psyche casta* (Pallas)

PROUTIA Tutt

Proutia Tutt, 1899, *Entomologist's Rec.J.Var.* **11**: 121.

A genus with two to four species in Europe and the Middle East, one of which occurs in England.

Imago.   Male. Antenna bipectinate, the pectinations devoid of scales (figure 68b); ocelli absent; labial palpus reduced. Foretibia with long spur, midtibia with apical spurs, hindtibia with medial and apical spurs. Forewing slender with apex somewhat acute; intercalary cell present.   *Female.* Apterous. Antenna simple, longer than in *Bacotia* Tutt. Legs without spurs.

PROUTIA BETULINA (Zeller)

Psyche betulina Zeller, 1839, *Isis, Leipzig* **1839**: 283.
Proutia eppingella Tutt, 1900, *Nat.Hist.Br.Lepid.* **2**: 295.
Type locality: Germany; Glogau (now Poland; Glogów).

Description of imago (Pl.7, figs 21,22)

Male. Wingspan 11–14mm. Antenna with 19–24 segments, bipectinate, the pectinations devoid of scales (figure 68b, p. 141). Foretibia with a long spur (figure 69b). Forewing slender, narrower and with apex more rounded than in *Bacotia sepium* (Speyer), with intercalary cell and nine veins emanating from cell (figure 70); broad-scaled, dark uniform brown. Hindwing somewhat narrower-scaled, brown shading to grey. *Female.* Head dark; antenna with 14 segments, simple; eyes small. Legs with number of tibial spurs varying from two to five, but remaining more or less constant within a population. Wings reduced to miniature appendages. Body reddish brown with dark dorsal plates on all abdominal segments; anal hair-tuft silver.

Life history

Ovum. Laid in the larval case and covered with hairs from the ventral tuft.

Larva (Pl.7, figs 49,50). Head black. Body purplish brown; prothoracic plate and more slender plates on thoracic segments 2 and 3 black. Case somewhat pointed, coated with fragments of bark, pine-needles, grass and lichen; length 8–10mm, width 2.5–3.0mm. The larva feeds on lichen, decaying plants and fresh leaves. August to May.

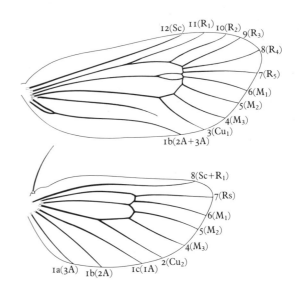

Figure 70 *Proutia betulina* (Zeller), wing venation

Pupa. In the case attached to a trunk or twig.

Imago. Univoltine. The adults emerge from late May to July and mate in the early morning; the pupal exuviae remain within the case.

Distribution (Map 27)

Local in south-eastern England from Kent to Hampshire and Huntingdonshire. Throughout Europe from Italy to Finland.

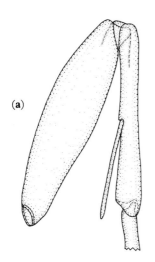

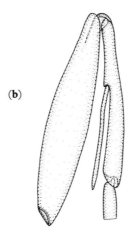

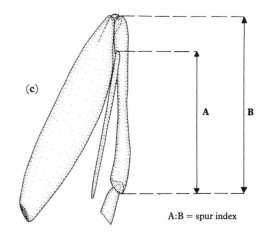

(a)   (b)   (c)

A:B = spur index

Figure 69 Foretibial spurs
(a) *Bacotia sepium* (Speyer)   (b) *Proutia betulina* (Zeller)   (c) *Psyche* spp.

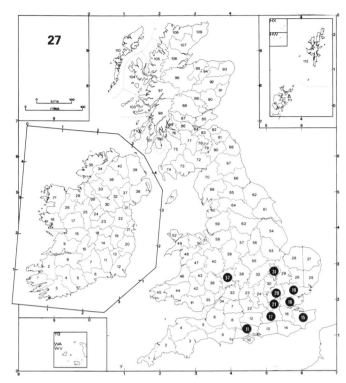

Proutia betulina

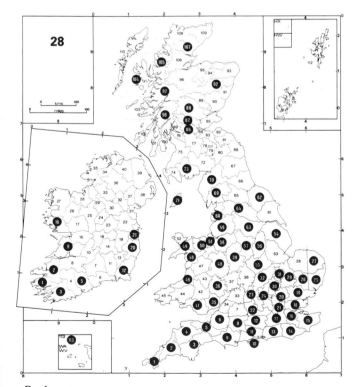

Psyche casta

PSYCHE Schrank

Psyche Schrank, 1801, *Fauna boica* 2: 156.

There are two species in Great Britain, one of which also occurs in Ireland.

Imago. *Male.* Head and body with sturdy brown hairs. Antenna pectinate, the pectinations with scales (figure 68c, p. 141); ocelli absent; labial palpus reduced, with a tuft of hair. Foretibia with a spur which is of systematic importance (figure 69c) (Dierl, 1964), midtibia with apical spurs, hindtibia with medial and apical spurs. *Female.* Apterous. Antenna simple. Legs normally developed.

PSYCHE CASTA (Pallas)

Phalaena casta Pallas, 1767, *Nova Acta Acad.Caesar. Leop.Carol.* 3: 437.
Type locality: not stated.

Description of imago (Pl.7, figs 16,17)

Male. Wingspan 11–15mm. Antenna with 18–21 segments, bipectinate, the pectinations with scales (figure 68c, p. 141); eyes large, the distance apart approximately equal to diameter of eye. Foretibia with spur index 0.76–0.82 (figure 69c) (Dierl, 1964). Forewing relatively broad, apex rounded; nine veins emanating from cell, 8 (R_4) and 9 (R_3) often stalked (figure 71); scales broad. Hindwing with somewhat narrower scales. All wings uniformly dark brown. *Female.* Apterous. Head dark brown; antenna with 10–15 segments, simple. Legs with 3–5 tarsal segments. Body whitish, thoracic segments dark brown, all abdominal segments with dark brown dorsal and ventral plates and dark scales laterally.

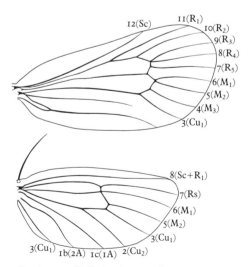

Figure 71 *Psyche casta* (Pallas), wing venation

Life history

Ovum. Laid in the larval case, hatching in 3–5 weeks.

Larva (Pl.7, figs 45,46). Head dark fuscous. Body pale reddish yellow, thoracic segments darker, often yellow-striped longitudinally. Case with pieces of grass-stem attached longitudinally, extending its entire length or beyond, constricted at fore end and splayed out at rear end (length 8–12mm). The larva feeds on grass, lichen and decaying plant matter. August to May.

Pupa. In the case, attached to a tree-trunk, fence or herbage. May to June.

Imago. Univoltine. The males emerge during the night, the females in the early morning. May to July.

Distribution (Map 28)

Occurs throughout the British Isles, commonly in the south but becoming more local in Scotland. Quite common throughout Europe from Spain to Finland, occurring from sea-level up to 2000m (6500ft). It was introduced into the U.S.A. where it occurs along the New England coast.

PSYCHE CRASSIORELLA Bruand

Psyche crassiorella Bruand, 1853, *Mém.Soc.Emul.Doubs* **3**: 29.

Type locality: France; Paris district.

Description of imago (Pl.7, figs 19,20)

Male. Wingspan 13–17mm, generally larger than *P. casta* (Pallas). Antenna with 21–25 segments, bipectinate. Foretibia with spur index 0.65–0.72 (figure 69c, p. 142) (Dierl, 1964). Forewing relatively broad, apex rounded; nine veins emanating from cell, 8 (R_4) and 9 (R_3) sometimes stalked; scales broad; glossy brown. Hindwing with scales narrower; brown with a greyish sheen. *Female*. Head light brown. Body whitish, prothoracic plate light brown; thoracic segments 2 and 3 and all abdominal segments each with a dark brown dorsal plate and a smaller ventral plate; white scales laterally.

Life history

Ovum. Laid in the case, hatching in 4–5 weeks.

Larva (Pl.7, figs 47,48). Similar to that of *P. casta*, but longitudinal stripes on thoracic segments more pronounced. Case similar to that of *P. casta* but slightly larger (length 10–15mm) and covered with more sturdy pieces of grass-stem. Feeds on grasses. A generation normally takes one year to develop but, depending on climate, may take two years. August to May.

Pupa. In the case. May to June.

Imago. Univoltine. The moths emerge in the early morning. June and July.

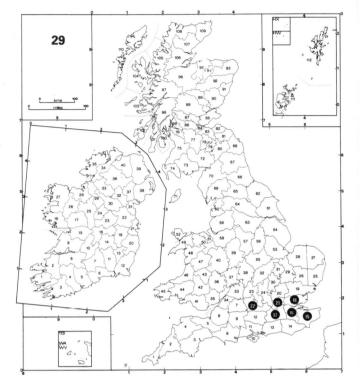

Psyche crassiorella

Distribution (Map 29)

Very local in the south-eastern counties of England. Throughout Europe from Spain to Finland.

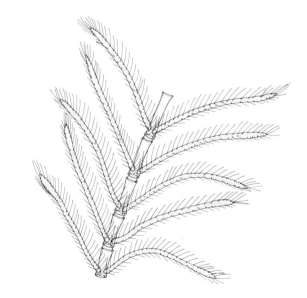

Figure 72 Part of antenna
Epichnopterix plumella ([Denis & Schiffermüller])

EPICHNOPTERIGINI

EPICHNOPTERIX Hübner

Epichnopterix Hübner, [1825], *Verz.bekannt.Schmett.*: 399.

A genus with more than a dozen Palaearctic species, one of which occurs in Great Britain and Ireland.

Imago. Male. Antenna with long pectinations (figure 72); eyes separated by more than one diameter of eye; ocelli absent; labial palpus reduced, with a tuft of hair. Foretibia without spur, midtibia with apical spurs, hindtibia with apical and medial spurs. Forewing broad, apex rounded, with eight veins emanating from cell and with intercalary cell; all wings dark brown to black, scales hair-like. *Female.* Apterous. Eyes minute; antennae and legs absent.

EPICHNOPTERIX PLUMELLA ([Denis & Schiffermüller])

Tinea plumella [Denis & Schiffermüller], 1775, *Schmett. Wien.*: 133.

Bombyx pulla Esper, 1785, *Schmett.* **3**: 232.

Type locality: [Austria]; Vienna district.

Description of imago (Pl.7, figs 23,24)
Male. Wingspan 10–14mm. Head and body with long black hairs; antenna with 17–21 segments, bipectinate, the pectinations long (figure 72). Wings with dense, narrow, hair-like scales, dark fuscous soon fading to brownish fuscous (Sieder & Loebel, 1954). *Female.* Apterous. Head sclerites much reduced; antennae absent; legs represented by short stumps; head and body reddish yellow; anal hairs whitish.

Life history
Ovum. Laid in the pupal skin within the case, hatching in 4–5 weeks.

Larva (Pl.7, figs 53,54). Head black. Body whitish, tinged purplish; thoracic segments with blackish plates (Meyrick, 1928). Case covered with flat pieces of grass placed longitudinally and attached at either end; length 8–12mm, width 2–3mm. Feeds on grasses. July to April.

Pupa. In the case affixed to a grass-stem. April to May.

Imago. Univoltine. The males emerge from April to June and fly at noon in sunshine. The female remains within the case and attracts the male by scent; the method of copulation is described in the introduction to the subfamily (p. 139).

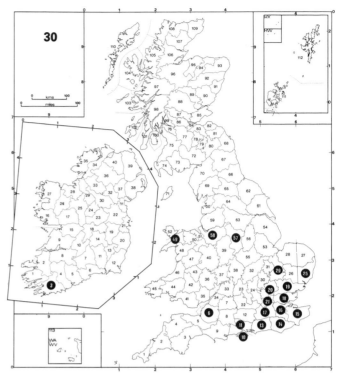

Epichnopterix plumella

Distribution (Map 30)
Local, occurring mainly in meadows and on moorland in England and Wales as far north as Derbyshire. In Ireland recorded from Co. Cork. In Europe known north of the Alps from the English Channel to Czechoslovakia. A few specimens have been captured in Paraguay where it was most probably accidentally introduced from Europe (Davis, 1964).

WHITTLEIA Tutt

Whittleia Tutt, 1900, *Entomologist's Rec.J.Var.* **12**: 20.

A genus with three to four species in Europe, one of which occurs in England.

Imago. Male. Antenna pectinate, the pectinations shorter than in *Epichnopterix* Hübner; eyes separated by approximately one and one-half times diameter of eye; ocelli absent; labial palpus reduced, with tuft of hair. Forewing broad, apex more rounded than in *Epichnopterix*, with eight veins emanating from cell and with intercalary cell; all wings white with grey to brown wavy striae, scales hair-like. *Female.* Apterous. Antennae and legs reduced to minute stumps.

WHITTLEIA RETIELLA (Newman)

Psyche retiella Newman, 1847, *Zoologist* **5**: xi, 1863.
Type locality: England; Isle of Sheppey, Kent.

Description of imago (Pl.7, fig.18)
Male. Wingspan 8–10mm. Antenna with 12–14 segments, bipectinate, the pectinations without scales. Body and tibiae covered with white hair. Fore- and hindwing strongly rounded; whitish with pattern of dark grey, hair-like scales, tending to take the form of irregular transverse lines. *Female.* Apterous. Antennae absent. Legs rudimentary.

Life history
Ovum. Laid in the pupal skin, hatching in 4–5 weeks.
Larva (Pl.7, figs 51,52). Head black. Body yellowish; narrow dorsal and broader lateral stripes paler; thoracic segments with black plates. Case with longitudinally attached pieces of grass, some of which project over the silken tube at the rear end; length 8–12mm. On common saltmarsh-grass (*Puccinellia maritima*) and other grasses growing in salt-marshes. September to April.
Pupa. In the case affixed to grass, the males near the roots, the females higher up. April to May.
Imago. Univoltine. The male flies in sunshine in late May and early June.

Distribution (Map 31)
Moderately common on most salt-marshes in southeastern England from Hampshire to Suffolk. Coastal in north-western Europe from the Netherlands to Denmark.

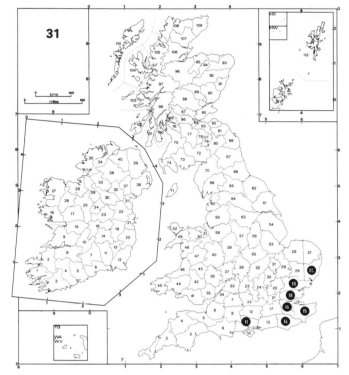

Whittleia retiella

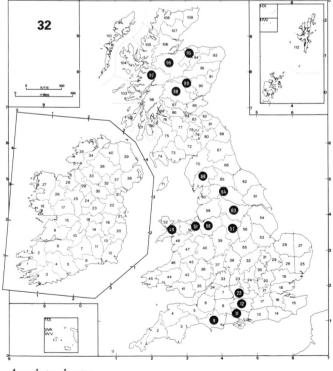

Acanthopsyche atra

Oiketicinae

The subfamily is divided into two tribes with similar behavioural characteristics.

ACANTHOPSYCHINI (*Acanthopsyche* Heylaerts, *Pachythelia* Westwood, *Lepidopsyche* Newman and *Thyridopteryx* Stephens)

The females have no wings, antennae or legs. Emergence of the female, copulation and oviposition are as described for the Epichnopterigini (Psychinae, p. 139).

PHALACROPTERYGINI (*Sterrhopterix* Hübner)

Female with characters as above. It never leaves the pupal skin and for copulation the male has to insert its abdomen through the closed rear opening of the case which was specially prepared for this purpose by the larva prior to pupation; further procedure as in the Epichnopterigini.

ACANTHOPSYCHINI

ACANTHOPSYCHE Heylaerts

Acanthopsyche Heylaerts, 1881, *Annls Soc.ent.Belg.* **25**: 66.

A genus containing several Palaearctic species and more than 12 in the Afrotropical and Oriental regions. One species occurs in Great Britain.

Imago. Male. Relatively large moths, wingspan 15–28mm. Antenna bipectinate to apex; eyes small and black, distance apart about one and one-half times diameter of eye; ocelli absent; labial palpus reduced, with hair-tuft. Foretibia with very long spur (figure 74), mid- and hind-tibiae without spurs. Wings with hair-like scales; veins 8 (R_4) and 9 (R_3) stalked (figure 73). *Female.* Apterous. Eyes reduced to dark spots; antennae and legs rudimentary.

Larva. The cases are covered longitudinally with plant matter.

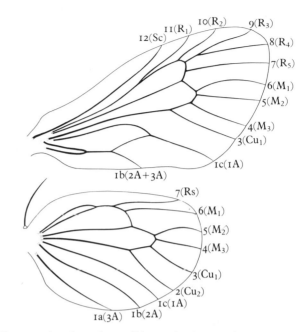

Figure 73 *Acanthopsyche atra* (Linnaeus), wing venation

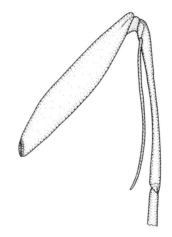

Figure 74 *Acanthopsyche atra* (Linnaeus), foretibial spur

ACANTHOPSYCHE ATRA (Linnaeus)

Phalaena (*Bombyx*) *atra* Linnaeus, 1767, *Syst.Nat.* (Edn 12): 823.
Psyche opacella Herrich-Schäffer, 1846, *Syst.Bearb. Schmett.Eur.* **2**: 20.
Type locality: Europe.

Description of imago (Pl.7, figs 25,26)
Male. Wingspan 16–22mm. Head and body covered with long black hairs; antenna with 28–30 segments, bipectinate. Foretibia with very long spur (figure 74). Wings

thinly scaled with dark grey hair-scales; veins and base of cilia darker. *Female*. Pale yellowish; head and thoracic segments dark brown.

Life history

Ovum. Laid in the pupal skin, hatching in 4–5 weeks.
Larva (Pl.7, figs 55,56). Head black. Body yellowish grey; thoracic segments black with yellow spots. Case cylindrical, generally dark in colour, covered longitudinally with grass-stems and twigs and leaves of heather; length 18–20mm, width 5–6mm.

Feeds on grasses and sallow (*Salix* spp.). August to April, but only a certain percentage of the larvae develop fully during the first year; depending on climate, a smaller or greater proportion overwinter for a second time and pupate after two years.
Pupa. In the case. April to May.
Imago. Univoltine. The males fly in May and June in the afternoon or, in the case of some populations, in the evening.

Distribution (Map 32)

Occurs locally on heathland and moorland in three main areas of Britain – Berkshire, Hampshire and Dorset; North Wales, the Pennines and the Lake District; and the Highlands of Scotland. Widespread in Europe from Italy to Finland.

Dispersal

Jørgensen (1954), quoted by Hoffmeyer (1970), states that he observed that some females left their cases and dropped to the ground a few days after pairing. Eleven of these were fed to a captive robin (*Erithacus rubecula*), and its droppings for the next 24 hours were retained. After two weeks between 30 and 40 larvae hatched from the droppings, made cases and started to feed. This experiment shows that ova can pass through a bird's gut unscathed and suggests a means whereby the dispersal of the species is effected.

PACHYTHELIA Westwood

Pachythelia Westwood, 1848, *Trans.ent.Soc.Lond.* **5**: 41.

A genus with only one species in the Palaearctic region.

Imago. *Male*. A large, stout moth. Antenna approximately half length of wing, bipectinate; eyes round, set at a distance apart of approximately one eye diameter; labial palpus with only one segment and hair-tuft. Foretibia with long spur, mid- and hindtibiae without spurs. Forewing broad with termen sinuate; veins 8 (R_4) and 9 (R_3) stalked (figure 75); scales hair-like, dark brown. *Female*. Apterous. All body appendages greatly reduced.

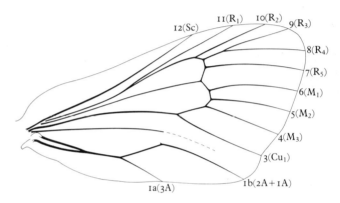

Figure 75 *Pachythelia villosella* (Ochsenheimer), forewing venation

PACHYTHELIA VILLOSELLA (Ochsenheimer)

Psyche villosella Ochsenheimer, 1810, *Schmett.Eur.* **3**: 180.
Type locality: Austria; Vienna.

Description of imago (Pl.7, figs 27,28)

Male. Wingspan 22–28mm. Antenna half length of wing with 35–39 segments, bipectinate. Foretibia with long spur. Forewing broad, costal margin straight, termen sinuate; dark brown, fading to paler, thin hair-scales giving a transparent appearance; generally a dark spot at base of radial veins; veins and base of cilia darker. Hindwing with similar scales. Body and head with pale greyish brown hairs. *Female*. Apterous. Cylindrical; pale yellow; body appendages much reduced.

Life history

Ovum. Laid in the pupal skin. After laying all its eggs, the female shrinks to a small ball of skin and remains dead in the head section of the exuviae, still in the case. The ova hatch in 4–5 weeks.
Larva (Pl.7, figs 57,58). Head whitish, marked with

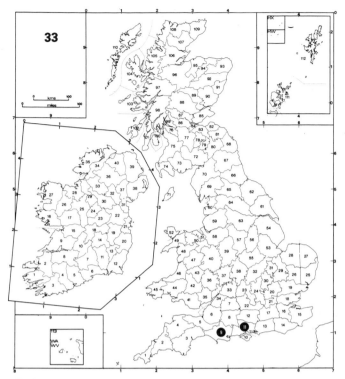

Pachythelia villosella

black. Body dark fuscous, laterally paler; thoracic plates yellowish with black longitudinal stripes and lateral spots. Case covered longitudinally with heather, grass, grass-stems and twigs; length 35–50mm, width 8–10mm; shortly before pupation the male larva lengthens the silken tube at the posterior end of the case without attaching any fragments thereto; the female case is covered with plant matter, usually heather, over the entire length. Development generally takes two years but descendants of the same female may complete development in the first or second year of the larval stage. July until May of the year of pupation.

Pupa. In the case attached, in the male, to the heather; in the female, to tree-trunks or posts (Heath, 1946). May to June.

Imago. Univoltine, emerging June to early August. The males fly in evening sunshine.

Distribution (Map 33)
Very local, confined to the heaths of Hampshire and Dorset. Widespread in Europe as far north as Finland, occurring from sea-level up to 2000m (6500ft).

Conservation. Because of its restricted distribution in Britain and vulnerability to heath fires and change in land usage, restraint should be exercised by collectors.

LEPIDOPSYCHE Newman

Lepidopsyche Newman, 1850, *Zoologist* **8** (Appx): 101.

A genus with two species in the Palaearctic region, one of which is included in the British list on the evidence of a single specimen supposed to have been taken in England.
   Generic characters differ little from those of *Pachythelia* Westwood.

LEPIDOPSYCHE UNICOLOR (Hufnagel)

Phalaena (Bombyx) unicolor Hufnagel, 1766, *Berlin.Mag.* **2**: 418.
Type locality: Germany; Berlin.

Imago (Pl.7, figs 29,30)
Barrett (1895: 343) writes, 'I have in my own collection a male example which was given me many years ago as a British specimen of *P. opacella*. By some misfortune the record of its locality is lost'. There is no other evidence for its occurrence in Britain. For a brief description of the imago and larval case (Pl.7, figs 59,60) see Barrett (*loc. cit.*). Occurs all over Europe northwards to Finland, from sea-level up to 2000m (6500ft).

THYRIDOPTERYX Stephens

Thyridopteryx Stephens, 1835, *Ill.Br.Ent.* (Haust.) **4**: 387.

A genus containing three Nearctic species, one of which was described from a supposedly British specimen.

THYRIDOPTERYX EPHEMERAEFORMIS (Haworth)

Sphinx ephemeraeformis Haworth, 1803, *Lepid.Br.*: 72.
Type locality: allegedly England; Yorkshire, but more probably U.S.A.; Georgia.

Described by Haworth from a specimen allegedly 'taken by the late Mr Bolton in Yorkshire'. Accidental introduction into Britain is a remote possibility. It is more likely that the moth was collected in Georgia by John Abbott and sent to a London dealer named Francillon, who is known to have mislabelled foreign specimens and sold them as British (Davis, 1964). It is, therefore, extremely unlikely that the species ever occurred in the British Isles.

PHALACROPTERYGINI

STERRHOPTERIX Hübner

Sterrhopterix Hübner, [1825], *Verz.bekannt.Schmett.*: 399.

A genus containing three species in the Palaearctic region, one of which occurs in Britain.

Imago. *Male.* Moth large with slender body. Antenna short, bipectinate (figure 77); eyes large, ocelli absent; labial palpus one-segmented, with tuft of hair. Foretibia without spurs, mid- and hindtibiae with very short spurs. Wings broad and rounded, scales hair-like; veins 4 (M_3) stalked, 8 (R_4) and 9 (R_3) stalked (figure 76). *Female.* Vermiform. All body appendages absent.

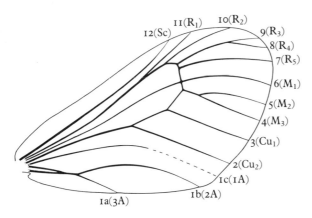

Figure 76 *Sterrhopterix fusca* (Haworth), forewing venation

STERRHOPTERIX FUSCA (Haworth)

Nudaria fusca Haworth, 1809, *Lepid.Br.*: 157.

Type locality: England; London district.

Description of imago (Pl.7, figs 31,32)

Male. Wingspan 18–25mm. Antenna with 22 segments, short, one-third to one-quarter of wing-length, bipectinate (figure 77); eyes large and round, approximately half eye diameter apart. Wings sparsely covered with short dark grey hair-scales, fading to paler grey. Head and body with long greyish hairs, body slender in relation to wings. *Female.* Apterous. Eyes reduced to small dark spots; all body appendages absent. Body short, length 5–6mm, diameter 3mm; a full ring of brown anal hairs.

Life history

Ovum. Laid in the pupal skin, hatching in 4–5 weeks.

Larva (Pl.7, figs 61,62). Head dark brown. Body brownish; thoracic plates dark brown, divided by a paler central line. Case somewhat narrowed towards ends, irregularly decorated with plant matter, sometimes placed transversely; length 16–20mm, width 5–7mm.

The larvae feed on grasses, oak (*Quercus* spp.), hawthorn (*Crataegus* spp.), sallow (*Salix* spp.), birch (*Betula* spp.), heather (*Calluna vulgaris*) and heaths (*Erica* spp.). August until May of the second year.

Pupa. In the case, affixed to heather, a tree-trunk or leaf. May to June.

Imago. Univoltine. The males emerge in the late afternoon during June and July and at night are attracted by light; pairing takes place at night.

Distribution (Map 34)

Very local in scattered sites in Britain as far north as Cumbria. Central Europe.

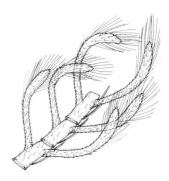

Figure 77 Part of antenna
Sterrhopterix fusca (Haworth)

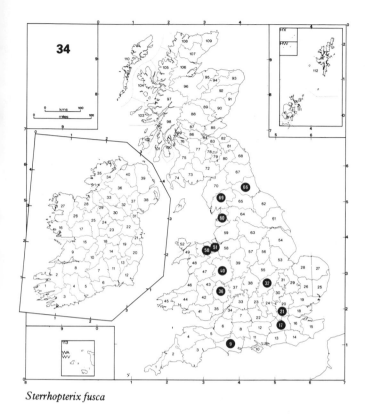

Sterrhopterix fusca

References

Barrett, C. G., 1895. *The Lepidoptera of the British Islands*, **2**: 333–369, pls 84–86. London.

Bradley, J. D., 1966. Some changes in the nomenclature of British Lepidoptera, 4. Microlepidoptera. *Entomologist's Gaz.* **17**: 213–235.

Davis, D. R., 1964. Bagworm moths of the Western Hemisphere. *Bull. Smith. Inst.* **244**: 1–233, 385 figs.

Dierl, W., 1964. Cytologie, Morphologie und Anatomie der Sackspinner *Fumea casta* und *crassiorella* sowie *Bruandia comitella* mit Kreuzungsversuchen zur Klärung der Artspezifität. *Zool. Jb.* (Syst.) **91**: 201–270.

Ford, L. T., 1946. The Psychidae. *Proc. Trans. S. Lond. ent. nat. Hist. Soc.* **1945–1946**: 103–110, 1 col. pl.; reprinted *in* [Agassiz, D. J. L.] (Ed.), 1978. *Illustrated papers on British Microlepidoptera*: 23–30.

Heath, J., 1946. The life history of *Pachythelia villosella* Ochs. (= *nigricans* Curt.) (Lep. Psychidae). *Entomologist's mon. Mag.* **82**: 59–63, 8 figs.

Hoffmeyer, S., 1970. Dispersal of *Pachythelia* species (Lep. Psychidae). *Entomologist's Rec. J. Var.* **82**: 33.

Jørgensen, P. L., 1954. Larver af *Acanthopsyche atra* L. klækket af fugleekskrementer. *Flora Fauna, Silkeborg* **1954**: 122–127.

Kloet, G. S & Hincks, W. D., 1972. A check list of British insects: Lepidoptera (Edn 2). *Handbk Ident. Br. Insects* **11**(2): viii, 153 pp.

Leraut, P., 1984. Mise à jour de la liste des Psychides de la faune de France. *Ent. gall.* **1**: 65–77.

Meyrick, E., 1928. *A revised handbook of British Lepidoptera*, vi, 914 pp. London.

Narbel-Hofstetter, M., 1964. La répartition géographique des trois formes cytologiques de *Luffia*. *Mitt. schweiz. ent. Ges.* **36**: 275.

Sauter, W., 1956. Morphologie und Systematik der schweizerischen *Solenobia* Arten. *Revue suisse Zool.* **623**: fasc. 3, no.27.

Sieder, L. & Loebel, F., 1954. Wissenwertes über die Gattung *Epichnopterix* Hb. (Lep. Psychidae). *Z. wien. ent. Ges.* **39**: 310–327, pl. 17.

Smith, F. N. H., 1983. *Luffia lapidella* Goeze (Lep.: Psychidae) in Cornwall. *Entomologist's Rec. J. Var.* **95**: 53–57, 1 pl.

Walsingham, Lord, 1899. *Talaeporia* (*Bankesia* Tutt) *staintoni* n.sp. and *montanella* n.sp. *Entomologist's Rec. J. Var.* **11**: 256–259.

TINEIDAE
E. C. Pelham-Clinton

A family of world-wide distribution of which about 350 species have been described in the Palaearctic region (Petersen, 1969). Of these 45 species may be considered as present or past residents in the British Isles. A number of others have been recorded but have not been shown to be permanent residents. The family name was formerly applied to a much wider range of species and the concept of the family was progressively reduced. Since Meyrick (1928) one genus, *Dryadaula* Meyrick, has been transferred to it from Lyonetiidae and four genera, *Narycia* Stephens, *Dahlica* Enderlein (*Solenobia* Duponchel), *Taleporia* Hübner and *Luffia* Tutt removed to Psychidae. *Ochsenheimeria* Hübner is now treated as a separate family, Ochsenheimeriidae. The genera treated in this volume in a separate family Hieroxestidae (formerly Oinophilidae) according to Davis (1978) should be included in the Tineidae.

The classification of the family is still in a fluid state and it is difficult to define subfamilies. However the grouping of genera into the subfamilies adopted by Bradley & Fletcher (1979) is, as far as possible, maintained here.

The family as now defined includes the clothes-moths (but not house-moths which belong to the Oecophoridae (*MBGBI* 4)) and some pests of grain. Their importance as pests has stimulated much taxonomic work on the family in the Palaearctic region. Of particular importance are works by Petersen (1957–58; 1969), Zagulyaev (1960; 1964; 1973; 1975; 1979), Hannemann (1977), Robinson (1979) and on larvae by Hinton (1956), all most useful for the study of British species.

Imago. Sexes similar, but the males of most species are on average smaller than females. Head rough-haired (figure 78) except in *Psychoides* Bruand. Antenna usually shorter

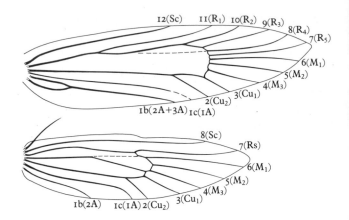

Figure 79 *Archinemapogon yildizae* Koçak, wing venation

than forewing, in male sometimes ciliate or pubescent, sometimes with raised whorls of scales which give a serrate effect; in those species with simple antenna, in the female usually more slender than in the male; scape with or without pecten. Ocelli absent. Labial palpus porrect or drooping; second segment with erect bristles, often conspicuous, on outer surface and at apex. Maxillary palpus of most genera long with up to five segments, the number of segments difficult to appreciate in dry specimens as the palpi are much folded. Haustellum reduced, shorter than labial palpus, or absent. Forewing normally with all veins present and unforked but reduced in some genera; vein 1b (2A+3A) usually forked at base, 11 (R_1) arising in basal half of cell; accessory cell normally present. Hindwing of most species rather broad, ovate, with all veins present and unforked (see figure 79). Exceptions to the venation pattern are noted in generic diagnoses. In the subfamily Tineinae the venation is somewhat unstable. Posterior tibia normally with long hairs and long spurs, the inner medial spur very long, almost reaching apex of tibia. Abdomen of female in most genera capable of great extension.

Moths of this family rest with body lying flat, the wings tent-like over the body, the forewings overlapping at the base but pressed together towards the apex. Against a suitable background they are most inconspicuous. When disturbed some species run rapidly instead of taking flight.

The rather broad ovate hindwings coupled with the rough-haired head will distinguish most members of the family at a glance, and the bristles of the labial palpi are characteristic of all genera, though they are also found in Hieroxestidae (*q.v.*).

Ovum. Translucent white; oval in all genera but *Dryadaula* in which it is subspherical. Laid singly or in small batches amongst the food material.

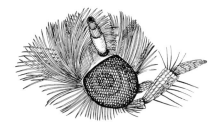

Figure 78 *Tinea pellionella* (Linnaeus), head

Larva. In most species of a rather uniform appearance, whitish with brown head and prothoracic plate. L group of prothorax trisetose in all genera except *Morophaga* Herrich-Schäffer. All legs are present and functional except in *Eudarcia* Clemens. Always concealed, either tunnelling in the food material or in a silken tube or portable case. In *Psychoides* feeding on ferns, but in all others as far as known feeding on fungi or lichens, on debris of animal or vegetable origin including material in the nests of birds, mammals or insects, on grain or other stored vegetable products or on animal substances. The Tineinae are mostly feeders on animal products such as hair, feathers, wool, skins and horn and are able to digest keratin.

The descriptions of larvae which follow are taken as far as possible from live specimens, but many have not been seen by me. Several descriptions are taken from Hinton (1956) or other authors.

Larval cases differ from those of the larvae of other families in being more or less flattened dorso-ventrally and open both ends (see figure 83, p. 195), but relatively narrower than those of Incurvariidae (*MBGBI* I).

Pupa. Abdominal segments 4–8 movable in male, 4–7 in female. Pupation either bare in the larval feeding place or in a cocoon. Pupa protruded on emergence of adult.

Collection. Moths of this family are mostly difficult to record or collect in the adult stage. Some of those species of which the early stages are unknown are extremely rare in collections; for instance, *Infurcitinea albicomella* (Herrich-Schäffer), *Ischnoscia borreonella* (Millière), *Stenoptinea cyaneimarmorella* (Millière) and *Nemapogon inconditella* (Lucas) are known in Britain from very few specimens although they are (or were) almost certainly resident. Some species are nocturnal and come into light-traps, but it is usually the most common species such as *Monopis laevigella* ([Denis & Schiffermüller]) and *Tinea semifulvella* Haworth that are taken this way. Probably the flight-time of nocturnal species is short.

The collection of larvae is a more reliable means of recording or obtaining specimens, but even with these it is only the species that feed on ferns and lichens that can be recorded and left *in situ*. It is not easy to identify the fungus-feeding species in the larval state and normally it is necessary to collect part of the fungus to see what moths appear. The increasing 'tidiness' of land management has led to a relative scarcity of bracket fungi in many areas and it is important to be conservation-minded when collecting fungi for this purpose.

Old birds' nests are a source of some species; these and the pellets of birds of prey probably provided the original food materials of those species which are now more dependent on man and his materials of wool, hair and feathers. Owing to the introduction of artificial fibres and of household insecticides some of these synanthropic species have become rare, and observations are needed to discover whether any are again taking to more natural materials. The most striking decline of a once abundant species is that of *Trichophaga tapetzella* (Linnaeus), *q.v.*, but most household species have declined in the last 30 years.

Distribution. The difficulty of setting out to look for and record many of the British species has meant that the maps produced for this family have had to be based largely on literature records. In some cases these are unreliable, especially in those groups of species which need examination of the genitalia for certain identification. In such cases two symbols have been used on the maps, open circles for unconfirmed records and filled-in circles for certain ones. The records provided by present-day collectors who appreciate the value of correct species determinations made by genitalia preparations have been a great help in preparing these maps.

Adventitious species. Notes on non-resident species have been provided by Dr Gaden S. Robinson of the British Museum (Natural History).

Key to species (imagines) of the Tineidae

1   Forewing with round or oval subhyaline spot at end of cell (*Monopis* spp.) 2

–   Forewing without such spot 8

2(1)   Forewing with large white costal blotch ....
.... *Monopis monachella* (p. 190)

–   Forewing without large white blotch 3

3(2)   Forewing with distinct yellow or ochreous dorsal streak
.... 4

–   Forewing without distinct yellow streak, sometimes indistinctly ochreous on dorsum 5

4(3)   Forewing with scattered yellow scales in apical third; hindwing pale grey .... *M. crocicapitella* (p. 189)

–   Forewing without scattered yellow scales; hindwing purplish fuscous .... *M. obviella* (p. 188)

5(3)   Forewing with indistinct triangular dull ochreous spots on costa and dorsum just beyond middle ....
.... *M. fenestratella* (p. 190)

–   Forewing without such spots 6

6(5)   Forewing narrowly edged ochreous on costa in apical half; hyaline spot in middle .... *M. imella* (p. 189)

–   Forewing not edged ochreous on costa; hyaline spot before middle 7

7(6)   Forewing purplish fuscous with a distinct yellow tornal spot .... *M. weaverella* (p. 187)

–   Forewing fuscous with yellow tornal spot vestigial or absent .... *M. laevigella* (p. 186)

8(1)   Forewing unicolorous and unmarked or almost so 9

–   Forewing at least mottled or with a discal stigma or tornal or apical spot, most species with numerous light or dark markings 13

9(8)   Forewing ochreous yellow 10

–   Forewing fuscous 11

10(9)   Wingspan 15–18mm; forewing deep ochreous yellow; in ants' nests .... *Myrmecozela ochraceella* (p. 167)

–   Wingspan 9–16mm; forewing pale ochreous yellow; in buildings .... *Tineola bisselliella* (p. 193)

11(9)   Forewing pale greyish brown ....
.... *Tinea columbariella* (part) (p. 195)

–   Forewing dark purplish fuscous 12

12(11)   Wingspan 9–10mm; head dark fuscous ....
.... *Psychoides verhuella* (p. 159)

–   Wingspan 11–13mm; head orange yellow; costa very inconspicuously edged ochreous posteriorly ....
.... *Cephimallota angusticostella* (p. 171)

13(8)   Forewing dark brown or fuscous, mottled or irrorate with paler scales but without distinct markings 14

–   Forewing with definite markings, or at least with a light or dark spot or spots 15

14(13)   Genitalia, see figure 80, p. 170 ....
.... *Haplotinea insectella* (p. 169)

–   Genitalia, see figure 80, p. 170 .... *H. ditella* (p. 169)

15(13)   Forewing basal two-fifths fuscous, apical three-fifths white .... *Trichophaga tapetzella* (p. 191)

–   Forewing otherwise 16

16(15)   Forewing mostly ferruginous or ferruginous orange; a black dot at tornus .... *Tinea semifulvella* (p. 204)

–   Forewing otherwise 17

17(16)   Forewing with pale markings on a dark ground 18

–   Forewing with dark markings on a paler ground 31

18(17)   Forewing unmarked except for a pale tornal spot ....
.... *Psychoides filicivora* (p. 160)

–   Forewing with more numerous markings 19

19(18)   Forewing with distinct white or silvery white spots and/or fascia on a blackish ground 20

–   Forewing otherwise; if with definite markings, then not white on a blackish ground 23

20(19)   Wingspan at least 17mm; forewing with conspicuous pale spots, two costal and two dorsal ....
.... *Triaxomera fulvimitrella* (p. 185)

–   Wingspan less than 11mm; forewing markings otherwise 21

21(20)   Forewing with a subterminal silvery white line parallel to termen; other markings more complex (see couplet 25) .... *Dryadaula pactolia* (part) (p. 158)

–   Forewing otherwise; a distinct straight rather oblique fascia before middle 22

22(21)   Forewing, besides the antemedian fascia, with three white spots, or a second fascia and one white spot ....
.... *Eudarcia richardsoni* (p. 162)

–   Forewing with more numerous small white spots beyond the antemedian fascia ....
.... *Infurcitinea argentimaculella* (p. 163)

23(19)   Forewing very narrow, with whitish spots only in apical quarter on costa and termen; three small scale-tufts in disc; two obscure dark fasciae ....
.... *Stenoptinea cyaneimarmorella* (p. 166)

–   Forewing with pale markings at least in basal half; without scale-tufts 24

24(23)   Forewing with a distinct pale antemedian fascia 25

–   Forewing without a distinct pale fascia 26

25(24)   Forewing with a dull whitish antemedian fascia, narrowed on fold; a costal and a dorsal spot of similar colour
.... *Triaxomasia caprimulgella* (p. 185)

–   Forewing with an irregular whitish fascia at one-quarter joined through an elongate spot in fold to a large median costal spot; markings in apical area silvery white ....
.... *Dryadaula pactolia* (part) (p. 158)

26(24)   Forewing markings formed of whitish irroration, more pronounced on dorsum; cilia unicolorous ochreous white .... *Infurcitinea albicomella* (part) (p. 164)
–   Forewing markings including numerous clear white spots; fringe chequered 27

27(26)   Wingspan 17–20mm; forewing with whitish markings inconspicuous, made up mainly of separate small dots; sometimes a larger whitish spot at tornus or indistinct blotch on dorsum at one-third; costal spots in apical half ochreous .... *Triaxomera parasitella* (part)(p. 184)
–   Wingspan 10–17mm; forewing with whitish markings more extensive including a series of distinct costal spots in apical third 28

28(27)   Forewing with clear whitish dot in disc at three-quarters 29
–   Forewing without this dot; usually some whitish irroration at this position, often with a dark brown dot below it; genitalia, see figure 81, p. 176 .... *Nemapogon granella* (part) (p. 173)

29(28)   Forewing with much pale reddish buff suffusion, mainly in apical area; genitalia, see figure 82, p. 177 .... *N. ruricolella* (part)(p. 179)
–   Forewing usually some shade of dark brown 30

30(29)   Forewing dark brown with markings including a blackish brown median costal spot, truncate on lower margin and narrowed in middle, this with other markings darker than average ground colour; genitalia, see figure 81, p. 176 .... *N. cloacella* (part) (p. 174)
–   Forewing with median costal spot appreciable only on costa, below merging with the general dark brown colour of wing; genitalia, see figure 81, p. 176 .... *N. wolffiella* (p. 175)

31(17)   Forewing whitish ochreous with small fuscous spot covering extreme apex .... *Ischnoscia borreonella* (p. 165)
–   Forewing with more extensive dark markings 32

32(31)   Forewing with ground colour white 33
–   Forewing with ground colour ochreous to fuscous 39

33(32)   Forewing pattern including a black sinuate streak, edged below with ferruginous brown, along fold from base .... *Archinemapogon yildizae* (p. 182)
–   Forewing without this streak 34

34(33)   Forewing pattern including an irregular blackish streak from base of costa to apex, fused with a median fascia and other markings .... *Nemapogon picarella* (p. 181)
–   Forewing otherwise 35

35(34)   Forewing with a blackish median angulated fascia sometimes broken into costal and dorsal spots, a basal costal spot and small apical markings; remainder of wing almost clear white .... *N. clematella* (p. 180)
–   Forewing otherwise 36

36(35)   Forewing markings composed entirely of dark brown irroration forming indefinite fasciae and spots; wingspan less than 10mm ... *Infurcitinea albicomella* (part) (p. 164)
–   Forewing pattern including distinct fasciae or spots; wingspan more than 10mm 37

37(36)   Forewing with dark brown or blackish costal spots indistinct and no clear median quadrilateral costal spot; yellowish brown markings in disc on a mainly whitish ground .... *Nemaxera betulinella* (p. 183)
–   Forewing with a dark median quadrilateral costal spot 38

38(37)   Wingspan 13–17mm; genitalia, see figure 82, p. 177 .... *Nemapogon inconditella* (p. 179)
–   Wingspan 10–14mm; genitalia, see figure 82, p. 177 .... *N. variatella* (p. 178)

39(32)   Forewing with plain ochreous ground colour; a series of dark brown marginal spots often united to form fasciae; without white or whitish spots .... *Tenaga nigripunctella* (p. 161)
–   Forewing otherwise; normally with whitish spots 40

40(39)   Forewing with a dark median quadrilateral costal spot, more distinct than other costal markings 41
–   Forewing without a median costal spot more distinct than others 44

41(40)   Wingspan more than 20mm; forewing median fascia distinct on dorsum .... *Morophaga choragella* (p. 156)
–   Wingspan less than 20mm; forewing median fascia represented only by a quadrilateral costal spot and a spot in fold 42

42(41)   Forewing with whitish postmedian dot in disc 43
–   Forewing without this dot; usually some whitish irroration in this position; genitalia, see figure 81, p. 176 .... *Nemapogon granella* (part) (p. 173)

43(42)   Forewing ground colour smooth pale reddish buff, especially evident in outer third; genitalia, see figure 82, p. 177 .... *N. ruricolella* (part) (p. 179)
–   Forewing ground colour ochreous or whitish ochreous with a varying amount of dark brown suffusion; genitalia, see figure 81, p. 176 .... *N. cloacella* (part) (p. 174)

44(40)   Forewing with numerous white and dark brown dots, mainly between veins, sometimes with irregular darker fasciae .... *Triaxomera parasitella* (part) (p. 184)
–   Forewing otherwise; few markings except for plical and discal stigmata 45

45(44)   Forewing with strong dark irroration, at least in apical area, and strongly chequered costal fringe (*Niditinea* spp.) 46
–   Forewing at most weakly irrorate and without chequering in costal fringe 47

46(45) Head brownish ochreous; forewing more densely irrorate; second discal and plical stigmata larger, first discal frequently absent; genitalia, see figure 84, p. 196 .... *Niditinea fuscella* (p. 194)

– Head whitish ochreous; forewing less densely irrorate; stigmata smaller and more discrete; genitalia, see figure 84, p. 196 .... *N. piercella* (p. 194)

47(45) Head bright yellow; forewing with plical stigma larger than the discals .... *Tinea trinotella* (p. 204)

– Head dull ferruginous or ochreous; forewing with a dark streak in fold or, if without a definite streak, at most a small plical stigma 48

48(47) Forewing usually with more extensive dark markings, a suffusion at base of dorsum, a streak along fold, two strong discal stigmata and some apical irroration; head brownish white .... *T. pallescentella* (p. 203)

– Forewing weakly marked; head dull ferruginous 49

49(48) Forewing markings very weak or absent; plical stigma, when present, represented by a slight suffusion along fold; genitalia, see figure 84, p. 196 .... *T. columbariella* (part) (p. 195)

– Forewing with at least second discal stigma fuscous and usually well defined (*T. pellionella* group, rarely distinguishable without dissection) 50

50(49) Forewing darker fuscous, scales more extensively tipped ochreous; hindwing grey; genitalia, see figures 85,86, pp. 200,201 .... *T. dubiella* (p. 202)

– Forewing scales paler fuscous or ochreous; ochreous-tipped scales more confined to apical area or absent; hindwing pale grey 51

51(50) Forewing pale glossy greyish ochreous; genitalia, see figures 85,86, pp. 200,201 ...... *T. flavescentella* (p. 203)

– Forewing darker 52

52(51) Forewing usually with some dark irroration and ochreous-tipped scales in apical area; genitalia, see figures 85,86, pp. 200,201 .... *T. pellionella* (p. 198)

– Forewing with dark irroration usually less obvious in apical area; genitalia, see figures 85,86, pp. 200,201 .... *T. translucens* (p. 199)

Scardiinae

MOROPHAGA Herrich-Schäffer

Morophaga Herrich-Schäffer, 1854, *Syst.Bearb.Schmett. Eur.* 5: 78.

Imago. Head densely rough-haired. Antenna about three-quarters length of forewing in male with cilia about four times diameter of shaft; scape with pecten. Labial palpus nearly twice as long as head, porrect or ascending; second segment densely rough-scaled beneath, bristles projecting on outer side; third segment thin, shorter than second. Maxillary palpus shorter than labial palpus. Haustellum reduced. Forewing with accessory cell, vein 7 (R_5) to apex, 8 (R_4) and 9 (R_3) stalked. Hindwing with vein 5 (M_2) and 6 (M_1) closely approximated or connate.

Larva. Unique in British Tineidae in lacking seta L3 of prothorax. Feeding in fungi and rotten wood.

The genus was formerly included in *Scardia* Treitschke, but is separated from it by the longer ciliate antenna and more densely scaled second segment of the labial palpus.

MOROPHAGA CHORAGELLA ([Denis & Schiffermüller])

Tinea choragella [Denis & Schiffermüller], 1775, *Schmett. Wien.*: 137.
Noctua boleti Fabricius, 1777, *Gen.Insect.*: 282.
Type locality: [Austria]; Vienna district.

Description of imago (Pl.8, fig.9)

Wingspan 20–30mm. Head whitish. Forewing whitish ochreous with much ochreous brown suffusion along veins especially towards apex, and more or less densely spotted and strigulated with blackish brown; more extensive blackish brown markings comprising a quadrate spot at one-quarter from base, a sinuate median fascia only distinct on dorsum and costa, a series of subbasal spots in cells 4–6 or 7 and other smaller costal and terminal spots. Hindwing dark fuscous with faint purple gloss. Fringes of both wings irregularly chequered.

There is some variation in extent of the markings, but this species is clearly distinguished from other resident British Tineidae by its large size.

Life history

Larva. Head reddish brown or dark brown. Prothoracic plate dark brown. Body whitish with conspicuous brown pinacula; dorsal plate of abdominal segment 10 brown. Feeding in galleries in various bracket fungi (*Laetiporus, Phellinus, Piptoporus*, etc.), or perhaps in rotten wood.

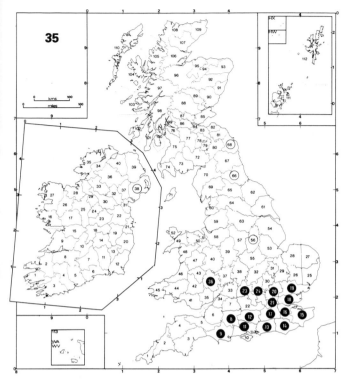

Morophaga choragella

Euplocaminae

EUPLOCAMUS Latreille
Euplocamus Latreille, 1809, *Gen.Crust.Ins.* **4**: 223.

EUPLOCAMUS ANTHRACINALIS (Scopoli)
Phalaena anthracinalis Scopoli, 1763, *Ent.Carn.*: 239.
Type locality: Carniola (Yugoslavia; Slovenija).

The inclusion of this large and conspicuous species on the British list is almost certainly erroneous: its citation as British is based on Turton's (1802) placement of an asterisk against the name, indicating it to be a British species. Stephens (1829; 1834–35) recorded that he had never seen specimens and that he doubted Turton's record (and that of Samouelle, 1819 – based on Turton). With a wingspan of 25–30mm, a conspicuous forewing pattern of large white spots on a black background and pectinate antennae in the male, this species could neither be overlooked nor confused with any other British species. Wood (1833–39) figures *E. anthracinalis* 'introduced by Turton without authority'. The distribution of this species encompasses central and south-eastern Europe and the Balkans.

Imago. Univoltine, flying from June to August. Crepuscular and nocturnal, but not often taken at light. Moths may be found by day sitting around their breeding sites.

Distribution (Map 35)
Very local in southern England and with more isolated records northwards to Northumberland. An old record from Co. Down requires confirmation. Distributed widely in the Palaearctic region.

Dryadaulinae

DRYADAULA Meyrick

Dryadaula Meyrick, 1893, *Proc.Linn.Soc.N.S.W.* (2)7: 559.

The genus was included by Meyrick (1928) in the Lyonetiidae. Bradley (1966) proposed a new subfamily Dryadaulinae for it and suggested that it might be transferred to the Tineidae. This placing was accepted by Morrison (1968). The genus is otherwise known principally from Australia and New Zealand.

Imago. Head rough with dense but fairly short hairs. Antenna about three-quarters length of forewing, simple in both sexes; scape without definite pecten. Labial palpus porrect, rather short; inner surface rough-scaled, bristles fine and inconspicuous; third segment longer than second. Maxillary palpus rather short. Haustellum absent. Forewing without accessory cell, veins 1b (2A+3A) not forked at base, 7 (R_5) and 8 (R_4) stalked. Hindwing with veins 6 (M_1) absent, 5 (M_2) and 7 (Rs) connate or stalked. Male genitalia strikingly asymmetrical (see Morrison, *loc.cit.*).

Ovum. Almost spherical.

Larva. On fungi.

DRYADAULA PACTOLIA Meyrick

Dryadaula pactolia Meyrick, 1902, *Trans.ent.Soc.Lond.* **1901**: 577.

Type localities: New Zealand; Nelson and Bealy River, Wellington.

Description of imago (Pl.8, fig.1)

Wingspan 8–11 mm. Head ochreous white, mixed darker on crown. Antenna whitish, ringed dark fuscous and with four broad dark bands in apical half. Forewing dark shining fuscous; a streak along fold white mixed with yellow, confluent with antemedian fascia in an enlarged longitudinal mark; other markings silvery white, comprising oblique antemedian and less oblique median fasciae, a slender transverse postmedian mark in disc and an irregular subterminal line, parallel to termen, expanded on costa; fringes irregularly chequered. Hindwing grey.

Life history

Described by Morrison (1968).

Larva. Head dark yellow-brown. Prothoracic plate dark brown. Body whitish with brown pinacula; dorsal plate of abdominal segment 10 brown. Feeding in a silk-lined gallery amongst the cellar fungus *Rhacodium cellare*.

Pupa. In a thin cocoon at the end of a larval gallery.

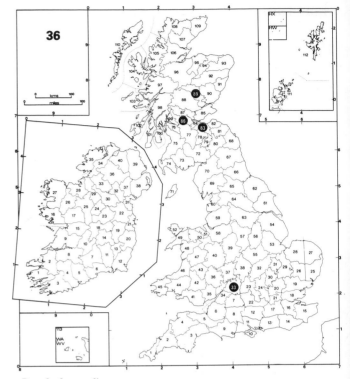

Dryadaula pactolia

Imago. Apparently in a continuous succession of broods, adults having been found or bred from larvae from March to November. The cellars in which larvae and adults have been found have a fairly constant temperature and humidity, the temperature of the warehouse from which Morrison's material was obtained varying little from 7° C.

History and distribution (Map 36)

In Britain first found at Gloucester by Clutterbuck (1916), the earliest specimen taken in a wine cellar in 1911 and another later found in a house. It was recognized by Meyrick as a species described by him from New Zealand. Later discovered by Morrison (*loc.cit.*) in a bonded whisky warehouse in Edinburgh. On 11 July 1981 a specimen was taken by Dr C. W. N. Holmes in a light-trap at Falkirk, a locality not far from a whisky distillery. No other British specimens are known. The species has been recorded in France from Bordeaux and from Normandy.

Teichobiinae

PSYCHOIDES Bruand

Psychoides Braund, 1853, *Mém.Soc.Emul.Doubs* (2)**3**: 109.

The moths are day-fliers and somewhat resemble Incurvariidae, in which family *Psychoides filicivora* (Meyrick) was originally described.

Imago. Head with procumbent scales, rough on crown. Antenna about half as long as forewing, ciliate in male; scape without pecten. Labial palpus porrect; second segment more or less rough-scaled beneath, bristles on outer side conspicuous at least at apex. Maxillary palpus very short. Haustellum absent. Forewing without accessory cell, veins 1b (2A+3A) not forked at base, 7 (R_5) to costa.
Larva. Feeding on ferns and unique amongst the British Tineidae in feeding on green plants.

PSYCHOIDES VERHUELLA Bruand

Psychoides verhuella Bruand, 1853, *Mém.Soc.Emul.Doubs* (2)**3**: 109, pl.2, fig.82.
Lamprosetia verhuellella Stainton, 1854, *Ins.Br.Lepid.*: 39.
Type locality: France; Besançon, Doubs.

Description of imago (Pl.8, fig.2)
Wingspan 9–12mm. Head with smoothly procumbent scales, crown with side tufts spreading inwards. Antenna of male shortly ciliate. Labial palpus first and second segments pale ochreous. Forewing dark fuscous with faint purple or violet gloss. Fringes of both wings paler towards margin.

Life history
Larva. Head and prothoracic plate black. Body yellowish white with brownish dorsal line. At first mining in a frond, the whitish mine visible on both sides; in early spring leaving the mine and burrowing into a sorus, in which it feeds on the sporangia; later spinning empty sporangia together to form a loose portable case; when it is fully grown in May this case on the underside of a *Phyllitis* frond resembles a misplaced sorus. Usually on hart's tongue (*P. scolopendrium*), but also recorded from spleenwort (*Asplenium* spp.) and rustyback (*Ceterach officinarum*).
Pupa. In the larval case, which is often fixed against a midrib.
Imago. Univoltine, occurring in June. Moths fly in daylight, especially in early morning sunshine and in late afternoon.

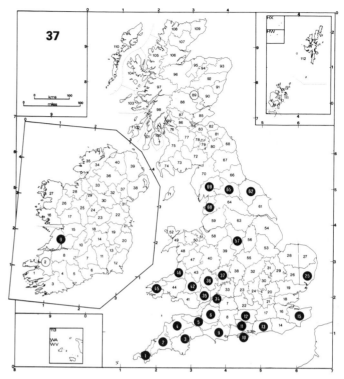

Psychoides verhuella

Distribution (Map 37)
Local in England and Wales and most numerous in the western counties, recorded as far north as Westmorland (Cumbria) and north Yorkshire. In Scotland there is an old record from Perthshire which seems doubtful. It has probably been overlooked in Ireland as there are records only from Cos Kerry and Clare. Abroad distributed from France through central Europe.

PSYCHOIDES FILICIVORA (Meyrick)
Mnesipatris filicivora Meyrick, 1937, *Entomologist* **70**: 194.
Type locality: Ireland; Seapoint, Co. Dublin.

Description of imago (Pl.8, fig.3)
Wingspan 10–12mm. Head ochreous or whitish with loosely procumbent scales, side tufts of crown brownish. Antenna of male serrate-ciliate, the cilia about three times diameter of shaft. Forewing dark fuscous with violet gloss; a small shining whitish tornal spot. Hindwing grey. Fringes of both wings paler towards margin.

Life history
First described by Beirne (1937).
Larva. Full-fed *c.*4.5mm long. Rather short and broad. Head pale brown. Prothoracic plate weakly sclerotized, with a pale brown posterior margin interrupted in middle. Body whitish with gut showing green. Feeds on a variety of ferns, commonly on soft shield-fern (*Polystichum setiferum*) or male-fern (*Dryopteris filix-mas*), but also on hart's tongue (*Phyllitis scolopendrium*), black spleenwort (*Asplenium adiantum-nigrum*), maidenhair spleenwort (*A. trichomanes*) and probably other species, eating sporangia or the lower frond surface, usually under a cover of sporangia spun into an irregular mass. The feeding causes extensive browning of the frond which is visible on the upper surface.
Pupa. In a tough white cocoon, usually embedded in the mass of sporangia, but sometimes spun in the open on the underside of the frond.
Imago. In a succession of generations from spring to autumn, the winter passed as a small larva. Moths fly at any time on dull days, but are particularly active in early morning sunshine and late afternoon.

History and distribution (Map 38)
First discovered in the type locality by Beirne (*loc.cit.*), but specimens from the same county were later found dating back to 1909. The first published record from England was made at Bournemouth in 1940 and since then the species has been found in Kent, throughout southwest England and in west and north Wales. In Ireland it is widely distributed. The apparent spread of the species could have been facilitated by the trade in ferns but, as the distribution is confined to maritime counties, it seems that inland it can only maintain a population indoors or under glass. There is a possibility that *P. filicivora* was present in the British Isles much earlier and misidentified as *P. verhuella* (Bruand), especially in the larval state; thus Stainton (1856) recorded Mr Drane finding *P. verhuella* larvae on wall-rue (*Asplenium ruta-mutaria*) at Caerphilly, Glamorgan – *P. filicivora* would be much more likely on this foodplant. There are no records from outside the

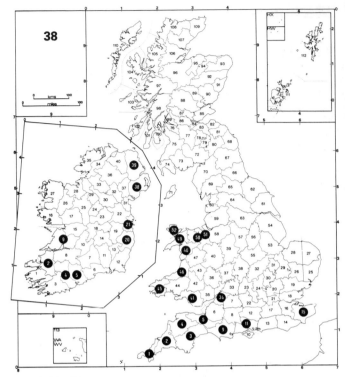

Psychoides filicivora

British Isles. Possibly the species was originally imported on ferns from the far east, since a related species occurs in Japan.

Meessiinae

TENAGA Clemens

Tenaga Clemens, 1862, *Proc.ent.Soc.Philad.* **1**: 135.
Lichenovora Petersen, 1957, *Beitr.Ent.* **7**: 344.

Only two western Palaearctic species have been described. *Lichenovora* was synonymized with *Tenaga* by Davis (1983).

Imago. Head densely rough-haired. Antenna nearly as long as forewing; scape with pecten. Eye small; greatest diameter less than length of scape. Labial palpus porrect, more than three times as long as diameter of eye; bristles of second segment strong. Maxillary palpus greatly reduced. Haustellum absent. Forewing without accessory cell, veins 1b (2A+3A) not forked at base, 5 (M_2) and 6 (M_1) stalked, 7 (R_5) and 8 (R_4) stalked, 7 (R_5) to costa, 9 (R_3) absent. Hindwing veins 5 (M_2) and 6 (M_1) stalked. Male genitalia remarkably reduced without definite uncus, gnathos or saccus.

Larva. Perhaps on detritus.

TENAGA NIGRIPUNCTELLA (Haworth)

Tinea nigripunctella Haworth, 1828, *Lepid.Br.*: 564.
Type locality: [England].

Description of imago (Pl.8, fig.4)

Wingspan 7–11mm. Head ochreous. Antenna ochreous. Labial palpus second segment with conspicuous dark apical bristles. Forewing ochreous with markings dark brown, comprising a median fascia, sometimes interrupted in fold, indistinct subbasal and subapical fasciae, sometimes reduced to costal and dorsal spots, and other indistinct markings and scattered dark brown scales. Hindwing grey.

A narrow-winged species of distinct appearance, no other British tineid being marked in this fashion. Much variation occurs in the extent of the dark markings.

Life history

Larva. Apparently undescribed. Continental authors have reported that it feeds in a portable case on lichens on fences, but this seems unlikely. Apparently there are no specimens of cases in British collections, but in the general collection of the BMNH is a Russian specimen from the Christoph collection: this is a rather irregular case which appears to be made from ground detritus including sand particles and without any lichen material. The habits of the imago suggest that the larva feeds in or around buildings.

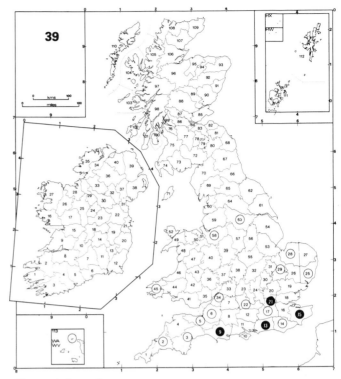

Tenaga nigripunctella

Imago. Formerly found in houses and outbuildings, the moths appearing in June and July. Barrett's (1878) note on the species in Pembrokeshire (Dyfed) is of interest: he writes 'At Tenby I have found it sitting on a house door, where it occurs rather frequently; but, from the description of building which it principally frequents – here and elsewhere – its larva may be suspected of tastes and habits which can hardly be described as decent, much less fastidious'.

Distribution (Map 39)

Formerly local and uncommon but recorded from Kent to Cornwall and northwards to Cheshire and Yorkshire. Records from north Northumberland and Ireland must be regarded as doubtful. The species appears not to have been seen in Britain for at least 50 years. Abroad the species has been found from Spain and Morocco to Anatolia and the Caucasus.

TENAGA POMILIELLA Clemens

Tenaga pomiliella Clemens, 1862, *Proc.ent.Soc.Philad.* **1**: 136.

Type locality: America.

Two specimens of this species were collected at Deal by H. W. Daltry in 1928 and 1930 (Daltry, 1929a; 1929b; 1931). The moth is yellowish, marbled with grey-brown and resembles vaguely *T. nigripunctella* (Haworth). The specimens were identified by Meyrick but have never been re-examined; their location is unknown. *T. pomiliella* is otherwise known only from the eastern U.S.A. and Sydney, Australia. It is not certain that the identification of Australian specimens is correct.

EUDARCIA Clemens

Eudarcia Clemens, 1860, *Proc.Acad.nat.Sci.Philad.* **12**: 10.

Meessia Hofmann, 1898, *Dt.ent.Z.Iris* **10**: 227.

A holarctic genus with several continental species. The synonymy of *Meessia* with the north American genus *Eudarcia* was introduced by Davis (1983) as new, but had previously been·noted by Forbes ([1924]). The north American and European species are undoubtedly congeneric.

Imago. Head densely rough-haired. Antenna about three-quarters length of forewing; scape with small pecten. Labial palpus short, drooping; second segment with conspicuous bristles. Maxillary palpus longer than labial palpus. Haustellum reduced. Forewing without accessory cell, veins 7 (R_5) and 8 (R_4) from 6 (M_1) (differing in this respect from continental species), 7 (R_5) to costa. Hindwing vein 4 (M_3) absent.

Larva. Feeding on lichens in a portable case; abdominal legs complete but short and useless for locomotion.

EUDARCIA RICHARDSONI (Walsingham)

Tinea richardsoni Walsingham, 1900, *Entomologist's mon. Mag.* **36**: 176.

Type locality: England; Portland, Dorset.

Description of imago (Pl.8, fig.5)

Wingspan 7–9mm. Head blackish brown, hairs around antennal base and face pale ochreous. Antenna blackish above, white below. Labial and maxillary palpi pale ochreous; bristles of second segment of labial palpus blackish. Forewing blackish with bronzy gloss; markings shining white, comprising an oblique slightly curved antemedian fascia, a median fascia oblique to tornus, usually broken into two spots, and a subapical mark from costa parallel to the fasciae; outer half of fringe white at apex. Hindwing grey, paler in female.

Similar species. Infurcitinea argentimaculella (Stainton) has similar colour and markings but more numerous white spots in the apical area and its wings are comparatively short and broad.

Life history

Described by Richardson (1895).

Larva. Head brown, fronto-clypeus and sides darker; back of head produced into a large two-lobed apodeme visible through the pale brown prothoracic plate; the latter with long setae directed forwards. Body yellow, broadest behind middle. Thoracic legs long, brownish. Abdominal segment 10 with numerous short setulae.

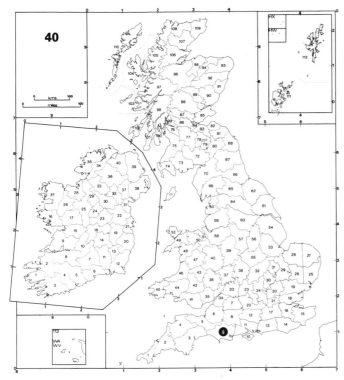

Eudarcia richardsoni

Feeds on lichens or algae on rocks in a portable case which lies flat against the rock. The full-sized case is about 6mm long, made of fine granules of rock and incorporating some lichen and alga particles; it is convex above and flattened beneath, and tapers slightly to a neck then widens to a broader valve at each end. The upper lamella of the valve is larger than the lower and meets the rock surface except when the larva extrudes its head and thorax; the lower lamella is drawn up to press against the inner surface of the upper when the valve is closed. If removed from its case the larva soon constructs a new one. Cases may be found in autumn on the lower surface of loose stones, but are fixed in a roughly vertical position before pupation, which takes place in spring; larvae perhaps live for two years.

Imago. Univoltine, the moth appearing in June and July.

History and distribution (Map 40)
Occurs at Portland, Dorset, where it was discovered by Richardson (*loc.cit.*) and at first supposed to be the continental *E. vinculella* (Herrich-Schäffer). It was later discovered in the Isle of Purbeck in the same county, but is not known from any other locality in Britain or elsewhere.

INFURCITINEA Spuler
Infurcitinea Spuler, 1910, *Schmett.Eur.* **2**: 461.

A western Palaearctic genus of numerous species. Closely related to *Eudarcia* Clemens but differing in venation and in the more complex male genitalia.

Imago. Head densely rough-haired. Antenna a little over half length of forewing, in male somewhat serrate with whorls of scales; scape with well-defined pecten. Labial palpus short, porrect or slightly drooping; bristles of second segment conspicuous. Maxillary palpus longer than labial palpus. Haustellum reduced. Accessory cell present in hindwing. Hindwing veins 5 (M_2) and 6 (M_1) stalked in *I. argentimaculella* (Stainton).

Larva. In a silken tube amongst lichens (larva of one British species unknown).

INFURCITINEA ARGENTIMACULELLA (Stainton)
Tinea argentimaculella Stainton, 1849, *Syst.Cat.Br. Tineidae and Pterophoridae*: 6.
Type locality: British Isles.

Description of imago (Pl.8, fig.6)
Wingspan 8–9mm. Head ochreous brown varying to light fuscous or dark brown; face ochreous whitish. Labial and maxillary palpi whitish; bristles of labial palpus second segment blackish. Forewing dark fuscous with coppery sheen; markings silvery white, comprising a curved oblique antemedian fascia, spots on tornus and opposite on costa, a transverse subapical mark on costa, a spot between this and tornal spot and other spots around apical margin; outer margin of fringe white. Hindwing fuscous. Tarsi black with white apical rings.

There is variation in extent of the white markings.

Similar species. Eudarcia richardsoni (Walsingham), *q.v.*

Life history
Larva. Head black. Prothoracic plate black; second and third thoracic segments each with two pairs, dorsolateral and lateral, of small brownish black plates. Body slender, brownish white. Pinacula slightly darkened, bearing rather long hairs. Plate of abdominal segment 10 brownish. The larva feeds on lichens of which *Lepraria incana* and *L.aeruginosa* have been identified (Brightman, 1965). It feeds in a long silken tube covered with lichen fragments spun amongst the foodplant, usually in shady situations on walls or rocks, sometimes on tree-trunks. Tubes containing larvae may be seen on the surface of the lichen from April to June, small larvae probably feeding through the winter below the surface. The larva pupates in the tube.

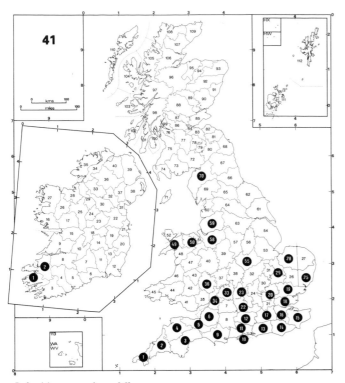

Infurcitinea argentimaculella

Imago. Univoltine. The moths fly in July and early August in sunshine.

Distribution (Map 41)

Most common in the south of England, but recorded in Britain as far north as Cumbria. In Ireland recorded only from Co. Kerry. It is a local species, but often common in suitable places and quite a small area of wall or rock can maintain a colony. Abroad from France to Scandinavia and central Europe.

INFURCITINEA ALBICOMELLA (Herrich-Schäffer)

Tinea albicomella Herrich-Schäffer, 1851, *Syst.Bearb. Schmett.Eur.* 5: 74.

Tinea confusella sensu auctt.

Type locality: Germany; Regensburg.

Description of imago (Pl.8, fig.7)

Wingspan 7–10mm. Head white, partly ochreous or fuscous on crown; face with some long fuscous hairs. Antennal scape and pecten dark brown. Labial palpus whitish; second segment with dark bristles; third segment brown-ish below. Maxillary palpus whitish. Forewing white, more ochreous towards apex, heavily irrorate with dark brown except towards dorsum; subbasal and broad median indefinite fasciae of dark brown irroration, not reaching dorsum; some indefinite costal spots towards apex and a more definite small tornal spot dark brown; fringe ochreous white. Hindwing light grey with purple gloss.

There is variation in intensity and extent of markings leading Meyrick (1895; 1928) to include it as two species under the synonyms given above.

Life history

Early stages. Unknown in the British Isles, but Hannemann (1977) states 'larva on lichens' without elaboration.

Imago. Probably univoltine, British specimens having been captured in July. All have been found on sea-cliffs. Birchall (1866) stated that at Howth 'it flits around and runs up the stems of the grass on the cliffs'.

History and distribution (Map 42)

Only known from four British localities. First recorded from Howth, Dublin (Birchall, *loc.cit.*), as *Tinea confusella*, then from Morecambe, Lancashire (Ellis, 1890), and from Black Hall Rocks, Durham (Robson, 1912), under the same name. In 1924 it was recorded as a new British species, *T. albicomella*, from Torquay, Devon (Metcalfe, 1924). It has not been recorded since that time.

This species has most probably been overlooked owing to its small size, obscure appearance and difficulties of access to its localities; it might well be widely distributed on southern and western cliffs. It has a wide distribution in Europe, where it is not necessarily maritime, from Spain to central Europe and Scandinavia and, according to Hannemann (*loc.cit.*), it requires hot dry situations.

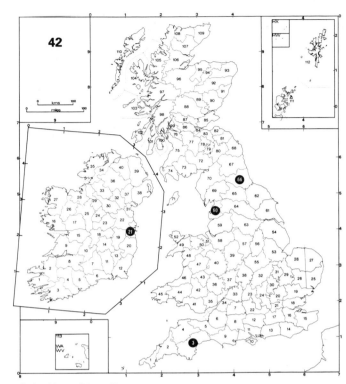

Infurcitinea albicomella

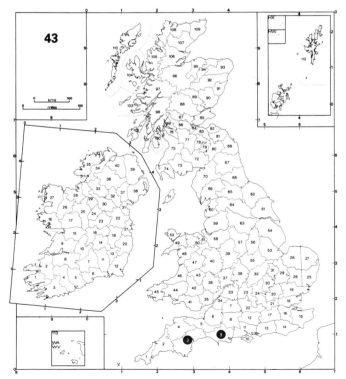

Ischnoscia borreonella

ISCHNOSCIA Meyrick

Ischnoscia Meyrick, 1895, *Handbk Br.Lepid.*: 783.

At least one other species in southern Europe.

Imago. Head rough, not densely haired. Antenna almost as long as forewing; scape without pecten. Labial palpus porrect, rather long and slender; bristles of second segment fine and inconspicuous; third segment longer than second. Maxillary palpus longer than labial palpus. Forewing veins 7 (R_5) to costa, 8 (R_4) and 10 (R_2) absent. Female genitalia unusual in the Tineidae in that the ovipositor is not greatly extensile.

Early stages. Unknown.

ISCHNOSCIA BORREONELLA (Millière)

Guenea borreonella Millière, 1874, *Revue Mag.Zool.* (3)2: 245.
Tinea subtilella Fuchs, 1879, *Stettin.ent.Ztg* 40: 341.
Type locality: France; Borréon, Alpes-Maritimes.

Description of imago (Pl.8, fig.8)

Wingspan 6–8mm. Head pale ochreous. Antenna pale ochreous fuscous. Labial and maxillary palpi ochreous white. Forewing whitish ochreous; a small fuscous spot covering extreme apex. Hindwing very pale grey. Fringes of both wings whitish ochreous.

Life history

Early stages. Unknown.

Imago. Probably univoltine; British specimens captured in July and August.

History and distribution (Map 43)

Known from only two British localities. Richardson (1891) first recorded it from Portland, Dorset, under the name *Tinea subtilella*. A number of specimens were taken but Richardson later (1896) described it as 'rather scarce'. On 13 August 1926 at about sunset Waters (1926) boxed a single specimen from a leaf on a cliff at Torquay, Devon. Since then the species has not been seen in Britain. Found abroad from France and northern Spain through central Europe to the Black Sea. Fuchs (*loc.cit.*) first found it in July 1878 on the walls of German vineyards.

STENOPTINEA Dietz

Stenoptinea Dietz, 1905, *Trans.Amer.ent.Soc.* **31**: 86.
Celestica Meyrick, 1917, *Exot.Microlepid.* **2**: 79.

A holarctic genus of three species, now including the single Palaearctic species formerly in the monobasic genus *Celestica* (Davis, 1983).

Imago. Head rough-haired. Antenna about three-quarters length of forewing; scape without pecten. Labial palpus porrect, rather long and slender; second segment bristles conspicuous; third segment shorter than second. Maxillary palpus longer than labial palpus. Haustellum reduced. Wings very narrow, the proportions similar to those of the Gracillariidae. Forewing with scale tufts. Hindwing with lower half of cell close to hind margin.

Larva. Reported on lichens or in rotten wood.

STENOPTINEA CYANEIMARMORELLA (Millière)

Argyresthia cyaneimarmorella Millière, 1854, *Annls Soc.ent.Fr.* (3)**2**: 64, pl.3.
Tinea angustipennis Herrich-Schäffer, 1854, *Syst.Bearb. Schmett.Eur.* **5**: 73.

NOMENCLATURE. Millière's name has five months priority over that of Herrich-Schäffer and the synonymy has recently been established by Leraut (1983).

Type locality: France; Mont Pilat, Haute-Loire.

Description of imago (Pl.8, fig.10)

Wingspan 12–14mm. Head ochreous yellow, orange-brown on crown. Forewing pale brown with varying amount of dark brown and blackish suffusion, mostly along costa and dorsum, sometimes forming diffuse oblique ante- and postmedian fasciae; whole wing sprinkled with bluish silver scales; three small scale tufts in disc, one antemedian blackish, two postmedian dull whitish anteriorly, black posteriorly, placed transversely; beyond scale tufts a broad orange-brown subapical area from costa to tornus; a series of dull whitish costal spots in apical half. Hindwing fuscous. Legs with all tarsal segments black with white apical rings.

Life history

Larva. Apparently undescribed. Reported by continental authors to feed on lichens on plum or in rotten wood of plum (*Prunus domestica*).

Imago. Recorded from May to September, perhaps bivoltine.

History and distribution (Map 44)

First found at Acton Green, Middlesex, by Sorrell (1876).

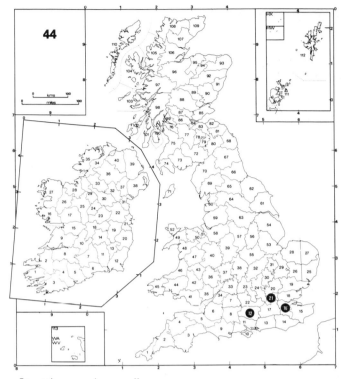

Stenoptinea cyaneimarmorella

This specimen was also recorded by Stainton (1876), and possibly the record from Richmond, Surrey, by Goss & Barrett (1902), 'A single specimen taken on a fence at Richmond by Mr. Sorrell', is an erroneous report of the same specimen. Meyrick's (1928) record from Kent may refer to a specimen labelled 'Bexley, 24.vi.1901' which was in the Walsingham collection and is now in the Studd collection in Exeter Museum. Another specimen from Bexley was collected by L. T. Ford on 14 June 1944 and is in his collection in BMNH. A Hampshire specimen not previously mentioned in British literature is in the Wocke collection in the Zoological Institute, Leningrad. Its label data are given by Zagulyaev (1979) as 'Anglia, Liss, 1 ♀ 4.viii.1893 (Wocke)'. The total number of British specimens is therefore perhaps no more than four. Evidently this is a species which is difficult to collect, but possibly other specimens have been collected and not recognized because of their narrow-winged non-tineid appearance. Recorded from most parts of Europe except the Iberian peninsula, as far east as Finland and Poland, and also from Lebanon.

Myrmecozelinae

MYRMECOZELA Zeller

Myrmecozela Zeller, 1852, *Linn.ent.* **6**: 103.

A genus of many species, mainly Palaearctic, but also with one species in south and east Africa.

Imago. Head rough-haired. Antenna about three-quarters length of forewing. Labial palpus porrect; second segment rough-scaled with inconspicuous bristles; third segment blunt, much shorter than second. Maxillary palpus very short. Haustellum reduced. Forewing vein 7 (R_5) to costa near apex.

Larva. In ants' nests.

MYRMECOZELA OCHRACEELLA (Tengström)

Tinea ochraceella Tengström, 1848, *Notis.Sällsk.Faun.Fl. fenn.Förh.* **I**: III.

Type locality: Finland; Uleåborg (now Oulu).

Description of imago (Pl.8, fig.11)

Wingspan 15–18 mm. Head ferruginous or ferruginous-ochreous. Forewing some shade of ochreous or brownish ochreous. Hindwing pale shining fuscous; cilia ochreous. Female abdomen with anal tuft of long wavy ochreous hairs.

Life history

Larva. Head yellowish brown with edges of sclerites darkened. Body yellowish white; dorsum matt with tergites more shining. No distinct prothoracic plate or pinacula. In a loose tube amongst debris at the edge of nests of the wood-ants *Formica aquilonia* Yarrow or *F. lugubris* Zetterstedt (on the Continent stated to be in *F. rufa* Linnaeus nests), feeding, according to Donisthorpe (1927), on the material of the nest. Young larvae feed through the winter and spring and pupation takes place in the larval tube in early June.

Imago. Univoltine, the moths appearing from late June to August. Usually nocturnal, becoming active about dusk, when moths may be found resting on grass blades around the ants' nests. During the day the moth remains inside the nest and may sometimes be made to fly out by tapping the nest with a stick.

Distribution (Map 45)

In Britain confined to the central Highlands of Scotland in open pine or birch forest. Very local and not always found where the ants occur, but sometimes very common. Records from the south of England appear to be erroneous and a record (King, 1901) from Glasgow also seems most

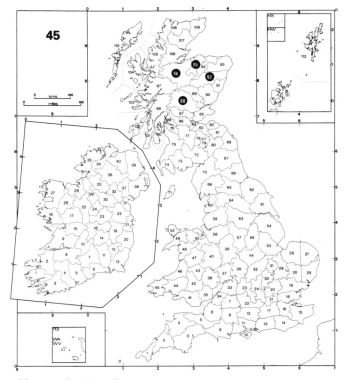

Myrmecozela ochraceella

unlikely. Central and northern Europe, from France to European parts of the U.S.S.R.

ATELIOTUM Zeller

Ateliotum Zeller, 1839, *Isis, Leipzig* **1839**: 189.

Metarsiora Meyrick, 1937, *Exot.Microlepid.* **5**: 76.

ATELIOTUM INSULARIS (Rebel)

Dysmasia insularis Rebel, 1896, *Annln naturh.Mus.Wien,* **II**: 125.

Metarsiora horrealis Meyrick, 1937, *Exot.Microlepid.* **5**: 76.

Type locality: Tenerife; Orotava.

One record of this species' introduction into Britain exists; a specimen was collected by S. N. A. Jacobs in a London warehouse in 1936 and was described as a new genus and species – *Metarsiora horrealis* – by Meyrick. Its synonymy with *Ateliotum insularis* was established by Bradley (1966). Pale grey-brown with a darker, marbled pattern, *insularis* has a wingspan of 11–15 mm. The male genitalia are figured by Petersen (1957–58), the female genitalia by Petersen (1960). *A. insularis* is otherwise known from the Canary Islands, Spain, southern France, Italy and Sicily.

Setomorphinae

SETOMORPHA Zeller

Setomorpha Zeller, 1852, *K.svenska VetenskAkad.Handl.*
1852: 94.

SETOMORPHA RUTELLA Zeller
Tropical Tobacco Moth
Setomorpha rutella Zeller, 1852, *K.svenska VetenskAkad.
Handl.* **1852**: 94.
Type localities: Botswana, Republic of S. Africa, Zimbabwe and Mozambique.

There is a single record of this species in Britain: a dead
female example was found in a cargo of sunflower seed
meal from Argentina in a London dock on 23 November
1944 by S. N. A. Jacobs (Corbet, 1945). As in *Lindera
tessellatella* Blanchard, the head-scales of *Setomorpha rutella* (erect in most other Tineidae) are directed forward; *S.
rutella* (wingspan 9–22mm) is smaller than *Lindera tessellatella* and is brown, speckled with dark brown or black.
The terminal segment of the labial palpus is markedly
spatulate whereas it is cylindrical and elongate in *L. tessellatella*. Detailed illustrations of this species and of *L.
tessellatella* are provided by Zimmerman (1978). *Setomorpha rutella* is widely distributed throughout the tropical
and subtropical Old and New World, the larvae feeding on
stored cereals, seeds, tobacco and other dried vegetable
material (Hinton, 1956).

LINDERA Blanchard

Lindera Blanchard, 1852, *Historia fisica . . . de Chile*,
Zoologia 7: 105.

LINDERA TESSELLATELLA Blanchard

Lindera tessellatella Blanchard, 1852, *Historia fisica . . . de
Chile*, Zoologia 7: 106.
Type locality: Chile.

This species was first found in Britain, breeding in floor-sweepings in a mill at Bootle, Lancashire, in January 1943
(Stringer, 1943). It is a large species with a wingspan of
20–30mm; the forewings are glossy, grey-brown, with a
speckled darker pattern. Uncharacteristically for a tineid
(but like *Setomorpha rutella* Zeller), the scales of the head
are directed forward over the vertex and down the frons,
giving *Lindera tessellatella* the superficial appearance of an
oecophorid.

Stringer (*loc.cit.*) figures the adult and genitalia; in the
male the eighth sternite is modified to form a slender,
stirrup-shaped structure. According to Hinton's (1956)
description of the larva of this species, its diet includes the
remains of mites and of lepidopterous larvae. The geographical range of *L. tessellatella* encompasses the subtropical regions of both the Old and New World. A subsequent British specimen was collected from a London
warehouse by S. N. A. Jacobs in June 1943.

Nemapogoninae

HAPLOTINEA Diakonoff & Hinton

Haplotinea Diakonoff & Hinton, 1956, *Entomologist* **89**: 31.

The original description of the genus includes full descriptions of the structure of larva and adult of the two British species. The distribution of the genus is Holarctic.

Imago. Head rough-haired. Antenna nearly as long as forewing; scape with pecten. Labial palpus drooping; second segment with conspicuous bristles. Maxillary palpus longer than labial palpus. Haustellum reduced. Forewing vein 7 (R_5) to costa near apex.

Larva. On stored products.

HAPLOTINEA DITELLA (Pierce, Metcalfe & Diakonoff)

[*Tinea*] *ditella* Pierce, Metcalfe & Diakonoff, 1938, *Genitalia of the British Pyrales with the Deltoids and Plumes*: 68.
Type locality: British Isles.

Description of imago (Pl.8, fig.13)

Indistinguishable superficially from *H. insectella* (Fabricius), *q.v.*, but the species may be separated by the genitalia of both sexes without dissection. In the male of *H. insectella* the uncus is longer than the valvae, ending in paired processes, inwardly concave; in *H. ditella* the uncus is much shorter than the valvae, ending in a pair of lobes with short points (figure 80, p. 170). The female of *H. insectella* may be recognized by a sclerotized process behind the ostium, absent in *H. ditella* (figure 80, p. 170)

Life history

Larva. Head pale to dark brown. Prothoracic plate pale yellowish brown. Body white, with pinacula indistinct. Feeds on stored vegetable products, especially grain, but also on rice and groundnuts.

Imago. Moths have been found from June to October, but there may be a continuous succession of generations.

Distribution (Map 46)

Local in the south of England, isolated records as far north as Cheshire and Yorkshire and an imprecise record by Hinton (1956) from 'Scotland'. As it was not recognized until 1938 some earlier records of *H. insectella* may refer to this species, but *H. ditella* appears to be much less common and has not been found in Britain away from warehouses, granaries and mills. Distributed abroad from France through central Europe to western Asiatic U.S.S.R.

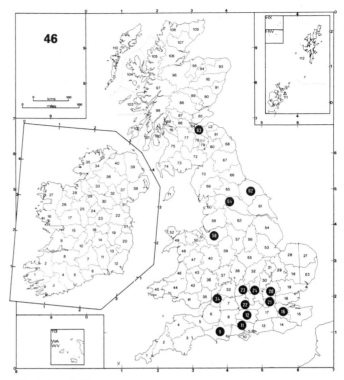

Haplotinea ditella

HAPLOTINEA INSECTELLA (Fabricius)

Tinea insectella Fabricius, 1794, *Ent.syst.* **3**(2): 303.
Tinea misella Zeller, 1839, *Isis, Leipzig* **1839**: 184.
Type locality: not stated.

Description of imago (Pl.8, fig.12)

Wingspan 11–20mm. Head ochreous brown or ferruginous brown. Forewing dark brown with slight coppery gloss, more or less densely irrorate or strigulate with ochreous; very indistinct dark markings comprising second discal stigma and a diffuse area along fold; fringe often obscurely chequered. Hindwing fuscous with purple gloss. A rare form has the forewing ground colour pale golden brown.

Similar species. *H. ditella* (Pierce, Metcalfe & Diakonoff), *q.v.*, separable by genitalia. *Niditinea fuscella* (Linnaeus) is more distinctly spotted. *Tinea pellionella* (Linnaeus) and related species are less irrorate and mostly have the markings more distinct.

Life history

Larva. Head brown. Prothoracic plate pale yellowish brown. Body white, with pinacula indistinct. Superficially similar to the larva of *Haplotinea ditella*. Feeds on a similar range of stored products to *H. ditella*, but also on fungus

Figure 80

Haplotinea ditella (Pierce, Metcalf & Diakonoff)
(**a**) male genitalia (**b**) female genitalia

Haplotinea insectella (Fabricius)
(**c**) male genitalia (**d**) female genitalia

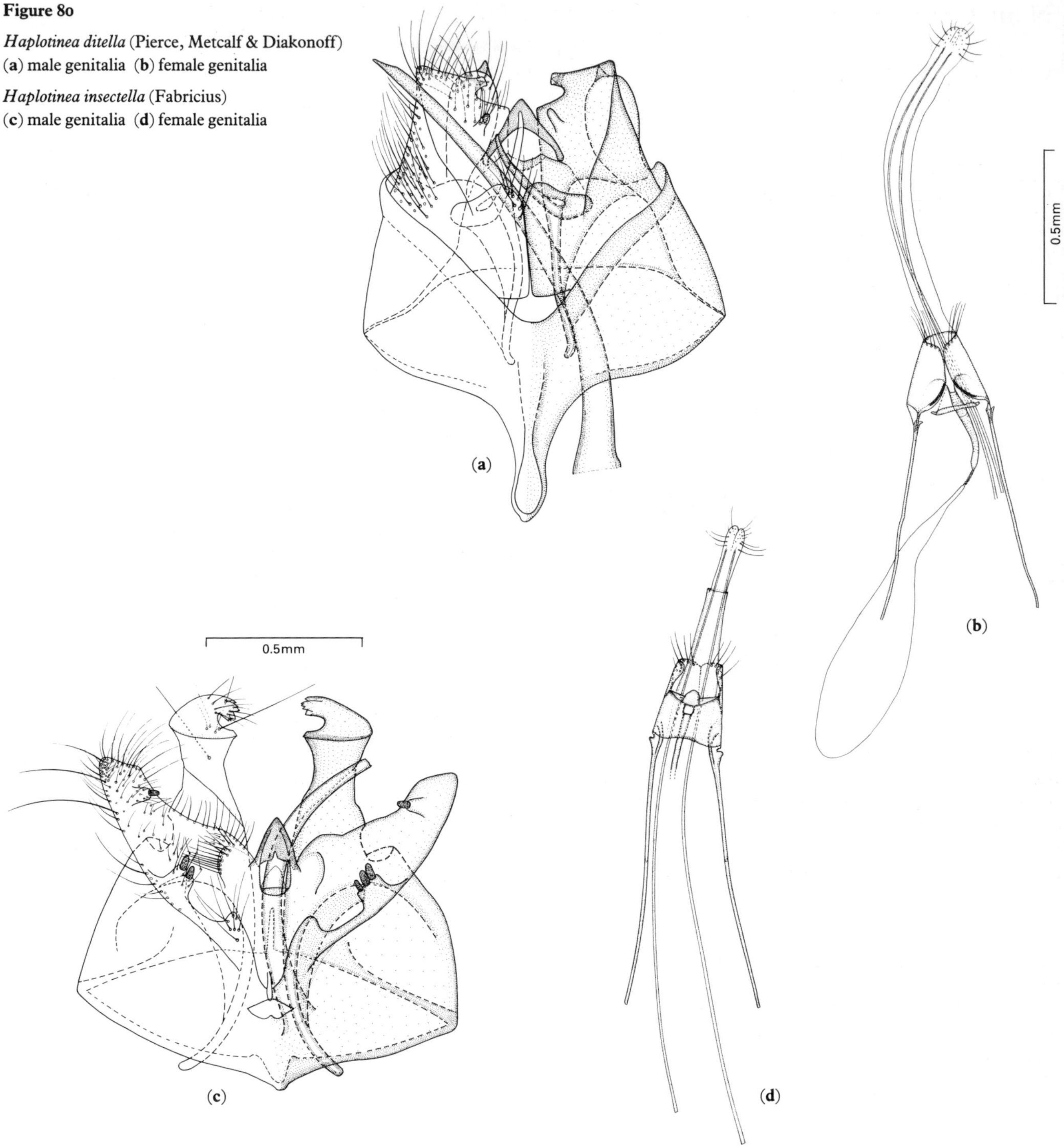

(**a**)

(**b**)

0.5mm

0.5mm

(**c**)

(**d**)

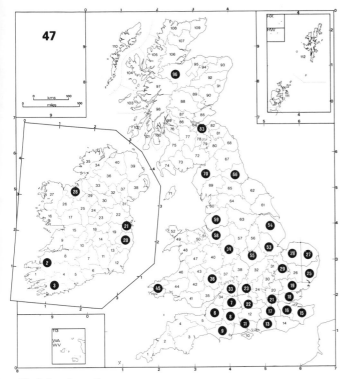

Haplotinea insectella

indoors and probably outside, as the moth has been found associated with fungus on tree-trunks. It has also been recorded from rotten wood and from poultry-house and stable refuse. Wakely (1958a) bred it from thick dead ivy stems.

Imago. Moths have been found from June to August and it seems likely that the species is univoltine. Moths may sometimes be found in houses or outbuildings. Ford (1931) found them in large numbers in farm-stables and reported that when disturbed they ran faster than any other 'Tinea' he had seen.

Distribution (Map 47)
Local over a large part of the British Isles including Ireland, but there are no records from south-west England and only one from Wales. Abroad it has much the same distribution in the Palaearctic region as *H. ditella*, but also occurs in north America.

CEPHIMALLOTA Bruand
Cephimallota Bruand, [1851], *Mém.Soc.Emul.Doubs* (1) **3** (1850) (3, livr. 5, 6): 32.

Several species occur in the Palaearctic region.

Imago. Head rough-haired. Antenna just over half length of forewing; scape with small pecten. Labial palpus drooping, laterally compressed; third segment blunt, shorter than second. Maxillary palpus longer than labial palpus. Haustellum reduced. Forewing with veins 7 (R_5) and 8 (R_4) approximated or connate, 7 (R_5) to costa near apex.

Early stages. Unknown.

CEPHIMALLOTA ANGUSTICOSTELLA (Zeller)
Incurvaria angusticostella Zeller, 1851, *Linn.ent.* **5**: 310.
Tinea simplicella Zeller, 1852, *Linn.ent.* **6**: 169.
Type locality: Hungary.

Description of imago (Pl.8, fig.14)
Wingspan 10–14mm. Head orange-yellow. Labial palpus orange-ochreous. Forewing dark brown with purple gloss; costa narrowly edged ochreous in apical half. Hindwing dark brown with purple gloss.

Life history
Early stages. Undescribed. Continental authors suggest that the larvae may develop in nests of bumble-bees or other aculeate Hymenoptera.

Imago. Probably univoltine, the moths appearing in June and July. A long lost species in Britain, recorded only from east and west Kent and from Surrey. Stainton (1854) knew it only from near Dover and from Mickleham and (1852) stated that it was to be found in chalky places and on sand. In the W. Tyerman collection in BMNH are a specimen each from South Foreland, July 1890, and from Deal, with no date, and one from Cuxton, west Kent, dated 26 July 1899 which appears to be the last specimen captured in Britain.

Hannemann (1977) states that it is found in Germany in dry meadows and herb-rich pine woods. The species has a wide distribution in western and central Europe.

CEPHITINEA Zagulyaev

Cephitinea Zagulyaev, 1964, *Ent.Obozr.* **43**: 680.

CEPHITINEA COLONGELLA Zagulyaev

Cephitinea colongella Zagulyaev, 1964, *Ent. Obozr.* **43**: 682.

A female specimen labelled 'Anglia' from the Wocke collection was made the holotype of this species by Zagulyaev. The only other known specimen, also a female, was a more recent specimen from the Crimea. Although the moth has a superficial resemblance to *Niditinea fuscella* (Linnaeus) it is scarcely possible that this is a species which has been overlooked here and probably the label of the holotype is erroneous (see Zagulyaev, 1975).

NEMAPOGON Schrank

Nemapogon Schrank, 1802, *Fauna boica* 2(2): 167.

The genus is basically Holarctic but includes a cosmopolitan species. About 20 species occur in Europe, several of them of close superficial appearance.

Imago. Head densely rough-haired. Antenna half to two-thirds length of forewing, in male of some species slightly pubescent or serrate due to whorls of scales; scape with pecten. Labial palpus porrect or drooping, about twice as long as eye diameter in most species; second segment usually with conspicuous bristles; third segment shorter than second. Maxillary palpus about or nearly as long as labial palpus. Haustellum reduced.

Larva. Usually in fungi, but some species have also become pests of stored products.

In the British Isles the earlier records of the group of species including *N. cloacella* (Haworth) are unreliable. These species are not easily recognized superficially and the genitalia of both sexes are therefore figured (figures 81,82, pp. 176,177). The males of most species may be recognized by brushing away a few scales from the genitalia and separated using the following key:

1   Tegumen broadly rounded without apical processes (figure 81a, p.176) .... *granella* (p. 173)
–   Tegumen with apical processes 2

2(1)   Tegumen rounded with small median projection with V-shaped notch (rarely this projection is folded down and inconspicuous) (figures 81c,81e, p. 176) 3
–   Tegumen with posterior margin emarginate, angled at sides 4

3(2)   Forewing ground colour dark brown without darker fascia or spots .... *wolffiella* (p. 175)
–   Forewing ground colour paler, with distinct darker markings .... *cloacella* (p. 174) (Difference in shape of gnathos arms is difficult to appreciate in dry specimens.)

4(2)   Tegumen posterior margin with shallow emargination (figure 82c, p. 177). Larger species (wingspan usually more than 14mm) with forewing ground colour white ... . *inconditella* (p. 179)
–   Tegumen posterior margin more deeply emarginate (figures 82a,82e, p. 177). Wingspan less than 14mm ... 5

5(4) Apex of valva with narrow blackish upcurved process (figure 82a, p. 177). Forewing ground colour white .... *variatella* (p. 178)

– Apex of valva with shorter inconspicuous process (figure 82e, page 177). Forewing ground colour mostly reddish buff .... *ruricolella* (p. 179)

Notes on the separation of four of these species and redescriptions of them are given by Pierce & Metcalfe (1934) and Corbet (1943).

NEMAPOGON GRANELLA (Linnaeus)
Corn Moth

Phalaena (Tinea) granella Linnacus, 1758, *Syst.Nat.* (Edn 10)**1**: 537.

Type locality: not stated.

Description of imago (Pl.8, fig.15)
Wingspan 9–16mm. Head ochreous or brownish ochreous mixed with dark brown on crown and at sides. Forewing white or ochreous white densely irrorate or strigulated with dark brown; darker brown markings comprising a basal patch in costal half, sometimes extended to dorsum, antemedian and median fasciae outwardly oblique from costa, both terminating in a large irregular patch on fold, numerous other spots sometimes linked to form irregular markings especially in apical area, and a series of spots along costa towards apex; fringe irregularly chequered, normally with a pair of white bars near apex and one towards tornus outside a line of dark-tipped scales. Hindwing fuscous with slight purple gloss. Genitalia, see figure 81, p. 176.

Considerable variation occurs in the extent of markings but even when the forewing is much darkened the median fascia and spot on fold remain conspicuous.

Similar species. *N. cloacella* (Haworth) has the forewing with patches or irroration of ochreous scales and a well defined postmedian white spot in the disc. *N. granella* is the only species of the group without a clear white postmedian spot in the disc of the forewing; in the corresponding position in this species is an area of irroration (light on dark or dark on a light ground).

Life history
Larva. Head pale yellowish to dark brown. Prothoracic plate pale brown. Body white, with pinacula indistinct. Feeds on fungi and on stored products. Out of doors it may feed on a variety of bracket fungi, including *Piptoporus betulinus*, *Coriolus versicolor*, *Laetiporus sulphureus* and

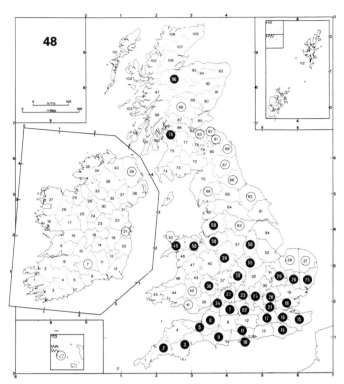

Nemapogon granella

Polyporus squamosus. Indoors on dry-rot fungus (*Serpula lacrymans*), on decayed wood, on corks in wine cellars and a great variety of stored vegetable products including dried mushrooms, grain, nuts and dried fruit. Probably in a continual succession of generations indoors.

Imago. Moths are found from March to September, and specimens of this group taken in buildings or outside early in the year are likely to be this species or *N. variatella* (Clemens).

Distribution (Map 48)
Local in Great Britain as far north as Inverness-shire. Records from many districts are unconfirmed and many specimens placed in collections as *N. granella*, especially from those northern localities, have been found to be *N. cloacella*. No Irish specimen has been seen, though there are old records from three vice-counties. The species has a cosmopolitan distribution through transport of stored products.

NEMAPOGON CLOACELLA (Haworth)
Cork Moth
Tinea cloacella Haworth, 1828, *Lepid.Br.*: 563.
Type locality: Britain.

Description of imago (Pl.8, fig.16)
Wingspan 10–18mm. Head ochreous white to brownish ochreous, mixed dark brown on crown and at sides; antenna of male subserrate. Forewing whitish mixed with ochreous and with a varying amount of dark brown suffusion; more definite dark or blackish brown markings as follows: a basal patch in costal half, sometimes extended to dorsum; a costal spot beyond this; a broad outwardly oblique mark from middle of costa, sometimes truncate beneath to form an anvil-shaped mark, sometimes continued inwards through a large spot on fold to form an almost complete angulated fascia to dorsum; costal strigulae before apex: white discal postmedian spot very clear; fringe irregularly barred with white, mostly from base. Hindwing fuscous. Genitalia, see figure 81, p. 176. An extremely variable species in size, colour and markings.
Similar species. N. granella (Linnaeus), *q.v. N. wolffiella* Karsholt & Nielsen has the ground colour more uniformly dark and the median spot or fascia is not clearly defined as in this species. *N. ruricolella* (Stainton) has a more uniform reddish ochreous ground colour especially in apical area. *N. variatella* (Clemens) has a clear white head and the forewing with white ground colour and less dark suffusion. *N. inconditella* (Lucas) is similar to this but much larger.

Life history
Larva. Head pale to dark red-brown. Prothoracic plate pale red-brown, crescentic, with very narrow median division. Body whitish, sometimes slightly tinged pink, pinacula inconspicuous. The larva feeds on a range of food materials similar to that of *N. granella, q.v.*, but this species is more frequently found feeding out of doors on various bracket fungi; it is especially common in *Piptoporus betulinus*, boring throughout the fungus and producing a little webbing. It has also been found feeding in the callus-tissue around wounds on birch (*Betula* spp.). Perhaps feeding in a succession of generations indoors, but outside appears to be bivoltine in the south and univoltine in the north.
Imago. Moths are found from May to September. They are on the wing in early morning sunshine and in late afternoon and are sometimes most numerous at dusk.

Distribution (Map 49)
Throughout mainland Britain, the Inner Hebrides, Isle of Man and Ireland. A common species, found in almost any type of locality but especially numerous in woodland and

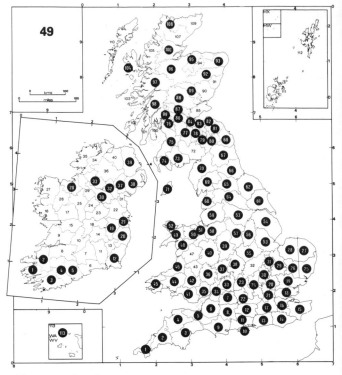

Nemapogon cloacella

in areas with much dead wood. Widely distributed in the Palaearctic region.

NEMAPOGON WOLFFIELLA Karsholt & Nielsen

Nemapogon wolffiella Karsholt & Nielsen, 1976, *Entomologica scand.* 7: 151.

Tinea albipunctella Haworth, 1828, *Lepid.Br.*: 564.

NOMENCLATURE. This species was re-named by Karsholt & Nielsen, since Haworth's name is a primary homonym of *Tinea albipunctella* [Denis & Schiffermüller], 1775; 319.

Type locality: England; near London.

Description of imago (Pl.8, fig.18)

Wingspan 10–14mm. Head pale ochreous with some dark brown hairs on crown and at sides; antenna of male subserrate and slightly pubescent. Forewing dark brown; prominent white spots antemedian on costa and at tornus; whitish spots, variable in extent, on costa near base and forming about four strigulae near apex, on dorsum near base extending to fold and a larger diffuse spot just before the tornal spot; a whitish postmedian spot in disc; fringe with five white bars between costal cilia and tornus. Hindwing dark fuscous. Genitalia, see figure 81, p. 176.

Similar species. N. wolffiella has a basically similar wing pattern to other species of this group and the genitalia show a close relationship to *N. cloacella* (Haworth), but the ground colour is darkened to such an extent that the dark markings corresponding to those of the other species are scarcely visible. In the male genitalia the shape of the gnathos arms distinguishes it from *N. cloacella* (figure 81, p. 176).

Life history

Larva. Undescribed. The moth has been bred from bracket fungi and from rotten wood.

Imago. Occurs in June and July and is presumably univoltine. Flies in late afternoon and at dusk.

Distribution (Map 50)

Very local in the south of England and as far north as Westmorland (Cumbria) and Durham. Frequents wooded localities but is not common. Found abroad from northern Spain to Scandinavia and the Caucasus.

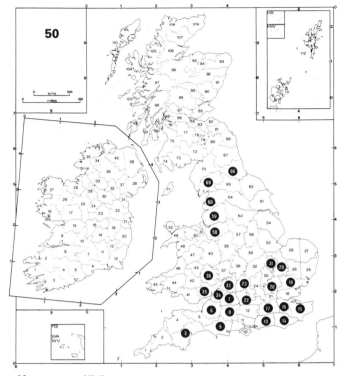

Nemapogon wolffiella

Figure 81

Nemapogon granella (Linnaeus)
(**a**) male genitalia (**b**) female genitalia

Nemapogon cloacella (Haworth)
(**c**) male genitalia (**d**) female genitalia

Nemapogon wolffiella Karsholt & Nielsen
(**e**) male genitalia (**f**) female genitalia

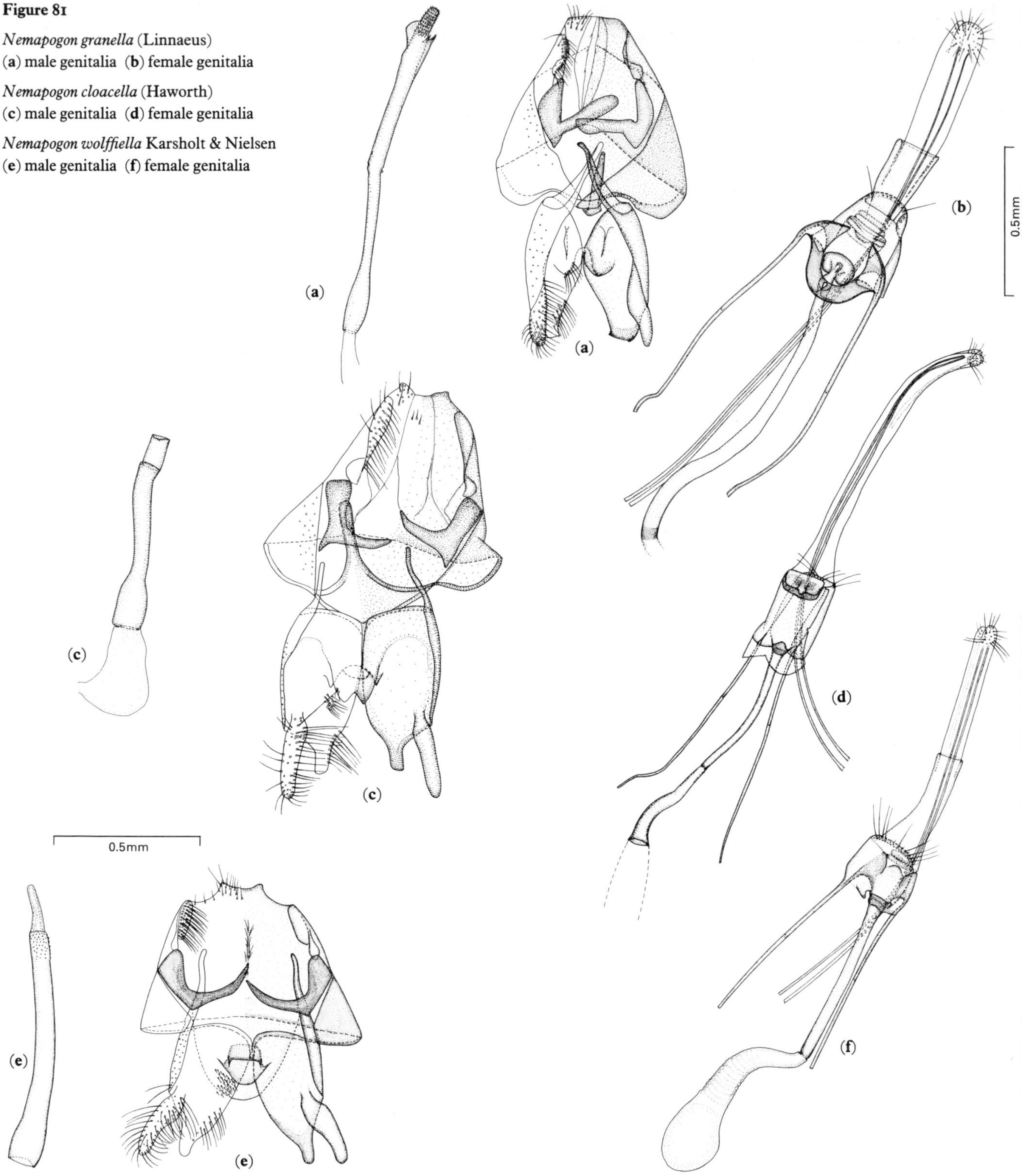

Figure 82

Nemapogon variatella (Clemens)
(**a**) male genitalia (**b**) female genitalia

Nemapogon inconditella (Lucas)
(**c**) male genitalia (**d**) female genitalia

Nemapogon ruricolella (Stainton)
(**e**) male genitalia (**f**) female genitalia

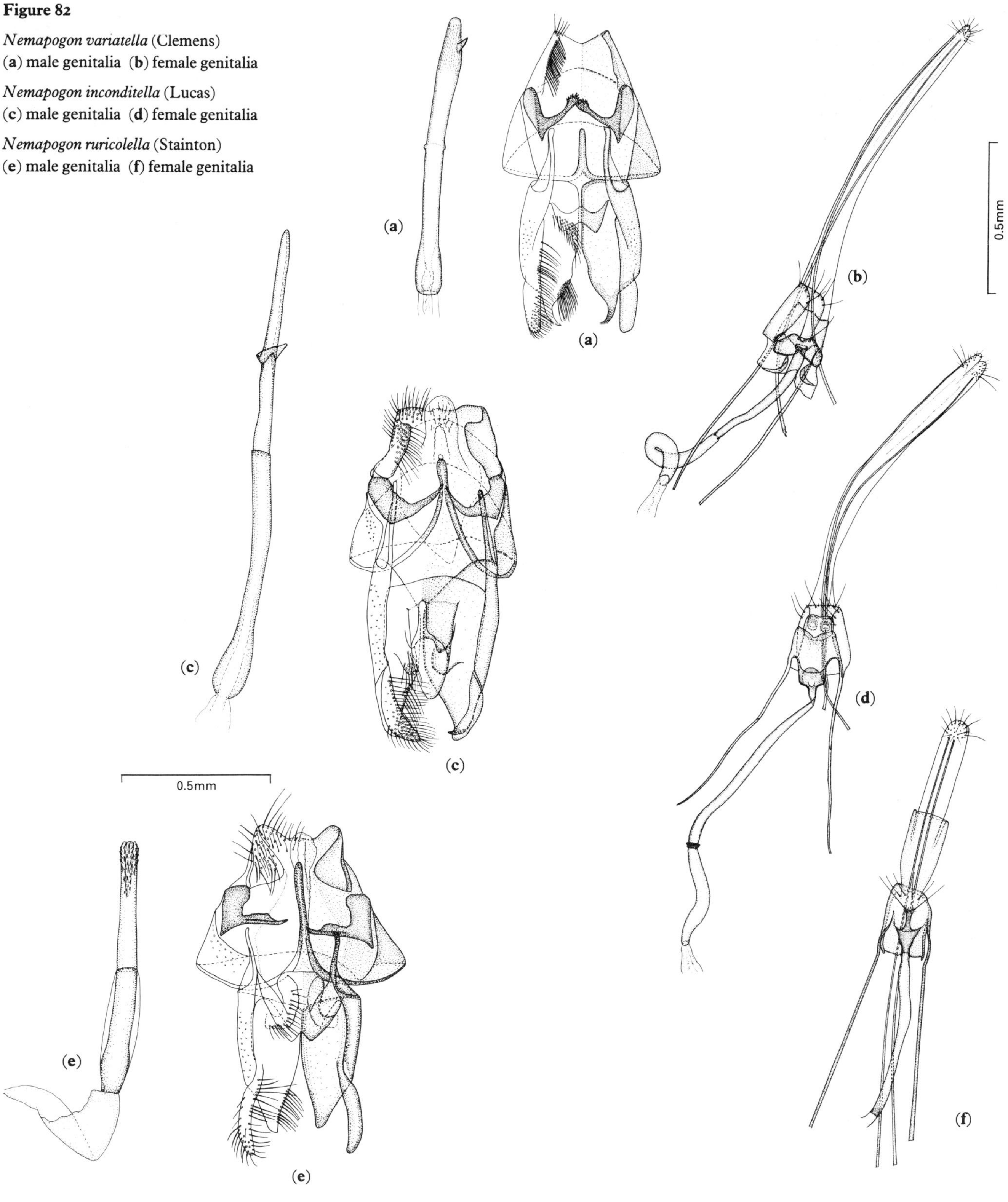

(**a**)

(**a**)

(**b**)

0.5mm

(**c**)

(**c**)

(**d**)

0.5mm

(**e**)

(**e**)

(**f**)

NEMAPOGON VARIATELLA (Clemens)

Tinea variatella Clemens, 1859, *Proc.Acad.nat.Sci.Philad.* **11**: 257, 259.

Tinea personella Pierce & Metcalfe, 1934, *Entomologist* **67**: 217.

Tinea infimella sensu Corbet, 1943, *Entomologist* **76**: 96.

NOMENCLATURE. After Corbet had published (*loc.cit.*) *Tinea personella* Pierce & Metcalfe as a junior synonym of *T. infimella* Herrich-Schäffer, 1851, the latter name was adopted by British authors in important works such as Ford (1949) and Hinton (1956). But the Herrich-Schäffer name was later recognized as a synonym of *T. cloacella* Haworth. Recently Davis (1983) has discovered the synonymy of *Nemapogon personella* with *N. variatella*.

Type locality: [U.S.A.; Philadelphia].

Description of imago (Pl.8, fig.19)

Wingspan 10–14mm. Head white with some dark brown scales at sides. Forewing white with a variable amount of dark brown suffusion, especially towards apex; markings dark brown as follows: a basal spot on costa reaching fold, a small antemedian costal spot; a broad mark outwardly oblique from middle of costa terminating in middle of wing, with lower margin somewhat produced towards termen; three costal spots in apical half; an outwardly oblique streak from middle of dorsum nearly reaching median costal mark, sometimes reduced to an elongate spot in fold; postmedian white discal spot bordered with dark brown posteriorly; fringe barred similarly to related species. Hindwing fuscous. Genitalia, see figure 82, p. 177.

There is much variation in the shape and extent of the dark markings.

Similar species. *N. variatella* has the forewing relatively narrower than in other species of the *N. cloacella* group, and the head is a purer white. Small specimens of *N. cloacella* (Haworth) can approach it very closely, but some ochreous suffusion is nearly always present on the forewing of that species.

Life history

Larva. Similar to that of *N. granella* (Linnaeus), *q.v.*, and like that species and *N. cloacella* stated to feed both on bracket fungi out of doors and on stored vegetable products. Fungus species recorded as foodplants include *Coriolus versicolor*, *Laetiporus sulphureus* and *Polyporus squamosus*. According to Hinton (*loc.cit.*) it largely replaces *N. granella* as a pest of grain in northern Europe, but if this is true it is surprising that it has not developed such habits in this country or in North America.

Imago. Possibly bivoltine, the moth having been found

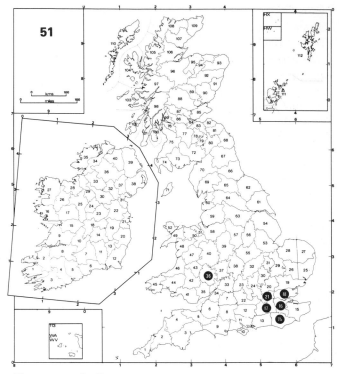

Nemapogon variatella

most commonly from March to May, but also as late as August.

Distribution (Map 51)

In Britain this species has been found mainly in the London area, but in the original description of *N. personella* Pierce & Metcalfe list a specimen from Brighton, Sussex (VC 14), and there is one in the Ffennell collection (Oxford University Museum) from Moccas Park, Hereford. Abroad it is found from northern Europe to North Africa and central Asia and in North America. It is recorded, rather surprisingly, from Iceland, to which it could have been imported with grain.

NEMAPOGON INCONDITELLA (Lucas)

Aristotelia inconditella Lucas, 1956, *Bull.Soc.Sci.nat.phys. Maroc* **35**: 255.

Nemapogon heydeni Petersen, 1957, *Beitr.Ent.* **7**: 73.

NOMENCLATURE. Introduced to the British list as *Nemapogon heydeni* but this was considered by Robinson (1982) to be a junior synonym of Lucas's species, described from North Africa in Gelechiidae.

Type locality: Morocco; Ifrane.

Description of imago (Pl.8, fig.17)

Wingspan 12–16mm. Head white, with a few dark brown hairs on crown and at sides of face. Forewing white with a variable amount of dark brown irroration or strigulation; dark brown markings as follows: costa with basal patch, extending to fold; a spot on costa beyond this; an almost rectangular slightly oblique median mark from costa to middle of wing; three spots on costa in apical third, the subapical large; an elongate spot in fold, a spot at base of dorsum and other marginal spots on termen. Hindwing fuscous. Genitalia, see figure 82, p. 177.

Similar species. N. *variatella* (Clemens) is similar in its white head and forewing, but *N. inconditella* is larger and the forewing has more extensive irroration or strigulation.

Life history

Larva. Undescribed. The moth has been bred on the Continent from *Coriolus versicolor* (Hannemann, 1977).

Distribution (Map 52)

The only known British specimen, a female, was collected on 7 July 1979 in south Devon (Pelham-Clinton, 1982). This is possibly an overlooked species which may yet be found in collections among specimens of the *N. cloacella* group. It seems likely that the British specimen came from a resident population. The species has a wide Palaearctic range, from Spain and Morocco to central Asia, but is nowhere common.

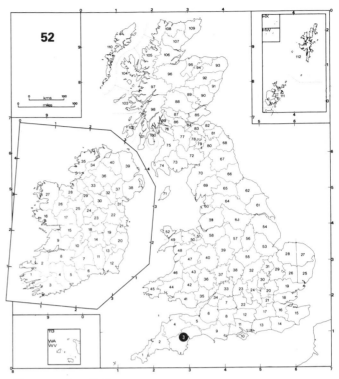

Nemapogon inconditella

NEMAPOGON RURICOLELLA (Stainton)

Tinea ruricolella Stainton, 1849, *Syst.Cat.Br.Tineidae and Pterophoridae*: 7.

Type locality: England; Lewisham.

Description of imago (Pl.8, fig.20)

Wingspan 10–14mm. Head creamy white with a few brown hairs at side of crown and of face; antenna of male subserrate and shortly pubescent. Forewing reddish buff or pale reddish brown, with whitish and brown irroration in basal half; dark brown markings as follows: a spot on costa not reaching fold; a small subbasal spot in fold; an antemedian costal spot; a large median subrectangular costal spot, narrowed in middle; a spot in fold, sometimes small and elongate, when larger almost forming an angulated fascia with median costal spot; other small marginal spots most distinct on both sides of dark spot in fold, in disc in postmedian area and around termen; fringe with median and terminal dark lines, with at least two whitish bars. Hindwing fuscous. Genitalia, see figure, 82, p. 177.

Similar species. Resembles *N. cloacella* (Haworth) most

closely, but is usually recognizable by the pale reddish brown subapical area, less mottled than in *N. cloacella*, in which the whitish postmedian spot stands out distinctly, this spot being not as clearly rounded as in *N. cloacella*.

Life history

Larva. Head yellowish brown, darkened below ocelli and posteriorly. Prothoracic plate pale yellowish brown. Body white with very indistinct pinacula. Feeds on bracket fungi including *Coriolus versicolor* and *Piptoporus betulinus*.

Imago. Univoltine, the moth appearing in June and July.

Distribution (Map 53)

Although described as a distinct species, *N. ruricolella* was later thought to be a form of *N. cloacella*, so that records have been confused until these species were clearly separated by Pierce & Metcalfe (1934). It appears now to be local in the south of England and more common in the west; farther north it occurs up to north Wales and Cheshire. In Ireland it is found in the extreme south-west. Abroad it occurs widely in Europe from France to European parts of the U.S.S.R. including Crimea.

NEMAPOGON CLEMATELLA (Fabricius)

Tinea clematella Fabricius, 1781, *Spec.Ins.* **2**: 297.

Tinea arcella sensu auctt.

Type locality: England.

Description of imago (Pl.8, fig.22)

Wingspan 12–15mm. Head white, sides of face dark brown; antenna of male subserrate. Forewing white with patches of ochreous irroration and some brownish irroration towards termen; blackish brown markings consisting of an elongate basal patch along costa, an angulate median fascia, sometimes interrupted, an irregular subapical costal mark or marks and a small dorsal spot at base. Hindwing pale fuscous.

Similar species. A species which is very distinct in appearance though closely related to species of the *N. cloacella* group according to characters of the genitalia. The female ductus bursae bears a ring of teeth otherwise in British species only found in *N. inconditella* (Lucas) (see figure 82, p. 177).

Life history

Larva. Head pale brown. Prothoracic plate not sclerotized. Body whitish. Recorded from fungal growths under bark of dead twigs of elm, oak, beech and hawthorn, the cushion fungus *Hypoxylon fuscum* on twigs of alder, the bracket fungus *Fomes fomentarius*, and from rotten wood (Petersen, 1969).

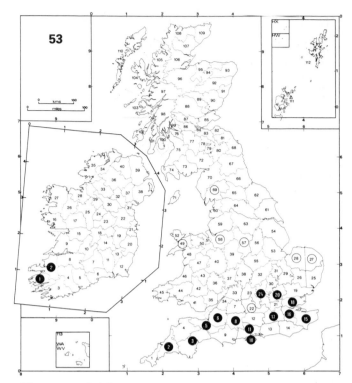

Nemapogon ruricolella

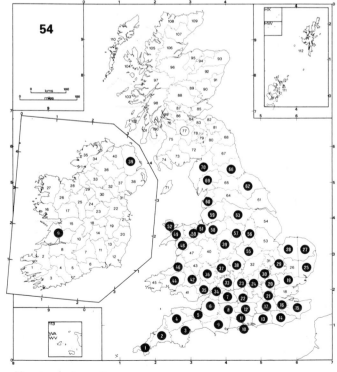

Nemapogon clematella

Imago. Univoltine, the moths flying from June to August. May be disturbed from hedges in the daytime more readily than the others of the genus – or perhaps it is more easily seen when so disturbed. The normal flight-time is at dusk or later and the moth is often taken at light.

Distribution (Map 54)

Locally common as far north as Westmorland (Cumbria) and Durham. In Ireland recorded from Cos Clare and Antrim. The latter was regarded by Beirne (1941) as doubtful as at that time it was the only Irish record. Abroad the range extends from Spain to Scandinavia, Anatolia and the Caucasus.

NEMAPOGON PICARELLA (Clerck)

Phalaena picarella Clerck, 1759, *Icones Insect.rar.* **1**: pl.10, fig.15.
Type locality: [Sweden].

Description of imago (Pl.8, fig.21)

Wingspan 12–19mm. Head white; labial palpus with outside of second segment dark brown. Forewing white; markings blackish brown with coppery gloss comprising a broad slightly sinuate streak from base of costa to near tornus and then to apex, adjoining this an antemedian costal spot and broad median mark outwardly oblique from costa, a wedge-shaped mark from base of dorsum to fold, an elongate or wedge-shaped spot obliquely across fold and marginal spots near base of dorsum, at tornus and around apex from costa to tornus, these often connected with the longitudinal streak; fringe with two dark lines and variable white bars between costa and tornus. Hindwing dark fuscous shading to whitish in basal half.

There is variation in extent of the blackish markings of the forewing, the oblique spot on fold sometimes linked to the median costal mark to form an angulated fascia.

Life history

Larva. Head shining brown. Prothoracic plate pale brown. Body whitish with broad pale grey pinacula. Feeds in bracket fungi, especially *Piptoporus betulinus*, and bores into the adjacent dead wood. Fully grown in April or May.
Imago. Univoltine, the moths flying in June and July.

Distribution (Map 55)

An uncommon species, though larvae may be found in numbers in one piece of fungus. Appears to have two centres of distribution, in Durham and in the central Highlands of Scotland, and to be rare elsewhere. Recorded from Monmouthshire (Gwent), Cheshire and Lancashire, so possibly overlooked in Wales. Not recorded from Ireland. Widespread in the Palaearctic region from Spain to east Siberia.

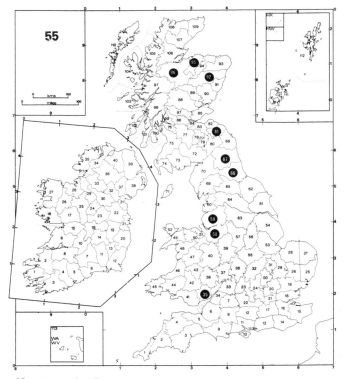

Nemapogon picarella

ARCHINEMAPOGON Zagulyaev

Archinemapogon Zagulyaev, 1964, *Zool.Zh.* **41**: 1041.

A small Palaearctic genus, separated from *Nemapogon* Schrank by the form of the genitalia, in the male by a peculiar elongate dorsal process on the aedeagus.

Imago. Head densely rough-haired. Antenna about half length of forewing, slightly serrate and pubescent in male. Labial palpus porrect or drooping, rather slender; second segment with distinct bristles; third shorter than second. Maxillary palpus nearly as long as labial palpus. Haustellum reduced.

Larva. The known larvae all feed on fungi.

ARCHINEMAPOGON YILDIZAE Koçak

Archinemapogon yildizae Koçak, 1981, *Priamus* **1**(1): 15.

Tinea laterella Thunberg, 1794, *Ins.Suecica* **7**: 94.

Tinea arcuatella Stainton, 1854, *Ins.Br.Lepid.*: 29.

NOMENCLATURE. *Tinea laterella* Thunberg is a junior primary homonym of *Tinea laterella* [Denis & Schiffermüller], 1775, *Schmett.Wien.*: 137. *T. arcuatella* Stainton is a junior primary homonym of *T. arcuatella* Schrank, 1802, *Fauna boica* **2**(2): 107, and Koçak's name is an objective replacement name for this.

Type locality: Scotland; Rannoch.

Description of imago (Pl.8, fig.23)

Wingspan 14–21mm. Head white, more or less mixed with ochreous on crown; outside of labial palpus dark brown except tip of third segment. Forewing white with variable amount of reddish or brownish ochreous irroration and fine longitudinal streaks; longitudinal blackish markings as follows: streaks from basal costal spot along and below costa; a sinuate streak in disc from before middle almost to costa near apex; a sinuate streak from base of dorsum partly below and partly in fold, ending below basal end of discal streak; all these mixed or bordered below with reddish brown; a median costal blackish mark sometimes connected to discal streak; other marginal spots blackish mixed with reddish brown; fringe with basal half brown-speckled, the whole length chequered brown and white. Hindwing fuscous.

There is much variation in extent of the dark markings.

Life history

Larva. Head red-brown. Prothoracic plate pale red-brown. Body whitish, tinged with greyish pink; pinacula inconspicuous. Feeds on bracket fungi on birch, *Piptoporus betulinus* or *Fomes fomentarius*, when on the latter normally in association with the tenebrionid beetle, *Bolitophagus reticulatus* (Linnaeus), which may be necessary to

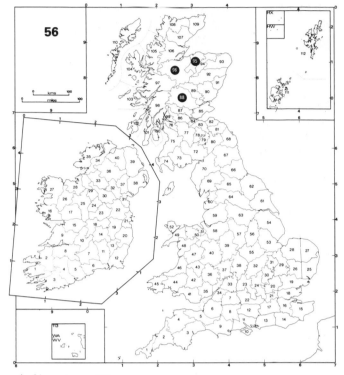

Archinemapogon yildizae

start the break-up of this very hard fungus before it can be used by the tineid. Full-grown in April.

Imago. Univoltine. The moths are on the wing from May to July and are rarely seen at large, most specimens in collections having been reared from larvae, which are sometimes abundant in a small area.

Distribution (Map 56)

In the British Isles the species is confined to the Scottish Highlands. In Glen Affric, Inverness-shire, the larvae are found mainly in *Fomes*, but in the east of the county, in Strathspey, they are found in *Piptoporus*. Abroad the species is found in north and central Europe, and eastwards to Siberia.

NEMAXERA Zagulyaev

Nemaxera Zagulyaev, 1964, *Fauna SSSR* Lepid. **4**(2): 186.

A monotypic genus, separated from *Nemapogon* Schrank by the structure of the male genitalia which have the costa of the valva produced and strongly spined.

Imago. Head densely rough-haired. Antenna about half length of forewing, shortly pubescent and subserrate in male; scape with pecten. Labial palpus porrect or somewhat drooping; second segment rather rough-scaled, the bristles inconspicuous. Maxillary palpus shorter than labial palpus. Haustellum reduced.

Larva. Similar to that of *Nemapogon*, feeding on fungi.

NEMAXERA BETULINELLA (Fabricius)

Alucita betulinella Fabricius, 1787, *Mant.Ins.* **2**: 255.
Tinea corticella Curtis, 1834, *Br.Ent.* **11**: 511.
Tinea emortuella Zeller, 1839, *Isis, Leipzig* **1839**: 184.

NOMENCLATURE. *Tinea corticella* Curtis is a primary homonym of *T. corticella* Haworth, 1828, *Lepid.Br.*: 566. Type locality: Sweden.

Description of imago (Pl.8, fig.24)

Wingspan 12–19mm. Head white. Forewing whitish, with patchy orange-brown irroration; a triangular orange-brown median mark on fold, its apex reaching or almost reaching dorsum; small dark brown spots around margin, the largest three in basal half of costa and one on costa before apex; small dark brown scales towards termen; fringe ochreous, irregularly barred with whitish. Hindwing dark fuscous.

Life history

Larva. Head yellowish brown. Prothoracic plate dark brown. Body white with very indistinct pinacula. Feeds on bracket fungi, especially *Piptoporus betulinus* and *Coriolus versicolor*, and occasionally on rotten wood. Fully grown in May.

Imago. Moths are found from May to August and a second generation has been recorded on the Continent.

Distribution (Map 57)

Very local in England as far north as Westmorland (Cumbria) and Northumberland. There are records from Co. Dublin which were accepted by Beirne (1941). Usually rare, but occasionally abundant in a small area. Abroad it is found from France through Scandinavia, northern and central Europe to central Siberia.

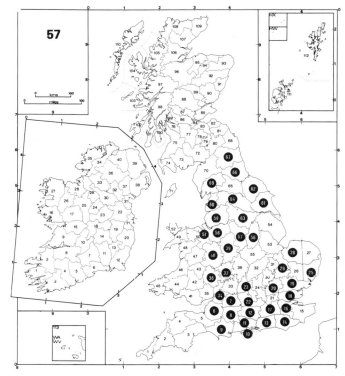

Nemaxera betulinella

TRIAXOMERA Zagulyaev

Triaxomera Zagulyaev, 1959, *Ent.Obozr.* **38**: 879.

A genus of three Palaearctic species, two of them British, separated from *Nemapogon* Schrank by the ciliate antenna and by the structure of the genitalia in the male.

Imago. Head densely rough-haired. Antenna half to two-thirds length of forewing, in male subserrate and with cilia as long as diameter of shaft; scape with pecten. Labial palpus porrect; second segment rough-scaled with long bristles. Maxillary palpus as long as labial palpus. Haustellum reduced.

Larva. Similar to that of *Nemapogon* but with only five instead of six ocelli on each side. Setae arise from well-marked pinacula. Feeds on fungi.

TRIAXOMERA PARASITELLA (Hübner)

Tinea parasitella Hübner, 1796, *Samml.eur.Schmett.* **3**: 16.
Type locality: Germany; Augsburg.

Description of imago (Pl.8, fig.25)

Wingspan 16–21mm. Head brownish ochreous, sometimes mixed with dark brown at side of crown. Forewing pale brownish ochreous, with many small white and dark brown spots mainly between veins; most clearly marked specimens with dark costal spots continued across wing as three irregular fasciae, the outer two merging in disc; a series of yellowish pre-apical costal spots; fringe chequered with white outside a dark median line. Hindwing fuscous with purple gloss. Tarsal segments dark with whitish apical rings.

Life history

Larva. Head reddish or dark brown, with margins of sclerites blackish. Prothoracic plate dark brown. Body white with well- marked pale brown pinacula. Feeds on a variety of bracket fungi, including *Coriolus versicolor*, and has been recorded from dead wood. It is fully grown in April.

Imago. Univoltine. Moths are on the wing from May to July, flying at dusk and after dark. Sometimes there is a flight about sunrise.

Distribution (Map 58)

Widely distributed and common in the south of England, but not recorded from Cornwall or from most of Wales. Further north it is more local, but is found as far north as Stirlingshire. Not recorded from Ireland. The continental distribution covers most of Europe and extends to Anatolia.

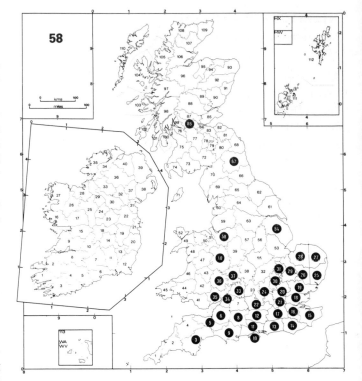

Triaxomera parasitella

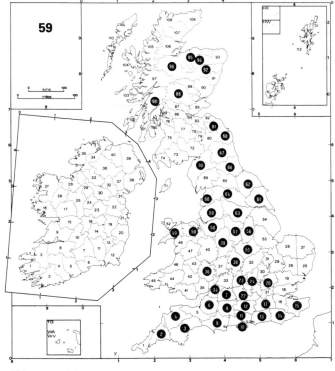

Triaxomera fulvimitrella

TRIAXOMERA FULVIMITRELLA (Sodovsky)

Tinea fulvimitrella Sodovsky, 1830, *Bull.Soc.Nat.Moscou* **2**: 74.

Type locality: Latvia; Livonia (now Latvian S.S.R.).

Description of imago (Pl.8, fig.26)

Wingspan 15–22mm. Head white to ochreous, with dark brown hairs at side of crown and face. Forewing blackish brown with slight coppery or purple gloss; four large white triangular or subquadrate marginal spots, each including blackish dots or strigulae, on costa antemedian and subapical, on dorsum near base and at tornus; sometimes additional white costal spots and smaller whitish spots towards termen; fringe chequered with white outside a dark median line. Hindwing rather dark purplish fuscous. Tarsal segments blackish with white apical rings.

Life history

Larva. Superficially identical with *T. parasitella* (Hübner), *q.v.* Feeds on various bracket fungi including *Inonotus radiatus* and *Piptoporus betulinus*. In the north it is particularly associated with *Piptoporus* but southern records suggest a preference for fungi on beech and oak. Larvae have also been found in dead wood and in callus-tissue around tree wounds. Fully grown in April.

Imago. Univoltine, the moths fly from May to July. This species is often found at rest on tree-trunks, its size and colouring making it more easy to see than others which no doubt rest undetected in similar situations.

Distribution (Map 59)

Found locally through most of mainland Britain except for the far north; it is perhaps most common in the central Highlands of Scotland. Not recorded from Ireland. Distributed abroad from France and Italy to Scandinavia, European U.S.S.R. from Crimea to the extreme north and Siberia as far east as Irkutsk.

TRIAXOMASIA Zagulyaev

Triaxomasia Zagulyaev, 1964, *Fauna SSSR* Lepid. **4**(2): 155.

A monotypic genus, sharing with *Triaxomera* Zagulyaev the presence of small spines at the apex of the tarsal segments – these are not easily seen without denuding a tarsus. Compared with *Triaxomera* the wings are narrow, the male antenna not ciliate and the genitalia of the male relatively simple.

Imago. Head densely rough-haired. Antenna about two-thirds length of forewing, subserrate in male; scape with pecten. Labial palpus rather short, length less than twice eye diameter, drooping; second segment with conspicuous bristles. Maxillary palpus longer than labial palpus. Haustellum much reduced. Hindwing veins 5 (M_2) and 6 (M_1) connate.

Larva. Reported to feed in dead wood.

TRIAXOMASIA CAPRIMULGELLA (Stainton)

Tinea caprimulgella Stainton, 1851, *Suppl.Cat.Br. Tineidae and Pterophoridae*: 2.

Type locality: Britain.

Description of imago (Pl.8, fig.27)

Wingspan 9–11mm. Head yellow-ochreous, brownish at sides. Forewing dark fuscous with coppery gloss; markings whitish, comprising a spot at base of dorsum, an indistinct antemedian fascia including dark strigulae, a spot at tornus and three costal spots in apical half; base of fringe pale coppery fuscous, outer half whitish, somewhat shining. Hindwing purplish fuscous; fringe shading to whitish on outer half.

Life history

Larva. Head shining brown. Prothoracic plate weakly developed. Body yellowish. Reported to feed till the spring on dead wood of oak, beech and elm. However the situations in which the moths have been found suggest that the larval food might include dead insects in spiders' webs.

Imago. Univoltine. The moth flies in June and July. A little-known species, obscure in appearance and habits. Warren (1880) wrote of it in Hyde Park, London, 'Very lazy, very local, and fond of dark corners: it may often be found hanging in cobwebs, where it remains perfectly still, apparently secure from the attacks of spiders.' Goss & Fletcher (1905) recorded it at Arundel, Sussex, from hollow trunks of ash. E. Bradford (*in litt.*) has found it in Kent near an old gnarled oak which probably harbours the species. An association with the insides of hollow trees

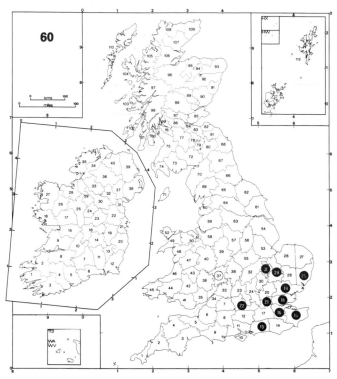

Triaxomasia caprimulgella

and, in Hyde Park, sheds, suggests that these might be breeding places and in this case dead insects might be the larval food (*cf. Diplodoma herminata* (Geoffroy), Psychidae, p. 131). However Wakely (1958b), recording moths resting on the smooth wood of scars on large elms in London parks, considered that the larvae had fed on the wood, perhaps on the scar tissue.

Distribution (Map 60)
British records are limited to the south-east of England, to Berkshire in the west and Cambridge and Suffolk in the north. Hyde Park, London, used to be the best-known locality and most specimens in collections are so labelled, but the most recent captures have been in Kent and East Anglia. Abroad it has been found locally in western Europe, as far east as Macedonia and northwards to Denmark.

Tineinae

MONOPIS Hübner
Monopis Hübner, [1825], *Verz.bekannt.Schmett.*: 401.

The genus is world-wide and includes species which have been carried by man to several geographical regions.

Imago. Head densely rough-haired. Antenna nearly as long as forewing, in male at most shortly pubescent; scape usually with pecten. Labial palpus porrect or drooping; second segment somewhat expanded with scales towards apex, its bristles mostly bunched at apex. Maxillary palpus longer than labial palpus. Haustellum reduced. Forewing with veins 3 (Cu_1) and 4 (M_3) and sometimes other veins stalked; a rounded or oval hyaline spot filling end of cell. Hindwing sometimes with veins 5 (M_2) and 6 (M_1) stalked.

Larva. Mostly feeding on animal products including dry faecal matter, but a few species on dry plant materials.

MONOPIS LAEVIGELLA ([Denis & Schiffermüller])
Skin Moth
Tinea laevigella [Denis & Schiffermüller], 1775, *Schmett. Wien.*: 139.
Tinea rusticella Hübner, 1796, *Samml.eur.Schmett.* 8: 61, pl.49, fig.339.
NOMENCLATURE. The long-established synonymy quoted above was forgotten until revived recently by Leraut (1983).
Type locality: [Austria]; Vienna district.

Description of imago (Pl.8, fig.31)
Wingspan 13–20mm. Head whitish or yellow-ochreous. Forewing pale violet-grey, densely irrorate with whitish and with blackish brown with coppery gloss; hyaline spot well before middle of wing (distance to base about 0.4 of wing length), surrounded with whitish scales; a series of indistinct whitish subcostal spots; fringe indistinctly chequered, with dark median line. Hindwing fuscous. Fore- and midtibia and tarsal segments blackish with white apical ring; hind tarsal segments less distinctly ringed.

Variation occurs in the amount of irroration, light or dark, of the forewing; sometimes a small whitish tornal mark is developed. Throughout most of the British and continental range the head colour is as described, but in St Kilda the head is invariably some shade of brown; this form occurs rarely on the main islands of the Outer Hebrides.

Similar species. M. weaverella (Scott), *q.v.*

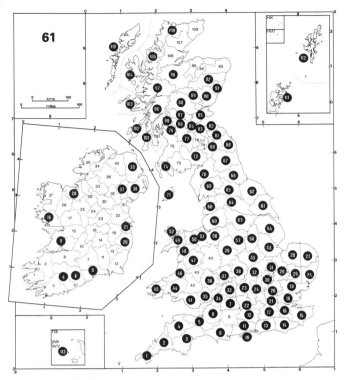

Monopis laevigella

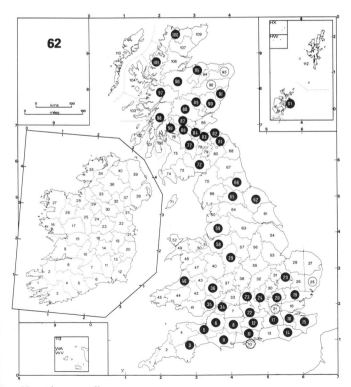

Monopis weaverella

Life history

Larva. Head light to dark brown with margins of sclerites usually strongly darkened. Prothoracic plate pale yellowish, divided medially. Body whitish; pinacula not developed. Feeding in a silken tunnel and making much webbing, on a great variety of foodstuffs of animal origin, mainly out of doors or in unheated buildings. It is one of the most common inhabitants of birds' nests of many kinds, and will feed on owl pellets and very often on dead birds and mammals; sometimes on various stored animal products including bird guano.

Imago. Found from May to September and perhaps bivoltine in the south. Forbes ([1924]) states that it has two broods in North America. The moth flies at dusk and after dark.

Distribution (Map 61)

The most universally distributed British tineid, found in all parts of the British Isles including Orkney, Shetland and the Outer Hebrides. It is the only tineid to have been recorded from St Kilda where the dark-headed race frequents cliffs and not the buildings of the village. There are gaps in the recorded distribution, especially in Ireland, but these are undoubtedly due to a lack of recording. The species has a Holarctic distribution, but it is stated by Forbes (*loc.cit.*) to be of sporadic occurrence in North America. It has a widespread distribution in Iceland and has been recorded from Greenland.

MONOPIS WEAVERELLA (Scott)

Tinea weaverella Scott, 1858, *Zoologist* **16**: 5964.
Type locality: Scotland; Rannoch, Perthshire.

Description of imago (Pl.8, fig.32)

Wingspan 13–18mm. Head whitish or yellow-ochreous. Forewing violet-grey irrorate with blackish brown with coppery gloss; hyaline spot just before middle of wing (distance to base about 0.45 of wing length), surrounded by whitish scales; slight whitish irroration along fold and below costa; a conspicuous yellowish white spot at tornus, continued into fringe; a variable number of yellowish white pre-apical costal spots and larger terminal spots, some continued into fringe; fringe pale ochreous outside a dark interrupted median line. Hindwing fuscous or dark fuscous with coppery purple gloss. All tarsi dark, variably ringed white at apex.

Similar species. Most easily distinguished from *M. laevigella* ([Denis & Schiffermüller]) by the large, usually roughly triangular pale spot at the tornus of the forewing, but *M. laevigella* sometimes has a pale spot in this position. When seen in a series the position of the hyaline spot, nearer the middle in *M. weaverella*, is a good recognition feature. The whole forewing of *M. laevigella* is more irrorate with rather transparent whitish scales or with the paler bases of dark scales so that the wing has a more mottled appearance compared with *M. weaverella*; the contrast between the darker wing and pale fringe of *M. weaverella* is more striking.

Life history

Larva. Apparently undescribed. Hannemann (1977) recorded it from birds' nests and dove-cotes, but if this were true the species should frequently have been bred in Britain, for it is not uncommon (see Pelham-Clinton, 1983). However Bland (1984) has recently recorded the species from fox faeces and Sterling (1984 and pers.comm.) from carcases of rabbit and fox.

Imago. Most frequently seen in June, but sometimes found from May to August.

Distribution (Map 62)

Widely distributed in mainland Britain and recorded from Orkney but not from Ireland. Much less common than *M. laevigella* in the south, but more frequent farther north, especially in the Scottish Highlands. Although described in 1858 the species was almost forgotten until Bankes (1910) clearly distinguished it from *M. laevigella*. Abroad it is local in northern Europe, from France to Scandinavia and European U.S.S.R.

MONOPIS OBVIELLA ([Denis & Schiffermüller])

Tinea obviella [Denis & Schiffermüller], 1775, *Schmett. Wien.*: 143.

Tinea ferruginella Hübner, [1813], *Samml.eur.Schmett.* 8: pl.51, fig.348, *nec* Thunberg, 1788.

NOMENCLATURE. Leraut (1983) has pointed out that not only is *Tinea ferruginella* Hübner a junior synonym of *T. obviella* [Denis & Schiffermüller] but also a junior primary homonym of *T. ferruginella* Thunberg, 1788.

Type locality: [Austria]; Vienna district.

Description of imago (Pl.8, fig.28)

Wingspan 10–13mm. Head yellow-ochreous. Thorax pale yellow; tegulae dark brown. Forewing dark brown, with copper and purple gloss; hyaline spot in middle, elongate; a broad pale yellow dorsal streak from base to tornus, expanded before middle almost to fold; a patch of shining whitish irroration between hyaline spot and costa and a number of scattered spots of same colour mostly forming a subcostal series; a few yellowish white spots, three or four pre-apical costal and one in terminal fringe; fringe pale coppery fuscous, paler outside a dark median line. Hind-

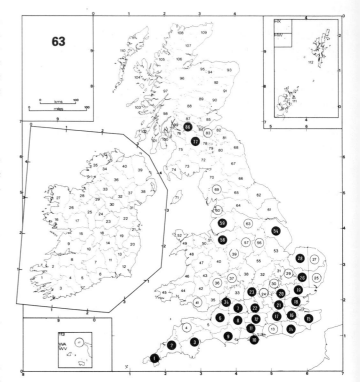

Monopis obviella

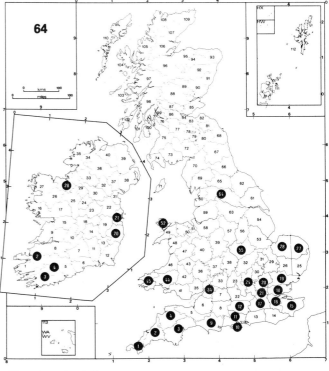

Monopis crocicapitella

wing dark fuscous with copper and purple gloss. Tarsal segments blackish with white subapical rings.

Similar species. Monopis crocicapitella (Clemens), *q.v.*

Life history

Larva. Similar to that of a small *M. laevigella* ([Denis & Schiffermüller]), but lateral dark margin of head narrower. More likely than *M. laevigella* to be found on manufactured woollen materials in outbuildings or out of doors, and less frequently in birds' nests or pellets of birds of prey. It has also been reported to feed on seeds (Bankes, 1912). Stainton (1854) recorded it as abundant in a coalmine at Campsie, Stirlingshire.

Imago. Found from May to October, probably in two generations.

Distribution (Map 63)

Locally common in southern Britain, but much scarcer northwards as far as Stirlingshire. As *M. crocicapitella* was not separated from *M. obviella* as a distinct British species until 1893 (Richardson, 1893, as *Blabophanes heringi*) and not clearly characterized until 1912 (Bankes, *loc.cit.*) the earlier records (and even some of the later ones) were confused. A two-symbol map has therefore been adopted, open circles for unconfirmed records. A number of Irish records were given by Beirne (1941) but all Beirne's specimens and other Irish specimens seen in the National Museum of Ireland, Dublin are *Monopis crocicapitella*. Abroad it is found throughout Europe, in North Africa and as far east as the Caucasus.

MONOPIS CROCICAPITELLA (Clemens)

Tinea crocicapitella Clemens, 1859, *Proc.Acad.nat.Sci. Philad.* **11**: 257.

Type locality: U.S.A.

Description of imago (Pl.8, fig.29)

Wingspan 10–16mm. Head yellow-ochreous. Thorax pale yellow-ochreous; tegulae dark brown. Forewing brown or dark brown with slight copper and purple gloss, more or less irrorate with pale ochreous or whitish, especially near costa and in apical third; hyaline spot in middle, oval; a broad pale yellow-ochreous dorsal streak from base to tornus, towards tornus meeting fold; fringe pale shining ochreous, whiter beyond an incomplete dark median line. Hindwing pale grey, darker towards apex. Tarsal segments blackish, variably ringed white at apex.

Similar species. M. obviella ([Denis & Schiffermüller]) has the forewing with a pale yellow, not ochreous, dorsal streak and pale irroration limited to a sprinkling of whitish scales near costa. The hindwing of *M. obviella* is dark fuscous, not pale grey.

Life history

Larva. Apparently undescribed, though reported as feeding on a variety of materials including flour, oats, other seeds and woollen refuse. Moths are sometimes common in poultry-houses, and have been bred from pigeons' nests.

Imago. Probably bivoltine, the moths occurring from June to October. In wild habitats as well as around outhouses, but rarely indoors. It flies in late evening and at dusk and sometimes comes to light.

Distribution (Map 64)

More restricted than *M. obviella* though sometimes abundant where it occurs. It is more common in maritime districts and especially so in south-western counties. Recorded as far north as Yorkshire and widely distributed in Ireland. The species has a very wide distribution around the world, including Asia, Australia, New Zealand, Africa, St Helena, North America and Hawaii.

MONOPIS IMELLA (Hübner)

Tinea imella Hübner, [1813], *Samml.eur.Schmett.* **8**: pl.51, fig.347.

Type locality: Europe.

Description of imago (Pl.9, fig.10)

Wingspan 11–14mm. Head whitish, ochreous at sides, to yellow-ochreous. Front of thorax yellow-ochreous. Forewing purplish fuscous to dark fuscous, shining, with sparse sprinkling of ochreous whitish scales, mostly near costa and in apical third; hyaline spot in middle, small, oval, often partially obscured with dark scales; costa finely edged yellow-ochreous from before middle to near apex; dorsum sometimes ochreous between margin and fold; fringe pale coppery fuscous with partial darker median line, becoming pale ochreous near tornus. Hindwing pale shining purplish fuscous, darker towards apex.

A rare form has the forewing entirely pale golden brown.

Similar species. The uncommon form with an ochreous streak along the dorsum of the forewing might be mistaken for *M. obviella* ([Denis & Schiffermüller]) or *M. crocicapitella* (Clemens), but those have a larger hyaline spot and lack the ochreous edge of the costa.

Life history

Larva. Apparently undescribed. Usually on woollen refuse on the ground in the open, but also recorded from dead animals and from birds' nests.

Imago. Moths are found from June to late September, probably in two generations, and fly in late evening and after dark, when they frequently come to light.

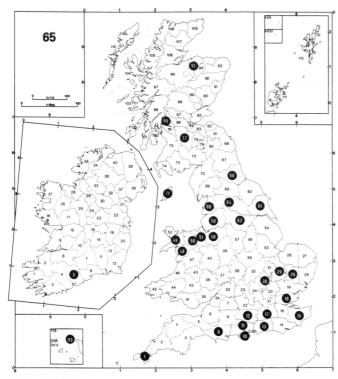

Monopis imella

Distribution (Map 65)

Found more in wild habitats than other British species of the genus and in the British Isles most common near the sea, but very local. Widely distributed in Britain as far north as Morayshire and recently found for the first time in Ireland in Co. Cork. Widely distributed in the Palaearctic region from Morocco through Europe to east Siberia.

MONOPIS MONACHELLA (Hübner)

Tinea monachella Hübner, 1796, *Samml.eur.Schmett.* **8**: 65, pl.21, fig.143.
Type locality: Germany; Augsburg district.

Description of imago (Pl.8, fig.30)

Wingspan 12–20mm. Head white. Antennal scape white; pecten absent. Labial palpus white mixed with brownish. Thorax creamy white; tegulae anteriorly dark fuscous. Forewing dark coppery fuscous with numerous small purplish grey spots composed of scales which form raised tufts towards wing-base, the two largest tufts in fold; a large subtrapezoidal creamy white costal blotch, the dorsal edge concave, including the rounded hyaline spot in its inner dorsal angle. Hindwing pale brassy or purplish fuscous.

Life history

Larva. Apparently undescribed. On skins, owl pellets and dead animals and in birds' nests.
Imago. Found from May to September.

Distribution (Map 66)

A rare and little-known species in the British Isles. Stainton (1854) referred to it in the Cambridgeshire fens as if it were resident, and this may have been so, as it was later recorded also from Yaxley Fen and Whittlesea Mere in Huntingdonshire. The most recent records, in 1973 and 1975, have been from east Suffolk and east Norfolk respectively. Not recorded from Ireland. This species has usually been quoted as having a world-wide distribution, but it has recently been found (K. R. Tuck & G. S. Robinson, unpublished) to be one of a complex of similar species. Our species is widespread in the Palaearctic region including the Himalayas; it also occurs in Sri Lanka.

MONOPIS FENESTRATELLA (Heyden)

Tinea fenestratella Heyden, 1863, *Stettin.ent.Ztg* **24**: 342.
Type locality: Germany; Wetterau district.

Description of imago (Pl.8, fig.34)

Wingspan 11–16mm. Head dull ochreous. Labial palpus ochreous; third segment dark brown, whitish at apex. Thorax dark brown, mixed with ochreous. Forewing dark coppery brown with purple gloss; hyaline spot in middle, rather elongate; base sprinkled with ochreous scales towards dorsum; a large triangular dull ochreous dorsal spot; a similar spot slightly beyond it on costa, produced apicad along costa; fringe ochreous. Hindwing brassy fuscous.

Life history

Larva. Undescribed. In birds' nests. Recorded by continental authors (*e.g.* Hannemann, 1977) from hornets' nests, dry plant material, dead wood and fungi.
Imago. Recorded from May to August.

History and distribution (Map 67)

First recorded in the British Isles from Chatteris, Cambridgeshire, in 1877 (Ruston, 1879), when it was thought to be breeding in rotten elm stumps. Since then it has been recorded from Surrey, Nottinghamshire and Lanarkshire, but these records seem unlikely and without specimens they cannot be accepted. The most recent record is one from Warwickshire (Simpson, 1981) of three specimens bred from a kestrel's nest. Abroad this is also a very little-known species, the few records extending from France to Ukraine.

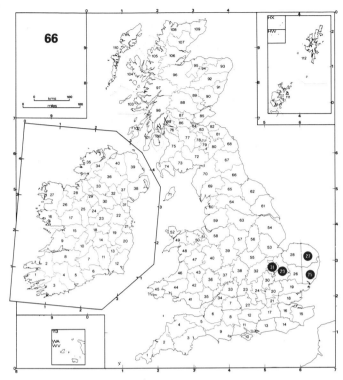

Monopis monachella

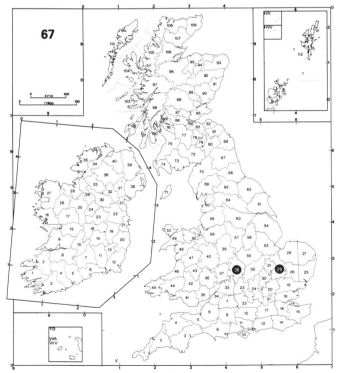

Monopis fenestratella

TRICHOPHAGA Ragonot

Trichophaga Ragonot, 1894, *Annls Soc.ent.Fr.* **63**: 123.

The genus includes a number of superficially similar species one of which is resident in Britain. The other species on the British list was overlooked until recently but is presumably a casual introduction. Yet other species might be found in British collections, particularly *T. scandinaviella* Zagulyaev which has the forewing apex with a larger dark area and the hindwing darker (Jalava & Kyrki, 1980).

Imago. Head rough-haired. Antenna about two-thirds length of forewing, in male pubescent; scape with pecten. Labial palpus porrect; second segment rough-scaled, with long bristles bunched at apex; third segment slender, as long as second. Maxillary palpus only half as long as labial palpus. Haustellum reduced. Forewing veins 9 (R$_3$), 10 (R$_2$), 11 (R$_1$) and 12 (Sc) not reaching costa, sometimes some of these anastomosing; space between veins 11 (R$_1$) and 12 (Sc) somewhat dilated and thinly scaled.
Larva. In structure closely related to *Monopis* Hübner. Feeds on hair, feathers and other animal materials.

TRICHOPHAGA TAPETZELLA (Linnaeus)
Tapestry Moth

Phalaena (Tinea) tapetzella Linnaeus, 1758, *Syst.Nat.* (Edn 10) **1**: 536.
Type locality: not stated.

Description of imago (Pl.8, fig.33)
Wingspan 15–22mm. Head white, brown at sides of crown and face; antennal scape white. Thorax white; tegulae pale fuscous. Forewing white, ochreous white on costa and towards apex, with sparse pale grey strigulae; basal third dark purplish fuscous, edge of dark area inwardly oblique from near middle of costa; a rounded postmedian spot in disc mixed pale grey and fuscous; apex and at least two small subapical spots blackish; fringe white, blackish around apex. Hindwing pale brassy fuscous.

Life history
Larva. Head pale to dark brown, anterior sclerites and lateral and occipital margins darker. Prothoracic plate pale brown. Body white, the pinacula not visible. Feeds in a silken tube amongst the food material and is fully grown in April. Food materials are normally fur, hair or feathers, either naturally occurring as in birds' nests or pellets of birds of prey, or in manufactured articles. It has never been common in heated dwellings but formerly was often a pest in unheated buildings.
Imago. Found from May to August.

Distribution (Map 68)

At one time widespread, abundant and troublesome. Since the 1939–45 war the species has suffered a remarkably sudden decline and is now rare in the British Isles. The introduction of more effective insecticides and of artificial fibres must have been the main agents of collapse of the populations in buildings, but probably the species can persist out of doors. It formerly occurred in most parts of mainland Britain and probably had a like distribution in Ireland. The most recent British record may be of specimens bred from barn owl pellets in Yorkshire in 1974 (H. E. Beaumont, *in litt.*). The species is distributed widely in temperate and warm temperate regions of the world.

TRICHOPHAGA MORMOPIS Meyrick

Trichophaga mormopis Meyrick, 1935, *Exot.Microlepid.* **4**: 575.

Type locality: Belgian Congo; Elizabethville (now Zaire; Lubumbashi).

This species was added to the British list on the strength of a single specimen introduced in a cargo of feathers from Formosa (Robinson, 1978). Since that record was published, a series of 11 specimens has been found in the L. T. Ford collection in BMNH; they are labelled 'I. of Wight, Sept. 1930, L. T. Ford/ex L. on fur'. This species is smaller than *T. tapetzella* (Linnaeus) and the outer margin of the dark part of the forewing is transverse, not oblique. *T. mormopis* is widely distributed throughout the Old World tropics.

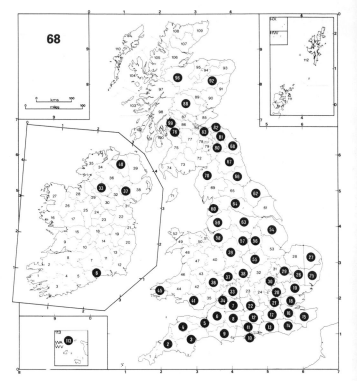

Trichophaga tapetzella

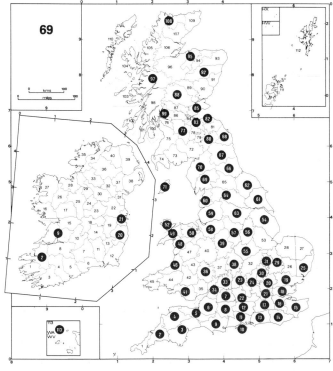

Tineola bisselliella

TINEOLA Herrich-Schäffer

Tineola Herrich-Schäffer, 1853, *Syst.Bearb.Schmett.Eur.* **5**: 23.

The genus contains two species, one of world-wide distribution and one restricted to southern Africa.

Imago. Head densely rough-haired. Antenna about three-quarters length of forewing; scape with pecten. Labial palpus rather short, porrect or drooping; second segment with conspicuous bristles. Maxillary palpus very short. Haustellum absent.

Larva. Head without convex ocellar lenses. Feeds in a thin tube on various animal and vegetable materials.

TINEOLA BISSELLIELLA (Hummel)
Common Clothes Moth
Tinea bisselliella Hummel, 1823, *Essais Ent.* **3**: 13.
Type locality: U.S.S.R.; St Petersburg (now Leningrad).

Description of imago (Pl.8, fig.35)
Wingspan 9–16mm. Head pale ochreous to reddish ochreous, sometimes brownish at sides. Forewing pale ochreous buff, somewhat glossy, sometimes darker at base; base of costa dark brown. Hindwing pale shining grey, tinted ochreous along veins and around margin.

Life history
Larva. Head pale brown with edges of sclerites darker. Prothoracic plate hardly darker than rest of integument. Body white. Pinacula very indistinct. Feeds in a flimsy white silken tube to which particles of food material adhere; much other webbing is produced and occasionally a flat mat of webbing is constructed covering several larvae. Fixed cylindrical cases may be made for moulting. The food is mostly of animal origin and includes a very wide range of materials, many of which are listed by Hinton (1956). Although this is a well-known pest of woollen fabrics and other animal fibres, as Hinton (*loc.cit.*) emphasizes, keratin is not only an unnecessary constituent of the diet, but the larvae actually develop faster with lower mortality on other materials such as fish meal or even flour. Indoors there is a continuous succession of generations, but periods of diapause may occur. Occasionally found out of doors in birds' nests, but usually in those around buildings; also in nests of social Hymenoptera.

Pupa. In a spindle-shaped cocoon in the larval spinning.

Imago. There is a natural flight of males in late afternoon, but otherwise both sexes remain mostly on the food material and run rapidly when disturbed. Moths may be found from February to September.

Distribution (Map 69)
Formerly an abundant and destructive household pest, found in most parts of the British Isles, though not recorded from the Orkney or Shetland Islands or the Outer Hebrides. It was often the most abundant clothes-moth and where it occurred it usually outnumbered *Tinea pellionella* (Linnaeus). It is now much more local than that species and is not often seen, but sporadic infestations still occur. It has a world-wide distribution, having been carried by man to all temperate and tropical regions.

NIDITINEA Petersen

Niditinea Petersen, 1957, *Beitr.Ent.* 7: 134.

A small Holarctic genus separated from *Tinea* Linnaeus by genitalia characters, in the male by the armature of the anellus and short saccus, and in the female by a pair of elongate signa.

Imago. Head densely rough-haired. Antenna about three-quarters length of forewing; scape with pecten. Labial palpus porrect or drooping; second segment with conspicuous bristles. Maxillary palpus shorter than labial palpus. Haustellum reduced. Forewing with hyaline patch near base below costa, between veins 12 (Sc) and base of 11 (R_1), covered with transparent scales on upper and lower surface of wing; the retinaculum is extended along vein 12 (Sc) and partially conceals the patch from below.

Larva. Differs only in chaetotaxy from those of the genus *Tinea*, but not constructing a portable case. Normally feeds in birds' nests.

NIDITINEA FUSCELLA (Linnaeus)
Brown-dotted Clothes Moth

Phalaena (Tinea) fuscella Linnaeus, 1758, *Syst.Nat.* (Edn 10) 1: 539

Tinea spretella [Denis & Schiffermüller], 1775, *Schmett. Wien.*: 142.

Tinea fuscipunctella Haworth, 1828, *Lepid.Br.*: 562.

Type locality: not stated.

Description of imago (Pl.8, fig.36)

Wingspan 11–17mm. Head brownish ochreous or ferruginous brown. Forewing pale brownish ochreous, more or less irrorate with dark brown or fuscous; a distinct dark brown rounded second discal stigma; other dark brown markings variable, usually including an elongate spot above fold at base often joined to a basal costal streak, a subbasal dorsal spot sometimes linked to a subbasal spot in fold, plical and first discal stigmata and a weak tornal spot; termen indistinctly chequered, the dark spots extending into basal part of fringe; outer half of fringe paler outside a weak dark median line. Hindwing pale bronzy fuscous. Tarsal segments dark with pale apices. Genitalia, see figure 84, p. 196.

Similar species. *Haplotinea* spp. have the forewing with very indistinct stigmata. *Niditinea piercella* (Bentinck) has a paler head and forewing with reduced but more contrasting markings; small pale specimens of *N. fuscella* are similar to *N. piercella* and the genitalia of these should be examined. *Tinea* spp. have the forewing with smooth ground colour without irroration.

Life history

Larva. Head dark brown, with edges of sclerites blackish. Prothoracic plate yellowish or reddish brown. Body white. Pinacula indistinct. In the nests of a variety of birds and in poultry-houses, feeding in a silken tube on feathers and other animal fibres, also perhaps on dry faeces. It has also been recorded as feeding on various stored vegetable materials.

Imago. Both sexes are active on the wing in late afternoon. The moths appear from May to September.

Distribution (Map 70)

Widely distributed in the British Isles but becoming more scarce northwards. The most northerly Scottish records require confirmation. Before 1943 *Niditinea piercella*, which is similar in appearance, was unrecognized in Britain, but this should have little effect on the records, *N. piercella* having specialized habits and being much scarcer than *N. fuscella*. The species has been recorded from most geographical regions but probably its true distribution is Holarctic, closely related species replacing it in other regions.

NIDITINEA PIERCELLA (Bentinck)

Tinea piercella Bentinck, 1935, *Tijdschr.Ent.* 78: 238.

Type locality: Netherlands; Overseen.

Description of imago (Pl.8, fig.37)

Wingspan 10–14mm. Head whitish ochreous. Forewing whitish ochreous, irrorate or reticulate with fuscous; plical, first discal and rounded second discal stigmata usually distinct, blackish brown; a blackish spot near base of costa; distinct fuscous chequering on apical margin; fringe with fuscous median line. Hindwing pale shining grey. Tarsal segments dark with pale apices. Genitalia, see figure 84, p. 196.

Similar species. *N. fuscella* (Linnaeus), *q.v.*.

Life history

Larva. Undescribed. The moth has been reared exclusively from nests in holes, mainly of various species of birds: it is commonly found in nest-boxes. On the Continent it has been recorded also from moles' nests.

Imago. The moths fly from June to August.

Distribution (Map 71)

First recognized in Britain by Wakely (1943) who reared moths from a sparrow's nest in a hole made by a woodpecker. Since then found locally in many localities in southern England and as far north as Yorkshire. Known in continental Europe in several countries from Italy to Scandinavia.

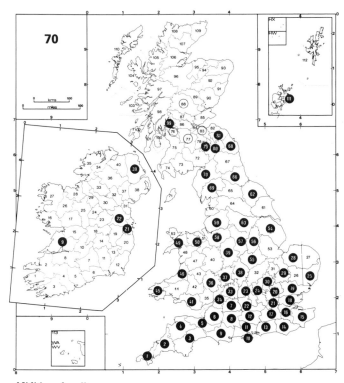

Niditinea fuscella

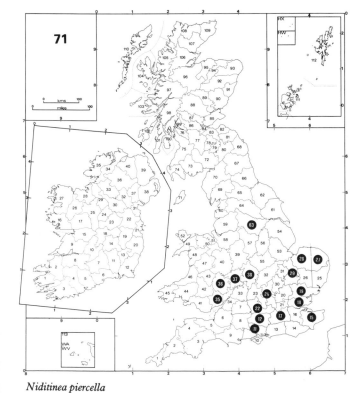

Niditinea piercella

TINEA Linnaeus

Tinea Linnaeus, 1758, *Syst.Nat.* (Edn 10) **1**: 496.

Many species of the family Tineidae were originally described in this genus and it still includes a large number which belong elsewhere but have not yet been transferred to other genera.

Robinson (1979) has revised the world species of the *T. pellionella* complex, which includes several clothes-moths of economic importance. The species of this complex are difficult to identify without dissection of the genitalia, and the distribution map of *T. pellionella* (Linnaeus) therefore differentiates records confirmed by dissection from other records under this name.

Imago. Head densely rough-haired (figure 78, p. 152). Antenna from two-thirds to as long as forewing; scape with pecten (without in some exotic species). Labial palpus porrect or drooping; second segment with conspicuous bristles. Maxillary palpus shorter than labial palpus. Haustellum reduced. Forewing with subcostal hyaline patch, as described for *Niditinea* Petersen, more or less developed, though hardly at all in *Tinea pallescentella* Stainton or *T. semifulvella* Haworth: this spot was described by Robinson (*loc.cit.*) only in the German species *T. steueri* Petersen in which it is particularly well developed, the costa convex around the spot.

Larva. Most species feed from a portable silken case, flattened and open at both ends (figure 83) which, except for *T. pallescentella*, *T. semifulvella* and *T. trinotella* Thunberg, is kept and enlarged throughout the larval life and used for pupation. The food is mainly animal fibres, either manufactured fabrics or the materials or detritus of birds' nests.

Figure 83 *Tinea* spp., larval case

Figure 84

Niditinea fuscella (Linnaeus)
(**a**) male genitalia (**b**) female genitalia

Niditinea piercella (Bentinck)
(**c**) male genitalia (**d**) female genitalia

Tinea columbariella Wocke
(**e**) male genitalia (**f**) female genitalia

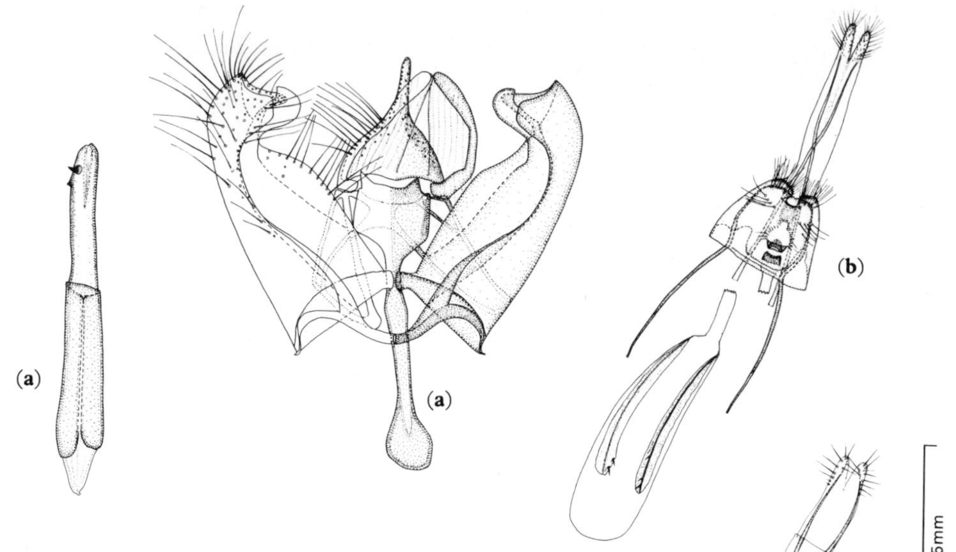

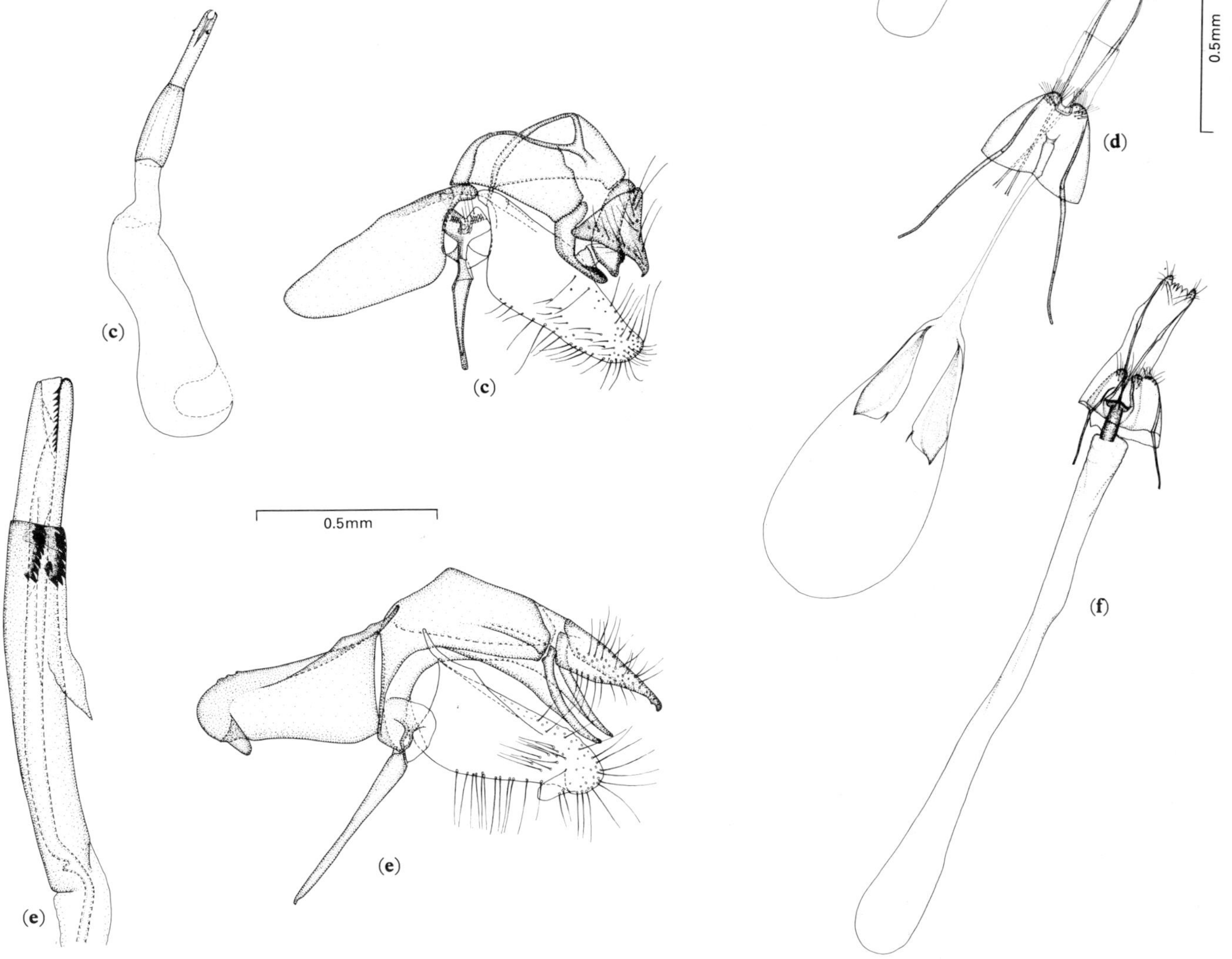

TINEA COLUMBARIELLA Wocke

Tinea columbariella Wocke, 1877, *Z.Ent.* **6**: 43.
Type locality: Germany (D.D.R.); Sömmerda.

Description of imago (Pl.9, fig.1)
Wingspan 8–16mm. Head ochreous brown. Forewing greyish brown, the scales tipped paler; base of wing slightly darker, at least on costa; second discal stigma dark fuscous, usually indistinct. Hindwing pale brassy fuscous. Genitalia,see figure 84, p. 196.

Similar species. Resembles darker species of the *T. pellionella* complex but the first discal and plical stigmata are lacking and the second discal is sometimes scarcely visible.

Life history
Larva. Head brown to dark brown, slightly darkened at edges of sclerites. Prothoracic plate paler brown. Body whitish, the pinacula scarcely visible. In a tough silken portable case incorporating food material, dorsoventrally flattened, up to 10mm in length. Usually in nests of birds of several species, most often those in and around buildings, such as house-sparrow (*Passer domesticus*), starling (*Sturnus vulgaris*) and swallow (*Hirundo rustica*); also in dove-cotes and poultry-houses. Occasionally found on woollen fabrics.

Pupa. In the larval case.

Imago. Moths appear mostly in summer but have been recorded in most months of the year.

Distribution (Map 72)
Local in the south of England and recorded as far north as Lancashire. The species was first recognized as British by Bradley (1950). Abroad it has a wide distribution in the Palaearctic region.

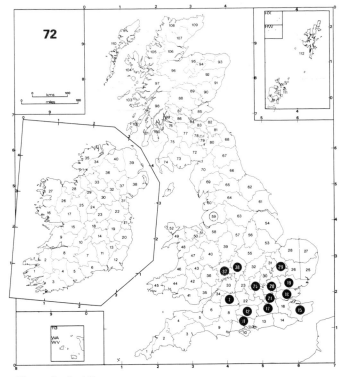

Tinea columbariella

TINEA FICTRIX Meyrick

Tinea fictrix Meyrick, 1914, *Supplta ent.* **3**: 59.
Type locality: Taiwan; Anping.

This species was added to the British list by Robinson (1978). There are two records of *T. fictrix* being introduced into Britain, the first with a cargo of Australian barley and the second with wheat residues, rodent droppings and coconut fibre in a ship's hold (Adams, 1979a). Superficially very similar to *T. columbariella* Wocke, *T. fictrix* has distinctive male and female genitalia which have been figured by Adams (*loc.cit.*) and Robinson (*loc.cit.*) respectively. Specimens from outside Britain have been bred from maize, copra dust, bat guano and pigeons' nests. *T. fictrix* is otherwise known from Nigeria, India, Taiwan and West Malaysia.

TINEA PELLIONELLA (Linnaeus)
Case-bearing Clothes Moth

Phalaena (Tinea) pellionella Linnaeus, 1758, *Syst.Nat.* (Edn 10) 1: 536.

Type locality: British Isles (neotype).

Description of imago (Pl.9, fig.2)
Wingspan 9–16mm. Head brownish ochreous to ferruginous ochreous. Forewing greyish ochreous to greyish brown, darker specimens with lighter-tipped scales towards apex; an irregular sprinkling of darker scales; wing-base, first and second discal and plical stigmata dark brown; subcostal hyaline patch usually distinct, contrasting with dark wing-base. Hindwing shining pale grey, darker and slightly brassy on margins and towards apex. Genitalia, see figures 85,86, pp. 200,201.

Similar species. T. translucens Meyrick in which the forewing has a smoother ground colour contrasting more with the dark markings. *T. dubiella* Stainton has the forewing darker with a sprinkling of ochreous scales and the hindwing is also darker. *T. lanella* Pierce & Metcalfe, *T. flavescentella* Haworth and *T. murariella* Staudinger usually have paler ochreous forewings.

Life history

Larva. Head dark brown with edges of sclerites very dark. Prothoracic plate brown. Body white without visible pinacula. In a portable case as described for the genus, normally feeding on wool, hair, fur or feathers. Occasionally found in birds' nests or among owl pellets, and also recorded from a great variety of stored vegetable products.

Pupa. In the larval case which is fixed some distance from the feeding site.

Imago. Flies in the late afternoon. Normally univoltine with moths appearing from June to October, but two or more generations a year could occur in heated buildings.

Distribution (Map 73)
This is probably still the most frequently occurring clothes-moth in the British Isles, though *Tineola bisselliella* (Hummel) can multiply more rapidly and produce more serious outbreaks. Formerly it was not clearly distinguished from *Tinea dubiella* and many records have to be relegated to the unconfirmed category; however the true distribution probably has at some time covered most of the British Isles. Now it is relatively uncommon and populations are probably maintained amongst outbuildings or in birds' nests around buildings. Abroad it occurs throughout the Holarctic region and in Australia and New Zealand.

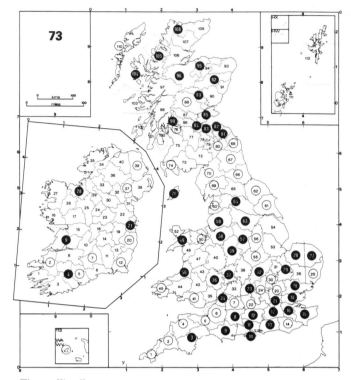

Tinea pellionella

TINEA TRANSLUCENS Meyrick

Tinea translucens Meyrick, 1917, *Exot.Microlepid.* 2: 78.
Tinea metonella Pierce & Metcalfe, 1934, *Entomologist* 67: 266.
Type locality: Pakistan; Peshawar.

Description of imago (Pl.9, fig.4)
Wingspan 9–18mm. Head pale orange-ochreous. Forewing pale shining greyish ochreous; basal area dark brown; stigmata dark greyish brown, first discal and plical elongate; subcostal hyaline patch less distinct than in related species, on upper surface mostly covered with scales of ground colour. Hindwing pale brassy grey, darker towards apex. Genitalia, see figures 85,86, pp. 200,201.
Similar species. T. *pellionella* (Linnaeus), *q.v.*

Life history
Larva. Similar to T. *pellionella.* In a portable case similar to that of T. *pellionella.* It has only been found feeding on manufactured materials or objects of animal origin.

Distribution
This species entirely replaces T. *pellionella* in the tropics and is frequently imported with goods from these regions, especially with 'bongo-drums' covered with zebra or cattle skin from East and South Africa. It can then establish itself in heated premises in the British Isles, but the infestations which have been discovered have probably always been eliminated. The species was first recorded in Britain by Cooke (1856) as T. *merdella.* Since that time records of '*Tinea merdella*' have been numerous, but most of them probably refer to T. *flavescentella* Haworth.

TINEA MURARIELLA Staudinger

Tinea murariella Staudinger, 1859, *Stettin.ent.Ztg* 20: 235.
Type locality: Spain; Cadiz.

Not certainly established in Britain. One specimen has been identified which was collected in 1946 from a cargo of hooves from Argentina in Glasgow docks, and in 1978 larvae were found damaging a carpet in Hackney, east London. The moth resembles T. *lanella* Pierce & Metcalfe and T. *flavescentella* Haworth but is usually more yellow in colour. Genitalia, see figures 85,86, pp. 200,201. For full accounts see Robinson (1979) and Adams (1979b).

TINEA LANELLA Pierce & Metcalfe

Tinea lanella Pierce & Metcalfe, 1934, *Entomologist* 67: 267.
Type locality: England; Liverpool.

Description of imago (Pl.9, fig.3)
Wingspan 11–17mm. Head pale ochreous. Forewing pale ochreous; base of costa suffused with blackish brown; first discal and plical stigmata indistinct, yellowish brown; second discal stigma pale greyish brown. Hindwing pale grey, tinted ochreous on margins. Genitalia, see figures 85,86, pp. 200,201.
Similar species. Very similar to T. *flavescentella* Haworth, which has the forewing with the dark suffusion at base of costa less extensive, and to T. *murariella* Staudinger in which the forewing is more yellow in tint.

Life history
Early stages. Undescribed.
Imago. Found in June and July.

Distribution
In the British Isles known only from moths collected in a wool warehouse at Liverpool in 1922. It has subsequently been found in Spain and Rumania. Most probably its true home is elsewhere.

Figure 85
after Robinson, 1979
Tinea pellionella group
male genitalia (**a**), with details showing tip of
aedeagi (**b–g**)

(**a,b**) *Tinea pellionella* (Linnaeus)
(**c**) *Tinea translucens* Meyrick
(**d**) *Tinea murariella* Staudinger
(**e**) *Tinea lanella* Pierce & Metcalfe
(**f**) *Tinea dubiella* Stainton
(**g**) *Tinea flavescentella* Haworth

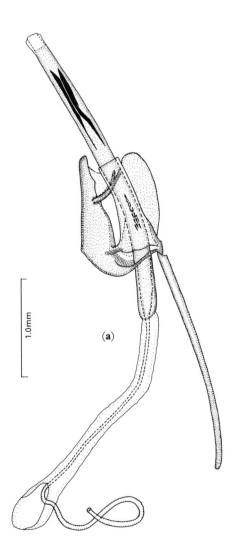

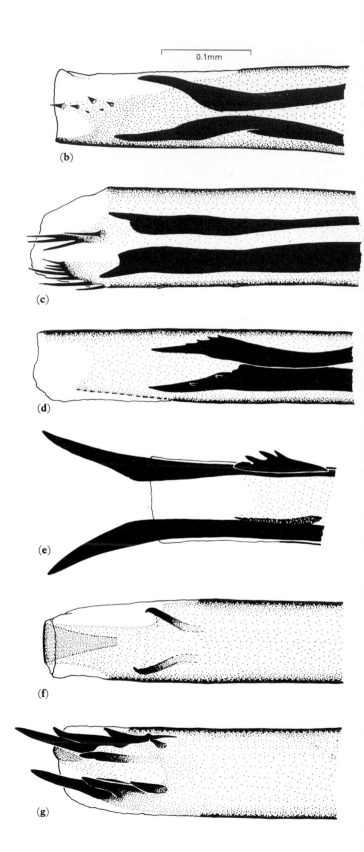

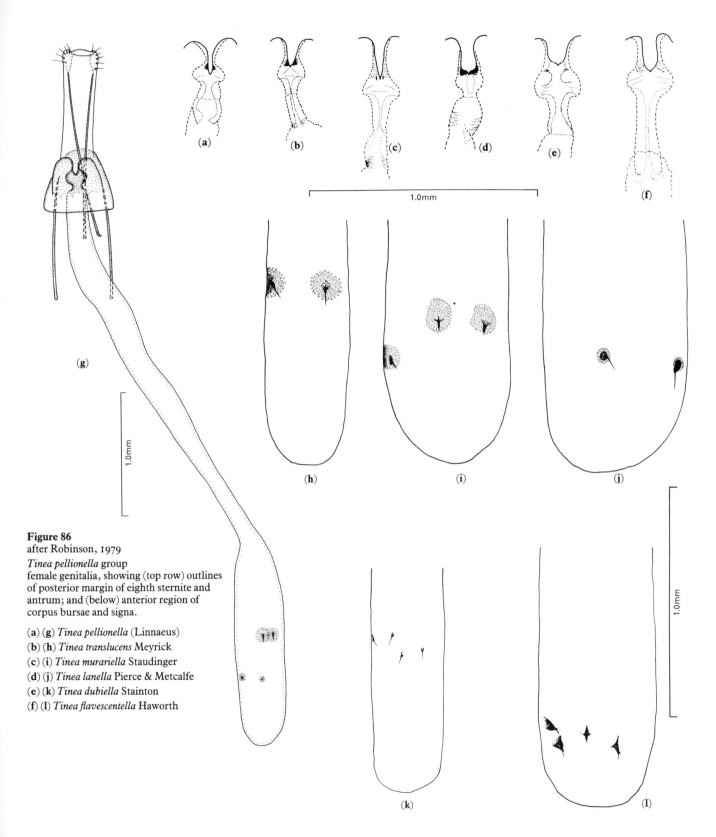

Figure 86
after Robinson, 1979
Tinea pellionella group
female genitalia, showing (top row) outlines
of posterior margin of eighth sternite and
antrum; and (below) anterior region of
corpus bursae and signa.

(**a**) (**g**) *Tinea pellionella* (Linnaeus)
(**b**) (**h**) *Tinea translucens* Meyrick
(**c**) (**i**) *Tinea murariella* Staudinger
(**d**) (**j**) *Tinea lanella* Pierce & Metcalfe
(**e**) (**k**) *Tinea dubiella* Stainton
(**f**) (**l**) *Tinea flavescentella* Haworth

201

TINEA DUBIELLA Stainton

Tinea dubiella Stainton, 1859, *Entomologist's wkly Intell.* **6**: 183.

Tinea turicensis Müller-Rutz, 1920, *Mitt.Ent.Zürich* **5**: 348.

Type locality: England; Liverpool.

Description of imago (Pl.9, fig.5)

Wingspan 9–15mm. Head ferruginous ochreous to brownish ochreous. Forewing fuscous, more or less sprinkled with ochreous or ochreous-tipped scales; stigmata blackish brown, first discal and plical sometimes very indistinct, second discal well developed; base of costa variably darker; subcostal hyaline patch narrow, inconspicuous. Hindwing grey, darker towards apex. Genitalia, see figures 85,86, pp. 200,201.

There is considerable variation in the degree of ochreous scaling.

Similar species. This species has a noticeably darker hindwing than other species of the *T. pellionella* group. The ochreous scales of the forewing are also characteristic, being brighter-coloured than the pale-tipped scales of *T. pellionella* (Linnaeus).

Life history

Larva. Head reddish brown. Prothoracic plate dark brown. Body white with inconspicuous pinacula. In a portable case similar to that of *T. pellionella*, up to *c.*10mm in length (figure 83, p. 195). Feeds on much the same range of materials and manufactured products of animal origin as *T. pellionella* and is found in birds' nests, especially those of swallows (*Hirundo rustica*), and also in owl pellets.

Imago. Flies in the late afternoon indoors. Moths are found throughout the summer, from May to September, probably in two generations.

Distribution (Map 74)

Widely distributed in the British Isles and formerly common. Almost entirely overlooked in the British Isles after its original description until Wakely (1962) published an account of it under the name *T. turicensis*. In his revision of the *T. pellionella* group Robinson (1979) found that about half the '*Tinea pellionella*' in British collections were *T. dubiella*. I have found *T. pellionella* to be more widely distributed than *T. dubiella*, but the latter seems to be more common in western Britain. It has been found as far north as Argyll, but from Ireland there is so far only one record, from Co. Cork (Robinson, *loc.cit.*). Abroad it has been recorded widely in the Holarctic region and in the southern hemisphere from St Helena, South Africa, Australia and New Zealand.

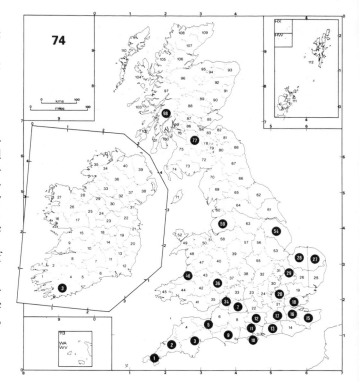

Tinea dubiella

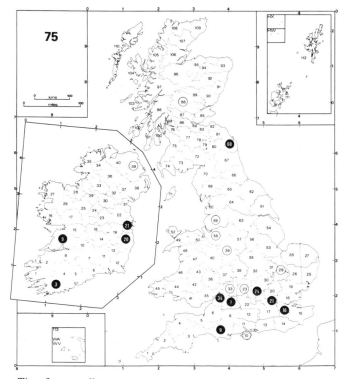

Tinea flavescentella

TINEA FLAVESCENTELLA Haworth

Tinea flavescentella Haworth, 1828, *Lepid. Br.*: 564.
Type locality: England.

Description of imago (Pl.9, fig.6)

Wingspan 8–17mm. Head whitish ochreous. Forewing whitish ochreous, becoming pale grey-brown towards base and towards apex; first and second discal and plical stigmata greyish brown, well defined; subcostal hyaline patch rather inconspicuous. Hindwing pale brassy fuscous, darker towards apex. Genitalia, see figures 85,86, pp. 200,201.

Similar species. T. lanella Pierce & Metcalfe, *q.v.*; *T. murariella* Staudinger also has the base of the costa of the forewing strongly darkened.

Life history

Larva. Not reliably described. Normally feeds on feathers indoors. Records of breeding in birds' nests were not verified by Robinson (1979) and neither were breeding records from other materials except possibly fur and insect remains.

Imago. Has been found from February to November.

Distribution (Map 75)

A fairly wide distribution in Britain and Ireland has been recorded but this has always been a local and uncommon species. According to Robinson (*loc.cit.*) it has usually been recorded in Britain as '*Tinea merdella*'. Abroad it occurs in western Europe and Algeria.

TINEA PALLESCENTELLA Stainton

Large Pale Clothes Moth

Tinea pallescentella Stainton, 1851, *Suppl.Cat.Br.Tineidae and Pterophoridae*: 2.

Type locality: England; Liverpool.

Description of imago (Pl.9, fig.7)

Wingspan 12–25mm. Head pale brownish ochreous. Forewing pale greyish brown variably speckled, irrorate or suffused with brown; base of wing darker brown, usually including a prominent dark brown mark below fold and sometimes connected to a distinct streak along fold; first and second discal stigmata and a more diffuse subapical discal spot dark brown; some dark brown chequering around apical margin; fringe paler towards margin, with indistinct dark median band; subcostal hyaline patch absent. Hindwing pale shining fuscous.

Similar species. Small weakly marked specimens might be mistaken for one of the *T. pellionella* group but can always be distinguished by the plical streak of the forewing.

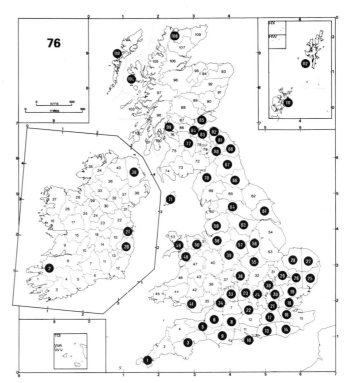

Tinea pallescentella

Life history

Larva. Head dark brown with edges of sclerites darker; a short black longitudinal stripe along ventral margin and a similar lateral stripe above it. Prothoracic plate reddish brown. Body whitish, with pinacula usually indistinct. Feeds on a variety of materials of animal origin, especially wool, hair and feathers, either in the open or in unheated buildings. It has also been bred from a wasp's nest. The larva makes a case only for resting while moulting, leaving it as soon as the integument is hard enough. When fully fed it leaves the feeding site and spins a tough cocoon for pupation.

Imago. Nocturnal, frequently coming to light. It may be found in any month of the year either indoors or out, and quite often makes mysterious appearances on the outside of windows in mid-winter.

Distribution (Map 76)

Recorded widely in the British Isles, including the Outer Hebrides and Shetland. It appears to be more common in some northern districts than in the south. Abroad it is known from Patagonia, U.S.A., Europe and New Zealand. Robinson (1979) comments on this distribution and considers that it may be a Patagonian species first imported to Europe in about 1840.

TINEA SEMIFULVELLA Haworth

Tinea semifulvella Haworth, 1828, *Lepid.Br.*: 562.
Type localities: England; Norfolk and near London.

Description of imago (Pl.9, fig.8)

Wingspan 14–22mm. Head bright ferruginous. Forewing pale grey or rosy grey shading to dull rose towards base; basal half of costa dark rose to blackish brown; apical quarter ferruginous, this colour extending inwards to a ferruginous or pink suffusion in disc; a clear dark brown tornal spot; subcostal hyaline patch scarcely evident. Hindwing fuscous. Outer half of all fringes whitish.

Life history

Larva. Head pale yellowish brown irregularly mottled with dark brown; a black longitudinal stripe along ventral margin and a black lateral stripe above it. Prothoracic plate dark brown. Body whitish with indistinct pinacula. Feeds in birds' nests of various species in the open, according to Hinton (1956) in a portable case, but if cases are indeed made they are probably only formed by small larvae or for moulting. For pupation a rough case of almost cylindrical section is constructed in the bird's nest. The larva has also been found on woollen material in the open and on the wool of a dead sheep.

Imago. Nocturnal, often found in light-traps. The flight period is from May to September; perhaps in two generations in the south of England.

Distribution (Map 77)

A common species throughout mainland Britain and Ireland, sometimes abundant in birds' nests. Abroad it is found throughout northern Europe and as far south as Austria, Rumania and Ukraine.

TINEA TRINOTELLA Thunberg

Tinea trinotella Thunberg, 1794, *Ins.Suecica* 7: 95.
Tinea lappella sensu auctt.
Tinea ganomella Treitschke, 1833, *Schmett.Eur.* 9(2): 263.
Type locality: Sweden; Vestrogothia (Västergötland).

Description of imago (Pl.9, fig.9)

Wingspan 11–17mm. Head yellow. Forewing of male pale yellow-ochreous, of female a little darker, the scales pale fuscous tipped with pale yellow-ochreous; basal third of costa dark brown; stigmata very dark brown, plical the largest, second discal smaller but distinct, first discal small or absent; fringe speckled or suffused with ochreous grey, margin pale yellow-ochreous; subcostal hyaline patch distinct. Hindwing bronzy fuscous, darker towards apex; margin of fringe whitish.

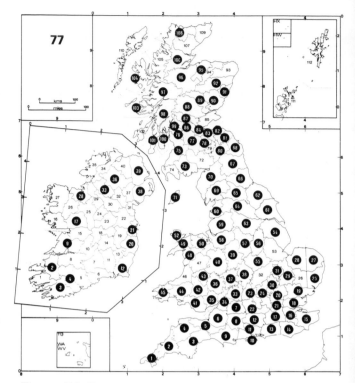

Tinea semifulvella

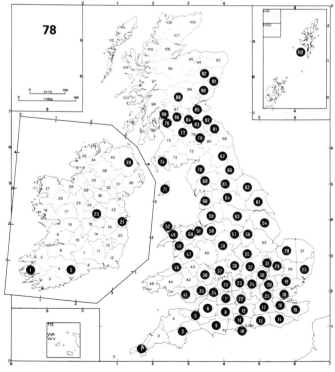

Tinea trinotella

Life history

Larva. Resembles that of *T. semifulvella* Haworth but the head is not mottled with dark brown. Habits similar to those of *T. semifulvella* and likewise stated by Hinton (1956) to feed from a portable case. Forms a similar case for pupation, almost cylindrical, up to 12mm in length, covered with material from the nest. Normally in birds' nests, but also recorded from woollen material in the open.

Imago. Nocturnal, often found in light-traps. The moth flies from May to August, probably in two generations in the south of England.

Distribution (Map 78)

An even more common bird's nest species than the last, recorded from most parts of mainland Britain and Ireland and from Shetland. Distributed abroad throughout Europe including European U.S.S.R.

CERATOPHAGA Petersen

Ceratophaga Petersen, 1957, *Beitr.Ent.* **7**: 130.

CERATOPHAGA ORIENTALIS (Stainton)

Tinea orientalis Stainton, 1878, *Entomologist's mon.Mag.* **15**: 134.

Type locality: England; Poplar, London.

This species has been imported into Britain twice. The first introduction was in 1878, with buffalo horns from Singapore. Stainton described the species from this introduced material. It was reintroduced in Glasgow in about 1946, possibly with a cargo of horn from India or Sri Lanka (Robinson, 1978). Specimens of *C. orientalis* are large (15–23mm), glossy, and ochreous grey; the species is known from Sri Lanka, India and Burma. *Ceratophaga* species feed, apparently exclusively, on the hooves or horns of dead mammals or on horns which have been shed.

CERATOPHAGA HAIDARABADI Zagulyaev

Ceratophaga haidarabadi Zagulyaev, 1966, *Trudȳ zool. Inst.Leningr.* **37**: 163.

Type locality: India; south Hydrabad [*sic*].

C. haidarabadi has been recorded once only from Britain, mixed with a series of *C. orientalis* (Stainton) collected in Glasgow in about 1946 (Robinson, 1978). It is similar in appearance to *C. orientalis* but larger (19–29mm) and more yellowish. It is otherwise known only from India.

References

Adams, R. G., 1979a. The male of *Tinea fictrix* Meyrick (Lepidoptera: Tineidae). *Entomologist's Gaz.* **30**: 198, 1 pl.

———, 1979b. *Tinea murariella* Staudinger in Britain (Lepidoptera: Tineidae). *Ibid.* **30**: 269–270, 1 pl.

Bankes, E. R., 1910. *Monopis weaverella*, Scott (n.syn. = *semispilotella*, Strand), specifically distinct from *M. rusticella*, Hb. *Entomologist's mon. Mag.* **46**: 221–228, 1 pl.

———, 1912. Stray notes on *Monopis crocicapitella*, Clms., and *M. ferruginella*, Hb. *Ibid.* **48**: 39–44, 1 pl.

Barrett, C. G., 1878. Notes on Pembrokeshire Tineina. *Ibid.* **14**: 268–272.

Beirne, B., 1937. Note on *Mnesipatris filicivora* Meyr. *Entomologist* **70**: 195–196.

———, B. P., 1941. A list of the microlepidoptera of Ireland. *Proc. R. Ir. Acad.* **47**(B): 53–147.

Birchall, E., 1866. The Lepidoptera of Ireland [part]. *Entomologist's mon. Mag.* **3**: 145–148.

Bland, K. P., 1984. *Monopis weaverella* (Scott), a mystery solved? *Entomologist's Rec. J. Var.* **96**: 37.

Bradley, J. D., 1950. On the occurrence of *Tinea columbariella* Wocke (Lep. Tineidae) in England, with a description of the species. *Entomologist* **83**: 169–172, 4 figs.

———, 1966. Some changes in the nomenclature of British Lepidoptera. Part 4, Microlepidoptera. *Entomologist's Gaz.* **17**: 213–235.

——— & Fletcher, D. S., 1979. *A recorder's log book or label list of British butterflies and moths*, [vi], 136 pp. London.

Brightman, F. H., 1965. Insects on lichens. *Lichenologist* **3**: 154.

Clutterbuck, C. G., 1916. *Dryadaula pactolia*, Meyr., in Gloucester. *Entomologist* **49**: 21.

Cooke, N., 1856. Capture of *Tinea merdella*. *Entomologist's wkly Intell.* **1**: 125.

Corbet, A. S., 1943. Observations on species of Lepidoptera infesting stored products. VI. The species of the *Tinea granella* (L.) complex (Tineidae). *Entomologist* **76**: 95–96.

———, 1945. Observations on species of Lepidoptera infesting stored products. XIV. *Leucania zeae* (Dup.) (Agrotidae) and *Setomorpha rutella* Zeller (Tinaeidae) found on ships in London Docks. *Ibid.* **78**: 88.

Daltry, H. W., 1929a. *Tenaga pomiliella* Clemens; a tineid new to the British list. *Ibid.* **62**: 34.

———, 1929b. *Tenaga pomiliella* Clemens, a tineid new to the British list: a description and further note. *Ibid.* **62**: 73–75.

———, 1931. *Tenaga pomiliella* Clemens in Kent. *Ibid.* **64**: 19.

Davis, D. R., 1978. The North American moths of the genera *Phaeoses*, *Opogona*, and *Oinophila*, with a discussion of their supergeneric affinities (Lepidoptera: Tineidae). *Smithson. Contr. Zool.* No. 282; 39 pp., 128 figs.

———, 1983. Tineidae *in* Hodges, R. W. *et al.*(eds), *Check list of the Lepidoptera of America North of Mexico*: 5–7. London.

Donisthorpe, H. St J. K., 1927. *The guests of British ants, their habits and life-histories*, xxiv, 244 pp., 16 pls, 55 figs. London.

Ellis, J. W., 1890. *The Lepidopterous fauna of Lancashire and Cheshire*, 136 pp. Leeds.

Forbes, W. T. M., [1924]. The Lepidoptera of New York and neighbouring states. Primitive forms, Microlepidoptera, Pyraloids, Bombyces. *Mem. Cornell Univ. agric. Exp. Stn* **68**: 729 pp., 439 figs.

Ford, L. T., 1931. Moths in stables. *Entomologist* **64**: 259.

———, 1949. *A guide to the smaller British Lepidoptera*, 230 pp. London.

Goss, H. & Barrett, C. G., 1902. Butterflies and moths. In *The Victoria history of the counties of England. Surrey*, **1**: 109–150.

——— & Fletcher, W. H. B., 1905. Lepidoptera Heterocera (Moths). In *The Victoria history of the counties of England. Sussex*. **1**: 170–210.

Hannemann, H. J., 1977. *Die Tierwelt Deutschlands* **63**. Kleinschmetterlinge oder Microlepidopteren III. Federmotten (Pterophoridae). Gespinstmotten (Yponomeutidae). Echte Motten (Tineidae), 275 pp., 17 pls. Jena.

Hinton, H. E., 1956. The larvae of the species of Tineidae of economic importance. *Bull. ent. Res.* **47**: 251–346, 216 figs.

Jalava, J. & Kyrki, J., 1980. Notes on the taxonomy and distribution of western palaearctic *Trichophaga* species (Lepidoptera, Tineidae). *Notul. ent.* **60**: 107–110, 3 figs.

King, J. J. F. X., 1901. Microlepidoptera. *In* Elliot, G. F. S., Laurie, M. & Murdoch, J. B., *Fauna, Flora & Geology of the Clyde area*, pp. 246–257. Glasgow.

Leraut, P., 1983. Quelques changements dans la nomenclature des Lépidoptères de France. *Ent. gall.* **1**: 35–36.

Metcalfe, J. W., 1924. *Tinea albicomella* H.S., a British species. *Entomologist* **57**: 113 (and note of exhibit by J. H. Durrant, *op.cit.*: 93).

Meyrick, E., 1895. *A handbook of British Lepidoptera*, vi, 843 pp. London.

———, 1928. *A revised handbook of British Lepidoptera*, vi, 914 pp. London.

Morrison, B., 1968. A further record of *Dryadaula pactolia* Meyrick (Lep., Tineidae) in Britain with notes on its life-history. *Entomologist's Gaz.* **19**: 181–188, 15 figs.

Pelham-Clinton, E. C., 1982. *Nemapogon heydeni* Petersen, 1957 (Lepidoptera: Tineidae) new to the British Isles. *Ibid.* **33**: 79–80, 3 figs.

———, 1983. *Monopis weaverella* (Scott), a continuing mystery. *Entomologist's Rec. J. Var.* **95**: 212.

Petersen, G., 1957–58. Die Genitalien der paläarktischen Tineiden. *Beitr. Ent.* **7**: 55–176, 338–379, 557–595; **8**: 111–118, 398–430.

———, 1960. Contribución al conocimiento de la distribuciòn geográfica de los Tineidos de la Península Ibérica (Lep. Tineidae). *Eos, Madr.* **36**: 205–236, 8 figs.

———, 1969. Beiträge zur Insekten-Fauna der DDR: Lepidoptera – Tineidae. *Beitr. Ent.* **19**: 311–388, 2 col. pls.

Pierce, F. N. & Metcalfe, J. W., 1934. *Tinea cloacella* Haw., *T. granella* Linn., *T. ruricolella* Staint., *T. cochylidella* Staint., and *T. personella* sp.nov. *Entomologist* **67**: 217–219 (and plate, 1935, **68**: pl. 2).

Richardson, N. M., 1891. Occurrence at Portland of *Tinea subtilella*, Fuchs, a species new to the British fauna. *Entomologist's mon. Mag.* **27**: 14 (with editorial note (H.T.S.): 15).

———, 1893. *Blabophanes heringi* at Portland: distinct from *B. ferruginella*? *Ibid.* **29**: 14–15.

———, 1895. Occurrence of *Tinea vinculella*, H.-S., at Portland, with notes on its life history. *Ibid.* **31**: 61–65.

———, 1896. A list of Portland Lepidoptera. *Proc. Dorset nat. Hist. antiq. Fld Club* **17**: 146–191, 1 pl.

Robinson, G. S., 1978. Four species of Tineinae (Lep., Tineidae) new to the British list. *Entomologist's Gaz.* **29**: 139–144, 3 figs.

———, 1979. Clothes-moths of the *Tinea pellionella* complex: a revision of the world's species (Lepidoptera: Tineidae). *Bull. Br. Mus. nat. Hist.* (Ent.) **38**: 57–128, 103 figs.

———, 1982. The Palaearctic Tineidae (Lepidoptera) described by Daniel Lucas. *Entomologist's Gaz.* **33**: 175–180.

Robson, J. E., 1912. A catalogue of the Lepidoptera of Northumberland, Durham, and Newcastle-upon-Tyne. Vol. II. Micro-lepidoptera. Part II – Tineina and Pterophorina. *Nat. Hist. Trans. Northumb.* **15**: 107–289.

Ruston, H., 1879. Occurrence of *Tinea fenestratella* (Heyden) in Britain. *Entomologist's mon. Mag.* **15**: 238–239.

Samouelle, G., 1819. *The entomologist's useful compendium*, 496 pp., 12 col. pls. London.

Simpson, A. N. B., 1981. *Monopis fenestratella* (Heyd.) in Warwickshire. *Entomologist's Rec. J. Var.* **93**: 45 (with ed. note).

Sorrell, T., 1876. New British *Tinea*. *Entomologist* **9**: 159.

Stainton, H. T., 1852. *The entomologist's companion*, iv, 75 pp. London.

———, 1854. *Insecta Britannica. Lepidoptera: Tineina*, viii, 313 pp., 10 pls. London.

[Stainton, H. T.], 1856. [Editorial note]. *Entomologist's wkly Intell.* **1**: 7.

Stainton, H. T., 1876. Occurrence of *Tinea angustipennis*, Herrich-Schäffer, in England. *Entomologist's mon. Mag.* **13**: 143–144.

Stephens, J. F., 1829. *A systematic catalogue of British insects*, xxxiv, 416, 388 pp. London.

———, 1834–35. *Illustrations of British entomology*. (Haustellata) **4**, 436 pp., 9 col. pls. London.

Sterling, P. H., 1984. *Monopis weaverella* (Scott), a solution to the mystery. *Entomologist's Rec. J. Var.* **96**: 37–38.

Stringer, H., 1943. Observations on species of Lepidoptera infesting stored products. X. *Lindera tessellatella* Blanchard, a tineid new to Britain. *Entomologist* **76**: 177–181, 12 figs.

Turton, W., 1802. *A general system of nature* **3**, 784 pp. London.

Wakely, S., 1943. *Arcedes (Tinea) piercella*, Benct. [*sic.*], in Britain. *Entomologist's Rec. J. Var.* **55**: 9.

———, 1958a. Notes on the Tineina. *Ibid.* **70**: 81–82.

———, 1958b. Notes on the Tineina. *Ibid.* **70**: 192–195.

———, 1962. Notes on *Tinea turicensis* Mull.-Rutz. (*metonella* Pierce). *Ibid.* **74**: 92–93.

Warren, W., 1880. Captures of Lepidoptera in the vicinity of London. *Entomologist's mon. Mag.* **17**: 137–138.

Waters, E. G. R., 1926. Micro-lepidoptera in south Devon, August, 1925. *Entomologist* **59**: 158–161.

Wood, W., 1833–39. *Index entomologicus*, xii, 266 pp., 54 col. pls. London.

Zagulyaev, A. K., 1960. *Fauna SSSR* Lepid. **4**(3): 267 pp., 3 col. pls, 231 figs. Moscow & Leningrad. [In Russian].

———, 1964. *Ibid.* **4**(2): 424 pp., 2 col. pls, 385 figs. Moscow & Leningrad. [In Russian].

———, 1973. *Ibid.* **4**(4): 127 pp., 2 col. pls, 99 figs. Leningrad. [In Russian].

———, 1975. *Ibid.* **4**(5): 428 pp., 11 pls (6 col.), 319 figs. Leningrad. [In Russian].

———, 1979. *Ibid.* **4**(6): 408 pp., 12 pls (4 col.), 332 figs. Leningrad. [In Russian].

Zimmerman, E. C., 1978. Microlepidoptera. *Insects Hawaii* **9**: i–xviii, 1–1903.

OCHSENHEIMERIIDAE
A. M. Emmet

A small Palaearctic family with only one genus. It is placed here according to the arrangement of Kloet & Hincks (1972); Kyrki (1983), however, proposes reassigning it to the Yponomeutoidea.

OCHSENHEIMERIA Hübner

Ochsenheimeria Hübner, [1825], *Verz.bekannt.Schmett.*: 416.

A genus with about 15 species, their status in some cases being still uncertain. Three of these occur in Great Britain and two in Ireland.

Imago. Head with vertex densely clad in long, shaggy, spatulate hair-scales with bifid or serrate apex; frons broad, smooth-scaled but usually covered by long scales drooping from vertex. Antenna slightly over half length of forewing; elongate scape and basal flagellar segments variably clad in dense hair-scales, the extent of these scales being of diagnostic value; the antennae are held extended in a weakly sinuate posture suggestive of the horns of a bull, a feature reflected in the nomenclature of species. Eye prominent but small, diameter less than half interocular distance; ocelli in pairs, distinct; haustellum developed; maxillary palpus short, two-segmented; labial palpus moderate, porrect, segment 3 short and almost concealed in a dense brush of broad hair-scales with bifid apex springing from apical half of segment 2. Forewing subrectangular with costa and dorsum almost parallel and the terminal cilia giving a truncate appearance; broad-scaled, the scales in discal area with serrate tips and suberect in some species; 11 or, less often, 12 veins present; 6 (M_1) separate or stalked with 8 (R_4) and 7 (R_5); discal cell elongate, accessory cell usually present; 1c (Cu_2) generally absent; 1b ($1A+2A$) separated at base, fused in outer half. Hindwing elongate-ovate, cilia slightly shorter than width of wing; wing-base thinly scaled and transparent in some species; six to eight veins present; 7 (Rs) and 6 (M_1) either forked near apex or completely fused; 1c (Cu_2) usually absent (figure 87). Foretibia without epiphysis, midtibia with one pair, hindtibia with two pairs of spurs. Abdomen stout in most species; often a pale band on dorsum of segment 6, more pronounced in female; male with eversible hair-pencil situated laterally on segment 8.

Larva. Thoracic legs and five pairs of prolegs present; crochets uniordinal and reduced in number to three or four in some species. Body weakly pigmented, without pattern except for dark spots surrounding spiracles; head, prothoracic plate and anal plate generally darker than body.

Pupa. Head furnished with a sclerotized 'cocoon-piercer'; caudal end with ten short, stout, black-tipped spines.

Life history and behaviour. The adults are diurnal, most species flying only in sunshine for an hour or two around midday. The wings are short in relation to the size of the body and the flight is most unlike that of other day-flying microlepidoptera, steady and direct, rather like that of a zygaenid. At other times of day and in dull weather moths rest low down on grass-stems and are seldom observed. *O. vacculella* Fischer von Röslerstamm has different habits; it roosts, often gregariously, behind loose bark, in sheds or even in occupied houses.

The ova are laid singly or in batches of two or three on leaves or in sheaths of many species of Gramineae; oviposition has also been recorded in haystacks (Davis, 1975). A few species including *O. vacculella* overwinter as a fully developed larva within the ovum, but the majority hatch in the autumn. Dispersal may then be effected by 'ballooning': long strands of silk are produced which are transported by the wind with the larvae suspended from them (Pavlov, 1961). Some then start feeding as leaf-miners (Pavlov, *loc.cit.*), but the mine does not seem to have been described. When still very small, they leave these mines and bore into the stems, a method of feeding adopted by other species from the outset. The young larvae mine downwards and overwinter in the stem just above the roots. In spring they start feeding again as stem-miners generally below the first node, causing distortion, blanching and withering of the upper growth. In some regions, but not in Britain, they are serious pests of cereal crops and pasture. Since the larvae change stems often, they do damage disproportionate to their numbers. Larvae have often been obtained by sweeping herbage; possibly they leave their mines to sun themselves and afterwards attack fresh stems. Pupation takes place in an oval, white silken cocoon spun between leaves of the foodplant or within a sheath. This stage lasts from two to three weeks.

All known species occur within an area extending from England across Europe and south-western U.S.S.R. to the Himalayas, and from North Africa to Finland. However, *O. vacculella* was accidentally introduced into the U.S.A. in the 1960s, where it now threatens to become an important pest of cereal crops.

Much of the information given above has been taken from Davis (*loc.cit.*) and Réal (1966).

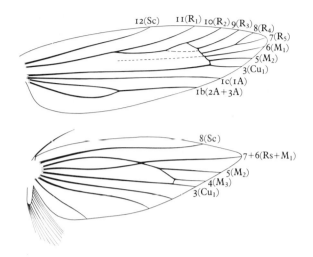

Figure 87 *Ochsenheimeria vacculella* Fischer von Röslerstamm, wing venation

Key to species (imagines) of the genus *Ochsenheimeria*

1 Antenna with long, shaggy scales in middle of flagellum 2
– Antenna without such scales 3

2(1) Scales thickening antenna more than three times width of shaft; forewing underside with terminal cilia dark-tipped; hindwing with cilia uniform fuscous ....*mediopectinellus* (p. 209)
– Scales thickening antenna less than three times width of shaft; forewing underside with terminal cilia not dark-tipped; hindwing with cilia grey, becoming whitish grey at inner angle .... *urella* ♀ (p. 210)

3(1) Antenna thickened at base and tapering, not annulated; hindwing with basal quarter thin-scaled ....*urella* ♂ (p. 210)
– Antenna not thickened at base, not tapering and with darker annulations; hindwing with basal third transparent .... *vacculella* (p. 210)

The antennae are figured by Karsholt & Nielsen (1984).

OCHSENHEIMERIA MEDIOPECTINELLUS
(Haworth)

Ypsolophus mediopectinellus Haworth, 1828, *Lepid.Br.*: 545.
Lepidocera birdella Curtis, 1831, *Br.Ent.* **8**: 344.
Type locality: England.

Description of imago (Pl.9, fig.11)
Wingspan 11–12mm. Head with long, spatulate, ochreous, dark-tipped scales, each with bifurcate apex; antenna slightly over half length of forewing, basal three-fifths thickened with shaggy fuscous and ochreous scales, longest in middle of flagellum where their length exceeds three times width of shaft. Forewing with scales in disc suberect and loosely attached, pale ochreous to fuscous, in paler specimens with scattered darker scales tending to form fasciae in median and subapical areas; cilia concolorous but dark-tipped on termen, more conspicuously on underside. Hindwing pale bronzy fuscous, paler and thinly scaled in basal one-quarter, the scales there tending to become detached in flight, so giving rise to a transparent patch; cilia fuscous. Abdomen with pale band on segment 6, more pronounced in female; male with lateral tuft of long scales on segment 8 and ochreous anal tuft.

Similar species. Female *O. urella* Fischer von Röslerstamm, in which the shaggy scales in the middle of the antennal flagellum are less than three times the width of the shaft, the forewing is generally more uniform pale ochreous and the terminal cilia are not or hardly dark-tipped; the hindwing has the cilia grey shading to whitish grey at the inner angle.

Life history
Ovum. Laid singly or in small batches in sheaths of cock's-foot (*Dactylis glomerata*), brome (*Bromus* spp.), meadow-grass (*Poa* spp.), foxtail (*Alopecurus* spp.) or other grasses.
Larva. Full-fed *c.*13mm long, slender, spindle-shaped. Head pale brown. Body whitish buff; a small dark spot on anterior edge of each segment just above spiracle. Mines the lower stem of its foodplant, causing wilting and discoloration, especially of the central leaf. It frequently changes stems and consequently may be obtained by sweeping herbage. September–May.
Pupa. Pale brown. In an elongate, fusiform white cocoon spun in a folded leaf or leaf-sheath or on a stem of the foodplant or an adjacent grass. May–July, though emergence may be delayed until August or September.
Imago. Univoltine, appearing from early July until September. The adult flies in sunshine between 12.00 and 14.00 hrs. At other times it rests generally low down in

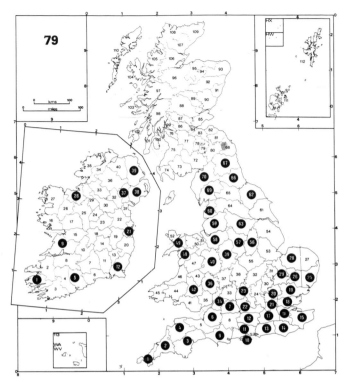

Ochsenheimeria mediopectinellus

herbage, though in warm weather sometimes high enough to be obtained by sweeping.

Distribution (Map 79)

Widespread in England, Wales and Ireland but not recorded from Scotland unless it is assumed that Scott (1854), who described the life history writing from Renfrewshire, did so from locally-taken larvae. Probably more common than the records suggest, since it passes unobserved unless the collector is at an appropriate locality at the right time of day in sunshine. Western Europe as far east as Germany and Switzerland.

OCHSENHEIMERIA URELLA Fischer von Röslerstamm

Ochsenheimeria urella Fischer von Röslerstamm, 1842, *Stettin.ent.Ztg* **1842**: 211.

Ochsenheimeria bisontella Lienig & Zeller, 1846, *Isis, Leipzig* **1846**: 274.

Type locality: Germany; Frankfurt am Main.

Description of imago (Pl.9, figs 12–16)

Wingspan 9–12mm. Sexually dimorphic. *Male.* Head with long, spatulate, ochreous, dark-tipped scales, each with bifurcate apex; antenna thickened to three-fifths with

recumbent scales, apical two-fifths simple. Forewing fuscous, obscurely irrorate darker; a few scattered, broad, suberect, pale-tipped scales; cilia concolorous, tips on termen not, or hardly, darkened. Hindwing grey with slight bronzy sheen; cilia grey, grading to whitish grey at inner angle. Abdomen with pale band on segment 6. *Female.* Antenna with basal three-fifths thickened with shaggy scales, longest in middle of flagellum but less than three times width of shaft. Forewing generally pale ochreous but sometimes fuscous as in male, or intermediate. Other characters as in male.

Similar species. Female *O. mediopectinellus* (Haworth), *q.v.*

Life history

Ovum. Laid singly or in small batches on leaves of couch (*Agropyron* spp.), brome (*Bromus* spp.) and probably other grasses.

Larva. Not described. First mines leaves but later bores into stems. April–May.

Pupa. In a white cocoon spun between leaves of the food-plant. Late May–June.

Imago. Univoltine. Flies in morning sunshine, later resting low down on grass-stems. Males will assemble to a freshly emerged female (Tutt, 1893). July–September.

Distribution (Map 80)

Widespread in Britain as far north as Shetland, occurring more freely in northern England and Scotland. Beirne (1941) recorded it doubtfully from Ireland but since its presence there has now been confirmed his records are accepted. Europe eastwards to south-western Russia and southwards to Italy.

OCHSENHEIMERIA VACCULELLA Fischer von Röslerstamm

The Cereal Stem Moth

Ochsenheimeria vacculella Fischer von Röslerstamm, 1842, *Stettin.ent.Ztg* **3**: 213.

Type locality: Germany; Frankfurt-an-der-Oder.

Description of imago (Pl.9, fig.17)

Wingspan male 11–12mm; female 12–14mm. Head with vertex clad in long, slender scales with spatulate, bidentate apex, ochreous at base grading to fuscous at apex; frons smooth-scaled, whitish to pale brown but concealed by shaggy scales drooping from vertex; antenna about one-half length of forewing, flagellum simple, pale fuscous with darker annulations, scape slightly thickened with scales. Forewing whitish ochreous mixed fuscous, all scales recumbent; often a pale dorsal blotch before middle; cilia consisting of dark-tipped, spatulate scales. Hindwing bronzy grey; basal area, ranging from one-third to

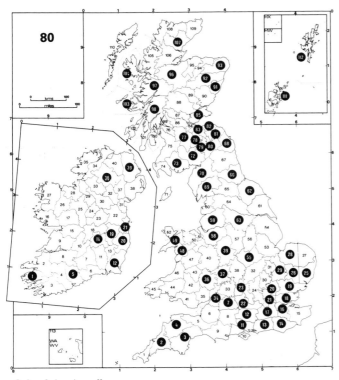

Ochsenheimeria urella

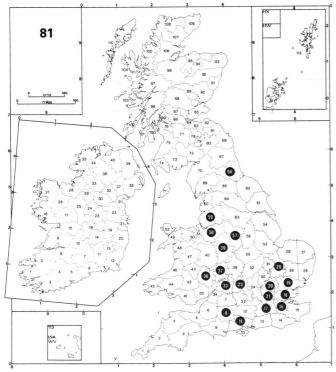

Ochsenheimeria vacculella

one-half, hyaline; cilia paler than wing with dark subbasal line. Abdomen with pale dorsal band on segment 6.

Life history

Almost unobserved in Britain but extensively studied abroad, especially in U.S.S.R. and the U.S.A. The information which follows is taken mainly from Pavlov (1961) and Davis (1975).

Ovum. Fusiform, length *c.*0.6mm, width *c.*0.2mm, micropylar end truncate; pale ochreous, somewhat shining, with irregular, coarse longitudinal furrows. Laid generally in pairs low down on dried leaves or in leaf-sheaths (Sich, 1908), most often on rye-grass (*Lolium* spp.), including cultivated rye, but also on couch (*Agropyron* spp.), brome (*Bromus* spp.), meadow-grass (*Poa* spp.), wheat (*Triticum aestivum*), meadow-fescue (*Festuca pratensis*) and cat's-tail (*Phleum* spp.). Laid also in haycocks, thatch and on cereals stored in barns. August, overwintering until April.

Larva. Full-fed 18–20mm long. Pale yellowish white; prothoracic plate brownish black, anal shield pale brown. Mines leaf-blades in the first instar; later bores into the lower stem, causing wilting. Abroad it is an important pest of cultivated rye and other cereal crops. April–May.

Pupa. In a white cocoon spun between leaves or in a sheath of the foodplant. Late May–June.

Imago. Univoltine, appearing in July and August. Flies for a brief period in midday sunshine; at other times of day it rests, often gregariously, under loose bark or in its crevices. When disturbed it is reluctant to fly and creeps deeper into its retreat or finds another crevice. It has also been found in sheds and houses. This habit is possibly associated with a period of aestivation, although this has not been suggested by any of the authors consulted.

Distribution (Map 81)

Although essentially an insect of grassland and arable country, it is more often recorded in open woodland, parks and gardens containing mature timber. Its reputation as a scarce species may in part stem from ignorance of its habits. Machin (1885) drew the attention of collectors to them after finding adults on old oak-trees in Epping Forest, Essex. A year later he wrote 'It now seems to have turned up in almost incredible numbers, not in any one particular place, but apparently all over the country' (Machin, 1886). Hearing that Stainton required specimens, Beaumont went to his house, took him out into his own garden and showed him moths behind the bark of an old willow-tree in such numbers that Stainton ran out of boxes (Porritt, 1905). It is, however, a species liable to great fluctuation in numbers (Stainton, 1885) and there is no evidence that it is as common today as it was in the 1880s. It is found in England as far north as Durham, but has not been recorded in Wales, Scotland or Ireland. Now Holarctic; its natural range extends across Europe to central and southern U.S.S.R., but since its accidental introduction into the north-eastern part of the U.S.A. it has there spread rapidly in cereal-producing areas.

References

Beirne, B. P., 1941. A list of the Microlepidoptera of Ireland. *Proc. R. Ir. Acad.* **47**(B): 53–147.

Davis, D. R., 1975. A review of Ochsenheimeriidae and the introduction of the cereal stem moth *Ochsenheimeria vacculella* into the United States (Lepidoptera: Tineoidea). *Smithson. Contr. Zool.*: no. 192, 20 pp., 31 figs.

Karsholt, O. & Nielsen, E. S., 1984. A taxonomic review of the stem moths, *Ochsenheimeria* Hübner, of northern Europe (Lepidoptera: Ochsenheimeridae). *Ent. scand.* **15**: 233–247, 36 figs.

Kloet, G. S. & Hincks, W. D., 1972. A check list of British insects: Lepidoptera (Edn 2). *Handbk Ident. Br. Insects* **11**(2): viii, 153 pp.

Kyrki, J., 1983. Adult abdominal sternum II in ditrysian tineoid superfamilies – morphology and phylogenetic significance (Lepidoptera). *Suom. hyönt. Aikak.* **49**: 89–94.

Machin, W., 1885. *Ochsenheimeria vacculella* in Epping Forest. *Entomologist* **18**: 264.

———, 1886. *Ochsenheimeria vacculella. Ibid.* **19**: 303–304.

Pavlov, I. F., 1961. Ecology of the stem moth *Ochsenheimeria vaculella* [sic] F.R. (Lepidoptera: Tineoidea). *Ent. Rev. Wash.* **40**: 461–466. English translation from *Ent. Obozr.* **40**: 818–827.

Porritt, G. T., 1905. Obituary: Alfred Beaumont. *Entomologist's mon. Mag.* **41**: 95–97.

Réal, P., 1966. Ochsenheimeriidae. *In* Balachowsky, A. S., *Entomologie appliquée à l'Agriculture* **2**: 254–255. Paris.

Scott, J., 1854. Larva and transformation of *Ochsenheimeria birdella. Zoologist* **12**: 4336–4337.

Sich, A., 1908. Ovum of *Ochsenheimeria vacculella* F.R. *Entomologist's Rec. J. Var.* **20**: 92.

Stainton, H. T., 1885. *Ochsenheimeria vacculella* – How does the larva live ? *Entomologist's mon. Mag.* **22**: 92–93.

Tutt, J. W., 1893. A few days' collecting in the western Highlands. *Entomologist's Rec. J. Var.* **4**: 285–288.

LYONETIIDAE
A. M. Emmet

A large heterogeneous family which has been variously interpreted. The grouping here adopted is that of the revised check-lists of Kloet & Hincks (1972) and Bradley & Fletcher (1979). In Britain four subfamilies, the Cemiostominae, Lyonetiinae, Bedelliinae and Bucculatriginae, are represented. The Bucculatriginae, which differ from the other subfamilies in several important respects indicated below, are given family status by some systematists (*e.g.* Chapman, 1902; Kyrki, 1983, who retains the Bucculatricidae in Tineoidea but transfers the Lyonetiidae to Yponomeutoidea).

Imago. Head usually rough-haired, at least on crown; ocelli present or absent; labial palpus moderate, short or rudimentary, porrect or drooping; maxillary palpus folded, short or, more often, vestigial; haustellum short and naked; antenna from two-thirds to just over length of forewing, with scape dilate with scales to form an eyecap, except in the Bedelliinae. Forewing usually narrow-lanceolate with apex acute and sometimes flexed upwards; venation much reduced, but seven to ten veins present; discoidal cell usually formed but open. Hindwing lanceolate to linear, with long cilia; five to six veins usually present.

Male genitalia: abdominal segment 7 or 8 with or developed into sclerotized or membranous flap-like lobes enveloping genitalia; uncus absent or weakly developed; gnathos usually present; valvae generally small and weakly sclerotized; aedeagus usually cylindrical or bulbous. Female genitalia: ovipositor adapted for piercing in *Lyonetia* Hübner and some *Bucculatrix* Zeller; apophyses anteriores usually weak or absent.

Larva. Typically with thoracic legs and five pairs of prolegs developed; crotchets uniordinal; prolegs sometimes absent in early instars. The larvae of most species feed entirely as leaf- or stem-miners, but later instars of *Bucculatrix* leave their mines and feed externally.

Pupa. Appendages fused with cuticle except sometimes at extremities; abdominal segments fused, without dorsal spines. Different characters are possessed by the pupae of the Bucculatriginae, *q.v.* Pupation takes place outside the mine either in an elliptical cocoon or suspended in strands of silk.

Key to subfamilies (imagines) of the Lyonetiidae

1   Antenna with scape expanded into eyecap 2
–   Antenna with scape elongate and with pecten, but without eyecap .... Bedelliinae (p. 226)

2(1)   Terminal cilia with one or more dark pencils or bars directed distad 3
–   Terminal cilia without such pencils; dark markings of cilia, if present, consisting of one or more lines subparallel to termen .... Bucculatriginae (p. 227)

3(2)   Terminal cilia with dark divergent bars not projecting beyond other cilia; a tornal spot of slightly arched, elongate leaden-metallic or violet scales ....
         .... Cemiostominae (p. 213)
–   Terminal cilia with a dark pencil projecting beyond other cilia; no tornal spot of specialized scales ....
         .... Lyonetiinae (p. 223)

Cemiostominae

A subfamily of world-wide distribution, represented in the British Isles by eight species placed in two genera.

Imago. Head with vertex rough-haired or smooth. Forewing broad-lanceolate with apex pointed and often slightly caudate; both wings with venation reduced but differing between genera. Male genitalia: eighth abdominal sternite represented by two developed processes incorrectly regarded as valvae by Pierce & Metcalfe (1935); valvae weakly developed; gnathos present or absent; aedeagus with broad, bulbous base. According to Kuruko (1964: 34), genitalic characters differ so much between species that it is difficult to use them as a basis for taxonomic grouping.

Forewing white or grey, with apical area variably yellow with a pattern of fuscous bars; a tornal spot consisting of elongate, prostrate but slightly arched leaden-metallic scales which overlie the normal scaling and may be lost in worn specimens; apical fringe with two to four pencils of dark cilia, whose relative angles are sometimes of diagnostic importance. The wing-tip tends to be flexed upwards, giving rise to the English (but almost unused) name of 'bentwing' (Heslop, 1964).

The cilia between the terminal pencils are interrupted in most species, notably those with grey forewings. This feature seems to have escaped the notice of both illustrators and authors, although it is an important part of a wing-pattern evolved to deceive predators. When the insect is at rest, its wing-tips project beyond the abdomen and are therefore its least vulnerable part. Their pattern mimics a small insect facing away from the moth itself. The costal and subapical bars represent its legs, the yellow subapical spot its thorax, its tornal spot its shining eye and the projecting cilial pencil its antenna. The combination of black and yellow may suggest a stinging hymenopteron. This pattern is analogous to the ocellus and tail found on the hindwing tornus of many lycaenids. The sluggish habits of the moths and their fondness for resting in conspicuous situations spring from their reliance on this protective adaptation. Stainton (1855: 286) writes of the genus *Leucoptera*, 'They are not easily roused from a state of repose, and have none of the restless activity which distinguishes the allied genus *Nepticula*. In calm weather, towards evening, they may be seen sitting on the terminal twigs of their foodplants, and flying leisurely from one twig to another. A broom bush, in a calm evening in June, becomes a beautiful sight, each twig being, as it were, illuminated by these brilliant little white insects.'

Ovum. Oval, flattened and slightly concave above.

Larva. Broadest in thoracic segments in most leaf-mining species but elongate and of even girth in *Leucoptera spartifoliella* (Hübner) which mines stems; intersegmental divisions usually deeply incised; prothoracic plate generally distinct; thoracic legs and five pairs of prolegs present in all species and best developed in *L. lotella* (Stainton), which alone among British species changes leaves; *L. malifoliella* (O. G. Costa) has fleshy lateral protuberances on the thoracic segments which become less pronounced with age.

Life history. The larvae chew parenchyma from the start; there is no sap-drinking phase as in the Gracillariidae. When the egg is laid on the underside of a leaf, the newly hatched larva mines upwards through to the upper surface. Thereafter the primary mine generally consists of a gallery and the secondary mine of a blotch. Pupation by all British species is in a white cocoon spun externally on a leaf or stem of the foodplant, or amongst leaf-litter. Except in *L. lotella*, the cocoon is pointed at each end, the ends being openable under pressure from within; the larval exuviae are ejected, but when the imago emerges, the pupal exuviae remain within the cocoon and there is no visible evidence that emergence has taken place. Except in *L. spartifoliella*, the winter is passed in the pupal stage. Pupae tend to overwinter more than once, and this habit may account for the virtual absence in some years of common species such as *L. malifoliella*.

Key to species (imagines) of the Cemiostominae

1   Forewing white 2
–   Forewing grey 7

2(1)   Hindwing white 3
–   Hindwing grey 6

3(2)   Inner margin of first costal spot of forewing extending to tornus .... *Paraleucoptera sinuella* (p. 222)
–   Inner margin of first costal spot of forewing reaching only halfway across wing 4

4(3)   Larger species (wingspan 7–9mm); antenna pale golden fuscous, apex white 5
–   Smaller species (wingspan 5–7mm); antenna dark golden fuscous, apex hardly paler .... *Leucoptera wailesella* (p. 217)

5(4)   Underside of forewing with reddish suffusion usually terminating in ill-defined diagonal line from middle of costa to tornus; a small, apical yellow spot .... *L. laburnella* (p. 215)
–   Underside of forewing with reddish suffusion reaching apex or ending in ill-defined vertical line from costa; apical yellow spot usually absent ....*L. spartifoliella* (p. 217)

6(2)   Hindwing underside with apical cilia white .... *L. orobi* (p. 218)
–   Hindwing underside with apical cilia golden grey .... *L. lathyrifoliella* (p. 219)

7(1)   Forewing with conspicuous purple-black apical spot .... *L. lotella* (p. 220)
–   Forewing without such an apical spot .... *L. malifoliella* (p. 221)

LEUCOPTERA Hübner

Leucoptera Hübner, [1825], *Verz. bekannt. Schmett.*: 426.
Cemiostoma Zeller, 1848, *Linn. ent.* **3**: 272.

A genus of world-wide distribution with seven species in Great Britain, five of which have also been recorded from Ireland.

Imago. Head smooth in white species but with rough-haired vertical tuft in grey species. Apex of forewing produced and generally slightly caudate; vein 1b (2A) simple; veins 3 (Cu_1) and 4 (M_3) absent, 5 (M_2) and 6 (M_1) stalked, with 6 (M_1) to apex or termen, 7 (R_5) and 8 (R_4) absent, 9 (R_3) and 10 (R_2) sometimes stalked and 11 (R_1) absent. Hindwing with vein 5 (M_2) present. See figure 88.

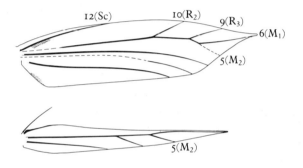

Figure 88 *Leucoptera laburnella* (Stainton), wing venation

LEUCOPTERA LABURNELLA (Stainton)
Laburnum Leaf Miner

Cemiostoma laburnella Stainton, 1851, *Suppl.Cat.Br. Tineidae and Pterophoridae*: 11.

Type locality: England; London.

Description of imago (Pl.9, fig.19)

Wingspan 7–9mm. Head, collar and eyecap shining white; antenna pale golden fuscous, grading to white at apex. Forewing shining white; a pair of outwardly oblique, more or less parallel, fuscous bars at three-fifths, extending from costa to near middle of wing; a similar, less oblique pair before apex; three divergent, straight fuscous lines in apical cilia; a triangular patch of long, sometimes black-tipped, leaden metallic scales at tornus, extending over the normal scaling into base of cilia and variably edged inwardly with black; the space enclosed by the first pair of costal bars completely filled with yellow and a white-centred, trifurcate yellow apical spot filling with one arm the lower half of the space enclosed by the anteapical bars; cilia shining white; underside basally suffused with pale reddish fuscous, usually terminating in an ill-defined diagonal line from middle of costa to tornus, and with a small, yellow apical spot. Hindwing and cilia white. Genitalia, see figure 89, p. 216.

Similar species. L. *wailesella* (Stainton) and L. *spartifoliella* (Hübner), *q.v.*

Life history

Ovum. Laid on the underside of a leaf of laburnum (*Laburnum anagyroides*); less often on dyer's greenweed (*Genista tinctoria*) or garden lupin (*Lupinus polyphyllus*).

Larva. Broadest in thoracic segments; intersegmental divisions deeply incised. Head transparent pale greyish brown, mouth parts darker. Prothoracic plate broad, with median sulcus; grey-brown. Abdomen greyish white, gut dark green. Legs pale brown, prolegs vestigial. In early instars, the head and prothoracic plate are darker. June to July; September.

Mine (Pl.1, fig.1). At first a circular brown blotch, diameter *c*.1.0mm, in which the larva mines to the upper surface of the leaf; then an irregular gallery, 5–10mm long, completely filled with greenish frass; finally a circular blotch, the black frass, which adheres to the upper cuticle, forming a spiral in the centre, leaving clear margins. Exit hole on upperside.

Pupa. Yellow-brown, in a flimsy, silken cocoon, pointed at the ends and spun under a web, usually on the underside of a leaf of the foodplant. The larval skin is ejected at pupation but the pupal exuviae remain in the cocoon after the emergence of the adult. July to August; September to May.

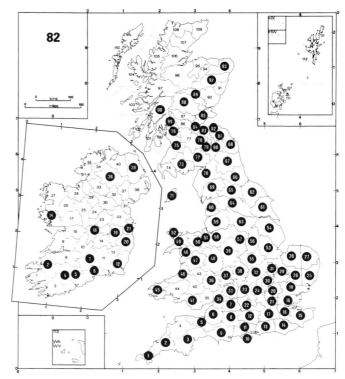

Leucoptera laburnella

Imago. Bivoltine; there may be three generations in favourable seasons. Flies at dusk in May and August.

Distribution (Map 82)

Predominantly an urban species, seldom occurring unless laburnum is plentiful. Widespread in Britain northwards to the Scottish Highlands; Ireland; Isle of Man. Throughout Europe; in southern Europe represented by subsp. *cytisella* Amsel, which feeds on *Cytisus* spp. North America.

Figure 89

Leucoptera laburnella (Stainton)
(**a**) male genitalia (**b**) female genitalia

Leucoptera wailesella (Stainton)
(**c**) male genitalia (**d**) female genitalia

Leucoptera spartifoliella (Hübner)
(**e**) male genitalia (**f**) female genitalia

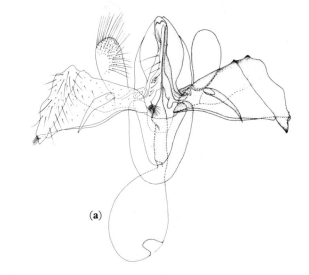

(**a**)

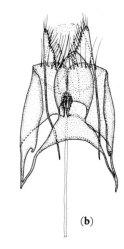

(**b**)

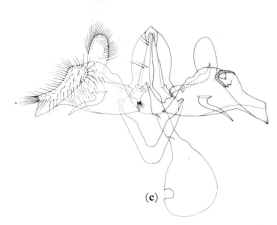

(**c**)

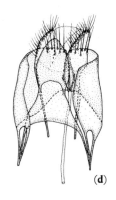

(**d**)

0.5mm

0.25mm

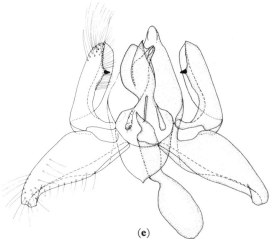

(**e**)

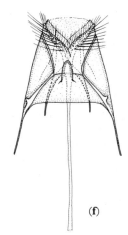

(**f**)

LEUCOPTERA WAILESELLA (Stainton)

Cemiostoma wailesella Stainton, 1857, *Entomologist's Annu.* **1858**: 115.

Type locality: England.

Description of imago (Pl.9, fig.20)

Wingspan 5–7mm. Differs from *L. laburnella* (Stainton) as follows. Antenna darker golden fuscous and not, or hardly, paler at apex. Forewing stated by Stainton (1859) to be obscurely bluish-tinged, but this is difficult to detect; costal and apical markings too variable to be used as distinguishing characters; underside with reddish fuscous suffusion deeper in hue and usually extending to apex. Hindwing in some specimens with faint golden sheen. The most reliable distinctive characters are the markedly smaller wingspan and the darker antenna. Genitalia, see figure 89, p. 216.

Life history

Ovum. Laid on the underside of a leaf of dyer's greenweed (*Genista tinctoria*).

Larva. Of even girth, intersegmental divisions deeply incised. Head transparent yellowish white, mouth parts pale brown. Prothoracic plate broad, almost colourless but with darker margins. Abdomen glossy pale yellowish white. Legs pale yellowish brown; prolegs moderately developed. June to July; September.

Mine (Pl.1, fig.5). At first a small brown blotch within which the larva eats through to the upper surface of the leaf; then a gallery filled with black frass leading to a blotch often extending over the whole leaflet, in which the black frass, which does not adhere to the upper cuticle, is scattered loose in the mine.

Pupa. Yellow-brown. The cocoon is spun on the underside of a leaf of the foodplant. First the leaf is slightly bent longitudinally by a belt of flossy silk, and then the pointed-oval cocoon is spun beside the belt. Both ends are openable and the larval exuviae are ejected and the pupal retained as in *L. laburnella*. In captivity an adult emerged successfully from a cocoon spun within its mine (Emmet, pers.obs.). July to August; September to May.

Imago. Bivoltine. Flies in May and August.

Distribution (Map 83)

The foodplant occurs mainly on clay and chalk. Local in England and Wales. Some records may be erroneous owing to confusion with *L. laburnella* when on the same foodplant. Western and central Europe, extending northwards into southern Sweden.

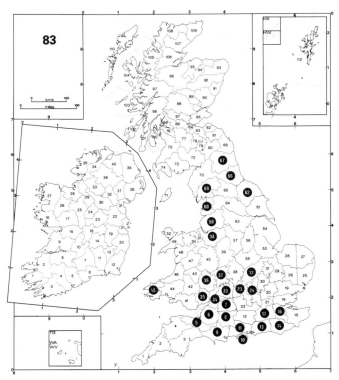

Leucoptera wailesella

LEUCOPTERA SPARTIFOLIELLA (Hübner)

Tinea spartifoliella Hübner, [1813], *Samml.eur.Schmett.* **8**: pl.49, fig.335.

Tinea punctaurella Haworth, 1828, *Lepid.Br.*: 578.

Type locality: Europe.

Description of imago (Pl.9, fig.21)

Wingspan 7–9mm. Differs from *L. laburnella* (Stainton) as follows. Forewing with second pair of costal bars sometimes convergent; first line in apical cilia slightly curving outwards towards termen; tornal spot more rectangular; apical yellow spot larger, often extending towards dorsum basad of tornal spot; terminal cilia with tips greyish golden; underside with reddish fuscous suffusion reaching apex or ending in an indefinite line at right angles to costa and without apical yellowish spot. All these characters vary; the most constant character is the normal absence of the yellow apical spot on the underside of the forewing. Genitalia, see figure 89, p. 216.

Life history

Ovum. Oval, with depression on upper surface. Laid on a first-year twig of broom (*Sarothamnus scoparius*) or dyer's greenweed (*Genista tinctoria*) (Emmet, 1976: 79). After the larva has hatched, the chorion is not filled with frass and so collapses and drops off; this readily distinguishes the mine from that of *Trifurcula immundella* (Zeller) (Nepticulidae) (Emmet, *MBGBI* 1: 210), where the chorion is filled with black frass, does not drop off and remains conspicuous even after the larva has vacated the mine.

Larva. Long and slender; intersegmental divisions moderately incised. Head from brown to black. Prothoracic plate T-shaped with median sulcus, two smaller bars on underside of first thoracic segment and small anal plate, all brown. Abdomen yellowish green, becoming greener as larva develops; gut somewhat darker green; an obscure chain of greyish brown spots along venter. Legs, which are well developed in later instars, dark brown or black. Starts feeding in October, but remains inactive in cold weather; feeds up rapidly in spring and is full-fed in April or May depending on the season.

Mine (Pl.1, fig.6). First usually mines upwards (*T. immundella* normally mines downwards), the early mine forming a sinuous brown ridge just under the epidermis; later it mines more deeply and its then greyish track fills the whole space between the angles of the stem. The larva mines through the angles and changes direction several times. In its later stages, the mine is indistinguishable from that of *T. immundella*.

Pupa. Yellow; appendages and abdominal segments 8–10 dark brown. Cocoon pointed-oval, generally spun on the stem or a leaf of the foodplant. Exuviae as for *Leucoptera laburnella*. April to June.

Imago. Univoltine. June and July.

Distribution (Map 84)

Widespread and common wherever the foodplant occurs throughout Britain; Isle of Man; in Ireland local or, more probably, under-recorded. Throughout Europe; north America.

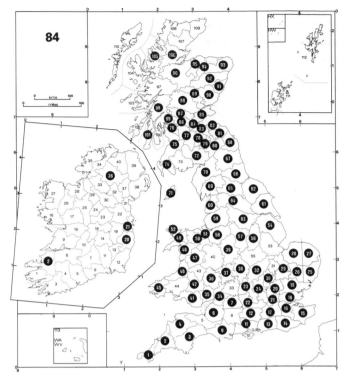

Leucoptera spartifoliella

LEUCOPTERA OROBI (Stainton)

Cemiostoma orobi Stainton, 1869, *Entomologist's Annu.* **1870**: 158.

Type locality: England; Scarborough, Yorkshire.

Description of imago (Pl.9, fig.22)

Wingspan 6–7mm. Head and eyecap shining white; antenna dark golden fuscous, apex paler. Forewing shining white; a pair of outwardly oblique, closely approximated fuscous bars at three-fifths, extending one-quarter of way across wing and sometimes reduced to a single bar; a similar, less oblique pair before apex; three fuscous bars in terminal cilia, the first usually curved and directed towards apex, the other two forming an angle of 15° and directed towards termen; a triangular patch of long, black-tipped leaden-metallic scales at tornus, extending over the normal scaling into base of cilia and inwardly edged black; when two costal bars are present, the space between may be filled with yellow or may be suffused fuscous; a variable, more or less linear yellow mark above tornal spot; cilia white. Hindwing purplish grey; cilia white. Underside of both wings with cilia white except for fuscous bars of forewing.

Similar species. *L. lathyrifoliella* (Stainton), *q.v.*

Life history

Ovum. Laid on either side of a leaf of bitter vetch (*Lathyrus montanus*); this plant was formerly known as *Orobus tuberosus* and the report that it also occurs on tuberous pea (*Lathyrus tuberosus*) seems to be due to a confusion of nomenclature.

Larva. Undescribed. Feeds in July; there is also a small second generation in September.

Mine (Pl.1, fig.4). At first a short, usually spiral gallery; later a blotch which absorbs and obscures the early gallery and often extends over the whole leaflet. The frass is at first concentrated in the centre of the mine but is later deposited haphazard.

Pupa. Undescribed. The cocoon is similar to that of *L. lathyrifoliella*, but is often spun on the foodplant. The larval exuviae are ejected as in *L. laburnella* (Stainton). August to May or August; September to May.

Imago. Mainly univoltine. Flies in May; a few moths of the next generation emerge in August, but the majority of the pupae overwinter, sometimes more than once. May be disturbed from its foodplant by day and is active shortly before dusk.

Distribution (Map 85)

Frequents moors, rough hillsides and open woodland. Very local but common where found. Recorded in England from north-eastern Yorkshire and Co. Durham; in Scotland from the Aviemore district; and in Ireland from Ballyeighter Wood and Newtown Castle in the Burren, Co. Clare, and from Ballynahinch, Co. Galway. Old records from Cambridgeshire (Fryer & Edelsten, 1938) are probably based on misidentifications since the foodplant does not occur there. Recorded in Europe only from Finland and Sweden.

LEUCOPTERA LATHYRIFOLIELLA (Stainton)

Cemiostoma lathyrifoliella Stainton, 1865, *Entomologist's Annu.* **1866**: 170.

Type locality: England; Shaldon, near Teignmouth, south Devon.

Description of imago (Pl.9, fig.23)

Wingspan 6–7mm. Differs from *L. orobi* (Stainton) as follows. Forewing with deeper and sometimes more extensive yellow markings at apex; dark markings deeper fuscous; bars in terminal cilia more divergent, forming an angle of 20–30°; terminal cilia broadly tipped golden grey, more easily seen on underside. Hindwing with apical cilia golden grey on both upper- and underside. Of these differences, the colour of the cilia seen from below is the most conspicuous.

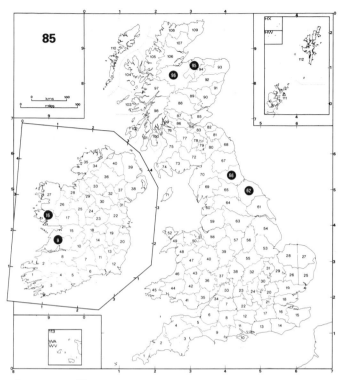

Leucoptera orobi

Life history

Ovum. Laid on the upperside of a leaf of narrow-leaved everlasting pea (*Lathyrus sylvestris*).

Larva. Broadest at metathorax; intersegmental divisions deeply incised. Head pale glossy yellow, darker in early instars; mouth parts pale brown. Prothoracic plate obscure but with two slightly darker anterior spots and median sulcus. Abdomen glossy pale yellow. Legs developed but small, concolorous with abdomen. June to late August.

Mine (Pl.1, fig.2). An oval blotch without any early gallery, but a small area surrounding the ovum is stained brown. The larva eats the palisade parenchyma only, leaving the spongy parenchyma intact. The greenish frass is deposited loose and haphazard. Sometimes several mines are in a leaf, and if these coalesce two or more larvae may be found in the same blotch.

Pupa. Yellow, turning brownish grey before the emergence of the adult. In a flimsy, white, oval-pointed cocoon, spun beneath strands of silk; exuviae as for *L. laburnella* (Stainton). The natural pupation site is probably leaf-litter, since cocoons have not been found even on heavily infested plants; in captivity the larvae usually pupate away from their foodplant. July or July to May.

Imago. Apparently bivoltine, but as adults and larvae in all stages of development occur simultaneously in July, the distinction between generations seems to be ill-defined. The moth flies from May onwards.

Distribution (Map 86)

Mainly confined in Britain to sea-cliffs and undercliffs; very local but common where found. It occurs on the coast of Devon from Torquay to Branscombe; at Luccombe Chine in the Isle of Wight; and in Stubby Copse, near Brockenhurst in the New Forest, Hampshire. Recorded in Europe only from Germany and Finland.

LEUCOPTERA LOTELLA (Stainton)

Cemiostoma lotella Stainton, 1858, *Entomologist's Annu.*
1859: 156.
Type locality: England; Scarborough, Yorkshire.

Description of imago (Pl.9, fig.24)

Wingspan 5–6mm. Head with vertex rough-haired, fuscous; frons shining grey; antenna shining purplish fuscous, eyecap grey. Forewing shining grey; an oblique, sometimes obsolescent, fuscous fascia extending from middle of costa to dorsum before tornus; wing beyond orange, enclosing two fuscous-edged, triangular, white costal spots; a large, triangular purplish black spot at apex, from which radiate four fuscous bars, the first forming the outer edge of the second white costal spot and the third directed more or less upward; an inwardly black-edged, purplish leaden-metallic spot at tornus, consisting of elongate scales which extend into base of cilia; costal cilia white, terminal and tornal shining grey; costal cilia on underside with fuscous bars absent or obscure. Hindwing, including cilia, grey.

Similar species. L. malifoliella (O. G. Costa), *q.v.*

Life history

Ovum. Laid on the upperside of a leaflet of common bird's-foot trefoil (*Lotus corniculatus*) or greater bird's-foot trefoil (*L. uliginosus*).

Larva. First thoracic segment broadest, head retractable; intersegmental divisions deeply incised. Head and small anal plate yellow-brown. Abdomen pale shining yellow. Legs well developed, enabling larva to walk easily when changing mines. July; sometimes a smaller second generation in August and September.

Mine (Pl.1, fig.8). A circular blotch without any early gallery, the blackish frass being arranged in a spiral and attached to the upper cuticle; in the later stages, spurs project from the blotch where the larva has been feeding. This is the only British member of the genus capable of moving to a fresh leaf; it may do this more than once. The mine is readily distinguishable from those of *Trifurcula* (= *Levarchama*) *cryptella* (Stainton) and *T. eurema* (Tutt)

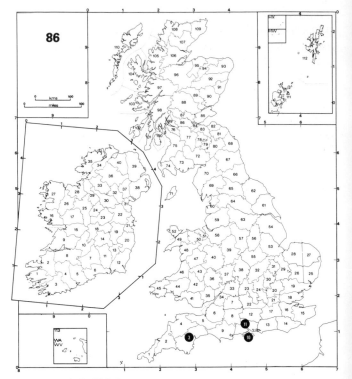

Leucoptera lathyrifoliella

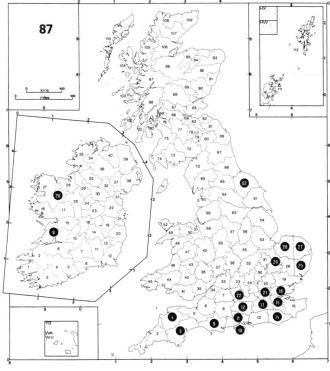

Leucoptera lotella

(Nepticulidae), which start with a long slender gallery containing linear frass (Emmet, *MBGBI* 1: 211–212).

Pupa. Yellow-brown, in a white, oval cocoon, less pointed than those of the other members of the genus; the ends of the cocoon are not openable under pressure from within and the larval exuviae are not ejected. It is generally spun beneath a leaf of the foodplant. August to May or August; September to May.

Imago. Generally univoltine, with the moth flying in May. In favourable seasons, at any rate in southern England, a small second brood occurs in August.

Distribution (Map 87)

Occurs on dry downland, where common bird's-foot trefoil is the foodplant, and in marshes on greater bird's-foot trefoil. More or less widespread in south-eastern England and also found in Yorkshire and the west of Ireland; probably much overlooked. On the Continent recorded only from Germany.

LEUCOPTERA MALIFOLIELLA (O. G. Costa)
Pear Leaf Blister Moth

Elachista malifoliella O. G. Costa, [1836], *Fauna Regno Napoli*, Lepidotteri: [293].

Opostega scitella Zeller, 1839, *Isis, Leipzig* **1839**: 214.

Type locality: Italy; Naples.

Description of imago (Pl.9, fig.25)

Wingspan 7–8mm. Differs from *L. lotella* (Stainton) as follows. Head with vertex almost smooth-scaled. Forewing broader; no purple-black apical spot; third cilial bar directed more or less outwards; tornal spot broadly dark-margined distally; underside with dark bars clearly visible in costal cilia.

Life history

Ovum. Laid on the underside of a leaf of a rosaceous tree or shrub, well away from the margin. The most usual foodplants are hawthorn (*Crataegus* spp.), apple (*Malus* spp.), rowan (*Sorbus aucuparia*) and pear (*Pyrus communis*), but it is also found on blackthorn (*Prunus spinosa*) and other *Prunus* spp., quince (*Cydonia oblonga*), *Cotoneaster* spp., other *Sorbus* spp., etc. It has been recorded on birch (*Betula* spp.) on the Continent but not in Britain.

Larva. Broadest in the thoracic segments, thence tapering towards the anus; intersegmental divisions moderately deeply incised; the young larva has large fleshy protuberances on abdominal segments 1–3, each ending in a seta, which become less prominent as the larva develops. Head and dorsal and ventral prothoracic plates blackish brown. Abdomen greyish white, gut dark green. Legs brown. August to September.

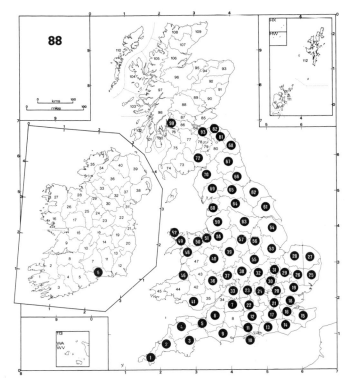

Leucoptera malifoliella

Mine (Pl.1, fig.3). A circular blotch, within which the larva feeds in a spiral, attaching its frass to the upper cuticle; in the final instar this pattern is less strictly followed and the frass is scattered loose. Often several mines in a leaf. When the mine occurs in the lobe of a hawthorn leaf, it somewhat resembles that of *Stigmella paradoxa* (Frey) (Nepticulidae), but in the latter the egg is laid at the apex of the lobe, within 1.0mm of the margin (Emmet, *MBGBI* 1: 238).

Pupa. 'Singularly flat' (Stainton, 1855). The cocoon, which is spun on leaf-litter, resembles that of *Leucoptera laburnella* (Stainton) and the larval exuviae are ejected as in that species. September to June.

Imago. Univoltine. Flies in June and July. Every few years, as in the hot summer of 1976, it occurs in profusion. Then the larval mines disfigure its foodplants; an infested hawthorn hedge may often be detected from afar, since the leaves look more brown than green. Between these 'population explosions' it is relatively scarce; probably parasites and/or disease are responsible for the sudden decline.

Distribution (Map 88)

Throughout England and Wales to southern Scotland. In Ireland observed sparingly in Co. Wexford in 1976 (Emmet, pers.obs.). Eurasiatic, reaching northern China.

PARALEUCOPTERA Heinrich

Paraleucoptera Heinrich, 1918, *Proc.ent.Soc.Wash.* **20**(1): 21.

A small Holarctic genus represented in Britain by a single species recorded only from Scotland.

Imago. Head with rough-haired vertical tuft. Forewing with apex produced but not caudate; vein 1b (2A) furcate; veins 4 (M_3) and 7 (R_5) present, 3 (Cu_1) and 8 (R_4) absent; veins 9 (R_3) and 10 (R_2) stalked. Hindwing with vein 5 (M_2) absent.

PARALEUCOPTERA SINUELLA (Reutti)

Leucoptera sinuella Reutti, 1853, *Beitr.rhein.Naturg.* **3**: 208.

Cemiostoma susinella Herrich-Schäffer, 1855, *Syst.Bearb. Schmett.Eur.* **5**: 342.

Type locality: Germany; Moosewalde, near Freiburg.

Description of imago (Pl.9, fig.26)

Wingspan 7–8mm. Head with vertex rough-haired, white; antenna golden fuscous, grading to white at apex, eyecap white. Forewing shining white; a moderately oblique yellow bar from costa at two-thirds, edged fuscous, the inner edging continued as a dark line to tornus; a similar, almost vertical bar at three-quarters, edged inwardly and sometimes partly outwardly with fuscous; a third vestigial bar at apex; an elongate yellow spot below the two outer bars; a tornal spot of elongate, leaden-metallic scales inwardly and sometimes outwardly edged black; cilia white with two parallel fuscous bars directed outwards subparallel to costa and sometimes one or two divergent bars below. Hindwing, including cilia, white.

Similar species. This is the only white British cemiostomine with the vertex of the head rough-haired, with a dark streak extending from the costa to the tornus and with the two upper cilial bars subparallel to the costa.

Life history

The description below is made mainly from living material received from the Netherlands.

Ovum. On the upper side of a leaf of aspen (*Populus tremula*), usually in a batch of about six laid in a row beside a vein. On the Continent, also on grey poplar (*P. canescens*), black poplar (*P. nigra*) and sallow (*Salix* spp.).

Larva. Head and abdomen whitish. July and September.

Mine. A blotch, at first brown but later turning blackish, with paler margins; the frass is deposited in fine grains in the centre of the mine. There are sometimes several larvae

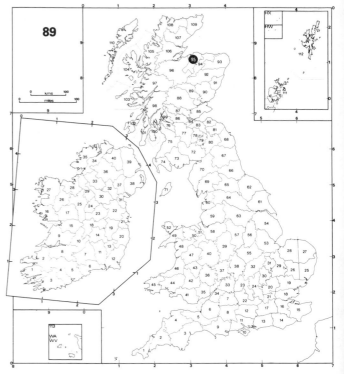

Paraleucoptera sinuella

in a mine, but seldom as many larvae as ova, some of which will be found to contain fully developed but dead larvae.

Pupa. Pale yellowish brown. Prior to pupation the larva spins a web measuring about 4×10mm on the upper surface of a leaf of the foodplant. Each of the shorter sides has a narrow, central V-shaped opening extending inwards for about 3mm. The cocoon is then spun beneath this web, its ends aligned with the openings. It is white, spindle-shaped and with a distinct cleft at each end, one for the ejection of the larval exuviae and the other for the emergence of the adult. July to August; September to May.

Imago. Bivoltine. June and August. The moths fly actively by day around their foodplant.

Distribution (Map 89)

Discovered at Aviemore, Inverness-shire, in a spinney of aspens near the railway station (Bankes, 1910). There it persisted until the 1950s when it disappeared suddenly; recent searching has failed to reveal any sign of its continued presence. The only other British record is a few taken at Grantown-on-Spey, Morayshire (Brown, 1945). It is strange that it was so restricted in Britain, as abroad it occurs commonly in varied climatic conditions in a range extending from northern Europe to north Africa and eastwards to Japan.

Lyonetiinae

A subfamily of world-wide distribution, represented in the British Isles by a single genus.

Key to species (imagines) of the Lyonetiinae

1   Forewing with outward oblique dark streak from middle of dorsum .... *Lyonetia prunifoliella*(p. 224)
–   Forewing without such a streak ...... *L. clerkella* (p. 225)

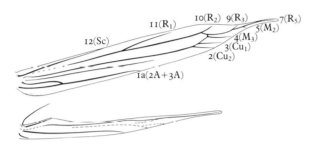

Figure 90 *Lyonetia clerkella* (Linnaeus), wing venation

LYONETIA Hübner

Lyonetia Hübner, [1825], *Verz.bekannt.Schmett.*: 423.
Argyromis Stephens, 1829, [June], *Nom.Br.Insects*: 49.
Argyromiges Curtis, 1829, November, *Br.Ent.* **6**: fasc. 284.

A large genus, two species of which have been recorded in the British Isles, although one has not been observed for over 50 years.

Imago. Head with vertex rough-haired; antenna slightly exceeding length of forewing; scape expanded with scales to form eyecap. Forewing narrow lanceolate, parallel-sided, with apex strongly produced; venation reduced; discoidal cell reaching two-thirds to three-quarters of wing; veins 4 (M_3), 8 (R_4) and sometimes 3 (Cu_1) absent; 7 (R_5) to costa near apex. Hindwing linear-lanceolate; venation reduced; transverse vein, 3 (Cu_1), 6 (M_1) and sometimes 4 (M_3) absent (figure 90). Male genitalia: eighth tergite with a longitudinal sclerotized ridge and a pair of sclerotized arms which project backwards from the posterior margin; a pair of extensible coremata on the ventral surface. Female genitalia: apophyses anteriores and apophyses posteriores well developed; ovipositor with file-like teeth adapted for piercing.

The predominant wing-pattern consists of a white wing, an elongate brown or orange discal spot, a similar subapical spot from which dark bars extend to the costa and costal cilia, an apical fuscous spot and a projecting fuscous pencil in the terminal cilia. As in the Cemiostominae, *q.v.*, this pattern, though simpler, may deceive a predator into mistaking the less vulnerable wing-tip for a smaller insect when the moth is at rest. Both British species have a melanic form.

Ovum. Undescribed.

Larva. Long and slender with intersegmental divisions deeply incised; thoracic legs and five pairs of prolegs developed.

Pupa. Appendages reaching extremity of abdomen and fused to it throughout their entire length.

Life history. The ovum is laid beneath the cuticle of a leaf. The larva feeds entirely on the parenchyma by mining; one of the British species can change leaves. Pupation takes place outside the mine in a very slight, flimsy cocoon suspended hammock-wise from silken threads, generally across a leaf surface drawn into a weak concavity by the supporting silken threads. Both British species overwinter in the imaginal stage.

LYONETIA PRUNIFOLIELLA (Hübner)

Tinea prunifoliella Hübner, [1796], *Samml.eur.Schmett.* **8**: pl.28, fig.191.

Tinea padifoliella Hübner, [1813], *ibid.* **8**: pl.46, fig.316.

Type locality: Europe.

Description of imago (Pl.9, figs 27,28)

Wingspan 9–10mm. Head and labial palpus white, vertex sometimes mixed pale fuscous. Thorax white. Forewing shining white; a reddish fuscous suffusion on costal half of wing from base to three-fifths, its lower edge darker and sinuate; an outward oblique, reddish fuscous streak from middle of dorsum meeting costal suffusion; a similar, shorter, sometimes interrupted streak from just before tornus; an elongate subapical fuscous streak terminating distally in a triangular black spot of dense, slightly raised scales; four divergent reddish fuscous bars extending from subapical streak to costa and three less well-defined streaks to tornal cilia; a pencil from apical spot projecting beyond cilia, curving slightly towards tornus and crossed near tips of cilia by a vertical bar, all reddish fuscous; apical and terminal cilia white, tornal and dorsal cilia pale reddish fuscous. Hindwing and cilia pale reddish fuscous.

Variation occurs in the development of the reddish fuscous markings of the forewing; additional small dorsal and subdorsal spots are sometimes present near base.

Life history

Ovum. Laid on the underside of a leaf of various rosaceous shrubs and trees, including blackthorn (*Prunus spinosa*), Japanese quince (*Chaenomeles japonica*), hawthorn (*Crataegus* spp.), apple (*Malus* spp. including cultivars), *Cotoneaster* spp. and *Sorbus* spp.; also on birch (*Betula* spp.). Sometimes several eggs are laid on the same leaf.

Larva. Pale green; July to August.

Mine. At first a fine gallery, hardly widening and almost filled with reddish brown frass; this leads abruptly to a large blotch, usually formed on the leaf-margin, in which the frass is dispersed, though some of it is ejected through one or more holes cut in the lower cuticle. Sometimes the blotch and gallery are separated, even being found on different leaves.

Pupa. Undescribed; in a flimsy cocoon suspended by silken threads on the lower surface of a leaf. August to September.

Imago. Univoltine, emerging in late September and overwintering; on the wing again in spring until May.

Distribution (Map 90)

Very local and not observed since about 1900. It was reported from the Midlands at Boughton, Worcestershire (Rea & Fletcher, 1901), Whittlebury Forest, North-

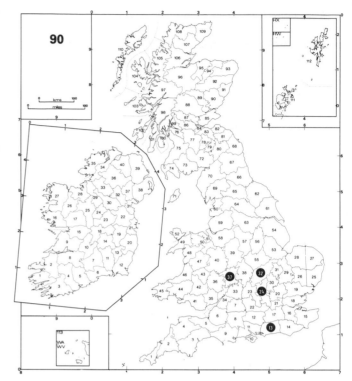

Lyonetia prunifoliella

amptonshire (Goss, 1902) and near Stony Stratford, Buckinghamshire (Stainton, 1859; Barrett, 1905); and also from Worthing, Sussex, where it was fairly common in the late nineteenth century (Goss & Fletcher, 1905). Palaearctic, occurring commonly throughout Europe and Asia Minor and in Japan.

LYONETIA CLERKELLA (Linnaeus)

Apple Leaf Miner

Phalaena (Tinea) clerkella Linnaeus, 1758, *Syst.Nat.* (Edn 10) **1**: 542.

Type locality: Sweden.

Description of imago (Pl.9, figs 29,30)

Wingspan 8–9mm. Head with vertex rough-haired, white, sometimes mixed fuscous; antenna as long as forewing, pale golden. Forewing elongate, costa and dorsum parallel; shining white; a golden brown, elongate, posterior discal spot above fold, edged or suffused fuscous; a similar subtriangular apical spot, the two separated by an angular fuscous fascia; costal and terminal cilia white, barred fuscous; a black dot at base of terminal cilia, from which emanates a projecting, slightly dorsally inclined, fuscous pencil; dorsal cilia golden fuscous. Hindwing dark grey, cilia golden fuscous.

A melanic form occurring as about 10 per cent of the population has the whole forewing except for the costal and terminal cilia suffused golden fuscous; intermediate forms are also found.

Life history

Ovum. The female pierces the lower epidermis to lay the egg in a leaf of various Rosaceae or birch (*Betula* spp.). Commonly chosen foodplants include hawthorn (*Crataegus* spp.), apple (*Malus* spp. including cultivars), *Prunus* spp., *Sorbus* spp. and *Cotoneaster* spp. Very occasionally *Salix* spp. are selected (Emmet, pers.obs.).

Larva. Long, flattened with intersegmental divisions deeply incised. Head pale shining brown; prothoracic plate with median sulcus and small anal plate darker brown; body pale greenish grey; thoracic legs black, prolegs vestigial. May, July and September to October.

Mine (Pl.1, fig.7). A long, narrow tortuous gallery which sometimes crosses the midrib; frass linear; exit hole on upperside. The larva mines venter upwards. The shape of the larva (see above) and the greater length of the final chamber when the mine is vacated readily distinguish this mine from that of a nepticulid.

Pupa. Pale green, wings darkening as the moth develops; a forked prominence on the head. In a slender, open-ended cocoon, suspended from silken threads on either side of a leaf of the foodplant or on vegetation below; there are usually two threads at each end of the cocoon which diverge at an angle of about 30°. May to June; July to August; October.

Imago. There are at least two generations and in favourable years three. The moth flies in June and August and those that emerge in October overwinter, often in thatch or evergreens, to reappear in April.

Distribution (Map 91)

Common throughout the British Isles northwards to the Caledonian Canal. Abroad throughout the Palaearctic region and in Madagascar.

Economic importance. In years of abundance, almost every leaf of rosaceous fruit-trees may be attacked by several larvae; leaf damage is chiefly due to necrotic areas caused by ringing and the crop-yield may be affected. Economic loss occurs in southern Europe but has rarely, if ever, been important in Britain.

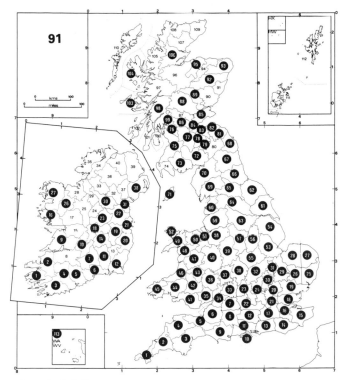

Lyonetia clerkella

Bedelliinae

A subfamily containing only one genus, under which its characters are described.

BEDELLIA Stainton

Bedellia Stainton, 1849, *Syst.Cat.Br.Tineidae and Pterophoridae*: 23.

A cosmopolitan genus, one species of which occurs in the British Isles.

Imago. Head with vertex rough-haired; labial palpus short, porrect; maxillary palpus rudimentary; antenna as long as forewing with scape elongate and somewhat thickened, without the eyecap characteristic of the other subfamilies but with a dense pecten. Forewing narrow-lanceolate with apex acute; veins 1b (2A) furcate, upper branch weak; 3 (Cu_1) and 4 (M_3) absent; 6 (M_1) and 7 (R_5) stalked; 8 (R_4) absent. Hindwing linear-lanceolate with very long cilia; transverse vein between 2 (Cu_2) and 5 (M_2) absent; 3 (Cu_1) and 4 (M_3) absent. Male genitalia: eighth tergite protruded beyond tegumen; eighth sternite with a pair of long-haired coremata; uncus absent; gnathos undeveloped; valva oblong. Female genitalia: corpus bursae elongate with crescent-shaped signum; ovipositor not adapted for piercing.

The imago rests with its anterior raised and the tip of the abdomen in contact with the substrate. This attitude is reminiscent of that adopted by a gracillariid, but *Bedellia* stands on its mid- and hindlegs only, whereas the gracillariid does so on all its legs. The British species overwinters as an adult.

Ovum. Oval. Laid on the surface of a leaf.

Larva. Spinneret functional throughout mining period (Hering, 1951: 132); legs and prolegs well developed. Moulting takes place within the mine with the larva curled in a semicircle, *i.e.* in the attitude adopted by *Bucculatrix* Zeller in their moulting cocoons. The larva readily makes fresh mines and changes leaves.

Pupa. All appendages fused. Pupation takes place without a cocoon, the pupa being suspended naked at the junction of a few strands of silk spun under a leaf, often the last in which the larva has fed.

BEDELLIA SOMNULENTELLA (Zeller)

Lyonetia somnulentella Zeller, 1847, *Isis, Leipzig* **1847**: 894.

Type locality: Sicily; Messina.

Description of imago (Pl.9, fig.18)
Wingspan 8–10mm. Head with vertex rough-haired, ochreous fuscous; antenna as long as forewing, ochreous ringed fuscous; scape with long pecten forming eyecap. Forewing ochreous with fuscous irroration becoming obsolescent towards dorsum; sometimes small fuscous dorsal spots at one-quarter, one-half and on tornus; cilia shining ochreous. Hindwing grey; cilia shining ochreous.

Life history
Ovum. Oval; laid, generally close to a vein, on either side of a leaf of field bindweed (*Convolvulus arvensis*), hedge bindweed (*Calystegia sepium*) or morning glory (*Ipomoea purpurea*); abroad other *Ipomoea* spp. and *Calystegia* spp. have been recorded as foodplants.

Larva. Body tapering towards anus; intersegmental divisions deeply incised. Head pale yellowish brown, sutures darker. Body greenish white; a subdorsal row of purplish brown spots, those on abdominal segments 1, 4 and 5 edged white; prothoracic plate brown, obscurely spotted darker; anal plate dark grey. Legs concolorous with body; anal prolegs marked dark grey. July to August; September.

Mine (Pl.2, fig.1). At first a narrow gallery, containing linear black frass and leading to a small, clear blotch; subsequently the larva frequently changes its mine, and sometimes also its leaf, making a series of irregular, clear blotches from which all the frass has been ejected. 'On the underside of the leaf beneath the mined areas [the larvae] attach a number of threads of silk and a certain amount of the frass they eject from the mine remains caught up in this web. There seems to be no explanation for this habit unless, perhaps, one assumes that they wish to make identification easy for the minologist' (Hering, 1951: 83). Possibly the purpose is to facilitate entry into a fresh mine. A larva with its jaws directed forwards has difficulty in piercing a leaf from a prone position on its surface. The silken threads provide a platform from which the larva can attack the leaf at an angle; compare the strategy adopted by *Aspilapteryx tringipennella* (Zeller) (Gracillariidae) for entering a new leaf (see p. 272). The angular attachment of the larval cases of phyllophagous Coleophoridae serves the same purpose. Kuruko (1964: 9) states that in Japan the larva rests on these threads when not feeding; this habit has not been observed in Europe.

Pupa. Grey-brown, with a dorsal keel and a prominent facial 'beak'. No cocoon is constructed. The prepupa rests

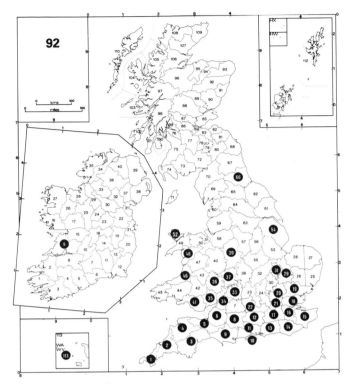

92

Bedellia somnulentella

on a few strands of silk spun above or below a leaf and after ecdysis the pupa remains attached to the larval exuviae, projecting at an angle. August; September to October.

Imago. Bivoltine, occurring in August and from October to May; the specific name probably refers to its winter 'sleep'. Numbers vary greatly from year to year, and it is sometimes locally abundant near the coast. This suggests that the native population may be reinforced by immigration; its almost universal distribution may result from a migratory tendency.

Distribution (Map 92)

Widespread but erratic in appearance in southern Britain and occurring with less frequency as far north as Co. Durham; in Ireland recorded only from the Burren, Co. Clare. It is found in every continent including Oceania, but in Europe its range northwards does not extend beyond Denmark and southern Sweden.

Bucculatriginae

A subfamily containing only one highly distinctive genus under which characters and behaviour are described.

Key to species (imagines) of the Bucculatriginae

1	Forewing unicolorous or almost so 2
–	Forewing not unicolorous 3
2(1)	Forewing grey .... *Bucculatrix cristatella* (p. 229)
–	Forewing shades of brown ... *B. maritima* (part) (p. 231)
3(1)	Forewing with ground colour yellow ....
	.... *B. thoracella* (p. 236)
–	Forewing otherwise 4
4(3)	Forewing with ground colour dark fuscous ....
	.... *B. cidarella* (p. 234)
–	Forewing with ground colour not dark fuscous 5
5(4)	Head black .... *B. nigricomella* (p. 230)
–	Head otherwise 6
6(5)	Head white, sometimes mixed fuscous 7
–	Head ochreous or brown 10
7(6)	Forewing with either one or two black streaks in disc subparallel to costa 8
–	Forewing without black streaks in disc 9
8(7)	Forewing with blackish apical streak ....
	.... *B. albedinella* (p. 234)
–	Forewing without black apical streak ....
	.... *B. crataegi* (p. 237)
9(7)	Antenna fuscous; eyecap white, edged fuscous ....
	.... *B. capreella* (p. 232)
–	Antenna white, ringed fuscous; eyecap uniform white ...
	.... *B. frangulella* (p. 233)
10(6)	Eyecap brown .... *B. maritima* (part) (p. 231)
–	Eyecap white, sometimes with a few darker scales ... 11
11(10)	Costa of forewing with four dark spots ....
	.... *B. ulmella* (p. 237)
–	Costa of forewing with three dark spots ....
	.... *B. demaryella* (p. 238)

BUCCULATRIX Zeller

Bucculatrix Zeller, 1839, *Isis, Leipzig* **1839**: 215.

A large, cosmopolitan genus, absent only from New Zealand. Of the 12 British species, one occurs only in Scotland, another, not included in the key, was recorded once over 100 years ago. The others have a relatively wide range within which they are locally common.

Imago. Head with vertex roughly tufted, frons smooth; ocelli absent; labial palpus minute; maxillary palpus rudimentary; haustellum short and naked; antenna two-thirds length of forewing, simple with scape dilated with scales to form an eyecap; first segment of flagellum deeply notched in male. Hindtibia clothed with long hairs above and short hairs below. Forewing lanceolate; veins 1b (2A) simple; discoidal cell elongate; 2 (Cu$_2$) and 5 (M$_2$) very short, from near apex of cell; 5 (M$_2$) sometimes absent; 3 (Cu$_1$) and 4 (M$_3$) absent; 6 (M$_1$) and 7 (R$_5$) stalked; 8 (R$_4$) sometimes absent; 9 (R$_3$) and 10 (R$_2$) very short; 11 (R$_1$) from before middle of discoidal cell. Hindwing two-thirds width of forewing, narrow lanceolate, cilia thrice width of wing; transverse vein between 2 (Cu$_2$) and 5 (M$_2$) absent; veins 3 (Cu$_1$) and 4 (M$_3$) absent; 6 (M$_1$) and 7 (Rs) stalked (figure 91). Male genitalia: uncus absent; socii developed to form two setose lobes; valvae tapered or rounded, sometimes bilobed, with basal angle of costa usually produced as a free arm, saccus sometimes semicircular; aedeagus rather elongate, more or less cylindrical. Female genitalia: apophyses anteriores absent; genital plate often densely set with scales; corpus bursae with signum consisting of spined ribs near juncture with ductus bursae.

Ovum. Oval. In British species laid on a leaf-surface. On hatching, the larva enters the leaf through the base of the egg which is filled with black frass.

Larva. Mouth parts in all instars adapted for chewing, directed forwards in the mining phase but downwards when the larva feeds externally. Body cylindrical, tapered little at extremities and without deeply incised intersegmental divisions; in most species unicolorous while mining but patterned and with the epidermis of a rough, shagreened texture when free-feeding; three pairs of thoracic legs and five pairs of prolegs fully developed in all instars.

Pupa. Pupa with abdominal segments capable of movement and not fused as in the other Lyonetiidae, herein approaching the Gracillariidae; however, important larval characters separate the Bucculatriginae from that family. The present position of the Bucculatriginae at the end of the Lyonetiidae and before the Gracillariidae has been hereby determined. Some systematists have regarded the pupal difference as sufficiently important to warrant the removal of the Bucculatriginae from the Lyonetiidae altogether (Chapman, 1902; see also Kuruko, 1964: 1–5).

Anal segments of abdomen free; lateral spines present; spiracles shortly stalked; cremaster consisting of two lateral spikes and a small bifid dorsal spike; head furnished with a chitinous toothed beak. The pupa ruptures the cocoon and protrudes before eclosion of the imago.

Life history. The larvae of all British species start feeding as leaf-miners, the duration of this phase differing between species. The mine takes the form of a gallery with linear frass. When tenanted it can be distinguished from that of a nepticulid by the presence of thoracic legs on the larva, and when vacated by its small size and the rough, matt appearance of the ovum, which in a nepticulid is smooth and shining. Small mines made by young yponomeutid or gelechiid larvae are more likely to cause confusion and these will be dealt with under the *Bucculatrix* species concerned.

On vacating its mine, the larva moults immediately; for this purpose it constructs a small, flat, circular, white 'cocoonet' on the surface of a leaf, generally on the underside at a junction of veins. Within this the larva curls itself in the shape of a horseshoe. When the moult takes place, the exuviae do not peel backwards in the usual manner but the larva emerges from its skin like a moth from its pupa, the skin remaining rigid and retaining its shape. Consequently, when a vacated cocoonet is held up to the light, it looks deceptively as if the larva were still present. Most species moult twice in this manner, the second cocoonet being considerably larger. In the free-feeding phase, the larva eats out 'windows' in the leaf, generally from below, *i.e.* it eats the lower epidermis and the mesophyll but leaves the upper epidermis intact; some species avoid even the finest veinlets. Species feeding on thick leaves, like *B. maritima* Stainton on sea-aster (*Aster tripolium*), may continue as leaf-miners in a series of short tunnels.

The remarkable manner in which the pupal cocoon (Pl.2, fig.10) is spun has long attracted the attention of naturalists. It was first observed in 1744 by the French entomologist Lyonet. He described the procedure of *B. ulmella* Zeller, an insect which was not to receive its name for over a hundred years. A few years later, in 1752, DeGeer described the pupation of *B. frangulella* (Goeze). Both these accounts are quoted *in extenso* by Stainton (1862: 56–64, 124–128). Jäckh (1955) gives a more detailed description, with a figure and photographs. The cocoon is spun almost entirely from the outside; this is rare in the Lepidoptera but is also the behaviour of *Nola aerugula* (Hübner) (Nolidae) (Tugwell, 1880; Revell, *MBGBI* 9: 119), though its larva follows a different procedure. The *Bucculatrix* larva starts by marking out the area within which the cocoon is to be constructed with a series of about 16 vertical silken posts; these are arranged in two semicircles about 2mm beyond the head and tail of the projected cocoon. Lyonet thought that the purpose of this palisade might be to protect the larva whilst it was at work, but it is much more likely that it is to guide the larva and help it keep its sense of direction. The larva starts work at the tail end of the cocoon, moving its spinneret in a series of short, dipping arcs, working, say, from left to right and then back again from right to left. As it spins it moves backwards until its anal end reaches its palisade. It has constructed a fluted tube, each of its ribs created by the upwards swing of its spinneret. It now enters its tube, turns round in the still elastic envelope and starts to work in the same manner from the head end. It works from outside as long as possible but the final junction has to be effected from within and at this point there is some blurring of the ribbed structure. Nevertheless, most larvae spin with such accuracy that the ribbing of the two halves maintains perfect alignment. After the 'outer cover' has been finished, the larva spins an 'inner tube' to complete its task. The palisade is only constructed when pupation is on a flat surface such as a tree-trunk or a broad leaf; it is dispensed with when the cocoon is spun on a blade or stalk of grass or a narrow leaflet. Some species are without a palisade even when the cocoon is spun on a suitable surface and in the descriptions which follow, the statement 'with' or 'without palisade' describes the larval behaviour in this respect.

The usual flight period of the adult is the evening, when such species as *B. crataegi* Zeller may fly in great abundance round their foodplant. At other times, the moths rest on trunks and fences. Free-feeding larvae drop on silken threads at the least disturbance and are best collected with the aid of a beating-tray.

Most species overwinter as pupae or prepupae, but *B. capreella* Krogerus does so as an adult and the first three species to be described probably pass the winter as very small larvae in their mines; *B. nigricomella* Zeller begins to feed again at the end of February if the weather is mild.

BUCCULATRIX CRISTATELLA Zeller

Bucculatrix cristatella Zeller, 1839, *Isis, Leipzig* **1839**: 214.
Type locality: Germany; Glogau (now Poland; Glogów).

Description of imago (Pl.9, fig.31)

Wingspan 6–7mm. Head with vertex ochreous grey; frons grey; antenna dark grey, eyecap ochreous grey. Forewing and hindwing uniform pale ochreous grey.

Similar species. Stigmella magdalenae (Klimesch) (*nylandriella* sensu auctt.) (Nepticulidae), which is smaller (3–5mm), has the vertex more ferruginous and the antenna shorter with white eyecap (Emmet, *MBGBI* **1**: 255).

Life history

Ovum. Laid on the upperside of a leaflet of yarrow (*Achillea millefolium*), close to the margin.

Larva. Head pale brown. Body dull green, dorsal line slightly darker; spots whitish green; prothoracic plate whitish green, minutely spotted black.

The young larva feeds in a gallery which follows the leaf-margin and contains linear black frass; it changes readily to a fresh mine. When too large for the leaflet to accommodate its body, it feeds as a partial miner, opening the side of a leaflet near its tip and excavating the parenchyma as far as it can reach while leaving the upper and lower epidermis intact. In later instars it feeds externally from above, leaving only the lower epidermis uneaten. The blanched and shrivelled leaflets fed on in this manner betray the presence of the larva. The white moulting cocoonets are sometimes spun on the upper surface of leaves and are then conspicuous. April to May; July. It is not known whether winter is passed as an ovum or a very small larva.

Pupa. Brown, with the chitinous beak vestigial. In a pure white cocoon without palisade spun on the foodplant or adjacent herbage. May to June; July to August.

Imago. Bivoltine, flying in June and August. Active in evening sunshine when it can be netted amongst its foodplant.

Distribution (Map 93)

Widespread and locally common on roadside verges, downland and grassland in England; not recorded from Wales; rare in Scotland; in Ireland noted only in the Burren, Co. Clare and in Co. Wicklow. Western and central Europe; Fennoscandia.

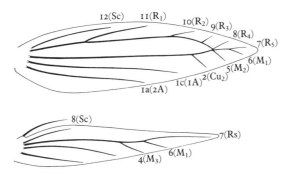

Figure 91 *Bucculatrix ulmella* Zeller, wing venation

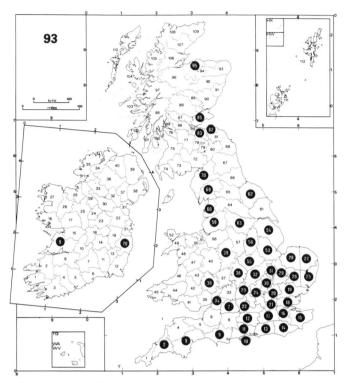

Bucculatrix cristatella

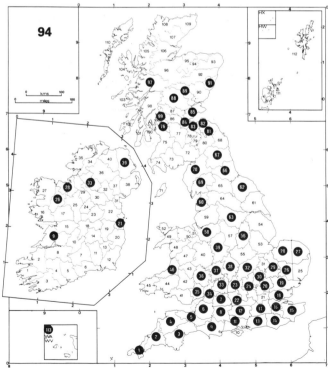

Bucculatrix nigricomella

BUCCULATRIX NIGRICOMELLA Zeller

Bucculatrix nigricomella Zeller, 1839, *Isis, Leipzig* **1839**: 215.
Bucculatrix aurimaculella Stainton, 1849, *Syst.Cat.Br. Tineidae and Pterophoridae*: 28.
Type localities: Bohemia [Czechoslovakia] and Germany; Glogau (now Poland; Głogów).

Description of imago (Pl.9, fig.32)

Wingspan 7–8mm. Head black, fawn brown laterally; antenna fuscous, shading to white at tip, eyecap white. Thorax yellowish metallic bronze, tegulae tipped white. Forewing yellowish metallic bronze; diffused whitish spots on dorsum at one-half and three-quarters and two similar spots on costa slightly beyond; a diffused whitish streak along fold from base to first dorsal spot; cilia bronzy grey. Hindwing bronzy grey, cilia slightly paler.

The continental form has the forewing uniform bronzy grey. The British form, described above, was at first regarded as a distinct species under the name *B. aurimaculella* Stainton. The continental form has not been observed in Britain but the British form occurs in the European population, especially in western Europe.

Life history

Ovum. Laid on the upperside of a leaf of oxeye daisy (*Leucanthemum vulgare*).

Larva. Head pale brown. Body greenish, gut darker, spots darker green; prothoracic plate paler green, speckled blackish.

The young larva feeds in a narrow gallery, starting from a small spiral; the frass is at first in a fine, continuous line but later is much interrupted and small in quantity. The mine is more commonly on the upperside but may be on the underside and sometimes passes from one side to the other; larger larvae make full-depth mines. The larva can change to a fresh mine, even in another leaf (Pl.2, fig.2). Sometimes the mine follows the leaf-margin or it may extend for a distance up the petiole before returning to the blade of the leaf. Moulting cocoonet white, generally on the under surface of a leaf. Subsequent feeding is mainly from the underside. March to April; July. It is not known whether the insect overwinters as an ovum or a very small larva in its mine.

Pupa. In a white cocoon on a stem of the foodplant or adjacent herbage. April to May; July to August.

Imago. Bivoltine. Late May to June and August. It flies in late evening sunshine low amongst its foodplant.

Distribution (Map 94)

Widespread, but local and seemingly absent from many places where its foodplant is common. It inhabits embankments, waste ground and grassy situations and occurs sporadically throughout Britain as far north as Invernessshire; probably under-recorded in Ireland from where there are scattered records; the Channel Islands. Europe, northwards to Fennoscandia.

BUCCULATRIX MARITIMA Stainton

Bucculatrix maritima Stainton, 1851, *Suppl.Cat.Br. Tineidae and Pterophoridae*: 28.
Type localities: England; St Osyth, Essex, and Gravesend, Kent.

Description of imago (Pl.9, figs 33,34)
Wingspan 8–9mm. Head with vertical tuft ochreous brown, mixed fuscous; frons shining brown; antenna brown with darker annulations, eyecap pale to dark ochreous brown. Forewing varying from pale to dark ochreous brown with darker irroration, more pronounced distally; small dark spots on fold at one-half and in disc at three-quarters; the following whitish markings are at best obscurely indicated and are frequently obsolete: a streak from base to one-half, outward oblique streaks from costa and dorsum at one-half and three-quarters and a subapical spot not quite reaching costa; the more distal costal spot sometimes dark-edged below; cilia greyish brown with a short basal and a longer subbasal line of dark-tipped scales. Hindwing and cilia pale ochreous grey.

Life history
Ovum. Laid on the underside of a leaf of sea-aster (*Aster tripolium*).
Larva. Head pale yellow. Body elongate, dull greyish yellow, gut darker, spots obscurely paler; prothoracic plate pale yellow, posterior region minutely spotted black; legs pale greenish yellow, spotted black.

At first in a long, narrow gallery which may be upper surface, lower surface or midway between the two; frass linear, black or reddish. On leaving its primary mine, it makes a series of shorter, full-depth mines and may continue feeding in this manner until full-grown; some larvae, however, eat out 'windows' from below in the usual manner of the genus. Occasionally mines are found in the epidermis of stems of the foodplant. The plants on which the larvae feed are liable to be submerged during spring tides and mining larvae may then have an advantage. The mines made in spring by *Scrobipalpa salinella* (Zeller) (Gelechiidae) are very similar but the larva of this species has a red dorsal stripe. April to May; July to August. It is not known whether the winter is passed as an ovum or an early instar larva.
Pupa. Pale olive brown. In a white cocoon, normally without palisade, on the foodplant, adjacent herbage or on

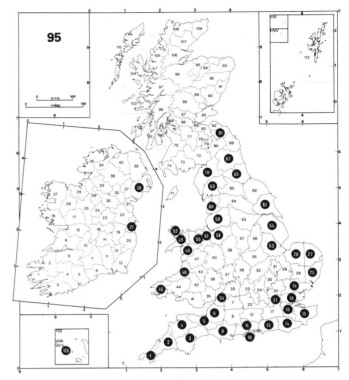

Bucculatrix maritima

debris which collects at the high-tide line. May to June; July to August.
Imago. Bivoltine. June and August. It flies low round its foodplant in evening sunshine mainly on the drier parts of salt-marshes; it has also been recorded at light.

Distribution (Map 95)
Common wherever the foodplant grows on saltings on the coast of England and Wales; just reaching Scotland in Berwickshire; in Ireland reported from Dublin by Kane in 1866 (Beirne, 1941: 128) and vacated mines found at Ballyconneely, Co. Galway, in 1973 were probably of this species (Emmet, pers.obs.); also recorded from Co. Down (H. G. Heal, pers.comm.). On coasts of western and northern Europe.

BUCCULATRIX CAPREELLA Krogerus

Bucculatrix capreella Krogerus, 1952, *Notul.ent.* **32**: 157.
Bucculatrix merei Pelham-Clinton, 1967, *Entomologist's Gaz.* **18**: 155.
Type locality: Finland; Punkasalmi.

Description of imago (Pl.9, fig.35)
Wingspan 8–9mm. Head with vertex white variably mixed fuscous and yellow-brown; frons shining white; antenna fuscous with paler annulations, eyecap white, edged posteriorly fuscous or yellow-brown. Thorax white, variably irrorate yellow-brown or dark brown. Forewing white with markings variably expressed; a yellow-brown suffusion, sometimes heavily irrorate with dark brown scales, may extend over whole wing except dorsum and pale costal spots at one-half, three-quarters and apex; this suffusion may be confined to the distal region or be almost wholly absent; a conspicuous black dorsal spot at one-half and a similar but more diffuse costal spot at three-quarters; sometimes also obscure subbasal and antemedian costal spots; apex with scattered black scales, often arranged as incomplete basal and subbasal lines in cilia; cilia shining ochreous grey. Hindwing ochreous grey, cilia paler.

Life history
Ovum. Laid on a leaflet of yarrow (*Achillea millefolium*).
Larva. Head yellowish brown. Body olive green; subdorsal, supraspiracular and spiracular rows of whitish, seta-bearing tubercles; legs yellowish brown.

Starts to feed by mining a leaflet, eating all the parenchyma and leaving a central line of black frass; later feeds externally from above in the same manner as the final-instar larva of *B. cristatella* Zeller, *q.v.* June to July.
Pupa. In a white cocoon, without palisade, spun on the foodplant or adjacent herbage. July.
Imago. Univoltine. The moths start emerging at the end of July and overwinter, after which they are on the wing until May. Two adults were beaten from alder on 9 September (Pelham-Clinton, 1967), suggesting that by this date they had already left the vicinity of their foodplant for hibernation. In the spring, specimens have been taken at rest on posts.

Distribution (Map 96)
First found at Aviemore, Inverness-shire, in 1966, it has since been found to be locally quite common in the valleys of the Scottish Highlands, where it frequents river-banks and roadside verges. Abroad it has been recorded in Finland and Sweden.

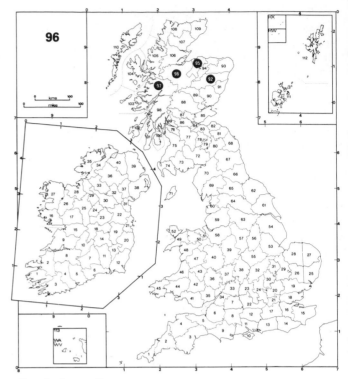

Bucculatrix capreella

BUCCULATRIX ARTEMISIELLA Herrich-Schäffer

Bucculatrix artemisiella Herrich-Schäffer, 1855, *Syst. Bearb.Schmett.Eur.* **5**: 340.
Type locality: Germany; Breslau.

A supposed example was reared from a larva found on yarrow (*Achillea millefolium*) at Folkestone, Kent, in 1865 (Knaggs, 1867). The usual foodplant of this species is field wormwood (*Artemisia campestris*) and it is therefore likely that the specimen was wrongly identified. Pelham-Clinton (1967: 158) suggests that it may have been *Bucculatrix clavenae* Klimesch which has somewhat similar wing-markings and feeds on yarrow.

BUCCULATRIX FRANGULELLA (Goeze)

Tinea frangulella Goeze, 1783, *Ent.Beyträge* **3**(4): 169.
Type locality: Europe.

Description of imago (Pl.9, fig.36)

Wingspan 7–8mm. Head white, vertex sometimes thinly mixed fuscous; antenna whitish with pale brown annulations, eyecap white. Thorax white, sometimes with one or two ochreous brown scales. Forewing white; plical spot at one-half, sometimes extending towards dorsum, and discal spot at three-quarters consisting of only one or two scales, black; the following markings pale ochreous brown, irrorate darker: rather broad ante- and postmedian outward oblique stripes from costa usually uniting in disc and extending to dorsum from just before plical spot almost to tornus, and a patch beyond discal spot almost filling apical area; a narrow, acute-angled terminal line of dark-tipped scales and a similar line in cilia; cilia shining ochreous grey, whiter at tornus. Hindwing and cilia ochreous grey, cilia shining.

The extent of the darker-irrorate, ochreous brown markings of the forewing differs between specimens; sometimes it extends almost to base of wing.

Similar species. B. capreella Krogerus, which has fuscous antenna with dark-edged eyecap, one or more black costal spots and less extensive dark irroration on the ochreous markings; according to present information the two species are allopatric in Britain.

Life history

Ovum. Laid on underside of a leaf of buckthorn (*Rhamnus catharticus*) or alder buckthorn (*Frangula alnus*).

Larva. Head very pale yellow-brown. Body pale yellowish green, thoracic segments faintly tinged purplish orange. During the mining phase, the young larva has a pale brown prothoracic plate and a chain of round, reddish brown spots on the venter; this is in contrast to most *Bucculatrix* which have the mining larva unmarked and the free-feeding larva with pattern.

The mine is at first a tightly wound spiral around the eggshell, in which the larva feeds venter upwards. This part of the mine is conspicuously stained blackish violet, according to Hering (1951: 94, 221) owing to the absorption by the plant tissues of liquid from the larval excreta. The last part of the mine straightens, and here the staining is absent (Pl.2, fig.4). After leaving the mine, the larva eats out windows from below, but sometimes eats the upper epidermis as well so as to make a complete hole. August, September.

Pupa. Undescribed. In a pale straw-brown cocoon with palisade spun on a leaf or amongst ground litter. October to June.

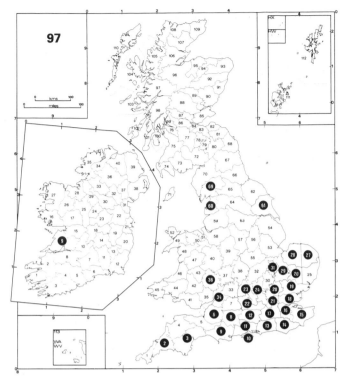

Bucculatrix frangulella

Imago. Univoltine. June to mid-July. Active in evening sunshine round its foodplants.

Distribution (Map 97)

Frequents buckthorn carr in fenland, open woodland and hedgerows. Widely but discontinuously distributed in southern England south of a line from the Wash to the River Severn; in some localities, such as Wicken Fen, Cambridgeshire, it can be very common. Further north, it is recorded from Skipwith, Yorkshire (Porritt, 1907), and Cumbria (N. L. Birkett, pers.comm.); in Ireland it has been recorded from the Burren, Co. Clare (J. M. Chalmers-Hunt, pers.comm.). Throughout Europe, including Fennoscandia; southern Africa.

BUCCULATRIX ALBEDINELLA Zeller

Bucculatrix albedinella Zeller, 1839, *Isis, Leipzig* **1839**: 216.

Elachista boyerella Duponchel, 1840, *in* Godart & Duponchel, *Hist.nat.Lépid.Fr.* **11**: 545, pl.309, fig.3.

Type localities: Germany; Frankfurt [-an-der-Oder] and Glogau (now Poland; Glogów).

Description of imago (Pl.9, fig.37)

Wingspan 8–9mm. Head white, mixed fuscous on vertex; antenna whitish with obscure fuscous annulations; eyecap white, often with a few fuscous scales. Thorax white, thinly irrorate fuscous. Forewing white, variably mixed with dark-tipped, ochreous scales which are more pronounced in certain areas to give the following predominantly ochreous markings: short antemedian, longer postmedian and subapical outward-oblique streaks and elongate dorsal spot extending beyond middle; the dark irroration more pronounced in certain areas to give the following black markings: spot of slightly raised scales basal to dorsal spot, a streak in disc marking outer edge of postmedian streak and subparallel to costa, and similar streak from lower edge of subapical streak to apex; terminal line black, narrow and evenly curved; cilia pale ochreous at apex, grading to white at tornus. Hindwing grey, cilia shining ochreous brown. Abdomen grey, anal tuft white.

Variation less pronounced than in related species, but the dark irroration sometimes extends to basal area.

Life history

Ovum. Laid on the underside of a leaf of English elm (*Ulmus procera*) or small-leaved elm (*U. carpinifolia*), usually close to the midrib and often at an angle of veins.

Larva. Head pale yellow-brown. Body greyish green; dorsal stripe olive-green, more faintly indicated on posterior segments; subdorsal stripe diffused and discontinuous. In the mining phase, the larva has a dark brown head and large oval dark spots on venter, thereby differing from most *Bucculatrix* in which the young larva is unmarked.

The young larva feeds in a slender gallery with linear frass, leaving clear margins; there are always two, usually three and often four diverticula, *i.e.* short, frass-free projections from the gallery which end abruptly; in some of these the lower epidermis is chewed through for an unexplained reason; no other British lyonetiid or nepticulid has a similarly formed mine (Pl.2, fig.7). In its final instars, the larva eats out windows from the underside of the leaf. July to early September.

Pupa. Black. In a cocoon differing from that made by other British members of the genus. It lacks the usual ribs and is spun under a flat web on bark or detritus. October to late May.

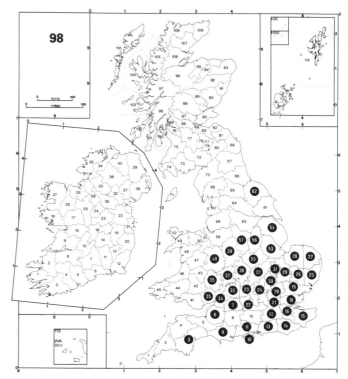

Bucculatrix albedinella

Imago. Univoltine; it flies in June in evening sunshine round elms, and earlier in the day and during dull weather it may be found resting on tree-trunks or fences.

Distribution (Map 98)

Frequents hedgerows and woodland borders and is widespread in England as far north as Yorkshire. It has not been reported from Wales, Scotland or Ireland, where its foodplants are rare. Southern and central Europe to southern Sweden; doubtfully recorded from Norway; Asia Minor.

BUCCULATRIX CIDARELLA Zeller

Bucculatrix cidarella Zeller, 1839, *Isis, Leipzig* **1839**: 216.

Type locality: Germany; Glogau (now Poland; Glogów).

Description of imago (Pl.9, fig.38)

Wingspan 8–9mm. Head with vertex ochreous mixed darker, frons white; antenna grey with dark brown annulations. Thorax dark fuscous. Forewing dark fuscous, the narrowly pale bases of the scales giving a roughened appearance; four whitish spots, the two on costa median and postmedian, and the two on dorsum antemedian and tornal; a small patch of slightly raised scales distal to inner dorsal white spot; cilia pale ochreous at apex, grading to grey at tornus. Hindwing and cilia pale grey.

When the larva has fed on *Myrica*, the ground colour is usually paler and the subdorsal patch of raised scales conspicuously darker than the ground colour.

Life history

Ovum. On the underside of a leaf of alder (*Alnus glutinosa*) beside a vein or on bog-myrtle (*Myrica gale*) beside the midrib. The matt, black, rough appearance of the egg helps to distinguish the ensuing mine from those of nepticulids.

Larva. Head pale brown. Body pale glossy green, thoracic segments with a slight purplish or orange tinge; gut obscurely darker green; pinacula whitish, bearing black setae; prothoracic plate concolorous with body, speckled brown. During the mining phase the head and prothoracic plate are dark brown.

The mine is a long, narrow gallery, yellowish brown when on bog-myrtle, almost filled with black, linear frass. When tenanted it can be distinguished from the similar mines of alder-feeding Nepticulidae by the more conspicuous larval legs, and by the dark brown head and prothoracic plate; the vacated mine differs in the distinctive ovum (see above), the smaller size and the larval exit hole which is on the upperside in *B. cidarella* and the underside in the nepticulids (Pl.2, fig.5). In its last two instars the larva feeds exposed on the underside of a leaf, eating out characteristic windows. August, September.

Pupa. Dark purplish brown; spiracles slightly raised; cremaster consisting of two lateral spines and a small bifid dorsal spine. In a pale yellow-brown cocoon without palisade, darkening with age, spun on a leaf-stalk or amongst detritus. October to May.

Imago. Univoltine. May and June. In favourable seasons some larvae feed up quickly to give rise to a small second generation in August. Rests by day on tree-trunks and fences and flies in evening sunshine. Although the coloration of adults resulting from *Myrica*-feeding larvae is usually distinctive, the populations on each foodplant are considered conspecific since there are no detectable differences in the genitalia of either sex or in the early stages. However, in many places where alder and bog-myrtle occur together *B. cidarella* has been found only on the former foodplant.

Distribution (Map 99)

Widespread and fairly common wherever there are alders throughout the British Isles as far north as Inverness-shire. It has been found on bog-myrtle only in north-west Wales on Anglesey (Michaelis, 1982) and Borth Bog, in south-west Ireland in Co. Kerry (K. G. M. Bond, pers. comm.) and in England on Arne National Nature Re-

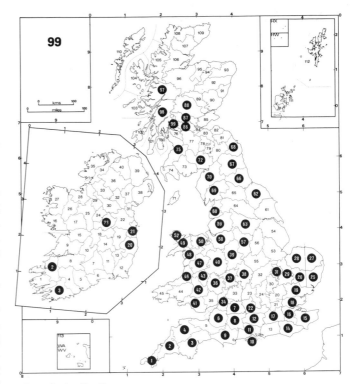

Bucculatrix cidarella

serve, Dorset (Emmet, 1982). When feeding on bog-myrtle, it appears to be very much more plentiful than it ever is on alder. Central and northern Europe, including Fennoscandia; on the Continent known only on alder.

BUCCULATRIX THORACELLA (Thunberg)

Tinea thoracella Thunberg, 1794, *Diss.Ent.* **7**: 88.
Elachista hippocastanella Duponchel, 1840, *in* Godart &
Duponchel, *Hist.nat.Lépid.Fr.* **11**: 530, pl.308, fig.4.
Type locality: Sweden; Delén.

Description of imago (Pl.10, fig.1)

Wingspan 6–8mm. Head with vertex ochreous, frons
white; antenna ochreous with darker annulations; eyecap
white. Thorax yellow; tegulae fuscous. Forewing yellow;
a broad subbasal fascia from which a narrow longitudinal
streak extends in disc to extreme apex, a large postmedian
costal spot not quite reaching discal streak and a smaller
median dorsal spot merging with discal streak, all fuscous;
a few black scales near anterior margin of dorsal spot; cilia
yellowish grey, darkened along axis of discal streak. Hindwing and cilia pale grey or grey.

Life history

Ovum. Laid on the underside of a leaf of small-leaved lime
(*Tilia cordata*), usually at an angle of veins and concealed
by their hairs. Certain populations lay on common lime
(*T.* × *vulgaris*), but where both trees are present the
small-leaved lime is strongly preferred.

Larva. Head pale yellow, mouth parts pale reddish
brown. Body pale greenish yellow, thoracic segments
sometimes with faint reddish tinge; gut green; prothoracic
plate pale brown.

The mine begins as an irregular gallery, often forming a
small blotch by fusion in the angle of veins; later it has a
short, more or less straight section which usually extends
along a vein before turning away at an angle; frass linear or
slightly dispersed (Pl.2, fig.3). The free-feeding larva eats
out windows in the leaf, generally from below, avoiding
even the finest veinlets. July to August; sometimes again
in September.

Pupa. In a pinkish brown cocoon spun on a trunk or
leaf-litter, usually without palisade. The two halves of the
cocoon are often joined more clumsily than is usual in
Bucculatrix. August to May.

Imago. Univoltine. June. Bivoltine on the Continent, and
in hot summers, at any rate in south-eastern England,
some adults emerge in August, giving rise to a second
generation of larvae in September. The main flight is in
evening sunshine but later it sometimes comes to light.

Distribution (Map 100)

Local, inhabiting ancient woodland or its fragmented re-
lics where small-leaved lime grows, preferring open rides
and the margins of woods. In localities where it has
adapted to common lime its habitat is more varied and it
may be found even in city gardens. It tends to occur in
great abundance in restricted areas, being completely ab-

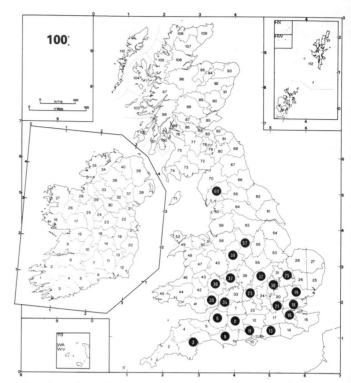

Bucculatrix thoracella

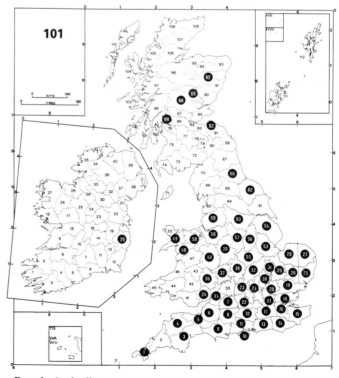

Bucculatrix ulmella

sent from apparently suitable nearby sites. Its headquarters are in the counties from Somerset along the Welsh border to Staffordshire. Elsewhere in its range its occurrence is sporadic. It has recently been found plentifully on common lime in and around London and there is evidence that it is extending its range (Emmet, 1984). Europe, including Fennoscandia.

BUCCULATRIX ULMELLA Zeller

Bucculatrix ulmella Zeller, 1848, *Linn.ent.* **3**: 288.
Argyromiges sircomella Stainton, 1848, *Zoologist* **6**: 2162.
Bucculatrix vetustella Stainton, 1849, *Syst.Cat.Br.Tineidae and Pterophoridae*: 28.

Type localities: Italy; Tuscany and Austria; Vienna.

Description of imago (Pl.10, fig.2)

Wingspan 7–8mm. Head with vertex ochreous, darker in centre of crown, frons white; antenna whitish, annulated fuscous; eyecap white, sometimes mixed fuscous. Thorax yellowish white mixed fuscous. Forewing yellowish white heavily irrorate with fuscous scales, the densest giving the following pattern: outward oblique subbasal, antemedian, postmedian and subapical spots on costa and a median blotch on dorsum, the anterior edge of which usually contains some raised black scales; cilia ochreous grey, shading to grey at tornus. These markings are variable in expression, the most strongly emphasized being the postmedian costal and the median dorsal spots. Hindwing and cilia pale grey.

Life history

Ovum. Laid on the upperside of a leaf of deciduous oak (*Quercus* spp.), generally close to the midrib.
Larva. Head pale brown. Body dull greyish green with numerous whitish dorsal and lateral spots, the posterior dorsal pair on each segment tending to unite to form a transverse band; prothoracic plate spotted grey-brown with an anterior transverse row of four blackish spots; legs very pale brown. In the mining phase, the head is pale brownish green and the body pale whitish green without markings.

The mine is a short, often contorted gallery close to the midrib, containing blackish frass (Pl.2, fig.9). The free-feeding larva eats out windows from the underside of the leaf. July; September, October.
Pupa. In a whitish cocoon, with or without palisade. The summer generation usually spins on a leaf and the autumn generation on a trunk; the larvae descend on silken threads and are often blown to a trunk not of their host tree. July, August; November to April.
Imago. Bivoltine. Late April until June and in August. Rests by day on trunks and fences.

Distribution (Map 101)

Widely distributed and common in Britain as far north as Perthshire; in Ireland recorded once from Co. Wicklow (Beirne, 1941). Europe, including Fennoscandia.
NOTE. Stainton (1862: 48) writes, 'The larva feeds on oak leaves (according to Mann also on elm leaves, but here I strongly suspect some mistake, as in various localities I have constantly observed the insect amongst oaks and never amongst elms; but Professor Frey assures me that he has repeatedly bred *B. ulmella* from *Ulmus campestris*)'. This species is no longer associated with *Ulmus* on the Continent and the present name may be misapplied.

BUCCULATRIX CRATAEGI Zeller

Bucculatrix crataegi Zeller, 1839, *Isis, Leipzig* **1839**: 216.

Type localities: Germany; Berlin and Glogau (now Poland; Glogów).

Description of imago (Pl.10, fig.3)

Wingspan 7–9mm. Head with vertex whitish, mixed ochreous fuscous; frons white, sometimes suffused ochreous; antenna whitish with fuscous annulations; eyecap white, usually with some ochreous scales. Thorax white, spotted ochreous. Forewing white, variably irrorate with blackish-tipped ochreous scales to give the following pattern: a diffused streak extending in disc from base to about one-third, antemedian, postmedian and subbasal outward-oblique streaks from costa, of which the postmedian is the most strongly expressed, and a subdorsal postmedian spot; the dark tips of scales most strongly emphasized on the postmedian and subdorsal markings, in disc at four-fifths where they form a short black streak subparallel to costa, and at apex where there is often a small black spot; cilia grey with blackish subbasal line. Hindwing grey, cilia with ochreous sheen.

Life history

Ovum. Laid on the upperside of a leaf of hawthorn (*Crataegus* spp.) or pear, including cultivars (*Pyrus communis*), generally beside the midrib or a major vein. It has also been recorded on wild service-tree (*Sorbus torminalis*), rowan (*S. aucuparia*) and apple (*Malus* spp. including cultivars).
Larva. Head light brown. Body dull greyish green; spiracular line and five raised spots on each segment yellowish white; prothoracic plate spotted grey. In the mining phase, it is uniform pale greenish yellow with darker head and prothoracic plate.

The mine is a short, contorted gallery with linear black frass, placed close to a major rib (Pl.2, fig.6). In the free-feeding phase, the larva eats out windows, generally from the upperside. July, August.

Pupa. In a white cocoon without a palisade spun on debris. September to April.

Imago. Univoltine. May and early June; in favourable seasons a few larvae feed up quickly to produce a small second generation in September. Flies in evening sunshine, sometimes in great abundance, around its food-plant.

Distribution (Map 102)

Widespread and locally abundant in scrub and along hedgerows and woodland margins. Occurs throughout England and southern Scotland but has not been recorded from Wales. In Ireland it was recorded by Birchall at Howth, Co. Dublin (Beirne, 1941), and in 1976 in east Cork (Emmet, pers.obs.). North Africa northwards to Finland and Sweden (but not recorded from Norway), and eastwards to Asia Minor.

BUCCULATRIX DEMARYELLA (Duponchel)

Elachista demaryella Duponchel, 1840, *in* Godart & Duponchel, *Hist.nat.Lépid.Fr.* **11**: 547, pl.309, fig.5.
Type locality: Latvia (now U.S.S.R.).

Description of imago (Pl.10, fig.4)

Wingspan 8–9mm. Head with vertex brown, darker-centred; frons yellowish white; antenna greyish white with darker annulations; eyecap yellowish white, mixed pale brown. Thorax yellowish white, variably mixed brown. Forewing yellowish white, densely irrorate with brown-tipped scales, the densest irroration giving the following pattern: a basal streak on fold to one-quarter, an outward-oblique antemedian spot, a postmedian fascia extending to tornus, an outward-oblique subapical spot, and a median subdorsal spot; of these markings, the fascia and subbasal spot are the darkest; the markings are generally clearly separated by areas in which the irroration is sparse or absent; sometimes a dark apical spot; cilia greyish ochreous with a dark brown, often irregular line. Hindwing and cilia pale ochreous grey.

Life history

Ovum. Laid on the underside of a leaf of birch (*Betula* spp.), also hazel (*Corylus avellana*), mainly in western localities, and sweet chestnut (*Castanea sativa*) (Chalmers-Hunt, 1966). On the last of these foodplants it has been described as a subspecies, *B. demaryella castaneae* Klimesch.

Larva. Head pale yellowish brown. Body dull grey-green; subspiracular stripe and spots white. In the mining phase, pale yellowish with the head darker.

The mine is a narrow gallery, often contorted at first, but later following a vein; the final chamber is often turned

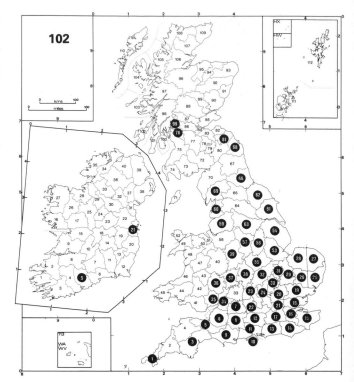

Bucculatrix crataegi

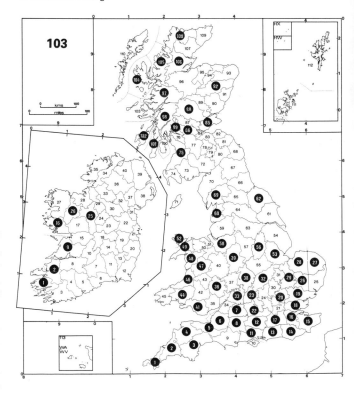

away at right angles; the frass fills the gallery (Pl.2, fig.8). In the free-feeding phase, the larva eats out windows from either side, leaving the veinlets. August.

Pupa. In a greyish ochreous cocoon spun amongst detritus. September to April.

Imago. Univoltine. May and early June. Rests on trunks and fences.

Distribution (Map 103)

Found in open woodland and on heaths, moors and bogs where birch grows throughout the British Isles, but more commonly in the western counties, Scotland and Ireland; it is local in south-eastern England. Central and northern Europe, including Fennoscandia.

References

Bankes, E. R., 1910. *Cemiostoma susinella*, H.-S., a tineid new to the British list, in Scotland. *Entomologist's mon. Mag.* **46**: 8–9.

Barrett, C. G., 1905. Lepidoptera. *Victoria County History of Buckinghamshire* **1**: 87–106.

Beirne, B. P., 1941. A list of the Microlepidoptera of Ireland. *Proc. R. Ir. Acad.* **47**(B) : 53–147.

Bradley, J. D. & Fletcher, D. S., 1979. *A recorder's log book or label list of British butterflies and moths*, [vi], 136 pp. London.

Brown, S. C. S., 1945. A new locality for *Leucoptera susinella* Herr.-Schäff. (Lep.). *J. Soc. Br. Ent.* **2**: 248.

Chalmers-Hunt, J. M., 1966. Spanish chestnut an apparently unrecorded natural pabulum of *Bucculatrix demaryella* Stt. (Lep. Lyonetiidae). *Entomologist's Rec. J. Var.* **78**: 138.

Chapman, T. A., 1902. The classification of *Gracilaria* and allied genera. *Entomologist* **35**: 81–88, 138–142, 159–164.

Emmet, A. M., 1976. Two species of Microlepidoptera reared from unusual foodplants. *Entomologist's Rec. J. Var.* **88**: 79–80.

——, 1982. *Bucculatrix cidarella* Zeller on *Myrica gale* in England. *Ibid.* **94**: 238.

——, 1984. *Bucculatrix thoracella* (Thunberg) (Lep.: Lyonetiidae). *Ibid.* **96**: 130–131.

Fryer, J. C. F. & Edelsten, H. M., 1938. Lepidoptera. *Victoria County History of Cambridgeshire* **1**: 139–161.

Goss, H., 1902. Lepidoptera. *Victoria County History of Northamptonshire* **1**: 94–100.

—— & Fletcher, W. H. B., 1905. Lepidoptera. *Victoria County History of Sussex* **1**: 164–210.

Hering, E. M., 1951. *Biology of the leaf miners*, iv, 420 pp., 1 col. pl., 180 text figs. 's-Gravenhage.

Heslop, I. R. P., 1964. *Revised indexed check-list of the British Lepidoptera*, vi, 145 pp. London.

Jäckh, E., 1955. Schutzvorrichtung zum Bau des Verpuppungskokons bei Arten der Gattung *Bucculatrix* Z. und *Lyonetia* Hb. (Lep., Lyonetiidae). *Z. wien. ent. Ges.* **40**: 118–121, 1 text fig., 4 pls.

Kloet, G. S. & Hincks, W. D., 1972. A check list of British insects: Lepidoptera. (Edn 2). *Handbk Ident. Br. Insects* **11**(2), viii, 153 pp.

Knaggs, H. G., 1867. Occurrence of a *Bucculatrix* (*B. artemisiella*) new to Britain. *Entomologist's mon. Mag.* **4**: 36.

Kuruko, H., 1964. Revisional studies of the family Lyonetiidae of Japan (Lepidoptera). *Esakia* **4**, 61 pp., 17 pls (1 col.), 4 text figs.

Kyrki, J., 1983. Adult abdominal sternum II in ditrysian tineoid superfamilies – morphology and phylogenetic significance (Lepidoptera). *Suom. hyönt. Aikak.* **49**: 89–94.

Michaelis, H. N., 1982. *Bucculatrix cidarella* Zeller (Lep., Lyonetiidae) on *Myrica gale*. *Entomologist's Rec. J. Var.* **94**: 102–103.

Pelham-Clinton, E. C., 1967. *Bucculatrix merei* sp.nov. (Lep., Bucculatricidae), a newly discovered Scottish species. *Entomologist's Gaz.* **18**: 155–158, 1 pl., 3 text figs.

Pierce, F. N. & Metcalfe, J. W., 1935. *The genitalia of the tineid families of the Lepidoptera of the British Islands*, xxii, 114 pp., 68 pls. Oundle.

Porritt, G. T., 1907. Lepidoptera. *Victoria County History of Yorkshire* **1**: 245–276.

Rea, C. & Fletcher, J. E., 1901. Lepidoptera. *Victoria County History of Worcestershire* **1**: 100–124.

Stainton, H. T., 1855. *The natural history of the Tineina*, **1**, xiii, 338 pp., 8 col. pls. London.

——, 1859. *A manual of British butterflies and moths* **2**, xi, 480 pp. London.

——, 1862. *The natural history of the Tineina* **7**, vii, 251 pp., 8 col. pls. London.

Tugwell, W. H., 1880. Life-history of *Nola centonalis*. *Entomologist* **13**: 42–45.

HIEROXESTIDAE
E. C. Pelham-Clinton

The two genera included in the British list that are at present placed in this family, formerly known as Oinophilidae, are maintained as a separate family here in spite of their inclusion by Davis (1978) in the Tineidae. The distinctive characters of the head, especially the peculiar ridge of scales between the antennae, are not found in any British Tineidae, though some exotic Tineidae have similarly flattened heads.

The family is mainly tropical and comprises at least 300 species, the majority in the genus *Opogona* Zeller.

This account owes much to Davis (*loc.cit.*) whose detailed descriptions of North American species include the British *Oinophila v-flava* (Haworth).

Imago. Sexes similar, except for some differences in wing markings in *Opogona sacchari* (Bojer). Head (figure 92) in side view tapering towards vertex, on which the antennae are connected by a compact rounded ridge of broad forwardly directed scales; this ridge sometimes bordered by dense vertical or frontal tufts; occiput and face long and smooth; antenna simple or slightly ciliate from three-quarters to length of forewing; ocelli absent; labial palpus three-segmented, porrect or drooping, directed outwards, the second segment with a few bristles at apex; maxillary palpus five-segmented, slender and inconspicuous in dried specimens; haustellum reduced, shorter than labial palpus. Forewing narrow; cell elongate, nearly three-quarters length of forewing; veins 6 (M_1), 7 (R_5) and 8 (R_4) stalked, 11 (R_1) absent. Hindwing narrow; venation greatly reduced. Frenulum with at least two bristles in female. Hindtibia densely haired.

Hieroxestidae are very diverse in appearance, the species occurring in Britain ranging in wingspan from 8 to 28 mm, but all may be recognized by the curious ridge of scales between the antennae. The labial palpi resemble those of Tineidae in having bristles on the second segment, but are always directed outwards. Moths rest with wings pressed close to the body, the tips of the wings slightly bent inwards.

Ovum. In *O. sacchari* ovoid and whitish, resembling those of Tineidae. Not described in other species.

Larva. Similar to those of Tineidae, in chaetotaxy most resembling Nemapogoninae, but ocelli reduced to no more than two pairs. On dead or living plant material in a silken gallery.

Pupa. In *O. sacchari* abdominal segments 5–7 are movable in the male, 5 and 6 in the female. Cremaster a pair of dorsally directed hooks. In the larval gallery, protruded on emergence.

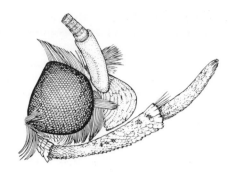

Figure 92 *Opogona sacchari* (Bojer), head

Occurrence in the British Isles. Only one of the three species on the British list is resident out of doors and that, *Oinophila v-flava*, only in the Isles of Scilly.

Key to species (imagines) of the Hieroxestidae

1   Forewing with yellowish fasciae ....
      .... *Oinophila v-flava* (p. 241)
–   Forewing without transverse pale markings 2
2(1)   Wingspan at least 20mm .... *Opogona sacchari* (p. 242)
–   Wingspan less than 20mm .... *O. antistacta* (p. 243)

OINOPHILA Stephens

Oinophila Stephens, 1848, *Trans.ent.Soc.Lond.* **5**: proc. xli.

About 50 species, mainly south African and Oriental, have been described in the genus but, as Davis (1978) points out few of them have been studied sufficiently to be sure that they are congeneric with *O. v-flava*, the type-species.

Imago. Head (figure 30, p. 66) as described for family, in dorsal as well as lateral view narrowed anteriorly, the interantennal scale-ridge bordered both anteriorly and posteriorly with tufts of long hairs. Antenna as long as forewing, slightly ciliate in male. Labial palpus short, length about one and a half times diameter of eye. Apex of both fore- and hindwing acuminate. Forewing veins 3 (Cu_1) and 4 (M_3) absent. Hindwing venation greatly reduced, only veins 1b (2A), 1c (1A) and 8 ($Sc+R_1$) easily distinguished (figure 93). Foretibia with epiphysis reduced to a minute spur.

Larva. Feeding on dry or decaying vegetable material.

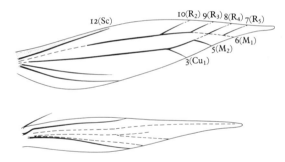

Figure 93 *Oinophila v-flava* (Haworth), wing venation

OINOPHILA V-FLAVA (Haworth)
Yellow V Moth

Gracillaria v-flava Haworth, 1828, *Lepid.Br.*: 530.
Type locality: [England].

Description of imago (Pl.10, fig.8)

Wingspan 8–12mm. Head with vertical and frontal hair-tufts ochreous orange; interantennal scale-ridge and face shining whitish; antenna whitish beneath scape and base of flagellum; occiput brownish fuscous. Forewing brownish fuscous, smoother in basal half, more or less irrorate with darker brown scales in apical half; a yellowish white median fascia, acutely angled outwards, the angle sometimes produced, narrowed and sometimes obsolete towards dorsum; irregular yellowish white markings in apical fourth, usually including a small spot at tornus, a larger spot beyond it on costa and a terminal subapical spot, sometimes linked to form a second angulated fascia. Hindwing fuscous, darker towards apex.

Life history

Larva. Head dark brown, paler at sides and posteriorly; ocelli reduced to a pair of single pigmented spots. Prothoracic plate pale yellowish brown. Body whitish with pale yellowish brown pinacula.

Feeds, in the open, on a variety of dry vegetable material such as palm-tree bark and old grass cuttings (Davis, 1978). Indoors it frequents wine-cellars and has been found on the fungus *Rhacodium cellare* (Morrison, 1968) and on wine corks.

Imago. Apparently univoltine in the open, the moths flying from July to September, but in California adults have been found mainly from April to June and in October and November. Moths fly mostly in the evening but may be found on the wing at any time of day.

Distribution

In the British Isles found in the open only in the Isles of Scilly where it is sometimes quite common. Larvae have not been found there and though moths are often found indoors they do not appear to frequent cellars. Elsewhere in the British Isles the species seems to be confined to cellars and warehouses. It is distributed widely in warm temperate regions of the world in central and southwestern Europe, the Azores, Madeira and Canary Islands, South Africa, California and Juan Fernandez Islands.

OPOGONA Zeller

Opogona Zeller, 1853, *Bull.Soc.Nat.Moscou* **26**: 504.

A genus currently including about 250 species, mainly of the Old World tropics, very diverse in superficial appearance. The two species on the British list are casual importations.

Imago. Head rounded in dorsal view; antennae widely separated; interantennal scale-ridge of long scales overlapping frons; vertical hair-tufts sometimes present. Antenna about three-quarters length of forewing; scape slightly flattened and concave ventrally. Labial palpus fairly long, length about twice diameter of eye. Forewing veins 5 (M_2) stalked with 6 (M_1), 7 (R_5) and 8 (R_4), these veins not well developed. Hindwing venation reduced, only veins 2 (Cu_2), 3 (Cu_1) and 8 ($Sc+R_1$) well developed. Foretibia with epiphysis well developed, about half length of tibia.

Larva. Feeding on a wide range of plants, primarily on decaying material but in a few species on living plants and in these cases sometimes attaining pest status.

OPOGONA SACCHARI (Bojer)

Alucita sacchari Bojer, 1856, *Report of the Committee on the 'Cane Borer'*: pl. 5.
Tinea subcervinella Walker, 1863, *List Specimens lepid.Insects Colln Br.Mus.* **28**: 477.
Type locality: Mauritius.

Description of imago (Pl.10, figs 5,6)
Wingspan 20–28mm. Head ochreous, irrorate with blackish brown; vertical tufts present, brownish ochreous; interantennal scale-ridge pale ochreous; face pale ochreous; labial palpus third segment slender, pale greyish ochreous; antenna greyish ochreous, paler beneath. Forewing pale ochreous irrorate with brown or blackish brown; discal stigmata of female distinct, blackish brown; in male stigmata less distinct, dark brown streaks usually present along fold and dorsum, and sometimes a whitish slightly sinuate streak from base above fold. Hindwing fuscous.

Life history
Described by Oldham (1928).

Larva. Head reddish brown; one pair of large ocelli and a second smaller pair posterior to the larger. Prothoracic and anal plates brown. Body whitish, with small yellowish pinacula.

Feeds on a wide range of plant material, both dead and living, and especially as a pest of bananas (*Musa* spp.) and sugar-cane (*Saccharum officinarum*). On the former it will feed on any part of the plant except the roots and leaf-blades, especially the inflorescences and fruits, and it is occasionally imported into the British Isles with bananas. The larva feeds in a silken gallery both internally and externally.

Pupa. In a cocoon, covered with frass and plant material, in the feeding position.

Imago. Moths are found in their native localities mainly from September to February. The natural flight time is from dusk till midnight.

Distribution
The species is a native of the islands surrounding Africa, from Seychelles and Mauritius to St Helena and the Canary Islands, from which it reaches this country in bananas. It could not maintain itself out of doors in the British Isles.

OPOGONA ANTISTACTA Meyrick

Opogona antistacta Meyrick, 1937, *Exot.Microlepid.* 5: 87.
Type locality: England; London (probably imported from Jamaica or Guyana).

Description of imago (Pl.10, fig.7)

Wingspan 14mm. *Female.* Head greyish ochreous, mixed brownish on crown; interantennal scale-ridge paler; vertical tufts absent; labial palpus third segment broad, widest in apical half. Forewing pale ochreous irrorate with brown; three small blackish brown spots, at apex and on costa and termen near apex. Hindwing grey, shading to fuscous towards termen.

Life history

Larva. 'Whitish . . . in a slight tubular web' on banana rind (Wakely, 1937).

History

The only known specimen was found at Finsbury, London, in the summer of 1936 as a larva on a banana in a bunch bought from a street stall and bred by Wakely on 4 September 1936. He subsequently sent it to Meyrick who described it and remarked that it was nearly related to *Opogona hemidryas* Meyrick and therefore probably imported from Jamaica or British Guiana (Guyana). The specimen is in BMNH.

References

Davis, D. R., 1978. The North American moths of the genera *Phaeoses*, *Opogona*, and *Oinophila*, with a discussion of their supergeneric affinities (Lepidoptera: Tineidae). *Smithson. Contr. Zool.* No. 282; 39 pp., 128 figs.

Morrison, B., 1968. A further record of *Dryadaula pactolia* Meyrick (Lep., Tineidae) in Britain with notes on its life-history. *Entomologist's Gaz.* 19: 181–188, 15 figs.

Oldham, J. N., 1928. *Hieroxestis subcervinella*, Wlk., an enemy of the banana in the Canary Islands. *Bull. ent. Res.* 19: 147–166, 15 figs, 2 pls.

Wakely, S., 1937. *Opogona antistacta*, Meyrick (Lep., Tineidae), bred from banana-feeding larva. *Entomologist* 70: 106–107.

GRACILLARIIDAE

A. M. Emmet, I. A. Watkinson and M. R. Wilson*

A very large, clearly defined family of world-wide distribution. There are 87 British species representing two subfamilies, the Gracillariinae and the Lithocolletinae.

The principal character separating the Gracillariidae from other families is the hypermetamorphosis of the larva. In early instars it mines the epidermal cells of leaves or, less often, tender bark as a sap-drinker; morphological adaptations, to be described below, are needed for this to be possible. After the second or third ecdysis the larva eats parenchyma in the normal manner of lepidopterous larvae, a change necessitating a major restructuring of the head-capsule and mouth parts. In the Phyllocnistidae the larva also undergoes hypermetamorphosis but for a different purpose; the second morph occurs after feeding has finished, its mouth parts are atrophied and it exists solely for cocoon-spinning. *Phyllocnistis* Zeller was formerly included in the Gracillariidae but has now been separated and raised to family status partly because of this important difference.

In general, the Gracillariinae have two sap-drinking instars and two phases involving several instars in the tissue-feeding stage, in the first of which they continue to mine and in the second of which they feed externally; exceptions will, however, be noted below. The Lithocolletinae have three sap-drinking instars and continue as leaf-miners throughout the two instars comprising the tissue-feeding phase.

Imago. Head with vertex smooth or rough-haired; antenna simple, as long as or slightly longer than forewing; scape with or without pecten; haustellum developed, naked; labial palpus moderate or long, porrect or ascending; maxillary palpus distinct but sometimes minute, filiform, porrect. Legs rather long, smooth-scaled or with bristles, but without long hairs; in some genera conspicuously ornamented. Forewing lanceolate with venation reduced and differing between genera. Hindwing narrow lanceolate, with long cilia. The characters given in the generic descriptions are mainly those used by Vári (1961).

The adults are mostly crepuscular but some, notably *Parornix* Spuler, fly also by night. Several species overwinter in this stage.

*The authorship within this family is divided as follows:

A. M. Emmet – Introduction and Gracillariinae

I. A. Watkinson – Lithocolletinae

M. R. Wilson – keys to and descriptions of *Phyllonorycter* pupae

Ovum. The structure has been little studied. The chorion collapses soon after the larva hatches and, even though it is possible to pin-point the exact spot of laying by tracing the mine back to its beginning, the egg itself can be found only with the aid of a powerful lens. The family differs in this respect from the Nepticulidae and Lyonetiidae in which the ova remain conspicuous. No investigation seems to have been made into how the minute larva, lacking the capacity to chew (see below), gains entry into the epidermis of hard leaves like those of evergreen oak (*Quercus ilex*), as in the case of *Caloptilia leucapennella* (Stephens), *Phyllonorycter messaniella* (Zeller) and, particularly, *Acrocercops brongniardella* (Fabricius) which must penetrate the very hard upper cuticle. Possibly it makes use of one of the stomata over which the egg has been laid. If such larvae are subsequently removed from their mines, they cannot effect re-entry (Hering, 1951: 117; Wakely, 1961: 84).

Larva. The structural changes which accompany the transition from sap-drinking to parenchyma-feeding amount almost to an additional metamorphosis and an understanding of the differences between the two larval morphs is essential in the study of the family (figure 94). The change takes place at the second or third ecdysis according to genus, the subsequent behaviour of the larva being correlated with this timing; where the transition is delayed until the third ecdysis the larva normally completes its growth as a leaf-miner.

An epidermal mine is, of necessity, very shallow and the young larva inhabiting it is strongly flattened dorsoventrally (figure 95a). The severe constriction of the

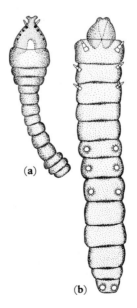

Figure 94 *Phyllonorycter corylifoliella* (Hübner), larva (**a**) first instar (**b**) final instar. After Hering

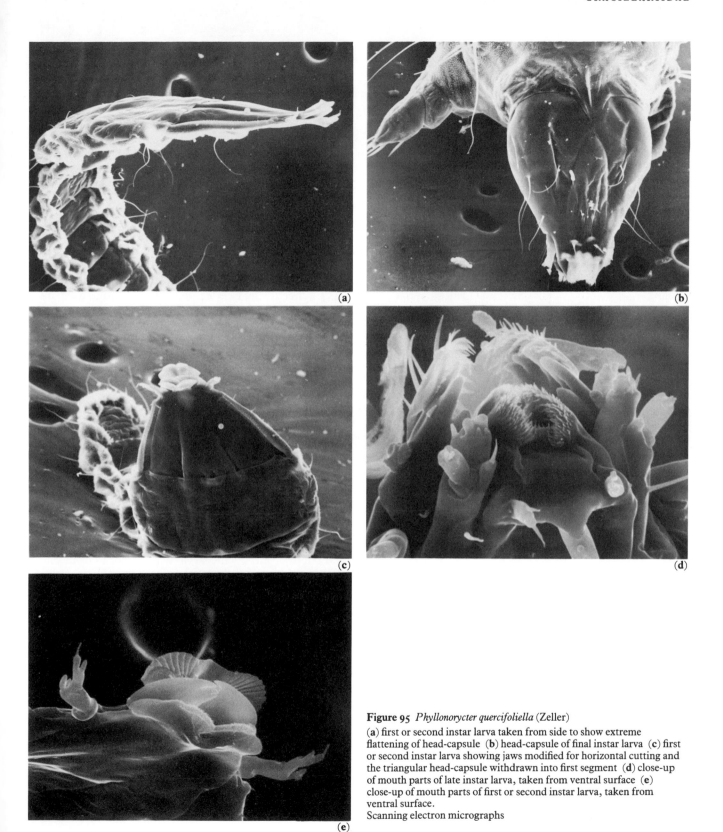

Figure 95 *Phyllonorycter quercifoliella* (Zeller)
(**a**) first or second instar larva taken from side to show extreme
flattening of head-capsule (**b**) head-capsule of final instar larva (**c**) first
or second instar larva showing jaws modified for horizontal cutting and
the triangular head-capsule withdrawn into first segment (**d**) close-up
of mouth parts of late instar larva, taken from ventral surface (**e**)
close-up of mouth parts of first or second instar larva, taken from
ventral surface.
Scanning electron micrographs

mine prevents the undulatory movement which is necessary for a larva walking with legs. Consequently, legs and prolegs being unusable tend to be greatly reduced or altogether absent. An alternative method of progression is adopted. Most species at first make a narrow gallery which is a close fit round the larva's body. By distending its posterior segments slightly to grip the walls, the larva can thrust forward its anterior segments; if the anterior segments are then distended, the posterior segments can be drawn up. This process is aided by the segments being somewhat broadened and the intersegmental divisions deeply incised so that the larva from above resembles a string of beads. Furthermore, the thoracic and posterior abdominal segments tend to be broader than those in between (figure 94a, p. 244). Gallery-feeding is usual in the first instar and blotch-feeding in the second, so it is natural for this shape to be more pronounced in the first instar.

The head-capsule of a free-feeding larva is roughly spherical and the mandibles are directed downwards towards a point just anterior to the legs of the first thoracic segment. In an epidermal miner the head-capsule is flattened into a very thin blade and the mouth parts are directed horizontally in front of the larva (figure 95a, p. 245). This requires a lengthening of the ventral and shortening of the dorsal components of the capsule. The dorsal apodemata are also withdrawn into the first thoracic segment to aid this realignment of the head. Seen from above, the capsule becomes elongate and more triangular (figure 95b,c, p. 245).

A normal larva has six stemmata or ocelli on each side of the capsule, arranged roughly in a semicircle so as to widen the range of vision. A gallery-miner needs to see only straight ahead and an eye must be situated on the knife-edge of the flattened capsule to achieve this. Consequently the stemma in this position is abnormally well developed and the other stemmata are more or less atrophied.

During the expansion of the epidermal mine silk is not produced and the spinneret is not developed in this stage.

The greatest changes, however, occur in the mouth parts. The mandibles of a tissue-feeding larva are adapted to chew out pieces of parenchyma (figure 95d, p. 245); a sap-drinking larva chews out nothing and its mouth parts are modified to cut the walls of the sap-filled cells so that it can imbibe the liquid contents. Hering (1951: 128–129) describes the modifications by means of the following analogy. He compares the mandibles of a tissue-feeding larva to cupped hands facing each other. The teeth, usually five, correspond to the fingers and thumb and interlock in chewing as the fingers can do if the hands are pressed together. Suppose now that the hands are rotated through 90° with the thumbs outwards and the fingers straight-

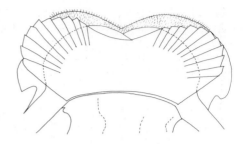

Figure 96 *Phyllonorycter* spp., labium. After Trägårdh

ened. In the same manner the mandibles of the sap-drinking larva are aligned horizontally instead of vertically and are flattened dorsoventrally. Each mandible consists of a laminar disc with a serrate cutting edge (figure 95e, p. 245). When the larva is feeding, the mandibles oscillate to and fro, cutting the tissues in front of them, the sap being imbibed as it is liberated from the severed cells.

Mandibles operating thus pose a danger of the severance not only of the epidermal cells but also of the leaf cuticle; if this were to happen, the larva would almost certainly perish. The danger is obviated by an adaptation of the labrum which is greatly enlarged and extends forwards over the mouth parts almost to their cutting edge (figure 96). The mandibles are thereby prevented from making contact with the cuticle, the labrum having a function similar to that of the protective head of a safety razor.

The labrum also has another function. After the sap-filled cells have been ruptured by the mandibles, their fibrous walls are still in position and would certainly inconvenience the larva, possibly even impeding its forward progress. The labrum has its anterior margin sharpened and often serrate. As the larva swings its head from side to side during feeding, the labrum acts as a saw, severing the cell walls close to the cuticle. The labium has no cutting capacity; it serves as a shovel collecting sap for swallowing.

In the moult which terminates the sap-feeding phase, far-reaching morphological changes occur. The body becomes more cylindrical and functional legs and prolegs appear; there are four pairs of the latter, the prolegs on abdominal segment 6 being absent in all gracillariids. The head-capsule becomes rounder and more similar to that of an external feeder. The mouth parts are adapted for chewing and not for cutting, and are directed downwards instead of forwards. The number of functional stemmata is increased, usually to four, and now give forward, downward and lateral vision. A functional spinneret appears, the labrum loses its specialized modifications and the larva turns into a conventional caterpillar (figure 94b, p. 244).

At the time of this ecdysis, the larva is inhabiting a blotch in which the cuticle has been detached like the skin over a blister; the epidermal cells have all been eaten but the parenchyma is still intact. The larva now eats downwards with its reorientated mouth parts (or upwards in a lower surface mine) into the parenchyma. The mouth, now directed downwards, can no longer be used to extend the area of the mine sideways, so the eventual size of the mine is dependent on the number of earlier sap-drinking instars. The Lithocolletinae have three such instars and a correspondingly larger blotch; the parenchyma it contains is sufficient for the whole of the larval phase and they complete their growth and pupate within the original mine. Most Gracillariinae, on the other hand, have only two sap-drinking instars and the parenchyma within their blotches is exhausted well before growth is completed. Accordingly they leave their mines and feed thereafter externally, though concealed in rolled or folded leaves. Among the British Gracillariidae there are three exceptions to this general rule: *Aspilapteryx tringipennella* (Zeller), *Parectopa ononidis* (Zeller) and *Leucospilapteryx omissella* (Stainton) are all capable of enlarging their mines and the first two can even start a fresh mine in a second leaf. There are modifications to the head-capsule and trophi correlated with these habits.

A new feature when tissue-feeding has started may be the presence of silk which modifies the shape and perhaps the structural strength of the mine. Silk spun over the fragile centrally detached cuticle shrinks and puckers the leaf, creating a cavity mine in which the larva is able to move more freely in greater seclusion. The amount of spinning depends on the duration of the mining phase. Where this is shorter, as in *Caloptilia* Hübner and *Parornix* Spuler, the spinning is slight and if the blotch is stretched the cuticle usually fractures easily. In *Phyllonorycter* Hübner, on the other hand, where both larval and pupal stages will be passed in the mine, the spinning is so extensive that the mine will often stretch considerably without splitting.

Mining in the tissue-feeding phase is obligate for the Lithocolletinae and if a mine is opened the larva usually dies, although minor repairs are sometimes made to the mine by larvae early in their fourth instar which is devoted to spinning rather than feeding. If the mine of a gracillariine is split open during tissue-feeding the larva will usually proceed at once to its next phase and spin the roll, cone or fold characteristic of its species. These spinnings are too varied to be covered here but the description of cone-spinning in *Parornix anglicella* (Stainton) (p. 280) is reasonably typical of the subfamily.

The information above is derived mainly from Chapman (1902); Trägårdh (1913); Hering (1951) and M. R. Shaw (pers.comm.).

Pupa. The first four abdominal segments are fused, but segments 5 and 6 are free in the female and segments 5–7 in the male. The appendages extend as far as or beyond the tip of the abdomen and are free distally. The head is furnished with a pointed 'cocoon-piercer'. Most species have the dorsal area of the abdominal segments darker and equipped with numerous small spines.

In many Gracillariinae pupation takes place in a smooth, shining, membranous cocoon, often spun on the surface of a living or dead leaf. Its cross-section resembles the letter D, flat side outermost, the curved component contiguous with the leaf-surface which is weakly bowed by silk. The flat surface sometimes has a central keel of puckered silk, resulting from contraction due to further internal spinning. In some species, notably of *Parornix*, the flat surface contracts so strongly that the cocoon becomes concealed in a tightly curved fold at the leaf-margin. The author has observed a gracillariid larva rubbing the inner surface of its cocoon with its anus, possibly applying an excretion which imparts to the silk its glossy, papery texture. Analogous behaviour has been observed in the Zygaenidae, which also construct parchment-like cocoons (p. 94; Beavis, 1973). Other pupation habits are described under the appropriate genus or species. In the Lithocolletinae pupation occurs in the mine, with or without a cocoon. In both subfamilies the pupae are capable of lively movement and protrude from the cocoon or mine before the emergence of the adult.

In captivity the larvae of some species tend to spin up in the angle between the wall and lid or floor of their container; pupal mortality is then high, probably through desiccation. The best remedy is to add damp sphagnum and to keep the box in a cool place. Other species like *Parornix* readily accept tissue-paper. Overwintering pupae should be kept out of doors.

Key to mines of the Gracillariidae

All gracillariid mines consist of an epidermal gallery which is later extended into an epidermal blotch. After the larval hypermetamorphosis and change of mouth parts, parenchyma is eaten and the blotch is developed into a tentiform mine by all British species except six of the Gracillariinae (four on *Acer* spp. and two on *Quercus* spp.); in these, the blotch lacks internal spinning, is not tentiform and consists of a small transparent window about 10sq.mm in area; this resembles a blotch made by a coleophorid but lacks the circular entrance hole and contains frass.

The majority of gracillariine larvae (29 species) leave their mines in the third or fourth instar and feed subsequently in a rolled leaf or a fold or cone on the margin of the leaf, using either the mined leaf or one nearby for this purpose. Determination is then best made by considering both patterns of feeding in conjunction. The larvae of the Lithocolletinae and a minority of the Gracillariinae (five species) continue mining throughout the larval stage.

Tentiform mines made by the Tischeriidae do not contain frass which is ejected through a hole in the epidermis. Hymenopterous mines on the same foodplants are neither tentiform nor fully transparent.

ACER (Maples and Sycamore)

1	Mine a full-depth, transparent blotch without internal spinning; larva later feeds in a rolled leaf or cone	2
–	Mine tentiform, opaque; larva mines throughout	5
2(1)	On *A. pseudoplatanus*	3
–	On *A. campestre*	4
3(2)	Mine larger, *c*.6mm long[1] .... *Caloptilia hemidactylella* (p. 268)	
–	Mine smaller, *c*.4mm long ...... *C. rufipennella*[2] (p. 261)	
4(2)	Mine tenanted in May, spinning in June; recorded only from the Isle of Wight .... *Calybites hauderi* (p. 273)	
–	Mine tenanted in June, spinning in July; recorded from southern England and Wales .... *Caloptilia semifascia* (p. 267)	
5(1)	On *A. campestre* .... *Phyllonorycter sylvella* (p. 356)	
–	On *A. pseudoplatanus* .... *P. geniculella* (p. 357)	
–	On *A. platanoides* .... *P. platanoidella* (p. 356)	

NOTES

[1] This character must be regarded as provisional owing to lack of material for study.

[2] *Caloptilia rufipennella* has been recorded on a wide range of *Acer* spp. on the Continent, but only on *A. pseudoplatanus* in Britain.

ALNUS (Alder)

1	Mine on upperside of leaf	2
–	Mine on underside of leaf	3
2(1)	Mine suboval; upper cuticle silvery flecked with brown frass .... *Caloptilia elongella* (p. 258)	
–	Mine subcircular; upper cuticle pale green, sometimes discoloured brown but not flecked with brown frass .... *Phyllonorycter stettinensis* (p. 351)	
3(1)	Mine on margin of leaf, small (*c*.10mm long); lower cuticle brownish; larva feeds later in folded leaf-edge .... *Caloptilia falconipennella* (p. 267)	
–	Mine usually away from leaf-margin and more than 10mm long; lower cuticle green; larva mines throughout	4
4(3)	Mine very large, extending from midrib almost to leaf-margin; larva grey; pupa in a cocoon without frass in centre of mine .... *Phyllonorycter froelichiella* (p. 352)	
–	Mine smaller, not exceeding 20mm in length; larva whitish	5
5(4)	Pupa in a cocoon edged with frass ..... *P. rajella* (p. 340)	
–	Pupa in a cocoon not edged with frass	6
6(5)	Always on *A. incana*; pupa usually in middle of mine .... *P. strigulatella* (p. 339)	
–	Usually on other *Alnus* species; pupa usually at one end of mine .... *P. kleemannella* (p. 353)	

BETULA (Birch)

1	Mine on upperside of leaf	2
–	Mine on underside of leaf	3
2(1)	Mine occupying most of the leaf which eventually almost closes over it; larva mines throughout .... *Phyllonorycter corylifoliella* f. *betulae* (p. 327)	
–	Mine small (*c*.12mm long); larva feeds later in a rolled leaf .... *Caloptilia betulicola* (p. 258)	
3(1)	Mine with lower cuticle brown; larva feeds later in a rolled or folded leaf	4
–	Mine with lower cuticle green; larva mines throughout	7
4(3)	Larva completes growth in a folded leaf-edge	5
–	Larva completes growth in a rolled leaf	6
5(4)	Bivoltine, feeding June and August – September; southern England to Caledonian Canal .... *Parornix betulae* (p. 278)	
–	Univoltine, feeding July – August; northern species .... *P. loganella* (p. 278)	

6(4) Final leaf-roll longitudinal ....
.... *Caloptilia populetorum* (p. 257)
– Final leaf-roll transverse .... *C. betulicola* (p. 258)
7(3) Mine large (15–20mm long); lower epidermis with 7–12 folds .... *Phyllonorycter cavella* (p. 333)
– Mine smaller (10–15mm long); lower epidermis with 1–6 folds 8
8(7) Mine almost exclusively on seedling birches; pupa without a cocoon .... *P. anderidae* (p. 342)
– Mine on seedling or mature birches; pupa in a cocoon ....
.... *P. ulmifoliella* (p. 347)

CARPINUS (Hornbeam)

1 Mine on upperside of leaf ....
.... *Phyllonorycter quinnata* (p. 338)
– Mine on underside of leaf 2
2(1) Mine subrectangular, lightly spun and little arched; both upper and lower epidermis brown with veins showing as reticulations; larva feeds later in folded leaf-edge .... *Parornix fagivora* (p. 279)
– Mine elongate between veins, strongly spun and arched, without reticulated appearance; lower epidermis green .... *Phyllonorycter tenerella* (p. 308)

NOTE. *P. messaniella* occasionally feeds on *Carpinus* (see *Quercus*).

CORYLUS (Hazel)

1 Mine on upperside of leaf ....
.... *Phyllonorycter coryli* (p. 337)
– Mine on underside of leaf 2
2(1) Mine subrectangular, lightly spun and little arched; smaller (c.10mm long); lower epidermis brown; larva feeds later in folded leaf-edge ....
.... *Parornix devoniella* (p. 282)
– Mine elongate between veins, strongly spun and arched; larger (15–20mm long); lower epidermis with strong central fold; larva mines throughout ....
.... *Phyllonorycter nicellii* (p. 353)

CRATAEGUS (Hawthorn)

1 Mine on upperside of leaf ....
.... *Phyllonorycter corylifoliella* (p. 327)
– Mine on underside of leaf 2
2(1) Mine smaller (c.5mm long); lower epidermis greyish brown; larva feeds later in a cone spun on a leaf-lobe .....
.... *Parornix anglicella* (p. 280)
– Mine larger (c.9mm long); lower epidermis green; larva mines throughout ... *Phyllonorycter oxyacanthae* (p. 314)

CYDONIA (Quince)

1 Mine on upperside of leaf ....
.... *Phyllonorycter corylifoliella* (p. 327)
– Mine on underside of leaf .... *P. cydoniella* (p. 320)

FAGUS (Beech)

1 Mine on upperside of leaf (rare aberration) ....
.... *Phyllonorycter maestingella* (p. 336)
– Mine on underside of leaf 2
2(1) Mine subrectangular, less than 9mm long, lightly spun and little arched; both upper and lower epidermis with veins showing as reticulation; larva feeds later in folded leaf-edge .... *Parornix fagivora* (p. 279)
– Mine oval or elongate, more strongly spun and arched; lower epidermis green without reticulation; larva mines throughout 3
3(2) Mine a broad oval, smaller (c. 12mm long); cocoon edged with frass .... *Phyllonorycter messaniella* (p. 312)
– Mine an elongate tube between veins or on leaf-margin; cocoon to one side of frass which is piled neatly in middle of mine .... *P. maestingella* (p. 336)

FRAXINUS (Ash) and LIGUSTRUM (Privet)

1 Mine with upper epidermis silvery; larva feeds later in a neatly constructed cone; pupa in last cone ....
.... *Caloptilia cuculipennella* (p. 256)
– Mine with upper epidermis yellow or brown; larva feeds later in an untidily constructed roll; pupa not in last roll
.... *C. syringella* (p. 270)

LEYCESTERIA (Himalayan honeysuckle) see LONICERA

LIGUSTRUM (Privet) see FRAXINUS

LONICERA (Honeysuckle), LEYCESTERIA (Himalayan honeysuckle) and SYMPHORICARPOS (Snow-berry)

1 Mine on upperside of leaf (rare aberration) ....
.... *Phyllonorycter trifasciella* (p. 355)
– Mine on underside of leaf 2
2(1) Mine large, occupying almost whole leaf and strongly inflated .... *P. emberizaepenella* (p. 348)
– Mine smaller, occupying only part of leaf which is often twisted into a cone .... *P. trifasciella* (p. 355)

MALUS (Apple)

1　Mine on upperside of leaf 2
–　Mine on underside of leaf 3
2(1)　Mine usually between veins, smaller (diameter *c*.10mm), and without central differently coloured patch; larva feeds later in folded leaf-edge .... *Callisto denticulella* (p. 289)
–　Mine usually over a vein, larger (diameter *c*.20mm) and with central differently coloured patch (p. 327); larva mines throughout .. *Phyllonorycter corylifoliella* (p. 327)
3(1)　Mine with lower epidermis silvery white .... *Callisto denticulella* (p. 289)
–　Mine with lower epidermis green or brown 4
4(3)　Mine subrectangular, both upper and lower epidermis brown; larva feeds later in a tight pleat resembling a mine in the centre of the leaf or in a folded leaf-edge .... *Parornix scoticella* (p. 283)
–　Mine more or less elongate, lower epidermis green; larva mines throughout 5
5(4)　On *M. domestica* .... *Phyllonorycter blancardella* (p. 319)
–　On *M. sylvestris* 6
6(5)　Mine larger (17–25mm); lower epidermis with strong central fold .... *P. cydoniella* (p. 320)
–　Mine smaller (13–19mm); lower epidermis usually with several distinct folds .... *P. blancardella* (p. 319)

ONONIS (Restharrow)

1　Mine an opaque ochreous brown gallery along midrib with clearer branches where the larva has fed; larva changes leaves and pupates externally .... *Parectopa ononidis* (p. 276)
–　Mine tentiform on underside of leaf; larva does not change leaves and pupates in the mine .... *Phyllonorycter nigrescentella* (p. 343)

POPULUS (Poplar and Aspen)

1　Epidermal gallery long, sometimes extending from midrib to leaf-margin; tentiform mine small (*c*.10mm long); larva feeds later in a cone or fold on leaf-margin .... *Caloptilia stigmatella* (p. 266)
–　Epidermal gallery short, usually obscured by later blotch; tentiform mine larger (*c*.13mm long) 2
2(1)　On *P. tremula* .... *Phyllonorycter sagitella* (p. 359)
–　On *P. alba*, *P. canescens* or occasionally other *Populus* spp. .... *P. comparella* (p. 359)

NOTE. Very occasionally *Salix*-feeding *Phyllonorycter* spp. attempt to mine *Populus*, but their mines are apparently always aborted.

PRUNUS (Blackthorn, Plum and Cherry)

1　On *P. padus* .... *Phyllonorycter sorbi* (p. 315)
–　On other *Prunus* spp. 2
2(1)　On *P. avium*, *P. cerasus* or cultivated cherry .... *P. cerasicolella* (p. 325)
–　On *P. spinosa* or *P. domestica* 3
3(2)　Mine larger (*c*.12mm long); lower epidermis green; larva mines throughout .... *P. spinicolella* (p. 322)
–　Mine smaller (*c*.8mm long); lower epidermis grey or whitish; larva feeds later in folded leaf-edge 4
4(3)　Mine elongate, more strongly arched by internal spinning; lower epidermis grey, opaque; larva grey, legs black .... *Parornix finitimella* (p. 286)
–　Mine subrectangular or triangular, only weakly arched by internal spinning; lower epidermis whitish and more or less transparent; larva whitish green, legs green .... *P. torquillella* (p. 288)

QUERCUS (Oak)

1　Mine an epidermal gallery on underside leading to a subquadrate blotch *c*.5mm across (triangular if in angle of veins); larva feeds later in cone on leaf-margin 2
–　Mine formed otherwise 3
2(1)　Univoltine; mine tenanted July – August, cone September – October .... *Caloptilia alchimiella* (p. 262)
–　Bivoltine; mine tenanted May and August, cone June and September – October; second generation indistinguishable from foregoing species ... *C. robustella* (p. 265)
3(1)　Mine upperside, large and extending over most of leaf 4
–　Mine underside 5
4(3)　Upper epidermis detached from parenchyma and silvery; mine slightly inflated .... *Acrocercops brongniardella* (p. 292)
–　Mine otherwise .... Hymenoptera spp.
5(3)　Larva mines only when young, feeding later in a cone on the leaf-margin .... *Caloptilia leucapennella* (p. 269)
–　Larva mines throughout 6
6(5)　Mine on *Q. ilex* .... *Phyllonorycter messaniella* (p. 312)
–　Mine on deciduous species 7
7(6)　Mine appearing macroscopically to be without creases in the lower epidermis 8
–　Mine with visible creases in lower epidermis 10
8(7)　Mine small (less than 10mm long), usually in lobe or at edge of leaf (autumn generation only) .... *P. heegeriella* (p. 308)
–　Mine larger (more than 17mm long) 9

9(8) Pupa in cocoon attached to the central green patch in the upper epidermis; mine 17–20mm long .... *P. roboris* (p. 307)

– Pupa without cocoon but in silken web; mine 22–28mm long .... *P. distentella* (p. 341)

10(7) Lower epidermis with numerous small creases 11
– Lower epidermis with at least one large crease 12

11(10) Very small mine usually in lobe or at edge of leaf, cocoon occupying most of mine (autumn generation only) .... *P. heegeriella* (p. 308)

– Mine usually on margin when leaf-edge folds right over almost concealing mine; pupa in flimsy, lace-like cocoon .... *P. saportella* (p. 309)

12(10) Cocoon incorporating no frass 13
– Cocoon incorporating frass 14

13(12) Mine small (less than 14mm long); cocoon attached to both upper and lower epidermis (summer generation only) .... *P. harrisella* (p. 306)

– Mine larger (more than 20mm long), almost always between veins extending from midrib; often several mines in a leaf .... *P. lautella* (p. 345)

14(12) Mine medium-sized or large (11mm or more long) 15
– Mine small (10mm or less long); cocoon attached to both upper and lower epidermis (summer generation only) ... . *P. heegeriella* (p. 308)

15(14) Cocoon attached to upper epidermis only 16
– Cocoon attached to both upper and lower epidermis 17

16(15) Cocoon completely frass-covered (summer only); mine irregular in shape, variously positioned in leaf .... *P. quercifoliella* (p. 311)

– Cocoon only lined with frass; a long mine between two veins and extending from midrib .... *P. muelleriella* (p. 313)

17(15) Cocoon flimsy and lined with only a little frass .... *P. messaniella* (p. 312)

– Cocoon strong with frass edging giving a distinct U or V shape 18

18(17) Mine with all the parenchyma attached to the upper epidermis usually consumed (autumn generation only) .... *P. quercifoliella* (p. 311)

– Mine with a patch of parenchyma on upper epidermis usually left uneaten .... *P. harrisella* (p. 306)

SALIX (Sallow and Willow)

1 Tentiform mine small (*c.*8mm long); larva feeds later in a folded leaf or cone 2
– Tentiform mine larger (over 15mm long); larva mines throughout 3

2(1) Rare montane species; larva feeds later in a folded leaf ... . *Callisto coffeella* (p. 290)
– Common, widespread species; larva feeds later in a cone .... *Caloptilia stigmatella* (p. 266)

3(1) Mine on *Salix repens* or *S. arenaria* .... *Phyllonorycter quinqueguttella* (p. 342)
– Mine on other *Salix* species. 4

4(3) Mine on smooth-leaved *Salix* species 5
– Mine on rough-leaved *Salix* species 6

5(4) Mine only on *S. viminalis*; mine long and narrow, often near petiole; pupa naked in mine without a cocoon .... *P. viminetorum* (p. 329)

– Mine on *S. viminalis, S. alba, S. fragilis* and occasionally other species; pupa in a cocoon ... *P. viminiella* (p. 328)

6(4) Cocoon white or pale yellow and loosely woven .... *P. salicicolella* (p. 330)

– Cocoon golden or light golden brown and strongly constructed 7

7(6) Outline of cocoon more obvious from outside mine .... *P. spinolella* (p. 332)

– Outline of cocoon not or hardly visible from outside mine .... *P. dubitella* (p. 331)

SORBUS (Rowan, Whitebeam and Wild Service-tree)

1 Mine on upperside of leaf .... *Phyllonorycter corylifoliella* (p. 327)
– Mine on underside of leaf 2

2(1) Mine small (5–8mm long); lower epidermis turns grey or brown; larva feeds later in a folded leaf or cone 3
– Mine long and narrow (20–30mm long); lower epidermis remains green; larva mines throughout 4

3(2) Larva feeds later in a cone on the leaf-margin; chiefly on *S. torminalis* .... *Parornix anglicella* (p. 280)

– Larva feeds later in a folded leaf-edge or, more often, in a centrally placed tight pleat which resembles a mine; on all *Sorbus* spp. .... *P. scoticella* (p. 283)

4(2) Mine on *S. torminalis* 5
– Mine on other *Sorbus* spp. 6

5(4) Mine with lower epidermis showing many longitudinal creases; pupa formed in a very pale brown loose silken chamber; frass disposed in a long loose line behind the cocoon (the most frequent species on *S. torminalis*) .... *Phyllonorycter mespilella* (p. 317)

– Mine with lower epidermis showing one major fold; pupa in a white silk-lined chamber, the frass heaped behind the cocoon (infrequently found on *S. torminalis*) .... *P. cydoniella* (p. 320)

6(4) Mine on *S. aucuparia* 7
– Mine on *S. aria* 9

7(6)   Pupa in a silk-lined chamber without any real cocoon   8

–   Pupa in a strong whitish cocoon, the frass heaped near the middle of the mine; mine extending along midrib or leaf-edge, strongly contorting leaf (most frequent species on *S. aucuparia*) .... *P. sorbi* (p. 315)

8(7)   For description see couplet 5a (infrequent on *S. aucuparia*) .... *P. mespilella* (p. 317)

–   Very rarely found on *S. aucuparia* .... *P. lantanella* (p. 326)

9(6)   For description see couplet 5a (most frequent species on *S. aria*) .... *P. mespilella* (p. 317)

–   For description see couplet 7b (rarely found on *S. aria*) .... *P. sorbi* (p. 315)

SYMPHORICARPOS (Snow-berry) see LONICERA

TRIFOLIUM (Clover)

1   Mine an opaque ochreous brown gallery along midrib with clearer branches where the larva has fed; larva changes leaves and pupates externally .... *Parectopa ononidis* (p. 276)

–   Mine tentiform on underside of leaf; larva does not change leaves and pupates in the mine 2

2(1)   Mine much inflated, distorting the leaf strongly; larva eats through to the upper epidermis .... *Phyllonorycter nigrescentella* (p. 343)

–   Mine little inflated and leaf only slightly distorted; feeding seldom reaches upper epidermis .... *P. insignitella* (p. 344)

NOTE. *P. nigrescentella* has been recorded on *Trifolium* on the Continent but not in Britain.

ULMUS (Elm)

1   Mine subcircular, strongly inflated; pupa in a cigar-shaped cocoon pointed at both ends and loosely spun to uneaten patch of parenchyma on upper epidermis .... *Phyllonorycter schreberella* (p. 346)

–   Mine elongate in the form of a narrow tube between veins; pupa in a cylindrical cocoon rounded at one end spun firmly to lower epidermis ..... *P. tristrigella* (p. 350)

OTHER FOODPLANTS

Only single gracillariid species or indistinguishable pairs are associated with the following plants:

Artemisia vulgaris (Mugwort) *Leucospilapteryx omissella*

Castanea sativa (Sweet chestnut) *Phyllonorycter messaniella*

Dryas octopetala (Mountain avens) *Parornix alpicola, P. leucostola*

Genista pilosa (Hairy greenweed) *Phyllonorycter staintoniella*

Hypericum (St John's wort) *Calybites auroguttella*

Lysimachia (Loosestrife) *C. phasianipennella*

Medicago (Medick) *Phyllonorycter nigrescentella*

Plantago lanceolata (Ribwort plantain) *Aspilapteryx tringipennella*

Pulmonaria (Lungwort) *Acrocercops imperialella*

Rhododendron (Azalea) *Caloptilia azaleella*

Rumex (Sorrel) *Calybites phasianipennella*

Sarothamnus (Broom) stems *Phyllonorycter scopariella*

Scabiosa (Small scabious) *P. scabiosella*

Symphytum (Comfrey) *Acrocercops imperialella*

Syringa (Lilac) *Caloptilia syringella*

Ulex (Gorse) stems *Phyllonorycter ulicicolella*

Vaccinium vitis-idaea (Cow-berry) *P. junoniella*

Viburnum (Wayfaring-tree, Guelder-rose) *P. lantanella*

Vicia sepium (Bush vetch) *P. nigrescentella*

NOTE. *Fomoria septembrella* (Stainton) and *F. weaveri* (Stainton) (Nepticulidae) pupate in inflated mines in the leaves of *Hypericum* and *Vaccinium vitis-idaea* respectively and these could be mistaken for gracillariid mines; there is no creasing of the lower epidermis in the nepticulid mines.

Key to subfamilies (imagines) of the Gracillariidae

1   Foretibia, apical half of midfemur and midtibia strongly thickened with projecting coloured scales, or a conspicuous series of bristles on hindtibia .... Gracillariinae

–   Fore- and midlegs not so thickened with scales and hindtibia without bristles .... Lithocolletinae (p. 294)

Gracillariinae

Morphological characters have either been discussed already in the introduction to the family or will be treated below under genera.

There are eight genera in Britain. Although the genera have been separated on the morphological characters of the adults, most also show clear distinctions in the biology of the early stages.

Key to species (imagines) of the Gracillariinae

NOTE. *Parornix* spp. (couplets 27–35) when taken as adults can seldom be determined with certainty without reference to the genitalia (figures 100,101, pp. 281,287).

1   Head smooth, vertex with appressed scales 2
–   Head rough-haired (*Callisto* and *Parornix*) 25

2(1)   Hindtibia above with conspicuous series of bristles 3
–   Hindtibia without bristles 5

3(2)   Forewing with pale markings irrorate brownish fuscous; a projecting dark pencil in terminal cilia ....
.... *Acrocercops brongniardella* (p. 292)
–   Forewing with pale markings not irrorate darker and without projecting dark pencil in terminal cilia 4

4(3)   Forewing with ground colour shining orange-brown; pattern silvery white .... *A. imperialella* (p. 291)
–   Forewing with ground colour pale fuscous, irrorate darker; pattern dull white ....
.... *Leucospilapteryx omissella* (p. 293)

5(2)   Midtibia smooth-scaled; forewing with four silvery white costal streaks and four similarly coloured dorsal spots .... *Parectopa ononidis* (p. 276)
–   Midtibia dilated with dense scales; forewing without dorsal spots or with fewer than four 6

6(5)   Labial palpus with segment 2 tufted beneath 7
–   Labial palpus with segment 2 not tufted 8

7(6)   Forewing with ground colour whitish, with patches of fuscous irroration forming fasciae ....
.... *Caloptilia cuculipennella* (p. 256)
–   Forewing very pale yellow or pale reddish or an intermediate colour, usually with scattered dark dots or rarely clouded with fuscous .... *C. leucapennella* (p. 269)

8(6)   Forewing with broad whitish costal streak ....
.... *Aspilapteryx tringipennella* (p. 271)
–   Forewing otherwise 9

9(8) Forewing with pattern clear yellow and sharply defined 10
– Forewing with pattern not clear yellow or, if yellow-ochreous, not contrasting sharply with ground colour 14

10(9) Forewing with large antemedian or median costal blotch 11
– Forewing with antemedian spot which does not extend to costa .... *Calybites auroguttella* (p. 275)

11(10) Forewing with terminal cilia distinctly paler than ground colour 12
– Forewing with terminal cilia concolorous with or darker than ground colour 13

12(11) Forewing with costal blotch produced along costa almost to apex .... *Caloptilia alchimiella* (p. 262)
– Forewing with costal blotch not so produced, terminating abruptly at about five-eighths .... *C. robustella* (p. 265)

13(11) Forewing with costal blotch subquadrate, hardly extending beyond middle; apical cilia yellow on underside .... *Calybites hauderi* (p. 273)
– Forewing with costal blotch elongate, almost reaching apex; apical cilia fuscous on underside .... *Caloptilia azaleella* (p. 262)

14(9) Forewing with white or whitish pattern 15
– Forewing otherwise 20

15(14) Forewing with one or more complete outward-oblique fasciae .... *C. syringella* (part) (p. 270)
– Forewing not fasciated 16

16(15) Forewing with white dorsal streak .... *Calybites phasianipennella* (p. 274)
– Forewing otherwise 17

17(16) Forewing with two costal and two dorsal spots .... *C. phasianipennella* f. *quadruplella* (p. 274)
– Forewing otherwise 18

18(17) Forewing with distinct triangular costal blotch, its posterior margin concave .... *Caloptilia stigmatella* (p. 266)
– Forewing with costal blotch formed otherwise 19

19(18) Forewing with costal blotch reduced to a whitish streak representing its anterior margin ... *C. semifascia* (p. 267)
– Forewing with costal blotch complete but ill-defined and heavily irrorate with ground colour .... *C. falconipennella* (p. 267)

20(14) Forewing with pattern, or traces of it, consisting of parallel, outward-oblique fasciae .... *C. syringella* (part) (p. 270)
– Forewing with pattern consisting of triangular costal blotch, longitudinal dark spots or altogether absent 21

21(20) Forewing with well-defined pale ochreous costal blotch anteriorly bounded by distinct reddish brown fascia; terminal cilia with three dark lines .... *C. hemidactylella* (p. 268)
– Forewing with costal blotch obsolete or ill-defined; anterior fascia, if present, only slightly darker than ground colour; terminal cilia without dark lines 22

22(21) Forewing with small, blackish median costal spot .... *C. populetorum* (p. 257)
– Forewing without this spot 23

23(22) Smaller species (11–12mm); forewing with ground colour chestnut or mahogany; foretibia with obscure, narrow pale median band .... *C. rufipennella* (p. 261)
– Larger species (14–16mm); forewing with ground colour reddish ochreous or yellowish ochreous; foretibia unbanded 24

24(23) Hindleg with trochanter and adjacent area white .... *C. betulicola* (p. 258)
– Hindleg with trochanter and adjacent area concolorous with forewing or yellowish .... *C. elongella* (p. 258)

25(1) Forewing dark brown with silvery white pattern 26
– Forewing white or whitish, irrorate with grey- or fuscous-tipped scales, often so heavily as to appear grey or fuscous with whitish pattern 27

26(25) Forewing with subbasal transverse silvery spot or strigula .... *Callisto coffeella* (p. 290)
– Forewing without subbasal spot .... *C. denticulella* (p. 289)

27(25) Forewing white with sparse fuscous irroration and without dark costal strigulae ..... *Parornix leucostola* (p. 285)
– Forewing heavily irrorate with dark scales and hereafter described as dark with pale markings; costa with numerous pale strigulae 28

28(27) Forewing blackish fuscous with strongly contrasting pale pattern .... *P. loganella* (p. 278)
– Forewing greyish fuscous with pattern not contrasting strongly 29

29(28) Forewing with pattern distinctly tinged ochreous .... *P. fagivora* (p. 279)
– Forewing with pattern not or only faintly tinged ochreous 30

30(29) Terminal cilia white-tipped at apex; third cilial dark line incomplete and starting below apex .. *P. betulae* (p. 278)
– Terminal cilia dark-tipped at apex; third cilial dark line complete 31

31(30) Labial palpus with segment 3 usually immaculate 32
– Labial palpus with segment 3 usually dark-banded ... 34

32(31) Forewing greyish fuscous extensively mixed white, especially basally where wing appears more white than fuscous; apical spot without conspicuous white anterior margin 33

– Forewing darker fuscous with purplish gloss, less extensively white-mixed, the basal area appearing more fuscous than white; apical spot with strong white anterior margin .... *P. torquillella* (p. 288)

33(32) Forewing with white subcostal streak from base to one-third .... *P. alpicola* (p. 284)

– Forewing without subcostal white streak ....
.... *P. scoticella* (p. 283)

34(31) Forewing dark fuscous; apical spot conspicuous and sharply margined clear white anteriorly; labial palpus with dark band on segment 3 interrupted ....
.... *P. finitimella* (p. 286)

– Forewing greyish or pale fuscous; apical spot less conspicuous, not sharply margined white anteriorly; labial palpus with dark band on segment 3 complete 35

35(34) Forewing grey; underside with pale line of terminal cilia continued round apex to reach subapical costal strigula
.... *P. devoniella* (p. 282)

– Forewing pale fuscous; underside with pale line of terminal cilia not reaching subapical costal strigula ....
.... *P. anglicella* (p. 280)

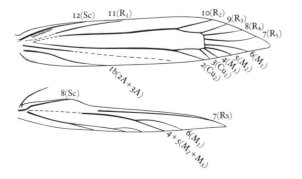

Figure 97 *Caloptilia alchimiella* (Scopoli), wing venation

CALOPTILIA Hübner

Caloptilia Hübner, [1825], *Verz.bekannt.Schmett.*: 427.

A very large genus of world-wide distribution; 13 species occur in Great Britain, seven of which are also found in Ireland.

Imago. Head with vertex and frons smooth; antenna about length of forewing with slight pecten; haustellum well developed; labial palpus long, slender, smooth, segment 3 as long as or longer than segment 2, pointed, ascending; maxillary palpus about three-quarters length of segment 2 of labial palpus, smooth, porrect or slightly ascending. Foretibia slightly thickened, apical half of midfemur and midtibia strongly thickened with projecting scales; hindtibia smooth, slender, tarsus one and a half to twice length of tibia. Forewing lanceolate, rather bluntly pointed; veins 1b (2A+3A) simple, 4 (M_3) and 5 (M_2) free, connate or short-stalked. Hindwing half to three-quarters width of forewing, lanceolate, cell open between veins 4 (M_3) and 5 (M_2), 1b (2A) usually present, 5 (M_2) and 6 (M_1) stalked (figure 97); cilia three times width of wing.

Life history

Larva. The larvae show virtually no distinctive characters between species; all are pale greenish white or yellowish white, even the head being sometimes more or less colourless. In the tissue-feeding phase, the thoracic legs and four pairs of prolegs are fully developed, the prolegs on abdominal segment 6 being absent.

There are two sap-feeding instars. In the first, the larva feeds in an epidermal gallery and in the second it expands the gallery into a blotch, the dimensions of which correlate with the time still spent as a miner in the tissue-feeding phase. This varies between species; in most the larva leaves the mine for external feeding soon after the third ecdysis, but the study of the transition from mining to external feeding is still incomplete. The method of feeding in the mine and the subsequent spinning is identical; the larva grazes downwards in relation to the larval attitude and through to the opposite epidermis which is left uneaten. A functional spinneret is acquired at the second ecdysis but the extent of spinning within the mine is variable. Species inhabiting small mines generally cause them to pucker very little if at all; others, such as *C. stigmatella* (Fabricius) make larger blotches which, as the result of more extensive spinning, take the form of strongly arched tentiform mines.

The spinning in which feeding is completed may consist of a cone, a roll or, in the case of *C. falconipennella* (Hübner), a folded leaf-edge resembling that made by a *Parornix* Spuler. Two or three such spinnings are made, the number depending on the size of the mine in which the larva formerly fed: for example, *Caloptilia rufipennella* (Hübner), which has a very small mine, nearly always spins three cones. One British species, *C. syringella* (Fabricius), is gregarious both as a miner and in its spinning.

Pupa. All British species except for *C. cuculipennella* (Hübner) and *C. falconipennella q.v.* pupate in membranous cocoons of the D-shaped pattern described on p. 247.

Imago. Many species are variable in the ground colour and markings of the forewing and one, *C. leucapennella* (Stephens), has two colour-forms. The characteristic attitude of the insect is to rest with its anterior parts raised high on its fore- and midlegs which are held so close together as to look like a single pair of limbs. These legs are splayed outwards, but the hindlegs are held close to the abdomen which slopes downwards so that its anal end rests on the substrate. The posture is similar to that adopted by *Bedellia somnulentella* (Zeller) (Lyonetiidae) which, however, uses only mid- and hindlegs (p. 226); it is the reverse of the attitude of *Argyresthia* Hübner and some other genera of the Yponomeutidae (*MBGBI* **3**). The legs of *Caloptilia* are thus prominently displayed and are ornamented to harmonize with the wing-pattern. The fore- and midtibiae are thickened with scales concolorous with the forewing, those on the inner side of the midtibia forming an expansible fringe. The hindlegs, which are almost concealed by the wings, are not thickened and are less richly coloured; however, in species which have a pale costal blotch on the forewing, the apex of the coxa, the trochanter and the base of the femur, the parts exposed when the moths are at rest, are coloured like the pale blotch to which they are adjacent and so do not disrupt the pattern. The pale tarsi and, in most species, the pale ventral surface of the abdomen contrast sharply with the prevailing dark colour. Clearly the insect's posture is adopted to display its pattern which is possibly disruptive, causing a predator to fail to recognize it as a moth.

Most species rest by day fully exposed on trunks or fences and fly in late afternoon sunshine and at dusk. The majority overwinter as adults, often in evergreen foliage, and these species may sometimes be taken at ivy-bloom in the autumn or sallow-bloom in the spring.

The graceful aspect of the moths, whether alive or in the cabinet, the interesting life history and the artistic skill of the larvae in making their spinnings render *Caloptilia* and related genera some of the most attractive of the microlepidoptera.

CALOPTILIA CUCULIPENNELLA (Hübner)

Tinea cuculipennella Hübner, [1796], *Samml.eur.Schmett.* **8**: 70, pl.28, fig.192.

Type locality: Europe.

Description of imago (Pl.10, figs 9,10)

Wingspan 11–12mm. Head pale ochreous, scales dark-tipped; antenna pale ochreous, annulated fuscous; labial palpus ochreous mottled fuscous, segment 2 shortly tufted beneath. Forewing whitish with indefinite ochreous or orange-ochreous markings irrorate darker; basal patch, subbasal fascia, antemedian fascia reaching fold and median fascia darker ochreous, all outward oblique and seldom complete; terminal cilia concolorous with forewing and with four dark lines; dorsal cilia dark brownish grey. Hindwing and cilia dark brownish grey.

Similar species. *C. syringella* (Fabricius), which lacks the tuft on segment 2 of the labial palpus.

Life history

Ovum. Laid on the upperside of a leaf of wild privet (*Ligustrum vulgare*) or ash (*Fraxinus excelsior*).

Larva. Head transparent with lateral black spot. Body greenish white, gut purplish; thoracic segment 1 with lateral black spot; in the mining phase, the head is relatively large and the intersegmental divisions deeply incised, the thoracic segments being broader than the abdominal.

Feeding commences in an epidermal mine on the upper surface of the leaf; the frass adheres to the cuticle in a brownish line. After the change of mouth parts, larval spinning causes the cuticle to contract and the leaf to fold upwards, almost concealing the mine; the frass, now black, is deposited in a mass. The mine may be distinguished from that of *C. syringella* on the same foodplants by its silvery colour, that of *C. syringella* being brownish or greyish green. After a further ecdysis, the larva leaves its mine and spins a neat cone formed by rolling the leaf downwards at a slight angle; the finished cone has pointed ends and is about one and a half times as long as the width of the leaf. Two such cones are constructed, the second, when on privet, occupying practically the whole leaf. July to early September.

Pupa. Pale brown. In a flimsy, spindle-shaped, silken cocoon of quadrate cross-section slung hammock-wise within the final cone. Prior to pupation the larva chews out a neat, round hole which it caps with silk at the point where the head of the pupa will be, and on eclosion the pupa projects through this hole. August to September.

Imago. Univoltine, emerging in September, overwintering and then flying until May. It is active in late afternoon around its foodplant.

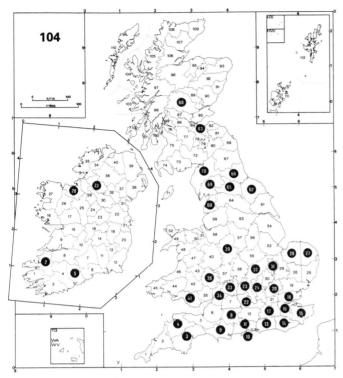

Caloptilia cuculipennella

Distribution (Map 104)

Frequents especially sea-cliffs and rough ground near the coast and woodland inland. Although widespread in Britain as far north as Perthshire and in Ireland, it is seldom common and is absent from many regions. Palaearctic.

CALOPTILIA POPULETORUM (Zeller)

Gracilaria populetorum Zeller, 1839, *Isis, Leipzig* **1839**: 209.

Type locality: Germany; Glogau (now Poland; Głogów).

Description of imago (Pl.10, fig.11)

Wingspan 11–14mm. Head ochreous to brown, collar darker; antenna pale ochreous annulated fuscous; labial palpus pale ochreous, segment 3 mottled fuscous. Forewing glossy pale ochreous to dark brown, variably irrorate darker; costa spotted fuscous; larger blackish spots on middle of costa and on fold before and beyond middle, the inner elongate and often bifid; usually some subapical dark scaling; all these markings may be obscured in dark specimens; cilia on costa pale ochreous, on termen very dark brownish fuscous obscuring the usual dark cilial lines and contrasting with rest of wing, and on dorsum grey. Hindwing and cilia grey.

Similar species. *C. falconipennella* (Hübner), which lacks the dark spots on the costa and fold and has the terminal cilia concolorous with the wing and the cilial lines distinct.

Life history

Ovum. Laid on either side of a leaf of birch (*Betula* spp.), young terminal leaves of seedlings or saplings being preferred.

Larva. Head pale whitish green to pale yellow-brown, mandibles darker. Body pale whitish green, becoming darker with age; gut bright green.

The mine, which may be on either side of the leaf, is long and rather broad, often extending from tip to base of the leaf, terminating in a blotch. After the change in mouth parts, the larva eats out the parenchyma from the blotch and spins the epidermis rather strongly, drawing the edges of the mine together. In the free-feeding phase, two rolls or folds are made. The first is of variable structure and may amount to little more than folding the edge of a leaf over as in *Parornix* Spuler. The second is of constant form, consisting of a whole leaf rolled longitudinally; this distinguishes it from the second roll of *Caloptilia betulicola* (Hering) which is made transversely on the same foodplant. July to August.

Pupa. Pale green, dorsum pale brown; legs spotted black. In a glistening, oval, pale whitish green cocoon in a rolled-up leaf, sometimes that in which the larva has been feeding, or beneath the edge of a leaf. Late July to September.

Imago. Univoltine, emerging from late August, overwintering and then flying until May. Earlier emergence may occur in hot summers, followed by a small second brood. It is not often observed as an adult until spring, when it flies round its foodplant in late afternoon.

Distribution (Map 105)

Occurs mainly on heaths and moors, and in open woodland where birch regenerates freely. Though widespread, it is uncommon and its apparently discontinuous distribution may be due to under-recording. It is found in southern Britain from Kent to Glamorgan and Herefordshire, in Cumbria and in western and northern Scotland. Rare in Ireland, being recorded only from Cos Wicklow, Clare and Galway. Palaearctic, its range extending through central and northern Europe to northern Asia.

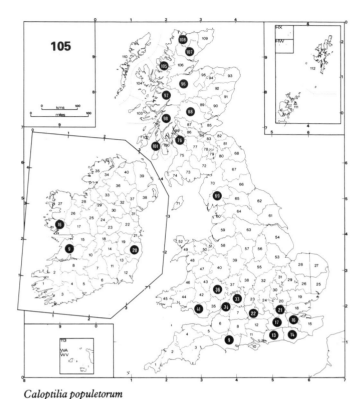

Caloptilia populetorum

CALOPTILIA ELONGELLA (Linnaeus)

Phalaena (Tinea) elongella Linnaeus, 1761, *Fauna Suecica* (Edn 2): 367.
Type locality: Sweden; Småland.

Description of imago (Pl. 10, figs 12–14)

Wingspan 14–16mm. Head concolorous with forewing; antenna ochreous annulated fuscous, the annulations obsolescent beneath. Hindleg with coxa, trochanter and femur concolorous with forewing or yellowish. Forewing with ground colour usually glossy reddish ochreous, but ranging to pale straw; often unicolorous but some or all of the following markings may be present; an obscure, slightly paler, subtriangular costal blotch, sometimes edged darker; costal and subdorsal dark spots, usually confined to basal half of wing; a series of large dark spots in disc, of which the antemedian, median and subtornal are the most prominent; costa beyond middle sometimes paler than ground colour; terminal cilia concolorous with wing, tipped slightly darker, tornal ochreous, dorsal grey. Hindwing dark grey, cilia paler. Genitalia, see figure 98, p. 260.

Similar species. *C. betulicola* (Hering), *C. rufipennella* (Hübner) and *C. hemidactylella* ([Denis & Schiffermüller]), *q.v.*

Life history

Ovum. Laid on the upperside of a leaf of alder (*Alnus glutinosa*), generally over a vein.

Larva. Head blackish in early instars, later pale brown. Body pale whitish green, gut greyish purple.

Feeds at first in an irregular gallery in the upper epidermis, with a central line of pale brown frass attached to the cuticle. This is developed into a blotch situated over a vein, within which, during the sap-feeding stage, the larva continues to attach brown frass to the cuticle. When tissue-feeding begins, spinning contracts the mine, closing it into a tube, the now black frass being packed at one end. On leaving the mine, the larva first rolls the edge of a leaf downwards into a small, elongate pocket; on changing its feeding place, it rolls half or even a whole leaf longitudinally and feeds on the lower surface from within. May; July.

Pupa. Green, dorsum varying from pale brown to smoky black. In a pale green, membranous cocoon, usually spun under a leaf. May to June; August.

Imago. Bivoltine. The first generation flies in June; the second emerges in September and overwinters, often in an evergreen such as yew. Moths reappear in the spring and continue on the wing until late April. The summer generation is relatively small and may not occur in unfavourable years.

Distribution (Map 106)

Occurs, often commonly, wherever its foodplant is found throughout the British Isles as far north as Orkney. Palaearctic.

CALOPTILIA BETULICOLA (Hering)

Gracilaria betulicola Hering, 1927, *Z.angew.Ent.* **13**: 168.
Type locality: Germany; Berlin.

Description of imago (Pl. 10, figs 15,16)

Wingspan 14–16mm. Head concolorous with forewing, frons yellowish; antenna ochreous annulated fuscous, the annulations obsolescent beneath. Hindleg with trochanter and adjacent areas of coxa and femur conspicuously whitish. Forewing glossy, ranging from pale to dark reddish ochreous; a paler, subtriangular, darker-edged costal blotch usually present; costa beyond middle often paler; occasionally darker antemedian and median discal spots; terminal cilia concolorous with wing, slightly darker-tipped, tornal ochreous and dorsal grey. Hindwing dark grey, cilia slightly paler. Genitalia, see figure 98, p. 260.

Similar species. Very similar to *C. elongella* (Linnaeus), but in *C. betulicola* the costal blotch is more often present,

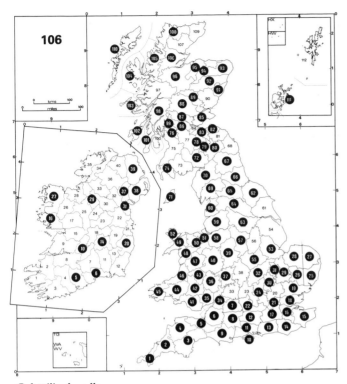

Caloptilia elongella

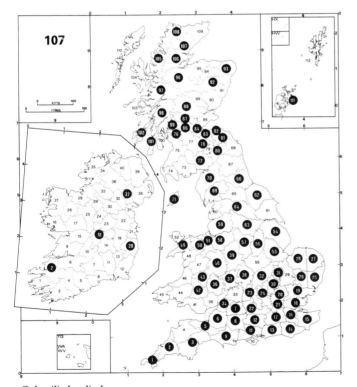

Caloptilia betulicola

more distinct and its edging darker; dark costal, subdorsal and discal spots are less common. However, the most constant difference is the whitish area round the trochanter of the hindleg, present only in *C. betulicola*; if this area is paler in *C. elongella*, it is yellowish. The absence of birch or alder from the habitat will eliminate the associated species with virtual certainty. *C. rufipennella* (Hübner) and *C. hemidactylella* ([Denis & Schiffermüller]), *q.v.*

Life history

Ovum. Laid on the underside, less often the upperside, of a leaf of birch (*Betula* spp.).

Larva. Head pale yellowish brown. Body pale greenish white, gut darker.

Feeding begins in a brownish epidermal minc on the same side of the leaf as the egg, with inconspicuous, red-brown linear frass; this is developed into an elongate blotch between veins. In the tissue-feeding phase spinning causes the blotch to arch strongly and the now black frass is stacked at one end. On leaving its mine the larva makes two successive rolls. The first may be little more than a *Parornix*-like fold, often on the same leaf as the mine and at the side if the mine has been at the tip. The second is a transverse roll starting from the tip and usually occupying almost the whole of the leaf. Within this the larva leaves the upper epidermis uneaten, a character which distinguishes the roll from those made by tortricid larvae. May; July.

Pupa. Green with the head, dorsum and tips of appendages blackish grey. Under a translucent, pale green silk membrane spun on the underside of a leaf. May to June; August.

Imago. Bivoltine. The first generation flies in late June and July; the second emerges in September or October and after overwintering continues until April. This species was formerly confused with *C. elongella*, although Wood (1890) suspected that the moths reared from alder- and birch-feeding larvae were specifically distinct. Brown (1947) was the first British author to confirm that this was the case.

Distribution (Map 107)

Widespread and common throughout Britain northwards to Orkney. Under-recorded in Ireland through confusion with *C. elongella*. Central and northern Europe; Japan.

259

Figure 98

Caloptilia elongella (Linnaeus)
(**a**) male genitalia (**b**) female genitalia

Caloptilia betulicola (Hering)
(**c**) male genitalia (**d**) female genitalia

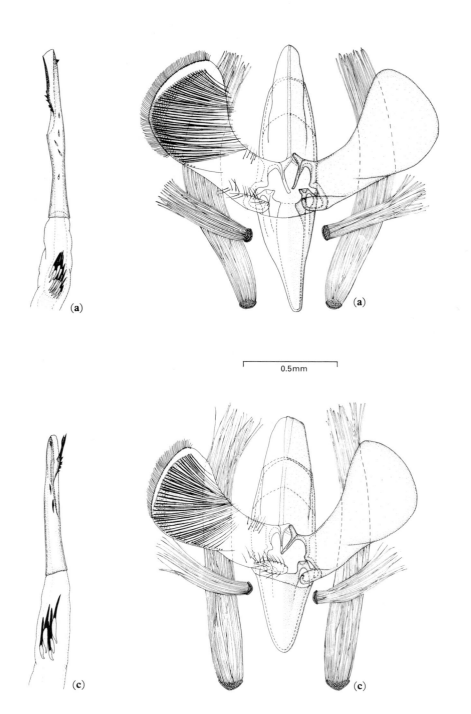

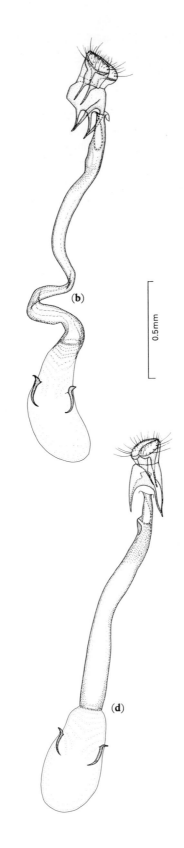

0.5mm

0.5mm

CALOPTILIA RUFIPENNELLA (Hübner)

Tinea rufipennella Hübner, [1796], *Samml.eur.Schmett.* 8: 67, pl.30, fig.204.

Type locality: Europe.

Description of imago (Pl.10, figs 17,18)

Wingspan 11–12mm. Head concolorous with forewing; antenna pale golden brown with obscure darker annulations. Foretibia clothed in dark chocolate-brown scales with an obscure pale central band; midfemur dark brown mixed white. Forewing glossy mahogany or chestnut-brown with violet sheen when viewed at an angle; sometimes a series of suffused fuscous costal, discal and subdorsal spots; cilia concolorous with wing on termen, dark grey on dorsum. Hindwing and cilia dark grey. Abdomen with dorsal surface grey and ventral white.

Similar species. C. elongella (Linnaeus) and *C. betulicola* (Hering), which are larger (14–16mm) and lack the pale band on the foretibia and the white admixture on the midfemur.

Life history

Ovum. Laid on the underside of a leaf of sycamore (*Acer pseudoplatanus*), close to a main rib; leaves of young trees from 2 to 3m up are preferred. Recorded on the Continent on other species of *Acer* (Hering, 1957).

Larva. Head pale green; mandibles reddish. Body green, gut darker.

Feeds first in a small epidermal mine on the underside leading into an angle of veins; in the tissue-feeding phase, the larva eats out a tiny, triangular full-depth mine in this angle with the black frass packed at the sides; there is no internal spinning. On leaving the mine it makes three successive cones, curling the leaf downwards so as to feed on the lower surface within. The first cone is very small, made by folding over the extreme tip of a lobe generally on the mined leaf. The second is much larger and often on a fresh leaf; the third is similar to the second. Sometimes the mine and all three cones are on a single leaf. Some individuals make only two cones (M. R. Shaw, pers.comm.). June, sometimes continuing into early July in East Anglia; in Scotland feeding occurs up to a month later.

Pupa. Green; abdominal segments with dorsal area pale brown. In a membranous, yellowish cocoon spun on the underside of a leaf, sometimes close to the final cone. July.

Imago. Univoltine; emerges in August and after overwintering, often in an evergreen such as yew, continues on the wing until April. It is probably a recent colonist, having been first reported in Britain in 1970 (Emmet, 1971b); however, its wide distribution and relative abundance suggest that it was already by then well established.

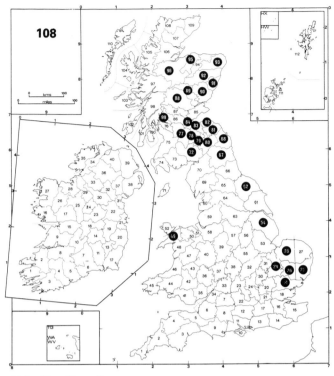

Caloptilia rufipennella

Distribution (Map 108)

Occurs mainly in two disjunct regions, in both of which it is common. The first is in eastern England from the extreme north of Essex to Lincolnshire; the second extends from Northumberland northwards to Moray (Emmet, 1979a; Shaw, 1981 and pers.comm.). A record from the Cleveland Hills, Yorkshire (Emmet, 1981), suggests either that the colonies are now merging or that it has been overlooked. In 1980 an apparently isolated breeding population was found in Caernarvonshire, north Wales (H. N. Michaelis, pers.comm.). Europe.

CALOPTILIA AZALEELLA (Brants)
Azalea Leaf Miner

Gracilaria azaleella Brants, 1913, *Tijdschr.Ent.* 56: LXII.

Type locality: Boskoop, Netherlands, the name having been given to moths reared from '*Azalea indica*' [*Rhododendron indicum*] imported from Japan.

Description of imago (Pl.10, fig.19)

Wingspan 10–11mm. Head with vertex leaden or purplish grey, frons whitish yellow; antenna ochreous, annulated fuscous; labial palpus whitish yellow, apex black. Thorax ochreous brown. Forewing glossy brown to ochreous yellow, variably irrorate purplish fuscous; an irregular, ochreous yellow costal blotch extending from one-quarter almost to apex, broadest anteriorly, on costa minutely dotted dark fuscous; terminal cilia purplish fuscous with two dark lines, dorsal cilia grey; underside with apical cilia purplish fuscous. Hindwing and cilia grey.

Similar species. *Calybites hauderi* Rebel, which has the yellow costal blotch trapeziform, hardly extending beyond the middle of the wing, and the apical cilia yellow on the underside.

Life history

Ovum. Laid on the underside of a leaf of azalea, especially *Rhododendron simsii*, *R. indicum*, the cultivar *R. hinomayo* and other cultivars.

Larva. Head pale yellow-brown. Body pale whitish yellow or greenish yellow.

Feeding starts in an irregular gallery in the lower epidermis which is developed into a small blotch, generally beside the midrib in the centre of the leaf, but sometimes close to the margin; in a small leaf it may occupy the whole of one side. In the tissue-feeding phase, spinning causes the lower epidermis to contract and the mine to arch upwards; the frass is packed at one end and the mine turns orange-brown. After leaving its mine, the larva makes two successive cones by rolling the tip of a leaf downwards, within which it feeds on the underside. June; September; sometimes also late in the autumn.

Pupa. Very pale brown. In a white, membranous silken cocoon spun beneath a leaf. July and September, generally overwintering until May.

Imago. Bivoltine. May; August; in mild seasons there is a small third generation in October. An accidental introduction, it is a pest on azaleas in greenhouses, where it is more or less continuous-brooded. It can also maintain itself in the open in southern England.

Distribution (Map 109)

Occurs, sometimes commonly, in sheltered gardens with extensive azalea beds in London (Buckingham Palace),

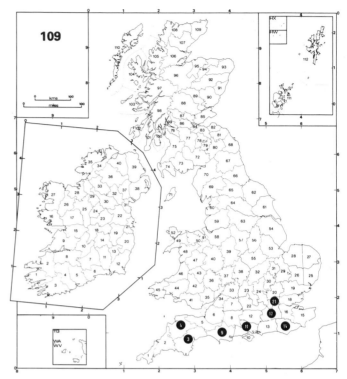

Caloptilia azaleella

Surrey (Royal Botanic Gardens, Kew and the Royal Horticultural Society Gardens, Wisley) and from Sussex to Devon; it is probably extending its range. The map shows only the vice-counties where it has been recorded in the open; it has been more widely reported in greenhouses. It is a native of eastern Asia but occurs as an introduction throughout Europe and also in north America and New Zealand.

CALOPTILIA ALCHIMIELLA (Scopoli)

Phalaena alchimiella Scopoli, 1763, *Ent.Carn.*: 254.

Tinea swederella Thunberg, 1788, *Mus.Nat.Acad.Uppsala* 6: 80.

Type locality: Carniola; Idria (now Yugoslavia; Slovenija).

Description of imago (Pl.10, fig.20)

Wingspan 10–13mm. Head with vertex pale purplish brown, frons yellow; antenna yellowish, annulated fuscous, the annulations obsolescent beneath. Thorax golden yellow, tegulae pale purple-brown. Forewing purplish brown with strong violet reflections; an elongate basal blotch on dorsum reaching one-quarter, a large subtriangular costal blotch not quite reaching dorsum and extending along costa almost to apex and a minute tornal spot all clear golden yellow; terminal cilia pale reddish brown, sometimes with faint indications of three darker lines, shading to grey on dorsum. Hindwing and cilia dark grey. Genitalia, see figure 99, p. 264.

Similar species. C. robustella Jäckh, *q.v.*

Life history

Ovum. Laid on the underside of a leaf of deciduous oak (*Quercus* spp.).

Larva. Head yellowish, mouth parts brown, ocelli black. Body uniform whitish yellow, gut purplish grey.

Feeding starts in an inconspicuous epidermal gallery on the underside which leads to an area where the larva continues to feed in 'a narrow thread-like mine, which it laces to and fro into the pattern of a small gridiron before throwing it into one square blotch' (Chapman, 1902: 139). When tissue-feeding begins, it makes this blotch full-depth, packing the frass at the sides. The blotch is triangular when in an angle of veins. After leaving its mine, it constructs successively up to three cones by curling the lobe of a leaf downwards, within which it eats through to the upper epidermis which it leaves intact. July to October, feeding up very slowly.

Pupa. Pale brown; in a symmetrical, oval cocoon, its upper surface consisting of a shining, whitish membrane with a weak central keel. It is often spun on the underside of the last leaf to be fed on, under the tip of a lobe loosely bent downwards. September or October, overwintering until late May or June.

Imago. Although it has always been regarded as bivoltine, M. R. Shaw (pers.comm.) found that a culture he was maintaining to provide hosts for parasites was univoltine, with the larvae developing much more slowly than those of the bivoltine *C. robustella*. This appears to be the regular regime, the adult having a long flight period extending from late May until July and the early stages also being protracted. The adult rests by day on trunks and fences, flies in the evening and later comes sparingly to light.

Distribution (Map 110)

Frequents oak woodland, especially where there are young trees. Occurs, sometimes commonly, throughout Britain as far north as Sutherland. The Channel Islands. Ireland. Only those records confirmed since the separation of *C. robustella* are shown on the map. The distribution of *C. alchimiella sensu lato* is Palaearctic.

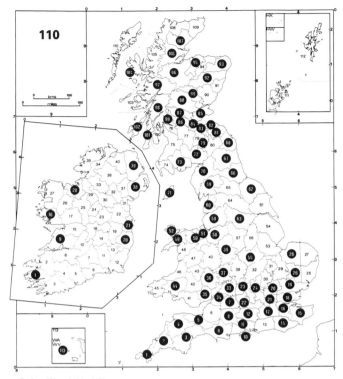

Caloptilia alchimiella

Figure 99

Caloptilia alchimiella (Scopoli)
(**a**) male genitalia (**b**) female genitalia

Caloptilia robustella (Jäckh)
(**c**) male genitalia (**d**) female genitalia

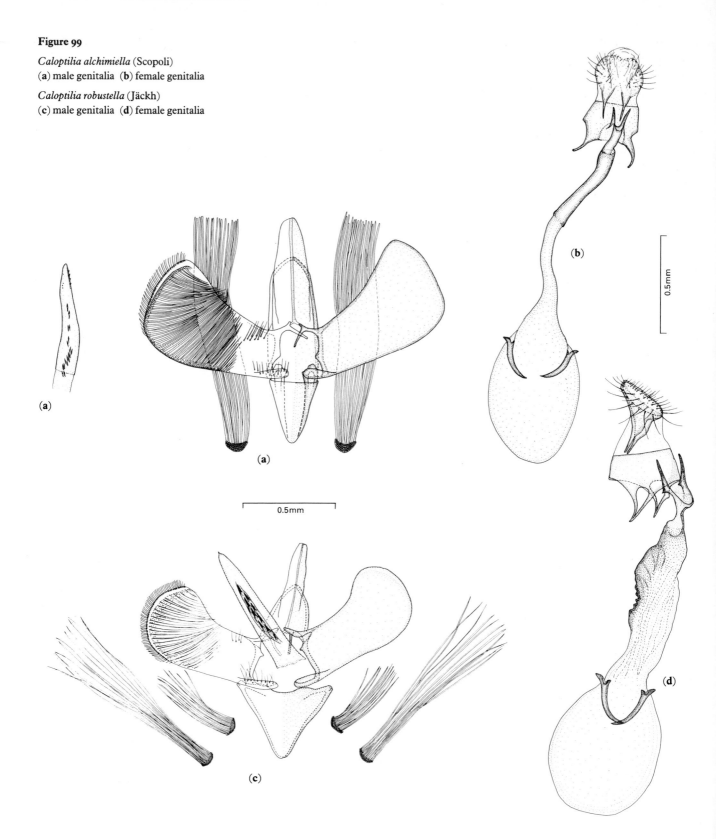

(**a**)

(**a**)

(**b**)

0.5mm

0.5mm

(**c**)

(**d**)

CALOPTILIA ROBUSTELLA Jäckh

Caloptilia robustella Jäckh, 1972, *Atti Accad.Sci.Torino*
106: 549.

Type locality: Germany; Kleinenkneten near Wildeshausen, Oldenburg.

Description of imago (Pl.10, fig.21)

Wingspan 10–13mm. Very similar to *C. alchimiella* (Scopoli), the points of difference being shown in the following table:

	C. alchimiella	*C. robustella*
Head	Vertex purplish brown, darker than thorax.	Vertex ochreous, concolorous with thorax.
Thorax	Golden yellow, tegulae pale purple brown.	Ochreous, including tegulae.
Forewing	Basal blotch sharply defined. Costal blotch with outer angle produced along costa almost to apex. Minute tornal spot present.	Basal blotch diffused. Costal blotch extends steeply to costa at about five-eights and is not or hardly produced along costa towards apex. No tornal spot.
Male genitalia (figure 99a,c, p. 264)	Aedeagus with many small cornuti arranged in two rows near base and a single, irregular row near apex.	Aedeagus with 6 to 7 large cornuti arranged in two rows.
Female genitalia (figure 99b,d, p. 264)	Ductus bursae a tube of even width, the granulation of its walls evenly dispersed over its surface.	Ductus bursae narrowing gradually towards bursa and abruptly towards ostium; granulation of its walls not evenly dispersed, being strongest at one side at widest section of ductus.

Life history

Except that it is bivoltine, no behavioural or morphological differences have been observed between this species and *C. alchimiella*, *q.v.*

Larva. Late May until July; September, with individuals feeding until early November.

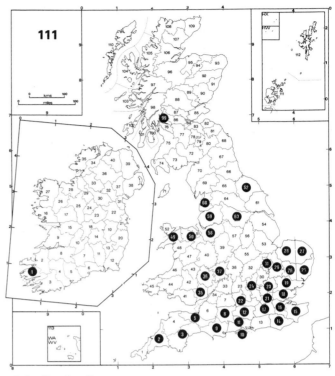

Caloptilia robustella

Pupa. July; October, overwintering until April or May.

Imago. Bivoltine. Mid-April to May; August. The first generation appears about three weeks before the earliest *C. alchimiella*.

Distribution (Map 111)

More restricted than *C. alchimiella*. Britain as far north as Aberdeenshire, being common in the south-east but rare in the west and north of its range; its apparent absence from much of the Midlands and central and southern Wales may be due to under-recording. In Ireland a single specimen has been taken near Killarney, Co. Kerry (Chalmers-Hunt, 1982). Central Europe from Italy to Sweden and eastwards to Asia Minor.

CALOPTILIA STIGMATELLA (Fabricius)

Tinea stigmatella Fabricius, 1781, *Spec. Ins* 2: 295.
Type locality: England.

Description of imago (Pl.10, fig.22)

Wingspan 12–14mm. Head with vertex red-brown, grading to yellow on frons; antenna yellowish white, annulated brown; labial palpus yellow, irrorate red-brown except at tip. Thorax red-brown. Forewing red-brown, darker immediately before a whitish, triangular costal blotch, thinly spotted red-brown and with its outer margin deeply excavate; terminal cilia red-brown, more ochreous on costa, and with three dark lines; dorsal cilia grey. Hindwing grey, cilia with reddish reflections. The red-brown ground colour is sometimes replaced by ochreous grey.

Life history

Ovum. Laid on the underside of a leaf of sallow or willow (*Salix* spp.), poplar or aspen (*Populus* spp.) or, very rarely, birch (*Betula* spp.). Broad-leaved sallows are less often selected. On the two occasions when the author has reared it from larvae found on birch, the trees were growing amongst sallows.

Larva. Head very pale yellowish brown, mouth darker. Body whitish yellow to whitish green.

Feeding starts in a relatively long gallery in the lower epidermis, which often follows the midrib before being directed outwards towards the leaf-margin, where it is expanded into a small blotch. When tissue-feeding begins, the blotch is puckered by spinning and resembles, and is often mistaken for, a small *Phyllonorycter* mine. On leaving its mine, the larva makes two or, less often, three successive cones or folds, the shape, at any rate of the first, being dependent on the foodplant. On willows and sallows it spins the tip of the leaf downwards into a cone, and on black poplar (*Populus nigra*) it likewise makes a cone at the tip or side of a leaf; however, on aspen (*P. tremula*), grey poplar (*P. canescens*) and white poplar (*P. alba*), it folds the edge of the leaf like the spinning of a *Parornix*, often misleading the entomologist. July to September.

Pupa. Green, dorsum grey. In a shining, membranous pale green cocoon on the underside of a leaf, often near the margin. September.

Imago. Univoltine, emerging in September and after overwintering continuing on the wing until May. Possibly occasionally bivoltine, since adults have been reared or taken in fresh condition in late June (Gregory, 1973). The moth may be beaten from evergreens in winter but is more often seen in spring, when it visits sallow catkins.

Distribution (Map 112)

Widespread and common throughout the British Isles northwards to Orkney. Holarctic.

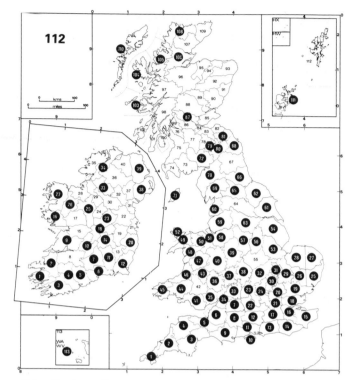

Caloptilia stigmatella

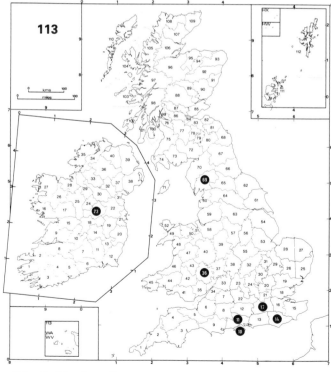

Caloptilia falconipennella

CALOPTILIA FALCONIPENNELLA (Hübner)

Tinea falconipennella Hübner, [1813], *Samml.eur.Schmett.* 8: pl. 46, fig. 317.
Type locality: Europe.

Description of imago (Pl. 10, fig. 23)
Wingspan 12–14mm. Head with vertex reddish to reddish fuscous, frons paler; antenna yellowish sharply annulated fuscous, the annulations obsolescent beneath. Thorax concolorous with base of forewing. Forewing with basal one-quarter reddish or brownish fuscous, this area bounded by an outward-oblique line from costa; wing beyond whitish heavily irrorate with colour of wing-base, more especially towards dorsum; terminal cilia concolorous and with three obscure darker lines, dorsal cilia grey. Hindwing and cilia grey.

Life history
Ovum. Laid on the underside of a leaf of alder (*Alnus glutinosa*).
Larva. Head pale yellowish brown. Body pale green, gut darker.

Although Hering (1957: 61) states that there is no preliminary gallery, my own notes from four rearings record a short preliminary gallery on the underside which tends to be absorbed in the subsequent blotch and thereby obscured; this is situated near the margin and is brownish in colour. On leaving its mine, the larva makes two or three successively larger folds on the margin of a leaf which closely resemble those of a *Parornix* species, their length ranging from 10 to 40mm. July to August.
Pupa. Green, dorsum grey to deep black. In a white silken cocoon spun under a leaf-margin, causing it to fold over as in *Parornix*. August to September.
Imago. Univoltine, emerging in September, overwintering and continuing on the wing until May. The adult has rarely been observed.

Distribution (Map 113)
A rare and elusive species. My personal opinion, supported by Wakely (1966), is that it is relatively widespread at very low density, this being based on vacated mines and folds, apparently of this species, observed at a number of localities. It has been reliably recorded from Sussex, Surrey, Hampshire, the Isle of Wight and Herefordshire. Hodgkinson, in the unpublished *Victoria County History*, recorded it from Cumbria. I have found vacated mines and folds, which can hardly have been of any other species, in Dunbartonshire, Scotland, and have had similar mines and folds submitted to me from Co. Meath, Ireland. Central and northern Europe, but not common.

CALOPTILIA SEMIFASCIA (Haworth)

Gracillaria semifascia Haworth, 1828, *Lepid.Br.*: 528.
Type locality: England; London district.

Description of imago (Pl. 10, figs 24, 25)
Wingspan 10–12mm. Head concolorous with forewing, frons sometimes mixed pale yellowish grey; antenna yellowish ochreous annulated fuscous, the annulations obsolescent beneath. Thorax concolorous with forewing. Forewing glossy reddish or blackish brown, sometimes mixed pale yellowish; costal blotch represented by a dark-edged ochreous white or yellowish streak from costa at one-third to fold where it is angled distad, this streak being often nearly or completely obsolete; some fuscous or dark reddish brown spots on costa and sometimes also minute whitish dots in area of costal blotch; sometimes a series of dark spots on dorsum; terminal cilia concolorous with forewing and with three obscure dark lines, dorsal cilia grey. Hindwing and cilia dark grey.

Life history
Ovum. Laid on the underside of a leaf of field maple (*Acer campestre*) close to a vein; less often on the upperside. Sycamore (*A. pseudoplatanus*) may also be selected where field maple is adjacent and heavily attacked (D. W. H. Ffennell, pers.comm.).
Larva. Head pale brown. Body pale green and rather transparent, gut darker.

Feeding starts in a narrow epidermal mine, generally on the underside of the leaf; this leads into a suboval blotch, often situated between veins near the petiole. In the tissue-feeding phase there is no internal spinning; the larva eats the parenchyma but leaves the veins so that the mine has a netted appearance when viewed against the light. Most of the frass is pushed to the margins. On leaving its mine, the larva feeds within cones or rolls, three of which are generally constructed. The first is usually on the mined leaf and is made by turning down the tip of a lobe; within this the larva again eats the parenchyma but leaves the veins. For the later chambers, it rolls down the whole of a large lobe into a tube and draws in the adjacent lobes to seal the ends. On sycamore, the larva makes a neat cone (D. W. H. Ffennell, pers.comm.). June to July.
Pupa. In a shining, yellowish white membranous cocoon spun on either side of a leaf, generally near the margin. July to August.
Imago. Univoltine; emerges from late July to September and remains active until October, being sometimes observed at ivy-blossom. After overwintering it reappears until May. Occasionally there is a small second generation of larvae in late August and September, producing adults in October.

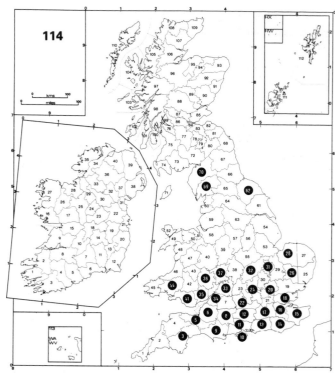

Caloptilia semifascia

Distribution (Map 114)

Frequents margins of woods and hedgerows where its foodplant is plentiful. Though widespread in southern England and Wales, it is very local, tending to occur in isolated but sometimes populous colonies. Apparently absent from the north Midlands, but reappears in Yorkshire and Cumbria. Central and northern Europe.

CALOPTILIA HEMIDACTYLELLA ([Denis & Schiffermüller])

Tinea hemidactylella [Denis & Schiffermüller], 1775, *Schmett.Wien.*: 144.

Type locality: [Austria]; Vienna district.

Description of imago (Pl.10, fig.26)

Wingspan 12–14mm. Head pale ochreous mixed reddish ochreous; antenna yellowish white, annulated fuscous. Thorax ochreous, mixed reddish. Forewing pale ochreous yellow, mixed and strigulated red-brown; a large, triangular, paler median costal blotch, on costa marked with several blackish dots, anteriorly margined by a well-defined dark reddish brown fascia; terminal cilia ochreous mixed reddish ochreous, with three reddish brown lines; dorsal cilia grey. Hindwing grey, cilia paler.

Similar species. Variegated forms of *C. elongella* (Linnaeus) and *C. betulicola* (Hering), from which it can be distinguished by its smaller size, the more distinct reddish brown fascia before the costal blotch and the three red-brown lines in the terminal cilia.

Life history

Ovum. Laid on a leaf of sycamore (*Acer pseudoplatanus*), probably on the underside. On the Continent recorded also on field maple (*A. campestre*) and Norway maple (*A. platanoides*).

Larva. Head very pale yellowish green, paler than body; mouth reddish brown. Body pale yellowish green, rather transparent; gut darker green (Stainton, 1864: 104).

The early feeding has not been described in detail. The Hering herbarium (BMNH) shows several cones and one mine on field maple; the latter consists of a short, tortuous epidermal gallery on the underside, leading to a full-depth blotch in which only the veins remain uneaten; the blotch is flat, without any sign of internal spinning. Stainton (*loc.cit.*) states that the larva then forms cones by rolling up a portion of the leaf on the underside and that these cones become of a chequered greenish grey. July to August.

The feeding can be distinguished from that of *C. rufipennella* (Hübner) by the larger size of the blotch, measuring 6mm as opposed to 3–4mm in its longer axis, if the single example studied is typical.

Pupa. Undescribed. August to September.

Imago. Univoltine, emerging in September, overwintering and flying again until May.

Distribution (Map 115)

Exceedingly rare and local. Stainton (*loc.cit.*) stated that it had been formerly not uncommon in Whittlebury Forest, Northamptonshire; three old specimens in the Bankes

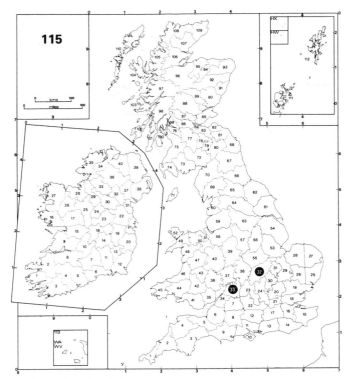

Caloptilia hemidactylella

collection and two in the Ford collection, all without data, could be from that locality. Wood (1890: 137) said it had not been taken in recent years. Later records have proved, if the specimens are available for examination, to be misidentified *C. elongella* or *C. betulicola* (Pierce & Metcalfe, 1935: 80; Brown, 1947), thereby casting doubts on other reports which cannot be checked. The only certain subsequent record is of several specimens beaten from dead bracken in November 1954 and 1955 near Cirencester, Gloucestershire (Newton, 1981). Records from Surrey, Norfolk, Monmouthshire, Staffordshire, Derbyshire, Cheshire, Cumbria and Perthshire are discredited or unconfirmed and are not shown on the distribution map. Central and southern Europe.

CALOPTILIA LEUCAPENNELLA (Stephens)

Gracillaria leucapennella Stephens, 1835, *Ill.Br.Ent.* (Haust.) **4**: 368.

Tinea sulphurella Haworth, 1828, *Lepid.Br.*: 564, nec Fabricius, 1775, *Syst.Ent.*: 670.

Type locality: England; New Forest, Hampshire.

Description of imago (Pl.10, figs 27–29)

Wingspan 12–14mm. Two colour forms occur, one very pale yellowish white (typical form) and the other reddish (f. *aurantiella* Peyerimhoff, fig.28); between these there is a full range of intermediate forms. Head, thorax and all appendages conform in general with colour of forewing. Head with frons paler in red specimens; antenna almost unmarked white in pale specimens, pale red with darker red annulations in red specimens; labial palpus yellowish above, even in red specimens, always with well-developed tuft on segment 2. Foretibia mottled fuscous in white form. Forewing often with cloudy fuscous spots in disc, in extreme examples this spotting so extensive as almost to obscure ground colour; terminal cilia concolorous with ground colour, but paler on tornus in red specimens; dorsal cilia reddish grey, even in white form. Hindwing grey, cilia with strong reddish sheen in all colour forms.

The typical form predominates in most of Britain, but Newton (1959) found f. *aurantiella* the more common in parts of Gloucestershire and in a series reared by the author from the Isles of Scilly only three out of 12 were of the pale form; Beirne (1941: 125) states that f. *aurantiella* occurs commonly with the typical form in Ireland.

Life history

'Little appears to be known about the early stages' (Brown, 1947); the species is not even mentioned by Hering (1957). Consequently I have had to rely almost entirely on my own notes which were made from larvae on evergreen oak; I have not observed its feeding on deciduous oak.

Ovum. Laid on the underside of a leaf of oak (*Quercus* spp.), including evergreen oak (*Q. ilex*).

Larva. Head brownish. Body greyish white (Meyrick, 1928).

Feeding starts in a long, narrow gallery in the lower epidermis which leads to an oval blotch between veins. In the tissue-feeding phase, the larva spins the lower cuticle causing arching and eats through to the upper epidermis. On leaving its mine, the larva first makes a small fold at the edge of a leaf. Later it constructs a cone at the tip of a leaf by rolling it downwards; at least two such cones are made, tender terminal leaves being selected. Meyrick (*loc.cit.*) and Brown (*loc.cit.*) give the larval season as June to July, but larvae were feeding in September in the Isles of Scilly.

Pupa. Under a pale greenish silk membrane on the underside of a leaf. August to October.

Imago. Univoltine, emerging from July but mainly in September or October, overwintering in evergreens and active again until May. Apparently the single generation extends its early stages over a long period.

Distribution (Map 116)

Frequents oak woodland, possibly with a preference for young trees. Very rare and not recently observed in southeast England, but progressively less uncommon westwards though always local; Wales; Scotland to Easter Ross; Isle of Man; Isles of Scilly. Locally common in Ireland, particularly in the south (Beirne, *loc. cit.*). Europe, including Fennoscandia.

CALOPTILIA SYRINGELLA (Fabricius)

Tinea syringella Fabricius, 1794, *Ent.syst.* **3**(2): 328.

Type locality: Germany.

Description of imago (Pl.10, figs 30,31)

Wingspan 10–13mm. Head ochreous, mixed leaden metallic on vertex; antenna dull yellow, annulated fuscous; labial palpus yellowish, banded fuscous. Thorax dull yellow, anteriorly fuscous. Forewing golden brown with variable black-edged ochreous white markings, consisting of very irregular fasciae at one-quarter, one-third and one-half, and two or three costal and three or four dorsal spots beyond; often a white-edged black apical dot; discal area beyond middle suffused fuscous; terminal cilia fuscous with whitish median line, sometimes obscurely double towards tornus; dorsal cilia dark grey. Hindwing and cilia dark fuscous.

The ground colour is sometimes dark golden fuscous and the pattern suffused with darker fuscous or obsolescent.

Similar species. C. cuculipennella (Hübner), *q.v.*

Life history

Ovum. Laid singly or in a row of two to eight against the midrib of a leaf of privet (*Ligustrum vulgare* or *L. ovalifolium*), ash (*Fraxinus excelsior*) or lilac (*Syringa vulgaris*); occasionally on white jasmine (*Jasminum officinale*) or *Phillyrea latifolia* f. *media* (Sich, 1911).

Larva. Head pale yellow-brown. Body yellowish white or greenish white, gut darker green. In early instars, the head is colourless, the body transparent and the segments deeply incised.

The young larvae start feeding in narrow, parallel epidermal galleries which merge to form a large blotch; during the sap-feeding stage the reddish frass is attached to the cuticle and the mine is greyish green. In the tissue-

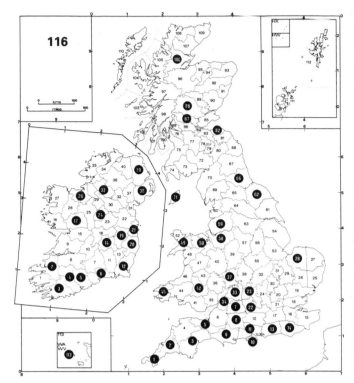

Caloptilia leucapennella

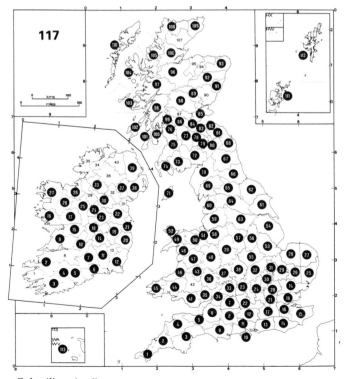

Caloptilia syringella

feeding phase, the frass is black and the mine is much distorted by internal spinning and turns orange-brown. After leaving their mine, the larvae construct an untidy cone by rolling a leaf downwards from the tip, within which they continue to feed gregariously; two such cones are usually made. June; August to September.

Pupa. Yellowish brown. In a greyish white, membranous cocoon, narrower and less glossy than those of other *Caloptilia*, spun on the underside of a leaf or, more often, on litter. June to July; October overwintering until April.

Imago. Bivoltine. April to May; July. Its presence can readily be detected by the unsightly discoloration made by the larva of the foliage of its foodplant, from which the adult may be dislodged by beating.

Distribution (Map 117)

Frequents woodland and gardens, where it is regarded as a pest because of the disfigurement made by the larvae of lilacs and privet hedges. Abundant throughout the British Isles to Shetland. Europe, Asia Minor and Canada.

ASPILAPTERYX Spuler

Aspilapteryx Spuler, 1910, *Schmett.Eur.* 2: 407.

A genus containing four species, three of which occur in southern Africa. The fourth, which is western Palaearctic, is found in Great Britain and Ireland.

Imago. Haustellum shorter than in *Caloptilia* Hübner. Forewing with veins 4 (M_3) and 5 (M_2) coincident. Other characters as in *Caloptilia*.

Life history

Larva. More strongly marked than in *Caloptilia*, especially with respect to the head and prothoracic plate. There are important ethological differences. *Aspilapteryx tringipennella* (Zeller), the only British member of the genus, spends the whole of the larval stage as a leaf-miner. In this and in pupating within the mine it resembles *Phyllonorycter* Hübner, but whereas that genus has three sap-feeding instars, *Aspilapteryx* has only two. It also differs from that genus in the tissue-feeding phase in that the trophi are adapted to chew in a horizontal plane rather than downwards, enabling it to enlarge the epidermal mine. Winter is passed as a small larva and in spring, if the host leaf has ceased to be palatable, it can start a fresh mine in a new leaf.

Pupa. In a cocoon in the mine.

Imago. Habits similar to those of *Caloptilia*. The British species is bivoltine.

ASPILAPTERYX TRINGIPENNELLA (Zeller)

Gracilaria tringipennella Zeller, 1839, *Isis, Leipzig* **1839**: 209.

Type localities: Bohemia and Germany; Glogau (now Poland; Glogów).

Description of imago (Pl.10, fig.32)

Wingspan 10–13mm. Head ochreous; antenna pale ochreous annulated pale fuscous. Forewing ochreous yellow to ochreous grey; a costal streak, extending from base nearly to apex, white or whitish grey; costa variably irrorate fuscous; subcostal and median longitudinal rows of fuscous spots, sometimes also a third incomplete row on fold; cilia ochreous, shading to grey on dorsum, sometimes with a fuscous line at base of terminal cilia. Hindwing pale grey, cilia tinged ochreous.

Life history

Ovum. Laid on the underside of a leaf of ribwort plantain (*Plantago lanceolata*).

Larva. Head brown or black. Body smoky whitish green, divided prothoracic plate and small anal plate brown. In early instars the head is pale brown and the body pale yellow, with the gut blackish grey.

Feeding starts in a long, tortuous gallery in the lower epidermis; its silvery white colour distinguishes it from dipterous mines on the same foodplant. After the change in mouth parts, the larva chews its way through the tissue to the upper surface where it makes a large blotch astride the midrib, the epidermis becoming very pale brown. Spinning causes the leaf to close progressively over the mine and finally almost to conceal it. When full-fed, the larva prepares an exit hole by chewing through to the lower epidermis which is left as a window covering the hole. June to July; October to April. In the overwintering generation, most of the feeding is done in the spring and is generally completed in a fresh leaf. To effect entry, the larva spins a slight web on the upperside of the leaf over the midrib. It then stands on the underside of this web with its dorsum towards the leaf surface and raises its head to chew its way in. This process is necessary because the mandibles are not adapted to chew downwards (see generic introduction, p. 271).

Pupa. Dark brown; in a white, spindle-shaped cocoon spun in the mine and attached at one end to the prepared exit hole. April to May; July to August.

Imago. Bivoltine. May; August. It is active in afternoon and evening sunshine.

Distribution (Map 118)

Frequents downland, rough grassland, waste ground, roadside verges and coastland. Widespread and locally common throughout the British Isles northwards to Shetland. Europe, Asia Minor, north Africa.

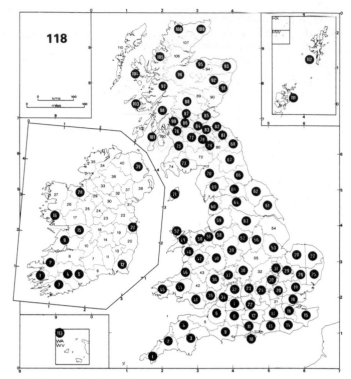

Aspilapteryx tringipennella

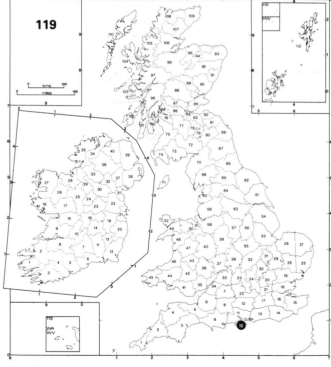

Calybites hauderi

CALYBITES Hübner

Calybites Hübner, 1822, *Syst.-alph.Verz.*: 66.

A genus as at present constituted comprising four species. Three are Palaearctic, of which all are found in Great Britain and two also in Ireland. The fourth species occurs in southern Africa.

Imago. Forewing with veins 4 (M_3) and 5 (M_2) coincident and stalked with 3 (Cu_1); venation otherwise similar to *Caloptilia* Hübner. Other characters as in *Caloptilia*.

Life history

There is no behavioural difference from *Caloptilia*. Fletcher (1942) cites two Indian species of *Caloptilia* which on the evidence of their biology and foodplants are very closely allied to *Calybites phasianipennella* (Hübner). The early stages of *Caloptilia semifascia* (Haworth) and *Calybites hauderi* Rebel likewise show them to be extremely closely related. It is therefore questionable whether *Calybites* should continue to be regarded as a distinct genus from *Caloptilia*.

Two of the British species of *Calybites* feed on herbaceous plants; two overwinter as adults and the third as a pupa.

CALYBITES HAUDERI Rebel

Calybites hauderi Rebel, 1906, *Verh.zool.-bot.Ges.Wien* **56**: 9.
Gracilaria pyrenaeella Chrétien, 1908, *Naturaliste* **30**: 246.
Type locality: Austria; Kirschdorf.

Description of imago (Pl.10, fig.33)

Wingspan 10–11mm. Head with vertex yellowish grey-brown to dark purplish brown; frons yellowish; antenna dull ochreous, annulated dark brown; labial palpus pale yellow, segment 3 with incomplete broad purplish brown median band, obsolete on upper and inner surfaces. Thorax yellow-brown to dark brown; tegulae brown, inwardly edged dull ochreous yellow. Forewing ranging from pale yellowish brown mixed darker to dark glossy chestnut with violet reflections; a large, clear yellow, subquadrate antemedian costal blotch, in paler forms sometimes darker-edged proximally and distally; some or all of the following yellow markings may also be present: a small basal spot, a subapical costal spot preceded by scattered scales on costa and a subdorsal patch of yellowish irroration near base; cilia concolorous with wing and with three darker lines, on underside yellow at apex. Hindwing brownish grey, cilia paler.

Similar species. Caloptilia azaleella (Brants), *q.v.*

Life history

Ovum. Laid on the underside of a leaf of field maple (*Acer campestre*), generally *c.*2–4m from the ground. Recorded on sycamore (*A. pseudoplatanus*) in Austria.

Larva. Head faintly tinged brown, mouth parts reddish. Body greenish white, gut blackish green. Closely resembles the larva of *C. semifascia* (Haworth), but the intersegmental divisions are rather more deeply incised.

Feeds at first in a long and rather broad epidermal gallery on the underside of the leaf. This is later expanded into a small blotch, always adjacent to a vein and often in the angle between a vein and the midrib. In the tissue-feeding phase there is no internal spinning; all the parenchyma within the blotch is eaten but the veins are left, giving the epidermis a reticulated appearance; the frass is stacked against the vein. After leaving its mine, the larva makes three successive cones by rolling down part of a lobe of a leaf; an adjacent lobe is sometimes drawn in to seal the larger end of the cone, but the spinning is nearly always cone-shaped and not cylindrical with both ends sealed by lobes as in *C. semifascia* (p. 267). In other respects the feeding pattern of the two species is indistinguishable. May to June, two to three weeks earlier than *C. semifascia*.

Pupa. Pale yellowish brown; in an oval, membranous, translucent, whitish green cocoon, conspicuously paler round its margin, spun usually on the underside of a leaf. June to early July.

Imago. Probably univoltine, with the moth emerging in July and overwintering. However, Chrétien (1908) suspected that it was bivoltine in southern France; if it is bivoltine in Britain, the second generation has not been observed and it is not known in that event whether the winter is passed as a pupa or adult.

Distribution (Map 119)

Occurs inside woods, differing in this respect from *C. semifascia* which prefers their margins and also hedgerows. Local in the Isle of Wight, Hampshire, especially near St Helen's where it was discovered by Ford (1933). Otherwise known only from the type locality and the western Pyrenees.

CALYBITES PHASIANIPENNELLA (Hübner)

Tinea phasianipennella Hübner, [1813], *Samml.eur. Schmett.* **8**: pl.47, fig.321.

Type locality: Europe.

Description of imago (Pl.10, figs 34–36)

Wingspan 10–11mm. Dimorphic. *Typical form.* Head yellowish brown to brown; antenna golden brown, annulated fuscous; labial palpus yellowish, outwardly irrorate brown, apex yellow. Thorax ochreous with brown central streak, tegulae brown. Forewing yellowish brown to chestnut brown; a narrow dorsal streak from near base to three-quarters white, its upper margin sinuate, dark-edged; costa with a subapical white spot, sometimes preceded by a whitish suffusion at three-quarters; terminal cilia concolorous and with three darker lines, becoming four at tornus; dorsal cilia grey. Hindwing dark fuscous, cilia grey with red-brown sheen. The f. *quadruplella* Zeller (fig. 35) differs as follows. Forewing with large, distinct, outward oblique white spots on costa at one-third and three-quarters; dorsum with elongate white spot at one-fifth and subtriangular spot at one-half; these spots variably dark-edged; preapical costal spot obsolescent or absent. Intermediate forms occur in which the white markings of f. *quadruplella* are obscurely indicated or represented only by intermittent dark margins.

Life history

Ovum. Laid generally on the underside of a leaf of water-pepper (*Polygonum hydropiper*), redshank (*P. persicaria*), black bindweed (*P. convolvulus*), common sorrel (*Rumex acetosa*), sheep's sorrel (*R. acetosella*), broad-leaved dock (*R. obtusifolius*), water-dock (*R. hydrolapathum*) or yellow loosestrife (*Lysimachia vulgaris*). At Wicken Fen, Cambridgeshire, the last of these is the most favoured choice.

Larva. Head whitish yellow with four black spots. Body dull greyish green with obscure darker spots; gut darker green; prothoracic plate anteriorly with two minute black dots and posteriorly with four larger spots.

Feeding starts in a gallery leading to and sometimes absorbed in a pale squarish blotch in the lower epidermis; on *Polygonum* spp. this mine is often on the upperside; brownish frass, due to be eaten later, is deposited in an indistinct line on the surface of the parenchyma. In the tissue-feeding phase, the larva eats the parenchyma and the frass adhering to it without enlarging its blotch. Spinning of the lower epidermis causes the leaf to pucker; the mine turns brown. The red-brown frass is packed at one end of the mine. A further ecdysis takes place within the mine. On leaving the mine, the larva cuts an irregular strip about 30mm long and 7mm wide from the edge of a leaf. The strip is rolled downwards slightly spirally, pulled over and secured with silk by its inner edge so that it projects

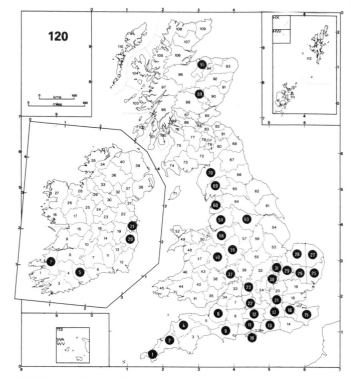

Calybites phasianipennella

downwards from the underside of the leaf like an untidy cone. Within this roll, the larva feeds on the underside of the leaf. Two such rolls are usually made. July to September.

Pupa. Dark brown and rather hairy. In a rather long, flimsy cocoon of white silk spun vertically in the last roll occupied by the larva. September.

Imago. Univoltine, emerging in September, overwintering and continuing on the wing until May. Active in the evening both in autumn and spring; it can be beaten from evergreen or withered vegetation in winter.

Distribution (Map 120)

Occurs in damp woodland and fens, but also in drier situations such as the edges of cornfields. Widespread and locally common in England, though absent from many apparently suitable areas; not recorded in Wales; rare in Scotland; local in southern Ireland. Palaearctic.

CALYBITES AUROGUTTELLA (Stephens)

Euspilapteryx auroguttella Stephens, 1835, *Ill.Br.Ent.* (Haust.) **4**: 363.
Type locality: England; Ripley, Surrey.

Description of imago (Pl.10, fig.37)

Wingspan 9–10mm. Head leaden metallic, frons silvery; antenna blackish grey with darker annulations, tip whitish; labial palpus ochreous grey, outwardly irrorate with leaden metallic to near tip. Forewing shining dark grey; subcostal spot at one-third, costal spot at three-quarters, and dorsal subbasal and pretornal spots golden; cilia dark grey. Hindwing and cilia dark fuscous grey.

Life history

Ovum. Laid on the underside of a leaf of perforate St John's wort (*Hypericum perforatum*) or slender St John's wort (*H. pulchrum*).

Larva. Head from pale whitish yellow to pale brown, mouth parts darker. Body pale whitish yellow or whitish green, posteriorly with faint reddish tinge.

Feeding starts in a tortuous gallery in the lower epidermis in which the red-brown frass is attached to the cuticle. In the second instar the mine is expanded into an epidermal blotch which absorbs the earlier working. In the tissue-feeding phase, spinning causes the mine to arch upwards to form rather a narrow tube. The frass is packed at one end and the mine turns brown. On leaving its mine, the larva forms a cone by spinning the tip of a leaf downwards; the cone is widest at its middle with the ends somewhat pointed. Later a second cone is made, nearly always higher up on the same plant. June; September to October.

Pupa. Rather dark brown. In a cocoon in a leaf rolled longitudinally and somewhat spirally, generally near the top of the plant in spring but often lower in the autumn, when the leaf may be secured to the stem by spinning. July to August; October, overwintering until late April.

Imago. Bivoltine. May; August. Flies low round its food-plant in the evening.

Distribution (Map 121)

Frequents coppiced woodland, rough grassland and roadside verges. Widespread and common throughout Britain to Sutherland; local and less common in Ireland. Europe; Asia Minor.

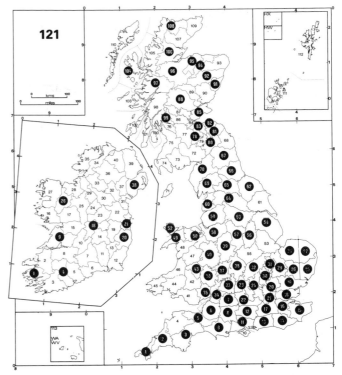

Calybites auroguttella

MICRURAPTERYX Spuler

Micrurapteryx Spuler, 1910, *Schmett.Eur.* **2**: 409.

A genus very close to *Parectopa* Clemens, with which it is synonymized by some taxonomists.

There is no reliable evidence that the single species it contains was ever taken in Britain.

MICRURAPTERYX KOLLARIELLA (Zeller)

Gracilaria kollariella Zeller, 1839, *Isis, Leipzig* **1839**: 209.
Type locality: Austria.

This species was placed on the British list on the evidence of a single specimen in the British collection at BMNH (Stainton, 1864: 134). There is no proof that the moth was taken in this country. The adult and its early stages are described and figured by Stainton (*loc.cit.*: 128–137, pl.3). Its larva feeds on broom (*Sarothamnus scoparius*). It occurs in central and southern Europe.

PARECTOPA Clemens

Parectopa Clemens, 1860, *Proc.Acad.nat.Sci.Philad.* **12**: 210.

A very large genus with world-wide distribution, but poorly represented in Europe. One species occurs in England.

Imago. Legs slender, smooth; hindtibia without bristly hairs above. Forewing with veins 2 (Cu$_2$) and 3 (Cu$_1$) coincident, 6 (M$_1$) and 7 (R$_5$) stalked, 8 (R$_4$) connate or short-stalked with stalk of 6 (M$_1$) and 7 (M$_5$). Hindwing with cell open between veins 3 (Cu$_1$) and 5 (M$_2$); 1b (2A) and 4 (M$_3$) absent, 5 (M$_2$) and 6 (M$_1$) stalked.

Resembles *Aspilapteryx* Spuler in mining throughout the larval stage, in enlarging the epidermal mine after hypermctamorphosis, in overwintering as a larva and in being capable of changing leaves; differs in ejecting the frass from the mine and pupating externally. The pupa lacks spines on the dorsal area of the abdominal segments; the cocoon is of the shape described for the family on p. 247 but the texture is less glossy and membranous.

PARECTOPA ONONIDIS (Zeller)

Gracilaria ononidis Zeller, 1839, *Isis, Leipzig* **1839**: 209.
Type locality: Germany; Glogau (now Poland; Glogów).

Description of imago (Pl.11, fig.13)

Wingspan 7–9mm. Head with forward-projecting scales on vertex, submetallic dark brown; frons silver; antenna ochreous, annulated fuscous; labial palpus brown, terminal segment white. Thorax dark brown with two posterior white spots. Mid- and hindlegs dark brown, prominently banded white. Forewing dark brown; costal streaks at one-quarter, one-half (both outward oblique), three-quarters and before apex (inward oblique), four dorsal streaks, each anterior to those on costa, and subapical dot all shining silver and edged blackish; first costal and second dorsal sometimes unite to form a fascia; cilia dark grey with basal and subapical dark lines at termen; underside uniform brown except subapical white spot on costa, which is conspicuous. Hindwing greyish fuscous, cilia paler.

Similar species. Acrocercops imperialella (Zeller), which lacks the two basal silver spots on the dorsum and has a conspicuous white apical spot, also clearly visible on the underside; the hindleg has a row of bristles on the tibia and is only obscurely banded with white.

Life history

Ovum. Laid on the underside of a leaf of red clover (*Trifolium pratense*), less often white clover (*T. repens*), possibly other clovers, or restharrow (*Ononis* spp.). On clover young leaves are preferred.

Larva. Head pale ochreous brown, mouth darker. Body golden yellow to yellowish green, gut darker.

Feeding starts in a tortuous gallery in the lower epidermis with a central line of red-brown or blackish frass. After the change in mouth parts, the larva makes a full-depth mine extending to the upper epidermis. An ochreous brown gallery is spun above the midrib, within which the larva rests concealed when not feeding. Paler brown, unspun branches emanate, where the larva has fed; these are later united to form a blotch, often occupying the whole leaflet which remains relatively flat since the spinning is confined to the midrib. All, or virtually all, the frass is ejected from the mine. Somewhat similar greenish white mines, lacking the gallery spun along the midrib, are dipterous; however, mines of this species on white clover almost lack the brown tinge which makes recognition easy on red clover. The larva changes leaves, often more than once; it usually emerges on the upperside near the petiole and enters the new leaf on the underside near the tip. The hairs on the leaves of red clover provide a platform from which the larva can select the appropriate angle of entry (*cf. Aspilapteryx tringipennella* (Zeller), p. 272); entry to the glabrous leaves of white clover has not been observed. July; September to April; overwintering larvae may begin to feed again as early as February (D. W. H. Ffennell, pers.comm.).

Pupa. Abdomen whitish; thorax and appendages blackish fuscous. In a white cocoon spun on either, but usually the upper, surface of a leaf of the foodplant and distorting it, or on adjacent herbage. April to May; July to August.

Imago. Bivoltine. Flies mainly from May to June and in August but the generations tend to overlap and there are probably three in favourable years. It is seldom seen as an adult.

Distribution (Map 122)

An inconspicuous and evasive insect which may have been overlooked in many areas. It has a preference for coasts and chalk downland, but is by no means confined to such situations. In some years it is common as a larva in suitable localities in southern England; in others it is very rare. It has been recorded from East Anglia and the south-eastern counties, the south Midlands, the counties bordering on south Wales, and Northumberland (Bolam, 1931). Hodgkinson (1891) reports rearing it at Ashton-on-Ribble, Lancashire, but does not state explicitly that the larvae were taken locally. Europe, including Fennoscandia.

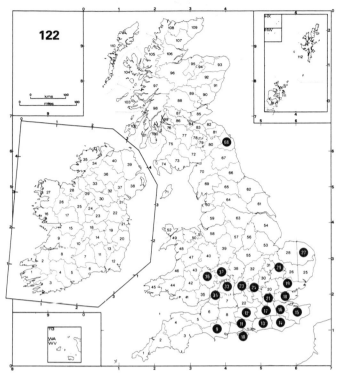

122

Parectopa ononidis

PARORNIX Spuler

Parornix Spuler, 1910, *Schmett.Eur.* **2**: 410.

A very large Holarctic genus. Ten species occur in Great Britain and six of these also in Ireland.

Imago. Head with vertex and frons rough-scaled; antenna about length of forewing, scape with pecten; haustellum well developed. Legs with ornamentation as in *Caloptilia* Hübner but hindtibia slightly more thickened with scales and tarsi shorter, about one and a quarter length of tibia. Wings more broadly lanceolate than in *Caloptilia*. Forewing with veins 4 (M_3) and 5 (M_2) short-stalked, 6 (M_1) and 7 (R_5) long-stalked. Hindwing with cell open between veins 3 (Cu_1) and 5 (M_2), 4 (M_3) absent, 5 (M_2) and 6 (M_1) short-stalked.

Life history

Larva. Sap-feeding in the first two instars. Legs in the tissue-feeding phase as in *Caloptilia* but differing from *Phyllonorycter* Hübner in the thoracic legs being four- as opposed to three-segmented. Instantly recognizable from other gracillariids by the prothoracic plate which is split into four black plates arranged in a transverse row.

The larval habits correspond with those of *Caloptilia* and show an equal degree of interspecific variation. The amount of spinning within the mine differs between species and may form a good means of distinction, as between *Parornix finitimella* (Zeller) and *P. torquillella* (Zeller). The mine is usually vacated after the third ecdysis and the majority of species then feed under a folded leaf-edge. An exception is *P. anglicella* (Stainton) which makes a cone like those of many *Caloptilia*; this species and *Parornix scoticella* (Stainton) both feed sometimes on wild service-tree (*Sorbus torminalis*) and may readily be separated by the difference in their spinnings.

Pupa. Pale brown; head with a pointed 'cocoon-piercer'; antennae longer than abdomen; dorsal area of abdominal segments with numerous short, dark, posteriorly directed spines. In a firm, opaque, usually ochreous cocoon, shaped like a segment of an orange but tapering in the converse direction in cross-section. It is generally spun under a narrowly folded leaf-edge in summer, but more often in leaf-litter in the winter generation. Old birds'-nests are a popular site and nests collected for tineids will often also yield a *Parornix*. All British species overwinter in their cocoons, most of them in the pupal stage.

Imago. Coloration duller than in most gracillariids. The ground colour is whitish but so heavily irrorate with grey or fuscous as to make it appear that the moths are dark with a pale pattern; traditionally this is how they have been described by authors and the same practice has been followed in this work. The posture adopted at rest is similar to that of *Caloptilia*. Most species are bivoltine with adults flying in May and August; however *P. torquillella* is an exception and those that are bivoltine in southern England may be univoltine in Scotland. The moths fly at dusk and remain active after dark, when they may sometimes enter light-traps in large numbers.

PARORNIX LOGANELLA (Stainton)

Argyromiges loganella Stainton, 1848, *Zoologist* **6**: 2162.
Type locality: Scotland; Luss, Dunbartonshire.

Description of imago (Pl.11, fig.1)

Wingspan 9–11mm. Head greyish ochreous, mixed dark brown, frons grey; antenna whitish with fuscous annulations; labial palpus white, with obscure grey subterminal spot. Thorax dark grey. Forewing blackish fuscous, paler at base; numerous costal strigulae, dorsal spots before and beyond middle and discal spot at three-quarters white; terminal cilia white, becoming grey on tornus and dorsum, with black subapical line. Hindwing and cilia dark grey.

Life history

Except that it is univoltine not distinguished in any stage from *P. betulae* (Stainton), *q.v.*

Ovum. Laid on birch (*Betula* spp.).

Larva. July to August.

Pupa. Overwintering from September to May.

Imago. Univoltine. May and June.

Distribution (Map 123)

Frequents woodland, hillsides and moors where there is a plentiful growth of birch. Occurs from the Lancashire mosses and Yorkshire northwards to Sutherland; more southerly records include Staffordshire, Herefordshire and an unlikely record from Hampshire (Goater, 1974: 38). The single Irish record is from Co. Kerry (Mere *et al.*, 1964). Sweden.

PARORNIX BETULAE (Stainton)

Ornix betulae Stainton, 1854, *Ins.Br.Lepid.*: 205.
Type locality: England.

Description of imago (Pl.11, fig.2)

Wingspan 9–10mm. Head pale grey, vertex ochreous fuscous; antenna whitish grey with sharply defined fuscous annulations; labial palpus white with subapical fuscous band, sometimes obsolescent. Forewing greyish fuscous; costa with numerous whitish strigulae, more strongly pronounced posteriorly; dorsal area paler with contrasting blackish spots before and beyond middle; an elongate

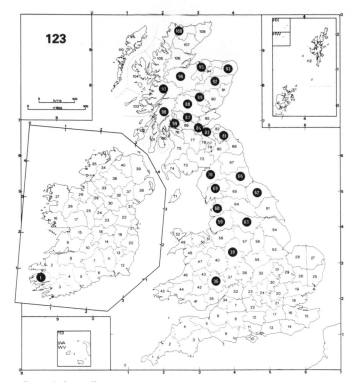

Parornix loganella

blackish streak in disc at three-quarters, broken by a whitish spot; an apical blackish spot; cilia grey with two complete darker lines, tips white at apex but dark-tipped in lower half of termen, forming an incomplete third dark line. Hindwing and cilia grey. Genitalia, see figure 100, p. 281.

Similar species. *P. betulae* is the only greyish *Parornix* with the tips of the apical cilia white.

Life history

Ovum. Laid on the underside of a leaf of birch (*Betula* spp.), with preference for seedlings.

Larva. Head brown with a darker spot on each lobe. Prothoracic plate with four blackish spots arranged transversely, the inner pair larger. Body greyish green, gut dark green; legs brown. In early instars, the head is almost transparent and the body yellowish grey.

Feeding starts in a gallery in the lower epidermis, in which the frass forms a faint reddish brown line on the cuticle; in the second instar this is developed into a blotch between veins. When tissue-feeding begins, spinning causes the lower epidermis to contract and the mine to arch upwards; both the upper and lower epidermes turn brown; the frass is packed at one end of the mine. The mine is hard to distinguish from that of *Caloptilia betuli-*

cola (Hering), and unless the larva with its characteristic prothoracic plate has been seen, records should be made only if the later feeding about to be described has also been observed. After leaving the mine, generally in the final instar, the larva makes a fold by turning the edge or tip of a leaf downwards and feeds therein; sometimes a second fold is made. Often the whole larval stage is completed on a single leaf. June; September to October.

Pupa. Pale brown. In an ochreous cocoon under a turned-down edge of a leaf. July to August; October overwintering until April or May.

Imago. Bivoltine. May; August. Active at dusk.

Distribution (Map 124)

Frequents heaths and open woodland where there are plenty of foot-high birch seedlings. Widespread and common throughout the British Isles northwards to Moray. Central and northern Europe.

PARORNIX FAGIVORA (Frey)

Ornix fagivora Frey, 1861, *Entomologist's wkly Intell.* **10**: 60.

Ornix carpinella Frey, 1863, *Linn.ent.* **15**: 19.

Type locality: Central Europe.

NOMENCLATURE. *P. fagivora* and *P. carpinella* are still regarded by some authors as distinct species (Hering, 1957: 250; Leraut, 1980: 61). *P. carpinella* is not listed by Kloet & Hincks (1972) even as a synonym. Stainton (1864: 210, 306) suggested tentatively that the *Parornix* feeding on *Fagus* and *Carpinus* were conspecific and this opinion is followed by Emmet ([1979]c: 55) and in this work, no differences having been observed in the genitalia of either sex. If there are two species, both occur in Britain.

Description of imago (Pl.11, fig.3)

Wingspan 9–10mm. Head ochreous white, mixed brown; antenna ochreous annulated fuscous; labial palpus whitish, segment 3 sometimes with fuscous median band. Thorax pale ochreous, tegulae brown. Forewing fuscous brown; costa rather obscurely strigulated ochreous; dorsal area more strongly mixed ochreous, with obscurely darker ante- and postmedian subdorsal spots; generally an ochreous spot in disc at three-quarters; an apical black dot, often obsolescent; cilia ochreous grey with three complete dark fuscous lines. Hindwing grey, cilia with ochreous tinge.

Similar species. The ochreous or fawn tinge of the pale markings and the lack of definition in the pattern help to distinguish *P. fagivora* from other *Parornix*.

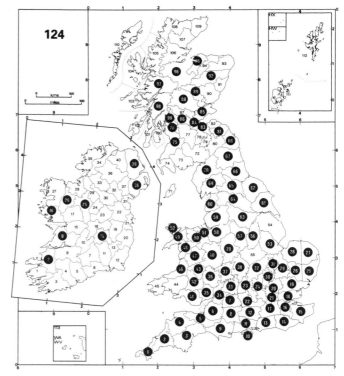

Parornix betulae

Life history

Ovum. On a leaf of beech (*Fagus sylvatica*) or hornbeam (*Carpinus betulus*), generally on the underside beside a vein.

Larva. Head pale brown with darker markings. Body whitish, gut dark green to purplish; prothoracic plate with four rather obscure dark spots.

Feeding starts in the lower epidermis, any early gallery being completely absorbed by the subsequent blotch in which the cuticle is at first marbled with reddish frass. In the tissue-eating phase, very little spinning is used and the mine remains almost flat. The larva eats through to the upper epidermis but leaves the veins, so that the mine has a reticulated appearance if held up to the light. The now black frass is deposited loose in the mine. After leaving its mine, the larva folds the tip or an edge of a leaf downwards and feeds within, still avoiding the veins. Each larva usually makes two such folds. July; September.

Pupa. In a white cocoon, usually spun under the edge of a leaf folded upwards. July to August; September, overwintering until April.

Imago. Bivoltine. May; August. Rests on trunks or may be disturbed from branches by beating.

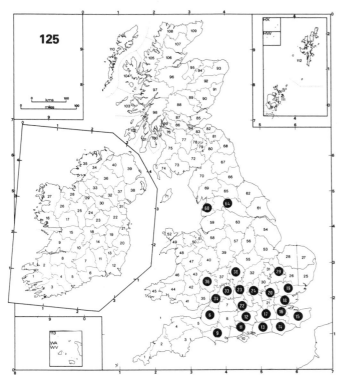

Parornix fagivora

Distribution (Map 125)

Frequents woodland where its foodplants predominate. Although fairly widespread, it occurs at low density and, being rather difficult to rear, is scarce in collections. It has been recorded in the area bounded by Kent, Dorset, Herefordshire and Cambridgeshire and also from Lancashire and west Yorkshire. Central and northern Europe, not extending to Finland.

PARORNIX ANGLICELLA (Stainton)

Ornix anglicella Stainton, 1850, *Trans.ent.Soc.Lond.* (2) **1**: 92.

Type locality: England; Lewisham, Kent.

Description of imago (Pl.11, fig.4)

Wingspan 9–11mm. Head with vertex ochreous white, mixed fuscous, frons fuscous; antenna ochreous, annulated fuscous; labial palpus white, segment 3 with median fuscous band. Thorax whitish ochreous, tegulae greyish fuscous. Forewing greyish fuscous mixed whitish, especially along dorsum; costa with numerous whitish strigulae, often obscure except towards apex; elongate dark fuscous ante- and postmedian subdorsal spots; a dark fuscous spot in disc at three-quarters followed by a small whitish spot, one or both of which may be absent; a blackish apical dot, obscurely bounded inwardly by an extension of the most distad whitish strigula on costa; cilia grey with three dark fuscous lines, the outermost seldom quite passing apex. Hindwing grey, cilia slightly paler. Genitalia, see figure 100.

Similar species. P. devoniella (Stainton) which has the ground colour paler and the dark markings more distinct. On the underside *P. devoniella* has a distinct white line in the terminal cilia which is continued round the apex to unite with the costal strigula nearest the apex; in *P. anglicella* this line is more obscure and stops at the apex, not reaching the anteapical strigula.

Life history

Ovum. Laid on the underside of a leaf of hawthorn (*Crataegus* spp.) or wild service-tree (*Sorbus torminalis*); exceptionally on rowan (*S. aucuparia*) or strawberry (*Fragaria* spp.) (Meyrick, 1928: 786).

Larva. Head pale yellowish brown with four small black spots. Prothoracic plate also yellowish brown with four black spots. Body greenish grey, gut blackish; abdominal segments each with six whitish spots; legs blackish. In the mining phase the head is thin, flat and transparent with the mouth parts and sutures pale brown; thoracic segments broader than abdominal segments which are deeply incised, flattened and tapering towards anus; body pale yellowish white, gut reddish; legs rudimentary.

Feeding starts in an epidermal gallery leading towards the leaf-margin where in the second instar it is developed into a blotch; the frass is at first attached to the lower cuticle which turns a dirty greyish brown colour. In the tissue-eating phase, the larva eats through to the upper epidermis and extensive spinning puckers the leaf strongly; both epidermes turn brown and the black frass is packed at one end of the mine. Its smaller size and the brown colour of the lower epidermis readily distinguish the mine from that of *Phyllonorycter oxyacanthae* (Frey), in which the lower epidermis remains green. On leaving the mine the larva constructs a cone after the manner of a *Caloptilia* rather than other *Parornix*. Its method is as follows. It spins a mat of silk on the underside of a lobe, covering the projected area of the cone. It then turns and spins a second mat above the first; meanwhile the first mat is contracting and flexing the leaf downwards. When the second mat is complete, the larva pushes its way under the mats and lifts the first and spins it to the second, applying more tension and thereby drawing the leaf over still further. The larva is now occupying, as it were, a segment of a circle, the surface of the leaf representing the arc and the silken mat the chord; as more spinning is applied to the latter causing further contraction, so the points of the arc are drawn closer together. As the new silk contracts, the

Figure 100

Parornix betulae (Stainton)
(**a**) male genitalia (**b**) female genitalia

Parornix anglicella (Stainton)
(**c**) male genitalia (**d**) female genitalia

Parornix devoniella (Stainton)
(**e**) male genitalia (**f**) female genitalia

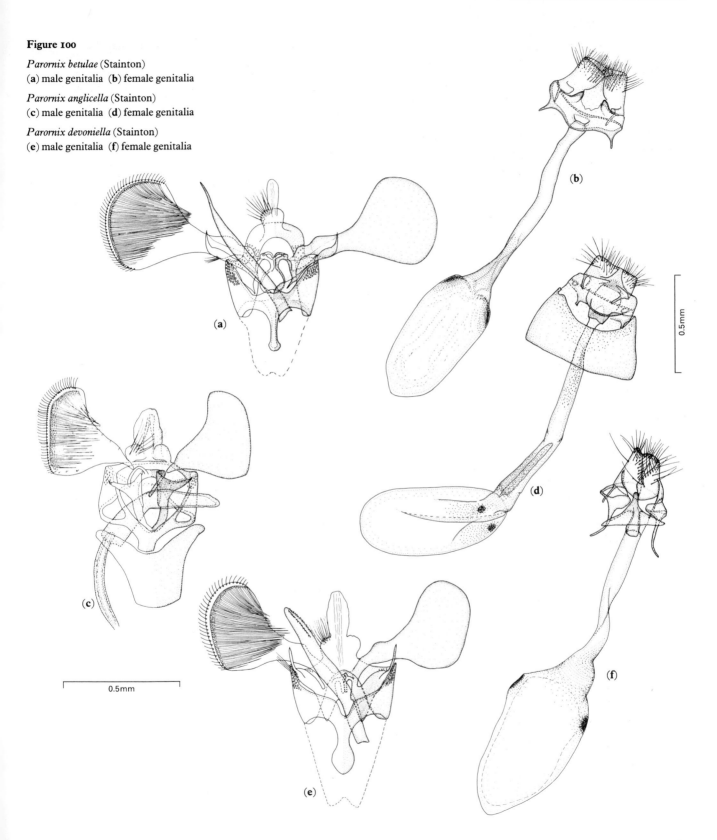

0.5mm

0.5mm

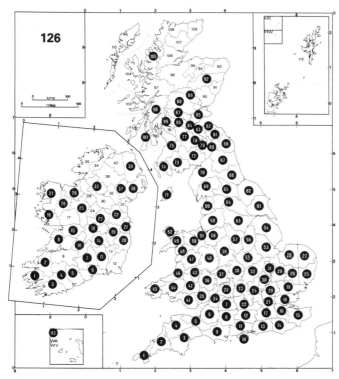

Parornix anglicella

old becomes baggy; some of this is worked into the mat and the rest is disposed of by eating. When the lobe has been drawn over far enough to form the cone, the larva comes out and adds a few external 'guy-ropes' to give rigidity. This complicated feat of engineering is completed within the space of about 15 minutes. Within the cone, the larva eats through to the upper epidermis which is left intact and later turns brown, a sign that the larva has gone. Often a second cone is needed. July; August to September.

Pupa. Pale yellow-brown; dorsal area of abdominal segments with numerous posteriorly directed, short dark spines. In a stoutly constructed, whitish cocoon, sometimes spun under a narrowly folded leaf-edge but often in leaf-litter or an old bird's-nest. July to August; September, overwintering until late April.

Imago. Bivoltine. Late April and May; August. Easily disturbed from its foodplants, round which it flies naturally from late afternoon onwards; it sometimes comes to light.

Distribution (Map 126)
Frequents gardens, hedgerows and woodland. Occurs abundantly wherever there is hawthorn throughout the British Isles to Perthshire, north of which it is more local and rare. Palaearctic.

PARORNIX DEVONIELLA (Stainton)

Ornix devoniella Stainton, 1850, *Trans.ent.Soc.Lond.* (2)**1**: 89.
Ornix avellanella Stainton, 1854, *Ins.Br.Lepid.* **3**: 204.
Type locality: England; Dawlish, Devon.

Description of imago (Pl.11, fig.5)
Wingspan 9–10mm. Head whitish, mixed fuscous; antenna pale ochreous, annulated fuscous; labial palpus white, segment 3 with median fuscous band. Thorax greyish white, sometimes irrorate fuscous, tegulae darker. Forewing pale grey, much mixed whitish, especially in dorsal area; costa with numerous whitish strigulae; ante- and postmedian subdorsal dark fuscous spots; an elongate fuscous streak in disc at three-quarters, interrupted by a whitish spot; an apical dark fuscous dot, sometimes obsolescent; cilia grey with three complete fuscous lines. Hindwing grey, cilia paler. Genitalia, see figure 100, p. 281.

Similar species. *P. anglicella* (Stainton), *q.v.*

Life history
Ovum. Laid on the underside of a leaf of hazel (*Corylus avellana*).

Larva. Head yellow-brown, with a darker mark on each lobe. Body pale greenish yellow, gut dark green; prothoracic plate pale greyish brown with four blackish spots.

Feeding starts in a gallery in the lower epidermis which is developed into a square blotch; occasionally the blotch is situated in an angle of veins and is then triangular. In the tissue-feeding phase, the larva first chews through to the upper epidermis round the margins of its mine, leaving the centre green. Later most of the parenchyma is eaten. The lower epidermis turns brown, but there is little spinning and the mine is only slightly arched. The usually smaller size, less elongate shape, lack of internal silk and the brown lower epidermis readily distinguish this mine from that of *Phyllonorycter nicellii* (Stainton). After leaving its mine, the larva folds the edge of a leaf either downwards or upwards and feeds therein; sometimes two such folds are made. July; September.

Pupa. In an ochreous cocoon, usually under a folded leaf-edge, more often that of a fallen leaf. July to August; September, overwintering until April.

Imago. Bivoltine. May; August. Rests on trunks and flies in the evening round its foodplant.

Distribution (Map 127)
Frequents woodland and, to a less extent, hedgerows. Widespread and abundant throughout England, Wales and Ireland; in Scotland, although extending up to Sutherland, it is rather rare and local; Isle of Man. Northern and central Europe.

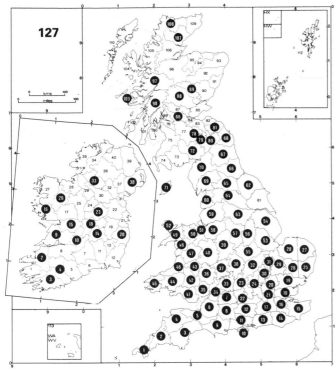

Parornix devoniella

PARORNIX SCOTICELLA (Stainton)

Ornix scoticella Stainton, 1850, *Trans.ent.Soc.Lond.* (2)**1**: 94.

Type locality: Scotland; Torwood, Stirlingshire.

Description of imago (Pl.11, fig.6)

Wingspan 9–10mm. Head whitish, mixed fuscous, frons white; antenna pale ochreous, sharply annulated fuscous; labial palpus white, segment 3 normally without dark median band. Thorax and tegulae whitish. Forewing fuscous, much mixed white, especially towards base and dorsum; costa with numerous white strigulae; dark fuscous ante- and postmedian subdorsal spots; a dark fuscous elongate streak in disc at three-quarters, often interrupted by a white spot which sometimes forms an extension of one of the costal strigulae; an apical blackish dot, margined inwardly by extension of costal white strigula nearest apex; cilia whitish with three entire dark fuscous lines. Hindwing and cilia grey. Moths reared from *Malus* spp. tend to be smaller and darker than those from *Sorbus* spp. Genitalia, see figure 101, p. 287.

Similar species. *P. alpicola* (Wocke), *q.v.* The white frons, usually immaculate white labial palpus and the less heavily marked forewing help to distinguish *P. scoticella* from other *Parornix*.

Life history

Ovum. Laid on the underside of a leaf of rowan (*Sorbus aucuparia*), common whitebeam (*S. aria*), Swedish whitebeam (*S. intermedia*), apple (*Malus* spp., preferring thick-leaved cultivars) or, less often, wild service-tree (*Sorbus torminalis*).

Larva. Head pale brown with two darker spots. Body pale whitish green or yellowish green; gut may be dark green, reddish or blackish, possibly depending on foodplant, tending to be dark green or blackish in larvae from rowan but reddish in those from apple; prothoracic plate concolorous with body with four irregularly shaped black spots.

Feeding starts in a rather broad gallery in the lower epidermis which is later expanded into a subrectangular blotch. In the tissue-feeding phase, the larva eats through to the upper epidermis leaving only the veins; the epidermis of both surfaces turns brown, a character which at once separates the mine from that of a *Phyllonorycter* and, when on apple, *Callisto denticulella* (Thunberg) which, when its mine is on the underside, has the lower epidermis white. There is relatively little internal spinning and the mine is only moderately arched, though more so on rowan than the other foodplants. After the larva has left its mine, it may make a typical *Parornix* fold at the edge of a leaf or spin the whole leaflet into a pod, but more often it spins a dense silken pad in a more central position on the

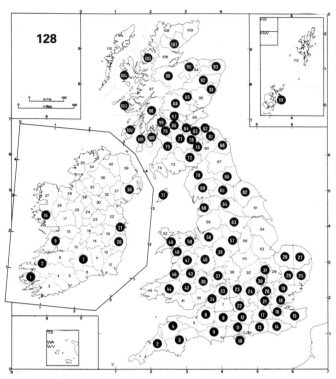

Parornix scoticella

underside, causing an abrupt pleat within which it eats through to the upper epidermis in the usual manner, causing it to turn brown. This pucker looks very much like a *Phyllonorycter* mine, but what appears to be the lower epidermis of the leaf is, in fact, the silken pad. Thus it was described as making a second *Phyllonorycter*-like mine beside its original mine on whitebeam by Bishop Whittingham (St Edmundsbury and Ipswich, 1937), an opinion accepted by Hering (1951: 31,36; 1957: 1009). July (south only); August to September.

Pupa. Pale yellow-brown; in a buff-coloured to white cocoon spun under a narrowly folded leaf-edge or on leaf-litter. July to August (south only); September, overwintering until late April.

Imago. Bivoltine in southern England, flying in May and August, univoltine in the north, flying only in August. Habits typical of the genus.

Distribution (Map 128)

Frequents gardens, orchards and woods in the south, but in the north mainly in woods and on moorland where there is rowan. Widespread throughout the British Isles, but more common in the north. Central and northern Europe.

PARORNIX ALPICOLA (Wocke)

Ornix alpicola Wocke, 1877, *Z.Ent.* (N.F.) **6**: 48.
Type localities: Austria; Tyrol and Germany; Bavaria.

Description of imago (Pl.11, fig.7)

Wingspan 8–10mm. Head with vertex white, thinly mixed fuscous; frons white; antenna fuscous with faint paler annulation; labial palpus white. Thorax white, sometimes mixed fuscous, tegulae grey. Forewing dark fuscous, very extensively mixed clear white, especially in basal two-thirds of wing, a subcostal white streak from base to one-third being a prominent character in most specimens; costa with numerous white strigulae; ante- and postmedian subdorsal fuscous spots; usually a white dot in disc at three-quarters; a well-defined apical black dot, margined inwardly by an extension of the costal white strigula nearest apex; cilia grey, with three complete fuscous lines. Hindwing and cilia pale grey.

Similar species. P. scoticella (Stainton), from which *P. alpicola* differs in its much darker antenna, in having an even greater admixture of white in the pattern of the forewing and in having only the apical dot darker than the ground colour. There is virtually no possibility of the two occurring on the same ground since the foodplants are not found together.

Life history

Ovum. Laid on the underside of a leaf of mountain avens (*Dryas octopetala*).

Larva. Head pale brown, margins of sclerites darker. Body pale green, gut darker; prothoracic plate concolorous, with four transversely placed blackish spots.

Feeding begins in an epidermal mine on the underside of a leaf which in the second instar is developed into a blotch which absorbs the earlier working and fills the whole of the leaf on one side of the midrib. In the tissue-feeding phase, the larva eats through to the upper epidermis which turns orange-brown, the mine being relatively little distorted by spinning. On leaving the mine, the larva changes leaves and spins the edges together to form a pod resembling that made by an *Ancylis* Stephens (Tortricidae) (*MBGBI* **5**); this spinning so closely resembles a mine that dissection is necessary to prove that it is differently formed. A second spinning is made if needed. Full-fed in late July or early August.

Pupa. Pale brown, in an ochreous cocoon spun under the folded edge of a leaf. The larva overwinters in its cocoon, the pupal stage lasting from March until May.

Imago. Univoltine. May and June. For its discovery in Britain, see Pelham-Clinton (1967).

Distribution (Map 129)

Confined to calcareous coastal hillsides on the north coast of Scotland. The only known localities are the Invernaver National Nature Reserve and a spot on the east side of Loch Eriboll, both in west Sutherland. The Alps of central Europe.

PARORNIX LEUCOSTOLA Pelham-Clinton

Parornix leucostola Pelham-Clinton, 1964, *Entomologist's Gaz.* **15**: 51.

Type locality: Scotland; Invernaver National Nature Reserve, Sutherland.

Description of imago (Pl.11, fig.8)

Wingspan 8–10mm. Head with vertex white mixed fuscous, frons white; antenna fuscous, scape and some basal flagellar segments partly white beneath; labial palpus white. Thorax white with faint pale fuscous median stripe; tegulae pale fuscous. Forewing white irrorate fuscous at least in basal quarter, this fuscous area being sometimes extended to one-half to connect with a fuscous patch and two short streaks bordering base of accessory cell; a small black apical spot; cilia white shading to fuscous on dorsum, tips at apex and a subapical patch on costa fuscous. Hindwing and cilia pale grey.

Status. The reasons for considering this a species distinct from *P. alpicola* (Wocke) are discussed by Pelham-Clinton (1967). He found differences in measurement but not in structure of the male genitalia; however, the sample studied was so small that only tentative conclusions could be drawn. There is no significant difference in the female genitalia. The ground colour of *Parornix* species is white and the pattern formed by fuscous irroration (see generic introduction, p. 277). It is possible that *P. leucostola* is a form of *P. alpicola* having this irroration greatly reduced in extent.

Life history

The foodplant is mountain avens (*Dryas octopetala*). No difference has been observed between this species and *P. alpicola*, *q.v.*, except possibly in timing, Pelham-Clinton (*loc.cit.*) having found that the adults of *P. leucostola* emerged marginally later than those of *P. alpicola*.

Imago. Univoltine. May and June.

Distribution (Map 130)

Known only from the type locality which is a coastal hillside with a plentiful growth of mountain avens.

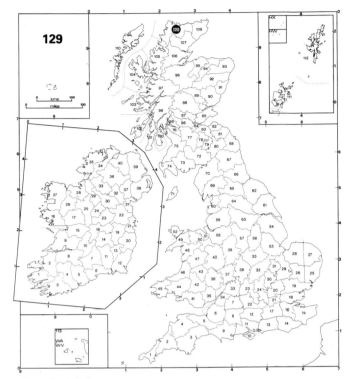

Parornix alpicola

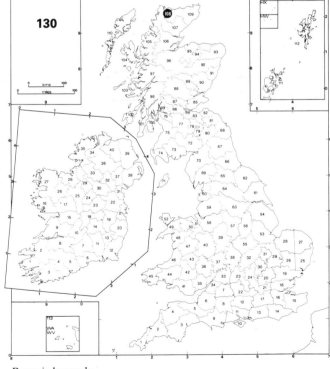

Parornix leucostola

PARORNIX FINITIMELLA (Zeller)

Ornix finitimella Zeller, 1850, *Stettin.ent.Ztg* 11: 162.

Type localities: Germany; Glogau (now Poland; Glogów) and Jena.

Description of imago (Pl.11, fig.9)

Wingspan 9–10mm. Head ochreous mixed fuscous; antenna ochreous annulated fuscous, labial palpus white, segment 3 with incomplete fuscous median band. Thorax whitish with fuscous median stripe, tegulae fuscous. Forewing fuscous mixed ochreous white, especially towards dorsum, this admixture becoming purer white towards apex; costa with numerous whitish strigulae; ante- and postmedian dark fuscous subdorsal spots; sometimes obscure dark streak in disc at three-quarters, interrupted by a white spot; an apical black dot, strongly margined inwardly by extension of white strigula nearest apex, which is continued hence through the terminal cilia at an angle hardly exceeding 90°; cilia grey with three complete dark lines. Hindwing and cilia pale grey.

Similar species. *P. torquillella* (Zeller), which usually lacks the dark band on segment 3 of the labial palpus and in which the costal white strigula on the forewing is continued round the apical dot and through the terminal cilia at an obtuse angle. Males of the two species can usually be separated by examination of the valvae without dissection (see figure 101).

Life history

Ovum. Laid on the underside of a leaf of blackthorn (*Prunus spinosa*) or wild plum (*P. domestica*).

Larva. Head pale brown with four black spots, outer pair larger. Body rather dark greenish grey, paler when full-grown, gut darker; paler grey subdorsal and lateral pinacula, obsolescent on abdominal segments; prothoracic plate concolorous with body, with four black spots, inner pair larger; legs ringed black.

Feeding starts in a gallery in the lower epidermis which is developed into a blotch between veins. When tissue-feeding begins, the larva eats through to the upper epidermis. Internal spinning causes the lower epidermis to become opaque like that of a *Phyllonorycter*; in colour it is greyish flecked with darker and it contracts strongly in a series of pleats, drawing the edges of the mine together. After leaving its mine, the larva folds the tip or edge of a leaf downwards and feeds therein through to the upper epidermis; at least two such folds are made. June to July; late August to October, feeding later than *P. torquillella* (M. R. Shaw, pers.comm.).

Both the larva and mine are readily distinguishable from those of *P. torquillella* and when they are being reared it is advisable to determine and segregate the species at this stage.

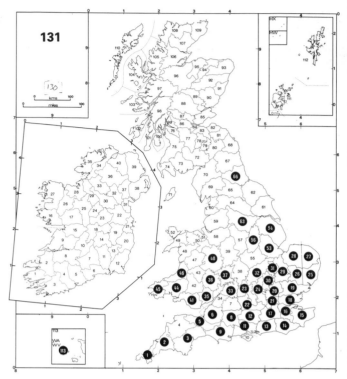

Parornix finitimella

Pupa. Pale brown; in an orange-yellow cocoon spun in a narrowly folded edge of a leaf or in detritus. July to August; October, overwintering until May.

Imago. Bivoltine. May; August. This species was not recognized as British until Pierce (1917) found it amongst specimens taken in Essex by C. R. N. Burrows; however, Stainton (1864: 296) had already suspected that it occurred here.

Distribution (Map 131)

Frequents scrub, woodland, hedgerows and gardens where there is blackthorn. Widespread in southern Britain to Yorkshire, though becoming progressively scarcer in the north of its range. In many places in the south it is more common than *P. torquillella*, but its full distribution is imperfectly known through confusion with that species. Europe northwards to Denmark and southern Sweden; Asia Minor.

Figure 101

Parornix scoticella (Stainton)
(**a**) male genitalia (**b**) female genitalia

Parornix finitimella (Zeller)
(**c**) male genitalia (**d**) female genitalia

Parornix torquillella (Zeller)
(**e**) male genitalia (**f**) female genitalia

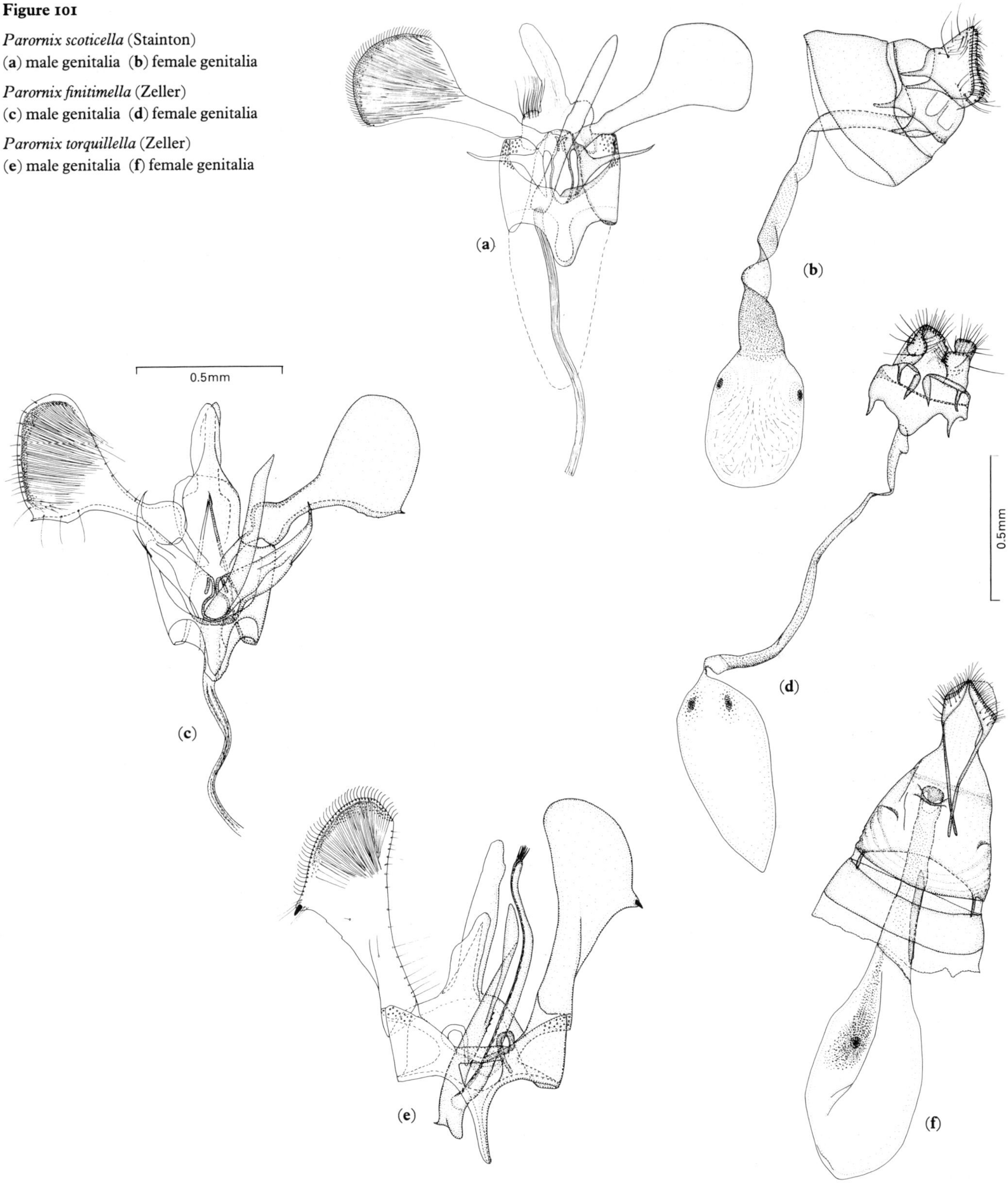

PARORNIX TORQUILLELLA (Zeller)

Ornix torquillella Zeller, 1850, *Stettin.ent.Ztg* **11**: 161.
Type localities: Italy; Florence, Pisa and Leghorn.

Description of imago (Pl.11, fig.10)

Wingspan 9–10mm. Head with vertex ochreous white, frons greyish white; antenna greyish ochreous annulated fuscous; labial palpus white, normally immaculate. Thorax whitish ochreous with fuscous median stripe, tegulae fuscous. Forewing dark fuscous with a purplish gloss; basal area paler, mixed whitish, especially towards dorsum; costa with numerous pale strigulae, whiter towards apex; elongate ante- and postmedian subdorsal dark spots; a dark streak in disc at three-quarters, generally interrupted by a white spot; an apical black dot, strongly margined inwardly by extension of costal white strigula nearest apex which is continued into terminal cilia at an obtuse angle; cilia grey with three complete fuscous lines. Hindwing grey, cilia paler. Genitalia, see figure 101, p. 287.

Similar species. P. finitimella (Zeller), *q.v.*

Life history

Ovum. Laid on the underside of a leaf of blackthorn (*Prunus spinosa*) or wild plum (*P. domestica*).

Larva. Head pale yellow-brown to brown with two cloudy and sometimes obsolescent dark spots. Body whitish yellow to yellow-green, gut darker green, blackish or reddish; prothoracic plate concolorous with body with four transversely placed black spots; legs yellow-green.

Feeding starts in the lower epidermis in a short gallery which is later expanded into a subrectangular or triangular blotch, often in an angle of veins. In the tissue-eating phase, the larva eats all the parenchyma through to the upper epidermis; there is little internal spinning and the broad, more or less white mine is only slightly arched and transparent if held up to the light, the frass, packed at one end, being clearly visible. This mine is nearly always easily distinguishable from that of *P. finitimella, q.v.* On leaving its mine, the larva folds the edge or tip of a leaf downwards and feeds therein. It may change its leaf several times. The larger folds often occupy a whole leaf which is spun into a pod, within which the larva consumes all the parenchyma starting from the tip which becomes blanched and distorted. Late July to September; *Parornix* larvae found on *Prunus* spp. in June and early July seem to be all *P. finitimella*.

Pupa. In an ochreous yellow cocoon spun under the narrowly folded edge of a leaf or in leaf-litter. September, overwintering until May.

Imago. Although hitherto described as bivoltine, it is probably entirely univoltine with an emergence period

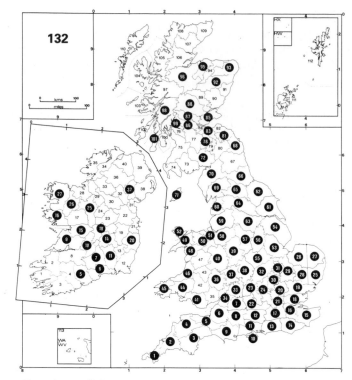

Parornix torquillella

extending from late May until July. The moths are active in evening sunshine and later come to light.

Distribution (Map 132)

Found in all situations where the foodplant occurs throughout the British Isles as far north as the Caledonian Canal; it is very common over most of its range but becomes more local in the north. Europe, not extending to Finland; Asia Minor.

CALLISTO Stephens

Callisto Stephens, 1834, *Ill.Br.Ent.* (Haust.) **4**: 276.

A genus containing several species in the Western Palaearctic Region and one in Japan. Two of them occur in Britain and one of them also in Ireland.

Imago. Antenna without pecten. Forewing with cell dilated in apical half; veins 2 (Cu_2) and 3 (Cu_1) coincident, 4 (M_3) present, 6 (M_1) and 7 (R_5) stalked. Hindwing with cell almost closed between veins 4 (M_3) and 5 (M_2). Other characters in general similar to those of *Parornix* Spuler.

Life history

There is no significant difference from *Parornix* in the early stages. The British species are univoltine. The adults are more robust-looking than *Parornix* species; the ground colour is dark brown and the pattern white or silvery.

CALLISTO DENTICULELLA (Thunberg)

Tinea denticulella Thunberg, 1794, *Diss.ent.* **7**: 97.
Gracillaria guttea Haworth, 1828, *Lepid.Br.*: 531.
Type locality: Sweden; Gedner.

Description of imago (Pl.11, fig.11)

Wingspan 10–12mm. Head with vertex deep ochreous mixed fuscous, frons paler ochreous; antenna deep ochreous fuscous with obscure darker annulations; labial palpus white. Thorax brown with two anterior and one posterior whitish spots. Forewing dark brown; costa with whitish subtriangular spots at one-third, one-half and three-quarters; dorsum with similar median and tornal spots; cilia brownish grey, whiter between two dark lines at apex, with a white bar on costa and a similar bar, not reaching base of cilia, in middle of termen, below which three dark lines extend to tornus. Hindwing brownish grey, cilia slightly paler.

Life history

Ovum. Laid on a leaf of apple (*Malus* spp.) including cultivars, generally on the upperside.

Larva. Head brown, sides darker. Body yellowish white or greenish white, gut dark green, reddish or blackish; prothoracic plate concolorous, with obscure darker spots. In the sap-feeding phase, the head is large, flat, triangular and pale yellow-brown; the body is glossy yellowish white, with the thoracic segments broader than the abdominal and all segments deeply incised.

Feeding begins in a fine epidermal gallery, generally on the upperside of a leaf, the frass being deposited in a slender red-brown line adhering to the cuticle. In the second instar the gallery is developed into a blotch situated

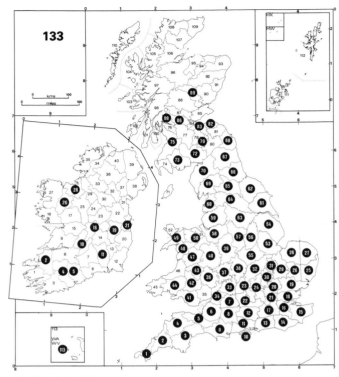

Callisto denticulella

between veins in which the frass, still adhering to the cuticle, is scattered, the blotch becoming conspicuously orange-brown. In the tissue-feeding phase, the larva seldom eats right through to the lower epidermis. The mine closely resembles that of *Phyllonorycter corylifoliella* (Hübner), *q.v.* (p. 327), but that is a double mine situated over a vein, not between veins, and as the result of more extensive internal spinning the cuticle tends to stretch under tension, not to tear as in *C. denticulella*. In certain years only, 1981 being the most recent example, many or even the majority of the mines in certain areas are on the underside of the leaf; possibly wet weather at the time of oviposition induces the female to select the drier lower surface. In underside mines, the raised cuticle is silvery white, almost completely lacking the rusty discoloration of upperside mines; the cause of this difference has not yet been studied (Hering, 1951: 219). After leaving the mine, the larva folds rather a long section of the leaf-edge downwards and feeds therein through to the upper epidermis which turns conspicuously brown. Two such folds are generally made. July to August.

Pupa. In a tough, oval, somewhat flattened ochreous brown cocoon, spun in leaf-litter. September, overwintering until the end of April.

Imago. Univoltine. May and June. Rests on tree-trunks and becomes active in the evening.

Distribution (Map 133)

Frequents gardens and orchards, and also all situations where there are crab-apples. Widespread and common throughout England and Wales; in Scotland it occurs as far north as Perthshire but is much more local. Widely distributed but not common in Ireland (Beirne, 1941: 124). Europe; north America.

CALLISTO COFFEELLA (Zetterstedt)

Ornix coffeella Zetterstedt, [1839], *Ins.Lapp.*: 1009.
Ornix interruptella Zetterstedt, [1839], *Ins.Lapp.*: 1009.
Type locality: Norway; Bjerkvik, Nordland.

Description of imago (Pl.11, fig.12)

Wingspan 10–12mm. Head with vertex blackish brown to ochreous, frons white; antenna dark brown with paler annulations; labial palpus white. Forewing glossy dark purplish brown with submetallic silver pattern consisting of a transversely elongate subbasal spot seldom reaching either costa or dorsum, a fascia at one-quarter angulate distad but sometimes broken into costal and dorsal spots, a large median strigula and four smaller strigulae distad on costa, a small discal spot, and scattered scales along dorsum from one-half to tornus which are sometimes grouped to form one or two small spots; a black apical dot; termen indented below apex; cilia purplish fuscous basally, grey distally, interrupted at indentation by a subtriangular white patch which is bisected by a fuscous bar. Hindwing purplish grey, cilia paler.
Similar species. C. denticulella (Thunberg) lacks the subbasal spot and has a white as opposed to a silvery pattern.

Life history

Not reared in Britain but vacated feeding places were found at the Scottish locality (see below) in September, 1984.
Ovum. Laid on the underside of a leaf of sallow (*Salix* spp.). Examples of the mine in the Hering Herbarium at BMNH are on mountain willow (*S. arbuscula*); those from Scotland on tea-leaved willow (*S. phylicifolia*).
Larva. Undescribed. Feeds at first in an epidermal gallery which is later developed into a tentiform blotch resembling that of a *Phyllonorycter*. After leaving its mine it folds part of a leaf downwards to form a pod. July to August.
Pupa. In a cocoon which is generally spun on a branch of the foodplant. August to May, but it is not yet known whether pupation takes place in the autumn or the larva overwinters in its cocoon.
Imago. Univoltine; the moth flies in June.

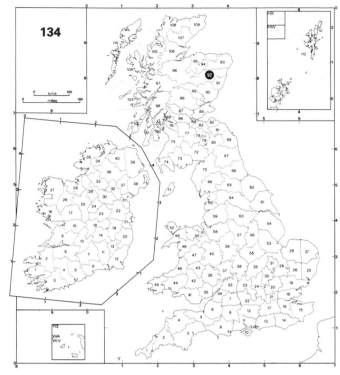

Callisto coffeella

Distribution (Map 134)

A montane species, known in Britain only from a single specimen found resting on a rock at about 500m (1600ft) near Braemar, Aberdeenshire (Palmer *et al.*, 1984). It is probably a long-established resident, having been hitherto overlooked owing to the inaccessability of its haunts and the rarity of its foodplants. The Alps; Fennoscandia.

ACROCERCOPS Wallengren

Acrocercops Wallengren, 1881, *Ent.Tidskr.* **2**: 95.

A very large genus of world-wide distribution, occurring especially in the southern hemisphere. Two species occur in Britain and a third was placed erroneously on the British list as a result of misidentification.

Imago. Head with vertex and frons smooth-scaled; antenna slightly longer than forewing, scape sometimes with tuft; labial palpus smooth, slender, ascending, segment 3 sometimes longer than segment 2, segment 2 sometimes tufted beneath. Fore- and midlegs smooth, hardly thickened with scales; hindleg slender, tibia with a row of bristly hairs above, tarsus slightly longer than tibia. Forewing with veins 1b (1A+2A) short, simple; 2 (Cu_2) often absent; 5 (M_2) and 6 (M_1) often connate; 7 (R_5) and 8 (R_4) usually stalked but 7 (R_5) sometimes absent; 11 (R_1) short, reduced or even absent. Hindwing half or less than half width of forewing, cilia very long, four to five times width of wing; cell open between veins 4 (M_3) and 5 (M_2), and 5 and 6 (M_1) stalked.

Life history

Larva. Has three sap-feeding instars like *Phyllonorycter* Hübner, herein differing from *Caloptilia* Hübner. The epidermal blotch is therefore relatively larger and enough parenchyma is exposed to satisfy the larva in the tissue-feeding phase; consequently growth is completed within the mine and no external spinning is needed. The trophi are such that the larva cannot enlarge the dimensions of the mine excavated in the sap-feeding phase and in this respect *Acrocercops* differs from *Aspilapteryx* Spuler, *Parectopa* Clemens and *Leucospilapteryx* Spuler, the other genera of the Gracillariinae in which larval growth is completed within the mine. Both the British species tend to be gregarious, with separate early galleries converging into a communal blotch. When full-fed, the larvae turn red.

Pupa. In a tough cocoon spun outside the mine. One British species overwinters in the pupal stage.

Imago. Both species are univoltine. Habits in general similar to those of *Caloptilia.* One species overwinters as an adult, often in thatch.

ACROCERCOPS IMPERIALELLA (Zeller)

Gracilaria imperialella Zeller, 1847, *Linn.ent.* **2**: 365.
Type locality: not stated.

Description of imago (Pl.11, fig.14)

Wingspan 7–8mm. Head leaden metallic, frons silver or white; antenna pale shining fuscous, apical one-sixth white; labial palpus golden brown or white. Thorax leaden metallic. Forewing orange-brown; extreme base of dorsum, an oblique fascia at one-quarter not reaching dorsum, and outward-oblique fasciae, which are often interrupted, at one-half, three-quarters and before apex, all silvery white more or less edged fuscous; cilia golden brown, containing a fuscous-edged, silvery white spot at apex. Hindwing brownish grey, cilia paler.

Similar species. *Parectopa ononidis* (Zeller), *q.v.*

Life history

Ovum. Laid on a leaf of common comfrey (*Symphytum officinale*) or lungwort (*Pulmonaria officinalis*), generally on the underside close to a rib.

Larva. Head transparent and colourless, or very pale brown. Body broadest in the thoracic segments, tapering in the abdominal; at first whitish, then very pale green and finally, when almost full-grown, bright crimson.

Feeding starts in a small spiral gallery in the lower epidermis which is soon developed into a blotch, absorbing the earlier working; the frass, still in linear arrangement even in the blotch, adheres to the cuticle. In the tissue-feeding phase, the larva eats irregularly through to the upper epidermis which turns brown; at this stage the mine is a large, squarish blotch, generally bounded by veins. There is little internal spinning and the mine remains almost flat and transparent; however, the larva or larvae, for there are sometimes two or three in the same mine, remain practically invisible until they turn scarlet. The frass is now scattered throughout the mine. Mines on the same foodplants with blackish discoloration of the upper epidermis are caused by Diptera. August to September, earlier in hot summers.

Pupa. In a flat, reddish cocoon spun in detritus or, occasionally, in the mine; in captivity, sometimes on tissue or polythene. August or September, overwintering until the end of April.

Imago. Univoltine. May and June. In the very hot summer of 1976, well-advanced mines were found in July and again in October, suggesting that it was bivoltine in that year (Emmet, 1977). The adult is seldom seen.

Distribution (Map 135)

Frequents fens and damp woodland. Occurs not uncommonly in the fens of Cambridgeshire and Huntingdon-

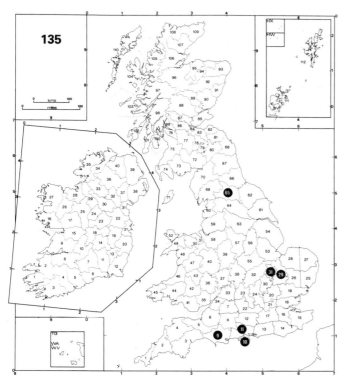

Acrocercops imperialella

shire; also recorded from Dorset, the Isle of Wight and the New Forest in Hampshire. There is an old record from north-west Yorkshire (Porritt, 1907). Central and southern Europe.

ACROCERCOPS HOFMANNIELLA (Schleich)

Gracilaria hofmanniella Schleich, 1867, *Stettin.ent.Ztg* **28**: 452.

Type locality: Germany; Stettin.

This species was formerly confused with *A. imperialella* (Zeller), and was redescribed by Stainton (1864: 194–203) under that name. Later he acknowledged his error (Stainton, 1868: 147; 1870: 12). Records of *A. hofmanniella* made before this correction almost certainly refer to *A. imperialella*, and since the differences between the two were recognized *A. hofmanniella* has not been recorded. There is, therefore, no evidence that it has occurred in Britain. The foodplant of *A. hofmanniella* is black pea (*Lathyrus niger*). The moth is recorded from central and south-eastern Europe.

ACROCERCOPS BRONGNIARDELLA (Fabricius)

Tinea brongniardella Fabricius, 1798, *Suppl.Ent.syst.*: 496.

Type locality: France.

Description of imago (Pl.11, fig.15)

Wingspan 8–10mm. Head ochreous; antenna ochreous, annulated fuscous; labial palpus ochreous, segment 3 black beneath, segment 2 strongly tufted. Thorax ochreous. Forewing ochreous mixed fuscous; an angulate fascia at one-quarter, outward-oblique streaks from costa to middle of disc at one-half, three-quarters and before apex, an elongate median dorsal spot and an elongate spot extending along termen to apex, all whitish irrorate fuscous and margined, especially anteriorly, fuscous; terminal cilia banded, counting from base, ochreous, fuscous, white, fuscous and grey, with a whitish or ochreous pencil emanating from apex and projecting beyond tips; two white spots in costal cilia; dorsal cilia grey. Hindwing and cilia grey.

Life history

Ovum. Laid on the upperside of a leaf of oak (*Quercus* spp.), including evergreen oak (*Q. ilex*) and introduced species.

Larva. Head pale yellowish brown, lobes darker. Body pale yellowish green; dorsum with an obsolescent chain of oval, greyish brown spots; prothoracic plate more or less colourless with obscure lateral and posterior grey-brown spots. When full-fed, the larva turns pinkish orange.

Feeding starts in a tortuous gallery in the upper epidermis with a central faint line of brown frass adhering to the cuticle. There are generally several mines in the same leaf and these converge to form a large blotch in which the larvae continue to feed in the epidermis, raising it slightly and causing it to take on a highly characteristic silvery hue. At this time they feed venter upwards, but after the change in mouth parts they turn over and eat down into the parenchyma, without enlarging the mine. The upper cuticle is lightly spun with silk which contracts, causing the mine to become inflated; the cuticle retains its silvery colour which at once separates the mine from the blotches made by the sawfly *Profenusa pygmaea* (Klug) in leaves of oak. The frass is scattered but collects mostly round the margins of the mine. The mine is not waterproof and tends to become waterlogged in heavy rain, apparently without detriment to the occupants. June, sometimes just feeding on into July.

Pupa. Pale yellow-brown; in a white cocoon similar in shape to those of *Caloptilia* Hübner but less glossy and translucent. The pupation site in the wild has not been observed but is probably leaf-litter. July.

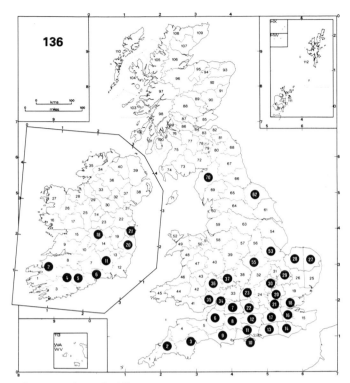

Acrocercops brongniardella

LEUCOSPILAPTERYX Spuler

Leucospilapteryx Spuler, 1910, *Schmett.Eur.* **2**: 408.

A Palaearctic genus of two species, one of which occurs in England.

Imago. Characters mostly as in *Acrocercops* Wallengren. Labial palpus loosely haired beneath, segment 3 as long as segment 2. Hindleg with a row of bristly hairs on tibia extending to first tarsal segment. Forewing with veins 2 (Cu$_2$) absent, 4 (M$_3$) and 5 (M$_2$) connate, 7 (R$_5$) absent. Hindwing with cilia three to four times width of wing; vein 1b (2A) absent.

Life history
Larva. Trophi adapted for enlarging the mine in the tissue-feeding phase as in *Aspilapteryx* Spuler and *Parectopa* Clemens, but whereas in those genera the larvae can change leaves, this is apparently not possible in *Leucospilapteryx*. As in *Acrocercops*, the larva turns red when fullfed. Growth is completed within the mine which is distended by internal spinning.

Pupa. In a tough cocoon outside the mine. Overwinters in the pupal stage.

Imago. Bivoltine.

LEUCOSPILAPTERYX OMISSELLA (Stainton)

Argyromiges omissella Stainton, 1848, *Zoologist* **6**: 2163.
Type locality: England; Charlton, Kent.

Description of imago (Pl.11, fig.16)
Wingspan 7–8mm. Head silvery grey; antenna pale fuscous, obscurely annulated darker; labial palpus white, rough-scaled beneath, segment 2 with apical and segment 3 with median fuscous band. Thorax silvery grey, irrorate fuscous, tegulae pale fuscous. Forewing pale fuscous, irrorate darker; an angulate fascia at one-quarter, outward-oblique fasciae at one-half and three-quarters, and a more direct fascia before apex white, edged fuscous, some or all fasciae often more or less interrupted in disc; an apical white spot extending into cilia, its centre bounded posteriorly by a short fuscous bar; remainder of cilia grey with an obscure fuscous line. Hindwing and cilia grey.

Life history
Ovum. Laid on the underside of a leaf of mugwort (*Artemisia vulgaris*).

Larva. Head pale yellowish brown. Body pale yellowish green, gut darker; prothoracic plate and small anal plate pale yellow-brown. When full-fed, the larva becomes orange-crimson.

Feeding starts in a long gallery in the lower epidermis

Imago. Univoltine, emerging at the end of July; in favourable weather there may be a small second generation in the autumn (Wakely, 1946). It overwinters in thatch, haystacks and evergreens and reappears in April and May.

Distribution (Map 136)
Frequents open woodland, especially where there are young trees. Very local, occurring in isolated colonies in England mainly south of a line from Lincolnshire to Herefordshire, but also in east Yorkshire and Cumbria; southwestern Wales; southern Ireland. Not recorded from Scotland. Palaearctic; very common throughout Europe, extending to north Africa, Asia Minor and eastern Asia.

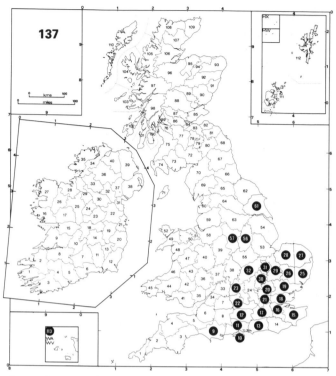

137

Leucospilapteryx omissella

which follows a vein or the margin of the leaf and leads into a blotch. When tissue-feeding begins, the larva can enlarge its mine which becomes conspicuously inflated as the result of internal spinning; the black frass tends to be massed in the centre. The upper epidermis becomes mottled with white where the parenchyma has been eaten away and turns first yellowish and finally purplish. July; August to September.

Pupa. Pale yellowish brown; in a dull, ochreous cocoon spun on leaf-litter. July to August; October, overwintering until April.

Imago. Bivoltine. May; August. Active in late afternoon.

Distribution (Map 137)

Frequents waste ground, downland and roadside verges where there is mugwort. Widespread and locally common in south-eastern England and also extending more sparsely northwards to south-eastern Yorkshire (Porritt, 1907: 273). Its incidence is puzzling, for in some areas it is extremely abundant, whereas in others, apparently equally suitable, it is absent. Central Europe.

Lithocolletinae

The subfamily Lithocolletinae is represented by only one genus in the British Isles. All members of the subfamily mine leaves or green bark, although species of different genera have different mining habits and biology.

PHYLLONORYCTER Hübner

Phyllonorycter Hübner, 1822, *Syst.-alph.Verz.*: 66.
Lithocolletis Hübner, [1825], *Verz.bekannt.Schmett.*: 423.

A very large genus of almost world-wide distribution, with 53 species found in Britain, of which 31 occur also in Ireland. Two species, *P. sagitella* (Bjerkander) and *P. staintoniella* (Nicelli), have only recently been added to the British list and it is possible that others remain to be discovered or will become established. Basic knowledge of the life-cycle of at least one species, *P. saportella* (Duponchel), has been discovered only recently, while aspects of the biology of some species are still unknown. In particular, the overwintering habits of the common species *P. messaniella* (Zeller) are not completely understood and the larval habits of the two bark-mining species, *P. scopariella* (Zeller) and *P. ulicicolella* (Stainton), are not fully described.

Imago. Head with vertex bearing a tuft of erect, rather long, rough scales, these being divided at their tips and frequently varying in colour from front to back of head; frons covered in appressed scales, nearly always white and showing some silvery or very pale golden sheen; antenna simple, nearly as long as forewing, scape short, slightly thickened and bearing a pecten on ventral side; labial palpus moderate, porrect, filiform, pointed; haustellum well developed, about twice length of labial palpus. Legs often with darker spots or markings on dorsal surface, those on hindleg sometimes aiding identification; fore-femur slightly thickened with scales; midtibia slightly thickened and bearing scales which project alongside a pair of short tibial spurs; hindfemur and hindtibia thickened, the latter more strongly and bearing two pairs of well-formed tibial spurs, hindtarsus smooth, slender and about one to one and a half times length of tibia. Forewing lanceolate, sharply pointed; veins 1c (1A) strongly represented, 2 (Cu_2) weak, simple, 5 (M_2) branching from median vein which meets radius to enclose large mesal cell, 7 (R_5), 8 (R_4), 9 (R_3) and 10 (R_2) short and branching from radial vein, latter weaker in basal third, 12 (Sc) weak and not reaching halfway; cilia long, particularly near tornus, there reaching width of wing. Hindwing three-quarters length of forewing, lanceolate; veins 1c (1A) short but distinct, 2 (Cu_2) weak, unbranched, 5 (M_2) and 7 (Rs) well developed and unbranched but anastomosed for proximal two-thirds in some species, 8 ($Sc+R_1$) weak (figure 102); cilia three times width of wing.

The adult is relatively constant in size, varying in wingspan from 6 to 9mm with only a very few species being outside this range. In almost all species the forewing markings are very attractive and striking, often showing clear white patterns contrasting sharply with metallic, brassy or orange-brown ground coloration. The pattern of the forewing generally follows one of three basic types shown in figure 103 which have been labelled with the terms used in the descriptions of individual species.

In figure 103a the forewing shows distinct white fasciae which usually have a smooth outline, as found in *P. kleemannella* (Fabricius), and in these species strigulae which occur distal to the fasciae are numbered as shown in

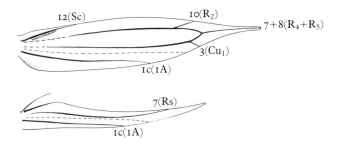

Figure 102 *Phyllonorycter harrisella* (Linnaeus), wing venation

figure 103a,b. Figure 103c shows the most commonly encountered wing-pattern, with the forewing bearing several white costal and dorsal strigulae, as found in *P. cydoniella* ([Denis & Schiffermüller]); this form usually possesses a pronounced basal streak and often a tiny dorsal patch. Figure 103b exemplifies species exhibiting an intermediate type of marking, with the first fascia appearing as if it is formed from two opposite strigulae which have merged. Indeed, in some variable species such as *P. cavella* (Zeller) both types illustrated in figure 103b,c may be found as normal extremes of the forewing pattern range. When this degree of diversity occurs within a species, the different forms have been treated differently in the key to species (imagines). In a number of species the white of the strigulae is so expanded as to give the forewing a white ground colour with dark strigulae superimposed; this colour form is found in the *Acer*-feeding and several of the *Quercus*-feeding species. In many species, the inner and less often the outer margins of the strigulae and fasciae have dark brown or black edging; the extent, form and

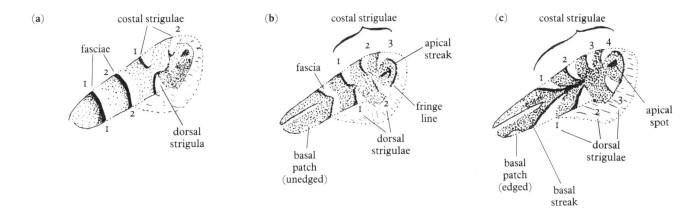

Figure 103 *Phyllonorycter* spp., forewing patterns

intensity of this edging is frequently of use in identification. The basal streak and dorsal patch, if present, may also show dark edging. A dark apical spot or streak is present in many species. The hindwing is devoid of marking, although areas of cilia may exhibit small changes in their shade of coloration.

In all species the forewing bears long cilia, giving the apex a rounded appearance, and the base of the cilia usually exhibits a dark thickening known as the fringe line. The apical cilia of several species bear thicker hairs or areas of thickening which form an obvious line or hook. As described by Le Marchand (1936), Jacobs (1945) and Bradley *et al.* (1969), this character may be one of three different types as shown in figure 104 and it has been used in the key to species (imagines) as an important distinguishing feature. Type 1, as found in *P. roboris* (Zeller), arises as a distinct line from the apical spot and extends directly out through the apical cilia. Type 2 may be of two forms, both with the thickening at the distal end of the apical cilia; in type 2a, found in *P. saportella*, the line is very strong and extends as a distinct projection out of the line of the cilia; type 2b lines are all weak as in *P. quercifoliella* (Zeller) or *P. quinnata* (Geoffroy) and do not noticeably interrupt the fringe line. The type 3 hook as found in *P. distentella* (Zeller) is very distinctive, arising from the distal edge of the last costal strigula, extending at a tangent to the fringe line and projecting beyond the other cilia.

According to Meyrick (1928), *Phyllonorycter* is a recent genus with groups of newly separated species showing wing-pattern and morphological similarities which make identification extremely difficult. For this reason, descriptions and figures of the genitalia are given for those species where identification cannot be made by external characters.

Ovum. Oval, *c.*4mm in length and of a flattened domeshape. The surface is reticulated and soft in texture; subsequent to hatching the egg-shell flattens against the leaf and is later frequently lost from its surface, having been dislodged either through the movement of the larva mining beneath or through the effect of weather. The larva eats through the lower surface of the egg and begins to mine directly into the leaf beneath. According to Pottinger & LeRoux (1971), newly hatched *P. blancardella* (Fabricius) larvae place up to 15 small faecal pellets in the empty egg-shell, arranging them in a semicircle around its inner edge opposite the emergence hole. It is not known whether this habit is shared by other *Phyllonorycter* species or whether this is the result of feeding while the anal end is still within the egg. The position on the leaf where the egg is laid is relatively constant for some species but the direction taken by the larva in developing its mine is variable.

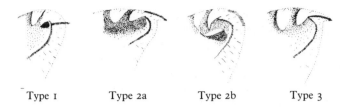

Type 1   Type 2a   Type 2b   Type 3

Figure 104 *Phyllonorycter* spp., types of forewing tips
(type 1) *P. roboris* (Zeller) (type 2a) *P. saportella* (Duponchel)
(type 2b) *P. quinnata* (Geoffroy) (type 3) *P. distentella* (Zeller)

Larva. There are only minor differences between species, and these only in the later instars. The colour ranges from pale greenish to whitish or pale yellow, and there may be a yellow to orange spot beneath the cuticle on the dorsal surface of abdominal segment 6, probably caused by a large area of fat body.

A full description of the sap-feeding to tissue-feeding hypermetamorphosis which the larvae undergo is given in the introduction to the family (pp. 244–247). In *Phyllonorycter* there are three sap-feeding and two tissue-feeding instars. The sap-feeding instars are dorsoventrally flattened with the thoracic segments distinctly broader than the abdominal. The first instar tends to have fewer setae and spiracles than the later instars and in all three sap-feeding instars the flattened appearance allows the gut to be seen very clearly. None of the first three instars has any developed legs, although vestiges of these may be seen in the third instar. Antennae are well developed.

Fourth and fifth instar larvae are cylindrical and have three pairs of thoracic legs and four pairs of prolegs on abdominal segments 3–5 and 10. The head is rounded and orientated for surface grazing; it often shows conspicuously darker markings than the rest of the larva. The antennae are relatively much reduced in size from the early instars. In these larger larvae a spinneret is present and setae are well developed, these being most pronounced in the fifth instar.

The first instar larva commences mining by cutting through the mesophyll cells underlying the epidermis, then feeds on the sap that is released by this process. It excavates a long narrow mine, usually alongside a tougher leaf vein, and at this stage the mine is very difficult to see. Gallery mining is continued by the second instar larva which, like the first, is almost devoid of setae and therefore uses the narrow mine as a means to anchor itself during the cell-cutting operation; presumably the thickened thoracic region helps considerably in this process, acting as a wedge against the sides of the mine. The third instar larva excavates a blotch, usually embracing the earlier gallery, and now, without the confines of a narrow mine, its small setae secure it during feeding. By this stage the mine is clearly visible as a flat, silvery blotch. In the fourth and fifth instars the larva feeds on the mesophyll cells within the mined area without further enlargement of the mine. These cells are eaten out in small islets down to the opposite epidermis of the leaf, the depressions thus formed bounded by small veins which are left uneaten as a raised lattice and lead to a mottled appearance on the opposite side of the leaf. In order to accommodate the larger size of larva, the mine is increased in depth by the fourth instar larva spinning silken threads across its inner surface. As these dry they contract, causing the epidermis to fold in longitudinal creases, thus shortening it and opening the mine into a tube or tentiform pucker or chamber. This folding and the pattern of the creases are characteristic for most species. In some species, such as the honeysuckle-feeders, the degree of folding is extreme and a tenanted leaf may be strongly distorted. The layers of silk confer considerable strength to the mine, providing protection not only from severe weather but also no doubt from certain predators. Within the mine small black faecal pellets are deposited; these may be scattered generally around the mine chamber, spun with silken threads into a neat pile, often with the larval exuviae, or incorporated into the cocoon. The deposition of faecal pellets is characteristic for each species although in several the summer and winter generations of the same species have different habits.

Most *Phyllonorycter* species overwinter as pupae, but some, such as *P. cerasicolella* (Herrich-Schäffer), pupate in the spring without further feeding. Other species such as *P. junoniella* (Zeller) overwinter as larvae which feed during warm periods and complete their growth in the spring.

Pupa. Some of the information in this section has been provided by M. R. Wilson who has made a comparative study of the pupae of *Phyllonorycter*.

Elongate, cylindrical, usually narrower in the last five segments, varying from light to dark brown. Abdominal segments 4–7 free in male, 4–6 in female, enabling the pupa to wriggle actively when disturbed; anterior of pupa furnished with a sharp keel or beak, enabling it to rupture the cocoon and mine before the emergence of the adult; wings long, extending to abdominal segment 6 and unattached at their distal ends; thoracic legs and antennae extending beyond wingtips to abdominal segment 7 and unattached to abdomen at extremities. Cremaster, dorsal and ventral spines on abdominal segments and setae differ between species and are of diagnostic value, see introduction to the key to pupae (pp. 302–305).

In all species pupation takes place within the mine and in the majority of species this stage is passed within the protection of some form of cocoon or pupation chamber. The colour, size, shape and position of the cocoon within the mine is usually constant for each species. In some species such as *P. distentella* the pupa is very loosely held in a fine net of silken threads, and a very few appear to have no obvious protection for the pupa. Bivoltine species show a tendency for the summer cocoons to be less robust than those formed by the autumn-feeding larvae to resist the harsher conditions of winter.

Behaviour. The adult usually emerges in the early morning and may frequently be found later resting on a tree-trunk, fence or other upright object near where the larva was feeding. This is particularly so in the spring emergence where the adults are coming from leaves which have overwintered on the ground. The adult is a crepuscular flier and may be found commonly on the wing at sunset. Most species are highly phagospecific, being found on only one or a small number of very closely related foodplants; the most polyphagous species is *P. corylifoliella* (Hübner) which feeds on a range of trees and shrubs, but since it has a very distinctive upperside mine it can be easily identified. The majority are associated with trees or shrubs, but a number such as *P. nigrescentella* (Logan) and *P. scabiosella* (Douglas) are to be found on low-growing herbaceous plants. Two species, *P. ulicicolella* and *P. scopariella*, feed on the fresh green bark of gorse (*Ulex* spp.) and broom (*Sarothamnus scoparius*) respectively. Some species such as *P. lautella* (Zeller) show a preference for young seedling trees, while others such as *P. saportella* feed on leaves situated high on the tree. With very rare exceptions, each species of leaf-feeder mines one surface of the leaf only, the underside in the majority of cases.

Collecting and rearing. Collecting adults is best carried out by netting them late in the day or at dusk, or by tubing them as they rest on tree-trunks. A few come to light but they are easily overlooked in a moth-trap containing many larger moths. The best method to secure a series is to rear them from collected mines. In the summer this is an easy task since leaves with fully developed mines contain last or penultimate instar larvae which, if kept moist and indoors in the warm, will feed up, pupate and emerge as adults in the space of one to three weeks. Autumn-collected mines need more careful treatment but have the advantage of greater abundance. In many species the winter pupal stage does not need to experience a cold spell before breaking diapause, and such mines, if brought into suitably warm conditions, can be forced to produce adults in a few days or weeks. However, better results are obtained if they are left out of doors for at least part of the winter. Many methods have been described for overwintering such mines, but in the experience of the author any method which allows excessive compression or disturbance of the leaves causes higher mortality. The most effective procedures permit natural fluctuations of temperature and humidity around the pupae in their mines. This can be achieved by storing the leaves in old stockings hung in an outhouse or shed, thus allowing good circulation of air around the leaves. Some entomologists prefer to lay the stockings on the ground, suitably anchored against wind dispersal, and consider that a covering of snow leads to better results. Protection against physical damage can easily be provided by housing the leaves in a disposable plastic cup or similar container within the stocking. Leaves stored in this way can be brought indoors in January or February and, if kept suitably moist, adult moths should start to emerge within two weeks or so. A high degree of parasitism is frequently experienced with this genus and therefore only a relatively low percentage of the mines will yield moths.

Key to species (imagines) of the genus *Phyllonorycter*

NOTES
1. Some species key out in more than one couplet.
2. For an explanation of the terms used, see figures 103,104, pp. 295,296.

1	Ground colour of forewing white or light grey 2
–	Ground colour of forewing not white or light grey 11
2(1)	Forewing with some form of apical hook in cilia 3
–	Forewing without apical hook in cilia 7
3(2)	Apical hook of type 1; base of forewing with broad oblique rich tan fascia from one-fifth costa to one-third dorsum .... *roboris* (p. 307)
–	Apical hook of type 2 4
4(3)	Forewing with black-edged basal streak .... *heegeriella* (p. 308)
–	Forewing with no such basal streak 5
5(4)	Pairs of strigulae joining or almost joining to form two or more angled fasciae .... *saportella* (p. 309)
–	Strigulae not forming two or more fasciae 6
6(5)	Forewing with four costal strigulae, all white, variously blended with light brown and edged with blackish brown .... *tenerella* (p. 308)
–	Forewing with three costal strigulae, the first visible only as the blackish edge .... *harrisella* (p. 306)
7(2)	Forewing with pairs of strigulae forming distinct fasciae or chevrons; ground colour clear white 8
–	Forewing with strigulae diffuse, not in distinct pairs and not forming fasciae; ground colour sprinkled with dark scales 10
8(7)	First distinct pair of strigulae meeting to form a sharply angled chevron, usually extending outwardly as a streak to meet the next pair of strigulae ...... *geniculella* (p. 357)
–	First distinct fascia not connected to second fascia 9
9(8)	First distinct fascia with inside edge forming an acute or occasionally a right angle, and outside edge mostly forming a right angle .... *sylvella* (p. 356)
–	First distinct fascia with inside edge forming an obtuse angle or smooth curve and outside edge smoothly curved .... *platanoidella* (p. 356)
10(7)	Forewing with four or five fuscous costal strigulae; fifth costal alone reaching or nearly reaching apical streak .... *comparella* (p. 359)
–	Forewing with five or six fuscous costal strigulae; fifth and sixth costal both reaching or nearly reaching apical streak .... *sagitella* (p. 359)
11(1)	Forewing showing at least one distinct transverse fascia, this being irregular, smoothly curved or angled, very occasionally barely broken in the middle 12

–	Forewing with no distinct transverse fascia; opposite strigulae when clearly visible may nearly join to form acute-angled chevrons 28
12(11)	Thorax metallic 13
–	Thorax not metallic 16
13(12)	Thorax silvery or leaden .... *schreberella* (p. 346)
–	Thorax brassy or golden 14
14(13)	Thorax dark brassy brown, forewing dark coppery brown; a short, sometimes obscure, dark-edged silvery basal streak .... *stettinensis* (p. 351)
–	Thorax light golden or brassy brown; basal streak absent 15
15(14)	Wingspan 9–10mm; first pair of strigulae beyond two fasciae, of equal size, smoothly curved and meeting to form a chevron; white markings with silvery sheen and very finely edged dark brown inwardly .... *froelichiella*(p. 352)
–	Wingspan 8mm; first pair of strigulae beyond two fasciae not of equal size and not meeting; white markings with brassy sheen and broadly edged dark brown inwardly .... *kleemannella* (p. 353)
16(12)	Thorax pale golden brown with white central band; first pair of strigulae very occasionally forming an acute-angled fascia .... *scopariella* (p. 334)
-	Thorax dark golden or orange-brown 17
17(16)	Forewing with at least two distinct, smooth or angulate transverse fasciae 18
–	Forewing with only one transverse fascia 22
18(17)	Second fascia slightly irregular or sinuate 19
–	Second fascia smoothly curved or straight 21
19(18)	Forewing without basal streak; first two pairs of strigulae strongly sinuate and irregular, broadly bordered blackish inwardly .... *trifasciella* (p. 355)
–	Forewing with short basal streak 20
20(19)	Forewing with first fascia straight or slightly outwardly curved; basal streak short, straight and not reaching first fascia; all fasciae and strigulae sharply edged with black; ground colour shining golden brown .... *scabiosella* (p. 348)
–	Forewing with first fascia irregular and varying in width; basal streak sinuate, reaching first fascia; fascia and strigulae only lightly edged with brown; ground colour light brown .... *emberizaepenella* (p. 348)
21(18)	Fasciae clearly edged inwardly with dark brown and outwardly with white fading smoothly into orange-brown ground colour; first pair of strigulae uniting to form a third black-edged fascia or chevron with another pair of strigulae beyond .... *nicellii* (p. 353)

–	Fasciae without clear dark edging, and outwardly with white grading rather coarsely through orange-brown into dark brown colour; first pair of strigulae barely touching and with a third white streak connecting their junction to the costa .... *tristrigella* (p. 350)
22(17)	Forewing shining orange-brown 23
–	Forewing golden, ochreous or light sienna brown 24
23(22)	One dorsal and one or two costal strigulae beyond fascia; fringe line uninterrupted by white from dorsal strigula .... *nigrescentella* (p. 343)
–	Two dorsal and three costal strigulae beyond fascia; white of second dorsal extending through fringe line .... *insignitella* (p. 344)
24(22)	Wingspan less than 6.5mm; forewing shining golden brown with black pencil of cilia on outer edge of fourth costal strigula opposite apical dot ..... *anderidae* (p. 342)
–	Wingspan greater than 6.5mm 25
25(24)	Thorax and tegulae with no trace of white; strigulae brassy white; basal streak narrow, black-edged on costal side .... *ulmifoliella* (p. 347)
–	Thorax with areas of white scales 26
26(25)	Thorax with white on anterior edge and sides but with no white central line; fascia and all strigulae broad and shining white, usually clearly edged with black; basal streak short, broad and unedged ...... *spinolella* (p. 332)
–	Thorax with white central line 27
27(26)	Strong central white line on thorax; tegulae white; forewing with broad dorsal patch; basal streak to one-third, finely black-edged on costal side ....*cavella* (p. 333)
–	Very fine central line of white or pale gold on thorax; anterior edge of thorax and posterior edge of tegulae also white or pale gold; forewing with thin, slightly sinuate basal streak; dorsal patch white or pale gold and only faintly visible .... *salicicolella* (p. 330)
28(11)	Thorax with lighter median line 29
–	Thorax without lighter median line 46
29(28)	Forewing showing darker hook in apical cilia 30
–	Forewing without darker hook in apical cilia 32
30(29)	Apical hook strong and of type 3; four costal and two dorsal strigulae; ground colour of forewing golden straw .... *distentella* (p. 341)
–	Apical hook weak and of type 2; four costal and three dorsal strigulae; ground colour not golden straw 31
31(30)	Ground colour of forewing pale golden brown; thorax pale golden brown; first costal strigula extended along costa towards base .... *quinnata* (p. 338)
–	Ground colour of forewing light sienna brown; thorax pale orange-brown; first costal strigula not extended along costa towards base .... *coryli* (p. 337)

32(29) Pale yellowish gold median line on thorax; forewing pale ochreous brown; basal streak sinuate; all strigulae pale whitish or golden .... *viminiella* (p. 328)
– White median line on thorax 33

33(32) Forewing with one costal strigula; ground colour rich matt chestnut-brown; basal streak long, fine and angulated, reaching two-fifths to one-half ....
.... *corylifoliella* (p. 327)
– Forewing with more than one costal strigula 34

34(33) Forewing with five costal strigulae 35
– Forewing with four costal strigulae 37

35(34) Forewing with three dorsal strigulae; first costal strigula reaching from base to one-quarter; ground colour shining golden or coppery brown; basal streak to one-third, finely dark-edged above and partly below with distal end angled slightly towards costa ..... *quinqueguttella* (p. 342)
– Forewing with four dorsal strigulae; first costal strigula at about two-fifths 36

36(35) Ground colour of forewing dark golden brown; basal streak pale, unedged and with distal end angled slightly towards costa .... *scopariella* (p. 334)
– Ground colour of forewing dark shining ochreous brown; basal streak strong, with costal margin straight and dark-edged, dorsal margin unedged and slightly sinuate .... *cavella* f. *milleri* (p. 333)

37(34) Basal streak straight with at least one side edged with black 38
– Basal streak not straight or with no strong black edging 41

38(37) Basal streak edged above and below with black; ground colour golden saffron brown; thorax dark orange-brown; distinct, black-edged dorsal spot below basal streak; white of second dorsal strigula extending through fringe line .... *insignitella* (p. 344)
– Basal streak black-edged on costal side only; basal patch absent or represented by only a few white scales unedged by black 39

39(38) First two dorsal strigulae strongly black-edged inwardly and for short distance round tips; fringe line strong; basal streak edged on costal side and round tip (occasionally linked to first dorsal); strigulae silvery white; ground colour golden ochreous .... *lantanella* (p. 326)
– First two dorsal strigulae weakly black-edged inwardly only; fringe line not strong; costal side of basal streak only faintly edged 40

40(39) Basal streak finely edged black on costal side; ground colour of forewing shining golden ochreous brown; head tuft mostly whitish; apical spot indistinct ....
.... *cavella* (p. 333)
– Basal streak almost unedged, with faint black line on distal end of costal edge; ground colour of forewing varying from sienna brown to greyish brown; head tuft pale yellowish brown; apical spot black and distinct ......
.... *maestingella* (p. 336)

41(37) Strigulae not dark-edged 42
– Strigulae dark-edged 43

42(41) Ground colour pale coppery or golden brown; pattern white .... *ulicicolella* (p. 334)
– Ground colour reddish brown; pattern with ochreous tinge .... *staintoniella* (p. 335)

43(41) First pair of strigulae not clearly defined, weakly edged and nearly meeting to form angled fascia; basal streak unedged .... *salicicolella* (p. 330)
– First pair of strigulae clearly defined and extended towards apex but not meeting; basal streak often showing some fine edging 44

44(43) Ground colour pale ochreous brown; dark edging to strigulae indistinct; only first pair of strigulae and basal streak clearly defined .... *dubitella* (p. 331)
– Ground colour dark orange; dark edging to strigulae distinct; all strigulae and basal streak well defined 45

45(44) Male genitalia with subapical spine surrounded by short stiff bristles and situated on a flat part of the valva; female genitalia with ductus bursae extending beyond anterior end of seventh segment (figure 108c,d, p. 324).
.... *cerasicolella* (p. 325)
– Male genitalia with subapical spine not surrounded by short stiff bristles and situated on a concave part of the valva; female genitalia with ductus bursae turning before anterior end of seventh segment (figure 108a,b, p. 324).
.... *spinicolella* (p. 322)

46(28) Forewing with darker hook in apical cilia 47
– Forewing without darker hook in apical cilia 50

47(46) Apical hook strong 48
– Apical hook weak 49

48(47) Apical hook of type 1; ground colour of forewing golden brown; basal streak to one-third; four costal and three dorsal strigulae .... *ulmifoliella* (p. 347)
– Apical hook of type 3; ground colour of forewing burnt umber; no basal streak; three costal and three dorsal strigulae .... *muelleriella* (p. 313)

49(47) Weak apical hook of type 2b; basal streak long, reaching to three-fifths and slightly curved towards costa; four costal and three dorsal strigulae; second dorsal strigula not reaching apex of basal streak ....
.... *quercifoliella* (p. 311)
– Very weak apical hook of type 2b; basal streak reaching to one-half; four costal and four dorsal strigulae; second dorsal strigula extending beyond apex of basal streak ....
.... *messaniella* (p. 312)

50(46) Thorax leaden metallic; ground colour of forewing bright coppery brown; three costal and three dorsal strigulae; basal streak and dorsal spot both strongly black-edged .... *lautella* (p. 345)
– Thorax not leaden metallic; ground colour of forewing not coppery brown; forewing with more or fewer than three costal strigulae 51

51(50) Forewing with five costal and three dorsal strigulae, outer ones often faint; ground colour golden brown ......
.... *scopariella* (p. 334)
– Forewing with four costal and three dorsal strigulae; ground colour not golden brown 52

52(51) Forewing with outer strigulae not clearly defined; basal streak indistinct, sinuate and unedged 53
– Forewing with strigulae clearly defined; basal streak distinct, usually straight and with some edging 54

53(52) Ground colour pale coppery or golden brown; pattern white .... *ulicicolella* (p. 334)
– Ground colour reddish brown; pattern with ochreous tinge .... *staintoniella* (p. 335)

54(52) Dorsal patch extended towards and often reaching basal streak, latter often reaching first pair of strigulae; strigulae generally not edged with darker scales ....
.... *viminetorum* (p. 329)
– Dorsal patch small if present, sometimes dark-edged; strigulae more or less clearly edged inwardly with black scales 55

55(54) Basal streak with costal edge arched towards costa and with middle section wider; ground colour dull light grey-brown to sienna brown, often mixed with white near base of wing .... *rajella* (p. 340)
– Basal streak narrow, straight or slightly sinuate; ground colour orange or golden brown 56

56(55) Basal streak and large dorsal patch clearly edged in black; first costal strigula blunt-ended and often curved over to form a hook; apical spot very distinct ....
.... *strigulatella* (p. 339)
– Basal streak not fully dark-edged along both sides; dorsal patch unedged if present; first costal strigula pointed and smoothly curved or wedge-shaped 57

57(56) Third pair of strigulae forming an almost unbroken inwardly curved fascia across wing; strigulae very lightly edged in black, ground colour light orange ....
.... *junoniella* (p. 322)
– Third pair of strigulae opposite, but with apices barely touching and not forming a distinct fascia; strigulae usually showing distinct dark edging; ground colour brownish orange 58

58(57) First dorsal strigula very smoothly sickle-shaped and ending in a fine point; first costal strigula wedge-shaped; basal streak fine, pointed and finely dark-edged on costal side and round tip .... *lantanella* (p. 326)
– First dorsal strigula not smoothly curved nor sharply pointed; first costal strigula oblique, usually pointed and angled towards apex 59

59(58) Wingspan greater than 8mm; ground colour shining coppery brown; first dorsal strigula very large, often sinuously curved and strongly black-edged; basal streak narrow, distal end black-edged on costal side 60
– Wingspan 8mm or less; first dorsal strigula moderate in size, more roughly outlined in black or unedged 61

60(59) White of strigulae silvery; apical streak represented by fine line of black scales surrounded by an area of coppery scales; hindtarsus often showing faint darker markings; male genitalia with costa of right valva much longer and stouter than left and bearing sharply hooked spine; female genitalia with sterigma strongly arched between apophyses anteriores but not forming a plate (figure 107c,d, p. 321) .... *cydoniella* (p. 320)
– Strigulae shining white; apical streak represented more by an elongate patch of black scales surrounded by a few scattered dark scales; hindtarsus not showing darker markings; male genitalia with right costa of valva thin, only slightly larger than left and bearing short straight spine; female genitalia with sterigma broad between two apophyses and arched up to form a square-ended plate (figure 106c,d, p. 316) .... *sorbi* (p. 315)

61(59) Strigulae very weakly dark-edged or unedged; male genitalia with right costa of valva much longer than left and bearing strongly hooked spine; female genitalia with sterigma smoothly arched between the two apophyses (figure 106a,b, p. 316) .... *oxyacanthae* (p. 314)
– Strigulae clearly dark-edged; male genitalia with right costa of valva longer than left and bearing straight or only slightly curved spine; female genitalia with sterigma not smoothly arched between the two apophyses (figures 106,107, pp. 316,321) 62

62(61) Basal streak broad, usually unedged and occasionally connected to first dorsal strigula; black edging to strigulae thick and not sharp-edged; ground colour with scattered dark scales; male genitalia with right costa long and narrow, terminal spine long and straight; female genitalia with the region between the two apophyses extending in a straight line (figure 107a,b, p. 321) .... *blancardella* (p. 319)

– Basal streak narrow, often finely black-edged on costal side; black edging to strigulae fine and sharply defined; ground colour usually clear; male genitalia with right costa short and stout, terminal spine long and slightly curved; female genitalia with region between the two apophyses containing an ovoid disc (figure 106e,f, p. 316) .... *mespilella* (p. 317)

Key to pupae of the genus *Phyllonorycter*

The main features of the pupae of *Phyllonorycter* are described in the introduction to the genus. The only characters dealt with here are those which are most useful in the identification of species, and are found on the dorsal and lateral surfaces of the abdominal segments (figure 105).

Cremaster. The form of the cremastral hooks varies considerably within the genus. The most common type has two pairs of hooks of approximately equal length, the outer pair stouter than the inner pair. Reduction of the outer pair frequently occurs, with this pair becoming smaller than the inner pair or being entirely absent. In some species both pairs are reduced in size. In these cases additional structures and spines may be present, possibly to provide effective anchorage in the mine.

Dorsal surface spines. The dorsal surface of the abdominal segments may be covered by small, posteriorly directed spines. In some species these are present extensively on each segment; in others they are reduced. Usually the spines towards the anterior margin are larger and they are often arranged in a distinct row or series of rows; in some cases they are reduced in number but greatly increased in size. Spines may either become progressively smaller towards the hind margin, or may become enlarged again to form a posterior row or rows; in this case there may be considerably smaller spines between the anterior and posterior rows. A few *Quercus*-feeding species, notably *Phyllonorycter muelleriella* (Zeller) and *P. messaniella* (Zeller), possess very large curved spines on abdominal segments 1–3 at the lateral margins (figure 105u,v). Smaller curved spines in the same position are found in *P. quercifoliella* (Zeller) (figure 105w).

Setae. The length of the setae on the dorsal surface of the abdomen is useful for separation of some species. It has been used here mostly to supplement other characters but in some cases serves to separate very similar species. Setae have been categorized as either 'short' or 'long'; this difference is most easily assessed by reference to the length of the lateral setae against that of the median pair on abdominal segments 2–4. In those species with 'short' setae the lengths are approximately the same (figure 105a,d,e); in those with 'long' setae the lateral setae are usually considerably longer than the median pair (figure 105b,f).

Preparation of pupae for examination. The pupal exuviae are extruded through the mine before the emergence of the adult. They may remain attached for some time and, even when the anterior portion has been damaged, the abdominal segments may remain intact and enable identification to be made.

Figure 105 (see facing page)
All dorsal views except where stated. Scales approximate.

(**A**) *Phyllonorycter messaniella*, whole pupa, lateral view (×15)
Abdominal segment 2 to show lengths of setae (×80)
(**a**) *tenerella* (short setae) (**b**) *quinnata* (long setae)
Cremaster and dorsal spines, showing typical appearance of 'rosaceous' group (×35)
(**c**) *oxyacanthae*
Dorsal spines of abdominal segment 2 (×80)
(**d**) *cavella* (distinct anterior and posterior rows)
(**e**) *harrisella* (anterior row distinct and with sparse spines)
(**f**) *ulmifoliella* (anterior rows only distinct)
Cremaster structure (×100)
(**g**) *mespilella* (two pairs, more or less equal in length – the most common form)
(**h**) *lantanella* (one pair, long)
(**i**) *comparella* (two pairs, very reduced)
(**j**) *emberizaepenella* (one pair, greatly reduced)
(**k**) *quinqueguttella* (two pairs, processes reduced in size)
(**l**) *spinolella* (**m**) *salicicolella* (one pair, processes illustrated to show differences between these two species)
(**n**) *harrisella* (one pair)
(**o**) *distentella* (two pairs, long)
(**p**) *sylvella* (two pairs, very long processes)
(**q**) *muelleriella* (two pairs, median pair reduced)
(**r**) *quercifoliella* (two pairs, one large but median greatly reduced)
(**s**) *scopariella*, lateral view (two pairs, upturned lateral pair longer than median pair)
(**t**) *trifasciella* (two pairs, reduced; elongated final segment; extensions laterally segment 7)
Large spines in intersegmental area of segments 1–3 (×140)
(**u**) *muelleriella* (**v**) *messaniella* (**w**) *quercifoliella*
Processes in intersegmental area (×100)
(**x**) *saportella* (**y**) *heegeriella*
Ventral view of segment 8 to show ventral spines (×100)
(**z**) *ulmifoliella*

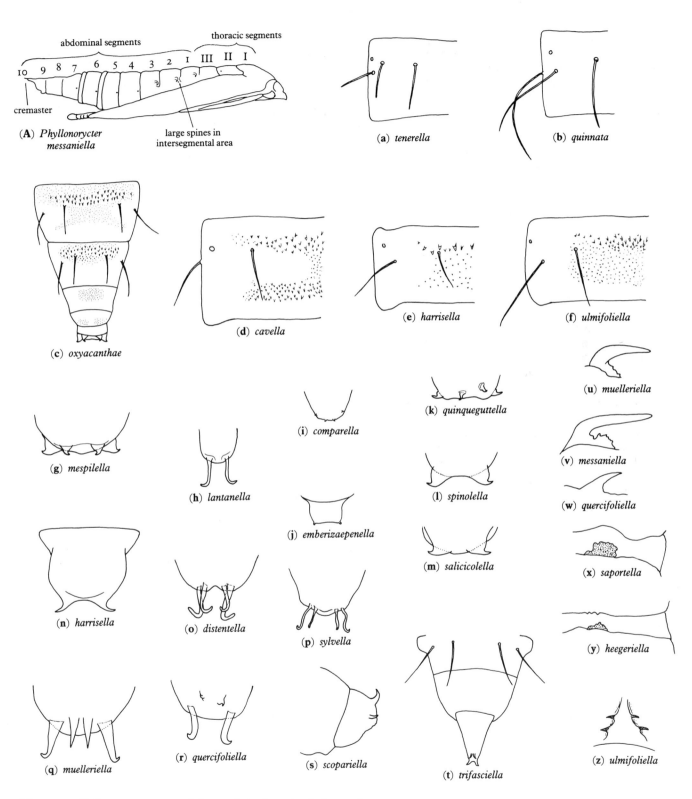

Figure 105 *Phyllonorycter* spp., pupal characteristics (legend on facing page)

The pupal exuviae should be carefully extricated from the mine, so as not to damage the cremaster. After some experience they may be examined dry under a good binocular microscope, both incident and transmitted light being useful. However, the simplest and perhaps best method is to mount them flat on a glass slide, when they may then be examined under the low power of a compound microscope. The exuviae should be warmed in 70 per cent alcohol for some minutes (in a tube heater or water bath) to soften the cuticle and help dispel trapped air bubbles. For temporary mounts, they may be carefully manipulated in a drop of glycerine to extend the abdominal segments, remove any attached debris, and to push out any trapped air bubbles. They may then be flattened under a glass cover-slip in glycerine with the dorsal surface uppermost. Permanent mounts may be made on slides using Euparal or Canada balsam as for genitalia preparations (*MBGBI* **I**: 129–131). Mounts may be made showing the lateral view, but this will obscure the cremaster. Dried pupal exuviae may be stored in gelatine capsules which should be labelled and pinned with the moths.

Use of the key. Since the majority of species are confined to a narrow range of host plants or are entirely monophagous, it was not thought appropriate to produce one large key attempting to cover all species. This would, in any case, involve many couplets and some difficult characters to separate species which are readily distinguishable by their host plants. When mines are collected, the foodplant can usually be identified and will give some indication of what species to expect. For this reason, keys have been made for host plants which have more than one species regularly found upon them. A few species are polyphagous and one of these species, *P. messaniella*, is recognizable by its large, lateral, abdominal spines (figure 105A,v, p. 303). In some cases (e.g. the difficult Rosaceae-feeding group) it has been found that very similar and perhaps closely related species have similar pupae and separation has not been possible.

ALNUS (Alder)

I   Cremaster with one pair of processes, the inner pair absent .... *froelichiella* (p. 352)

–   Cremaster with two pairs of processes 2

2(1)   Dorsal spines in distinct anterior and posterior rows with smaller spines between them ...... *kleemannella* (p. 353)

–   Dorsal spines with only anterior row distinct, spines gradually becoming smaller towards posterior margin 3

3(2)   Anterior rows of dorsal spines consisting of large spines in two rows .... *stettinensis* (p. 351)

–   Anterior rows of dorsal spines in several rows of smaller spines .... *rajella* (p. 340)/*strigulatella* (p. 339)

NOTE. No reliable character has yet been found to separate the pupae of *rajella* and *strigulatella*, *q.v.*, but they are unlikely to be confused because of the different foodplants.

BETULA (Birch)

I   Cremaster with processes greatly reduced and very small .... *anderidae* (p. 342)

–   Cremaster with processes not reduced and clearly visible 2

2(1)   Spines on dorsal surface of abdominal segments 1–3 in two distinct rows; cremaster with inner pair of processes shorter than outer pair .... *cavella* (p. 333)

–   These spines in only one distinct row; cremaster with inner and outer pairs of processes of equal length .... *ulmifoliella* (p. 347)

CARPINUS (Hornbeam)

I   Dorsal surface of abdominal segments 1–3 with a dense patch of long spines placed centrally towards anterior margin .... *quinnata* (p. 338)

–   Spines in this situation in a sparse patch .... *tenerella* (p. 308)

CORYLUS (Hazel)

I   Cremaster with two pairs of processes and very long setae .... *coryli* (p. 337)

–   Cremaster with one pair of processes and short setae .... *nicellii* (p. 353)

CRATAEGUS (Hawthorn)

I   Dorsal surface of abdominal segments with several rows of long spines towards anterior margin; setae short .... *oxyacanthae* (p. 314)

–   Spines in this area very short, setae very long .... *corylifoliella* (p. 327)

LONICERA (Honeysuckle)

1 Anal segment elongate, cremaster with two pairs of processes .... *trifasciella* (p. 355)
– Cremaster with one pair of minute lateral processes ......
.... *emberizaepenella* (p. 348)

MALUS (Apple)

No character has been found to distinguish the pupae of *blancardella* and *cydoniella*.

QUERCUS (Oak)

1 Cremaster with only one distinct pair of processes; inner pair, if present, very small 2
– Cremaster with two distinct pairs of processes; inner pair may be slightly shorter than outer pair 3

2(1) Abdominal segments 1–3 each with a single large lateral spine; cremaster with processes narrow, inner pair minute .... *quercifoliella* (p. 311)
– Large lateral spines absent; cremaster with outer processes short, their bases joined by a concave margin ....
.... *harrisella* (p. 306)

3(1) Cremaster with inner pair of processes distinctly shorter than outer pair, tapering to a narrow point; abdominal segments 1–3 with single massive lateral spines 4
– Cremaster with inner pair of processes as long as outer pair; abdominal segments 1–3 without massive lateral spines 5

4(3) Abdominal segments 1–3 with spines not arranged in distinct rows; massive spines larger, with wide bases and directed laterally almost at right angles ....
.... *messaniella* (p. 312)
– Abdominal segments 1–3 with spines arranged in distinct anterior and posterior rows; massive spines smaller, with narrow bases and directed laterally in a curve, not a right angle .... *muelleriella* (p. 313)

5(3) Cremaster with processes long and narrow 6
– Cremaster with processes shorter, bases of outer pair wide 7

6(5) Cremaster with processes widely divergent, outer pair directed ventrally .... *roboris* (p. 307)
– Cremaster with processes parallel ..... *distentella* (p. 341)

7(5) Abdominal segments 1–3 with spines on dorsal surface in distinct anterior and posterior rows ... *lautella* (p. 345)
– Spines in this position larger on anterior margin, gradually becoming smaller towards posterior margin 8

8(7) Lateral intersegmental area of abdominal segments 1–3 with distinct sclerotized raised process ....
.... *saportella* (p. 309)
– These sclerotized raised processes smaller ....
.... *heegeriella* (p. 308)

SALIX (Sallow and Willow)

1 Cremaster with two pairs of processes 2
– Cremaster with one pair of processes 3

2(1) Cremaster with inner pair of processes much shorter than outer pair .... *quinqueguttella* (p. 342)
– Cremaster with inner pair of processes almost as long as outer pair .... *viminiella* (p. 328)

3(1) Setae on anal segment long .... *dubitella* (p. 331)
– Setae on anal segment short 4

4(3) Cremaster with processes narrow, apical portion long and thin .... *viminetorum* (p. 329)/ *salicicolella* (p. 330)
– Cremaster with processes broader, apical portion shorter .... *spinolella* (p. 332)

NOTE. *viminetorum* and *salicicolella*. No reliable character separates these species, but they are unlikely to be confused because of their different foodplants.

SORBUS (Rowan, Whitebeam and Wild Service-tree)

No character has been found to distinguish the pupae of *sorbi*, *mespilella* and *cydoniella*.

ULMUS (Elm)

1 Pupa dark brown; spines on dorsal surface of final segments large, sparse and scattered ... *schreberella* (p. 346)
– Pupa light brown; spines on dorsal surface of final segments smaller, more numerous and less scattered ....
.... *tristrigella* (p. 350)

PHYLLONORYCTER HARRISELLA (Linnaeus)

Phalaena (Tinea) harrisella Linnaeus, 1761, *Fauna Suecica* (Edn 2): 363.
Tinea cramerella Fabricius, 1777, *Gen.Insect.*: 296.
Type locality: Sweden.

Description of imago (Pl.12, fig.1)

Wingspan 7–9mm. Head with vertical tuft white, mixed with a few widely dispersed dark brown scales; frons and labial palpus shining white; antenna white, terminal segment dark grey, scape white. Thorax and tegulae white; legs whitish, tarsi banded greyish fuscous, more pronounced on foreleg. Forewing white, pale orange-brown apically; three costal and two dorsal strigulae, all inwardly edged dark brown, first pair visible only as the dark edging, remainder blending gradually into pale orange-brown on distal side; first pair with dark edging oblique, that of dorsal 1 longer and more curving than costal 1 and nearly meeting at three-fifths to form a sharply angled fascia; second pair almost opposite with dark edging usually meeting in middle of wing to form an obtusely angled fascia, the white of dorsal 2 sometimes forming a very wide-based triangle; costal 3 wedge- or arc-shaped; apical spot clear, black; fringe line strong blackish, extending from costal 3 to dorsal 2; cilia whitish, black-tipped from costal 3 to level of apical spot, forming type 2 apical hook. Hindwing whitish, tinged fuscous; cilia white. Abdomen greyish fuscous, lightly scaled white; anal tuft yellow and white.

In Scotland forms occur in which the whole forewing is more or less suffused with pale ochreous grey; sometimes the suffusion takes the form of an indistinct basal streak as in *P. heegeriella* (Zeller), but the latter may be distinguished by its four costal strigulae.

Life history

Ovum. Laid on the underside of a leaf of deciduous oak (*Quercus* spp.).

Larva. Head pale brown. Body pale whitish green, more yellowish from abdominal segments 5–7; dorsal line very weak or absent. Late June to July; August to early November.

Mine. Underside. Small, oval or irregular in shape, less than 14mm in length, variously positioned on the leaf; if at the edge or in a lobe, the upperside may be strongly curved over but placed otherwise it does not normally contort the leaf unduly; the parenchyma cells are usually all consumed; underside with a single fine crease extending nearly the full length of the mine.

Pupa (figure 105e,n, p.303). Light brown; setae short; abdominal segments each with an anterior row of large, sparsely placed spines on dorsal surface; cremaster with a

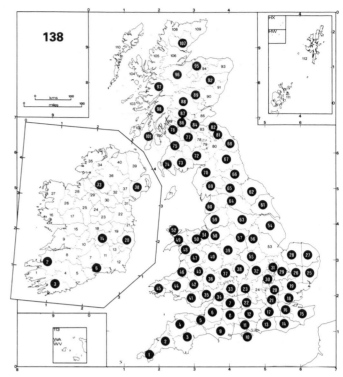

Phyllonorycter harrisella

pair of short processes, their bases joined by a concave margin. In a tough, white silken cocoon, attached to both upper and lower epidermis; in the autumn generation the frass is incorporated along the sides and posterior end of the cocoon giving it a distinctive U- or rounded V-shape when viewed against the light; in the summer generation the frass is distributed at random. The pupa is extruded through the lower surface of the mine before emergence of the adult. July to August; October to April.

Imago. Bivoltine. May and June; August and early September.

A useful account of this and all the other European oak-feeding species is given by Gregor (1952), while accounts of the British species are given by Emmet (1975b), Ffennell (1975) and Harper & Langmaid (1978). A further study dealing with the life histories of *P. harrisella* and *P. quercifoliella* (Zeller) is given by Miller (1973).

Distribution (Map 138)

Widespread and common throughout Britain; widespread but apparently less frequent in Ireland. Throughout northern and central Europe.

PHYLLONORYCTER ROBORIS (Zeller)

Lithocolletis roboris Zeller, 1839, *Isis, Leipzig* **1839**: 217.
Type localities: Germany; Berlin, Frankfurt am Main and Glogau (now Poland; Glogów).

Description of imago (Pl.12, fig.2)

Wingspan 7–9mm. Head with white vertical tuft; frons and labial palpus white; antenna white marked brown on outer edge of each segment, less so distally, scape white. Thorax light tan with white central line; tegulae white, lightly marked tan anteriorly; foreleg dark greyish tan marked white on tarsus, mid- and hindlegs whitish banded pale fuscous at joints, more so on midleg. Forewing shining white becoming pale orange-brown in apical third; a rather broad rich tan fascia extending from basal fifth of costa, narrowing and reaching dorsum at nearly one-third, both edges clearly defined but more strongly dark-bordered outwardly; a small tan basal patch on dorsum to one-sixth, unedged; middle third of wing devoid of markings; apical third with four costal and two dorsal strigulae, all white and edged dark brown inwardly, more so on costals; first pair wedge-shaped, opposite at two-thirds, with dark edging meeting on midline of wing to form an obtusely angled fascia shading into pale orange-brown inwardly; second pair opposite, each slightly curved with dark edges meeting to form an inwardly curved fascia across wing; dorsal 2 sometimes very weak or absent, not cutting fringe line, the dark edge then present only as a few individual scales; costal 3 narrow, inwardly curved with strong dark edging extending into cilia; costal 4 narrow, not always present and with dark edging represented only by a short line between costals 3 and 4 and not extending into cilia; apical spot black, very distinct and directly adjoining white of costal 4; fringe line strong from costal 4 to dorsal 1, enclosing dorsal 2 if present; strong apical hook of type 1, represented by a pencil of long black scales extending from apical spot beyond rest of cilia, surrounded by a few purplish scales at its junction with the apical spot; cilia white with slightly darker tips around apex. Hindwing pale grey, cilia white. Abdomen dark fuscous dorsally, whitish fuscous ventrally; caudal tuft pale fuscous above, whitish below.

Life history

Ovum. Laid on the underside of a leaf of deciduous oak (*Quercus* spp.).

Larva. Not described. July to August.

Mine. Underside. Large, oval, often up to 20mm in length, variously positioned and strongly contorting the leaf; the parenchyma of the upper surface is consumed with the exception of a small central patch which stays green; lower surface uncreased, but close examination shows it to be covered in very fine longitudinal ridges.

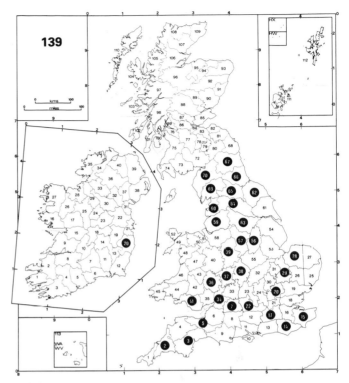

Phyllonorycter roboris

Pupa. Brown; setae short; abdominal segments with spines on dorsal surface very short; cremaster consisting of two pairs of long, narrow processes, the outer pair directed ventrally. In a rounded, whitish cocoon attached solely to the arched green patch on the upper epidermis and with no connections to the lower surface; it is supported by a fine sheet of silk which is spun from the edges of the green patch and makes contact with the lower surface of the cocoon; the frass is stacked in a heap behind the cocoon without a silken pad to retain it.

Imago. Univoltine (Emmet, 1976; 1979b). The moth flies in June. In Europe the species is reported to be bivoltine and may indeed be so in the British Isles if conditions are favourable.

Distribution (Map 139)

Found mainly in ancient woodland. Widely distributed but usually scarce throughout England and eastern Wales; in Ireland, Beirne (1941) recorded it from Co. Wicklow and gave an unlocalized record from the south. Throughout Europe.

PHYLLONORYCTER HEEGERIELLA (Zeller)
Lithocolletis heegeriella Zeller, 1846, *Linn.ent.* 1: 232.
Type locality: Germany; Glogau (now Poland; Glogów).

Description of imago (Pl.12, fig.3)
Wingspan 6.5–7.5mm. Head with white vertical tuft; frons shining white; labial palpus whitish; antenna and scape white. Thorax and tegulae pure white; legs white, very lightly banded fuscous on tarsi, more pronounced on foreleg. Forewing white to two-fifths, thence pale golden, with four costal and three dorsal strigulae strongly edged inwardly brownish fuscous, shading to golden brown; a fine brownish fuscous basal streak to two-fifths; first pair of strigulae with dark edging oblique, not meeting, and that of costal 1 not quite reaching as far as that of dorsal 1; second pair with dark edging often meeting to form an obtuse-angled or slightly outwardly curved fascia; dorsal 3 usually between costals 3 and 4, faintly edged and often indistinct; apical spot strongly black; fringe line strong from costal 4 to dorsal 2, enclosing dorsal 3 if present, becoming less strong dorsally; cilia white, weakly black-tipped in quadrant above apical spot, forming type 2 apical hook. Hindwing and cilia white. Abdomen in male light greyish dorsally, light fuscous ventrally, caudal tuft pale ochreous above, whitish below; in female whitish fuscous dorsally, white ventrally, caudal tuft yellowish white above, white below.

A form occurring regularly in the north-western counties of England and in Scotland has the white scales of the ground colour dark-tipped, giving the wing a coarse, roughened, greyish appearance; the basal streak and strigulae are clear white and their brownish fuscous edging is diffused (Emmet, 1975b).
Similar species. *P. tenerella* (Joannis), which lacks the distinctly dark-bordered basal streak, although it occasionally has an ochreous base to the wing with a white central streak extending through this coloration.

Life history
Ovum. Laid on the underside of a leaf of deciduous oak (*Quercus* spp.).
Larva. Head very pale greenish brown. Body very pale whitish green with the gut showing through as a darker green dorsal line on abdominal segments 1–6 and again on segment 8. July; September to late October.
Mine. Underside. Smaller than the mines of other oak-feeding species of *Phyllonorycter*, usually less than 10mm in length. About 85 per cent of the mines are situated on a leaf-edge, causing the edge or lobe to fold over, often completely so; oval in shape when in the centre of the leaf; usually most of the parenchyma attached to the upper epidermis is consumed; underside in the summer generation with a fine central crease, in the autumm generation with no crease but a series of fine longitudinal ridges strongly contracting the lower surface.
Pupa (figure 105y, p. 303). Pale brown; setae long; cremaster with two pairs of short processes, outer pair stout with wide bases. Summer generation, in a small, distinct, white, frass-lined cocoon attached to both surfaces of the mine; autumn generation, in a fine, white, slightly more open silken cocoon which occupies the greater part of the mine, the frass either stacked in a rough heap behind the cocoon or more generally distributed. July to August; October to May.
Imago. Bivoltine. May; August.

Distribution (Map 140)
Widely distributed throughout England and Wales, and in Scotland except in the extreme north, being found commonly in southern England; rare in Ireland. Throughout northern and central Europe.

PHYLLONORYCTER TENERELLA (Joannis)
Lithocolletis tenerella Joannis, 1915, *Annls Soc.ent.Fr.* **84**: 121.
Lithocolletis tenella Zeller, 1846, *Linn.ent.* **1**: 236. [Junior primary homonym of *Lithocolletis tenella* Duponchel, 1843.]
Type locality: France.

Description of imago (Pl.12, fig.4)
Wingspan 6–8mm. Head with vertical tuft white, sometimes with a few brown scales anteriorly; frons and labial palpus shining white; antenna white, marked above very pale brown on each segment, scape white. Thorax white to ochreous creamy white; tegulae white, sometimes narrowly edged pale brown at base; legs whitish, tarsi white banded grey, more strongly on foreleg. Forewing ranging from white to pale brownish ochreous, ochreous specimens with a white longitudinal median area from base to about two-thirds which may either nearly fill the disc or be reduced to an almost imperceptible white line; four costal and three dorsal white strigulae, broadly edged dark brown inwardly; costal 1 straight and very oblique, extending from one-third to one-half and often also along costa to base; costal 2 less acute; costals 3 and 4 at right angles to costa, then curving outward; dorsal 1 short, very slightly curved, commencing at one-third and stopping well before end of costal 1; dorsal 2 with dark edging very pronounced and joining that of costal 2 to form an acute-angled chevron and often linked by a further dark brown line with the apex of dorsal 1 so as to form an arrow-like

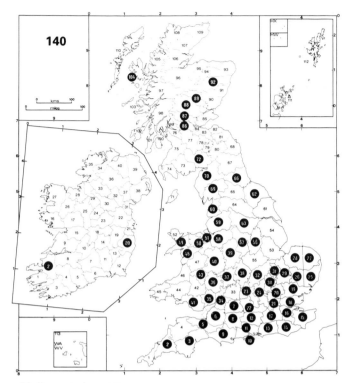

Phyllonorycter heegeriella

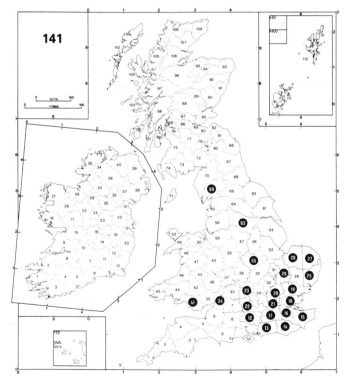

Phyllonorycter tenerella

marking in centre of wing; dorsal 3 indistinct, sometimes absent; apical spot black, distinct and slightly elongate, continuing inwards as a dark line to the apex of the chevron formed by the second pair of strigulae; fringe line not usually strong, extending from costal 4 to dorsal 2 and enclosing dorsal 3; cilia whitish, tipped fuscous from costal 4 to apex to form weak type 2 apical hook. Hindwing pale greyish fuscous; cilia whitish with light brown sheen. Abdomen greyish fuscous marked with a few light-coloured scales dorsally, white ventrally; caudal tuft pale brown above, white below.

Similar species. P. heegeriella (Zeller), *q.v.*

Life history

Ovum. Laid on the underside of a leaf of hornbeam (*Carpinus betulus*).

Larva. Head light brown. Body pale greenish yellow. June to July; September to October.

Mine. Underside. Long and narrow between two veins, but not usually extending to the edge of the leaf; the underside with one large central fold contracting the mine so that it forms a tube, the upperside with the parenchyma almost totally consumed.

Pupa (figure 105a, p. 303). Setae short; dorsal surface of abdominal segments 1–3 with central area containing few spines (*cf. P. quinnata* (Geoffroy)); cremaster consisting of two pairs of well-developed processes of equal length, the outer pair stouter than the inner pair. In a loose, pale brown cocoon at one end of the mine with the frass heaped in the central portion, or occasionally alongside the cocoon. The pupa is extruded through the underside before the emergence of the adult. July to August; October to May.

Imago. Bivoltine. May; late July and August.

Distribution (Map 141)

Widespread and locally common in south-east England; elsewhere a rare introduction in hornbeam plantations. Throughout central and north-eastern Europe.

PHYLLONORYCTER SAPORTELLA (Duponchel)

Elachista saportella Duponchel, 1840, *in* Godart & Duponchel, *Hist.nat.Lépid.Fr.* **11**: 539.

Tinea hortella Fabricius, 1794, *Ent.syst.* **3**(2): 327. No. 174, nec no. 43.

Type locality: France; Paris district.

Description of imago (Pl.12, fig.5)

Wingspan 7.5–9.0mm. Head with vertical tuft white mixed with scattered brown scales anteriorly; frons and labial palpus shining white; antenna white, proximal four-fifths with segments ringed pale fuscous at apex, terminal segment brown, scape white. Thorax white with a transverse band of dark brown scales at one-third from anterior margin; tegulae white tipped dark brown posteriorly; legs white banded dark fuscous on tarsal segments, more strongly on foreleg. Forewing pure white with five costal and three dorsal ochreous brown strigulae; first pair forming a slightly angular fascia before one-quarter, dark-edged on both sides but the outer more strongly, especially in paler specimens; sometimes a small dark brown mark on costa at about one-third; second pair of strigulae almost meeting in middle of wing to form a right-angled chevron, dorsal half dark-edged on both sides, costal half similarly edged but more strongly on outer side with a narrow extension along midline of wing often just reaching costal 3; third pair of strigulae nearly forming an acute-angled chevron, the dorsal short, truncate, dark-edged on both sides, the costal arising at about three-fifths, strongly dark-edged outwardly, curving towards apex of wing and continuing as a parallel-sided band irrorate with an increasing number of black scales to terminate in a diffuse apical patch of blackish scales; costals 4 and 5 distinct, dark-edged outwardly and curving at inner extremities into extension of costal 3 and inwardly blending smoothly into the pure white ground colour; fringe line black and distinct but only round apex, fading rapidly below this area; a very distinct blackish brown line at outer edge of cilia from above apex to opposite apex, forming type 2 apical hook; cilia below pure white, distinctly shorter immediately below hook. Hindwing light grey in male, whitish in female; cilia light fuscous on costa, white on dorsum. Abdomen in male fuscous dorsally, paler ventrally, caudal tuft ochreous fuscous; in female whitish grey dorsally, white ventrally, caudal tuft yellowish white.

Life history

Ovum. Laid on the underside of a leaf of deciduous oak (*Quercus* spp.), more commonly on a leaf situated on a high branch (Emmet, 1983).

Larva. Not described. June to early July; September to October.

Mine. Underside. Small, slightly larger than that of *P. heegeriella* (Zeller), and like that species usually situated on the leaf-edge, the under surface with many tiny creases of equal size, the contraction these induce causing the upper epidermis to curl under and in some cases virtually touch the lower epidermis; the parenchyma attached to the upper epidermis variously consumed, in some cases completely so but in others showing only a slight marbling of the green without any pronounced central green patch. When the mine is in the central part of the leaf, it is more oval and does not contort the leaf so strongly.

Pupa (figure 105x, p. 303). Pale brown; setae long; abdominal segments 1–3 each with large raised processes later-

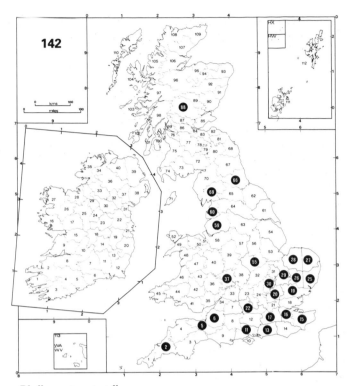

Phyllonorycter saportella

ally in intersegmental area; cremaster consisting of two pairs of short processes, outer pair wide-based. In a discrete but extremely flimsy, lace-like cocoon occupying a large part of the mine but located towards one end and attached solely to the upper epidermis; most of the frass is heaped in the other part of the mine. July to August; October to May.

Imago. Bivoltine. May; late July and early August. The moths rest, sometimes in numbers, on the trunks of oak-trees.

Distribution (Map 142)

In the past there have been scattered records from various parts of England and from Perthshire in Scotland where it was taken at Rannoch in 1938 by W. Mansbridge, his specimen now being in the Michaelis collection at Manchester Museum. Until recently the last records of this species were from Madingley and Gamlingay, Cambridgeshire, in 1949 (E. C. Pelham-Clinton, pers.comm.). In 1982 it was by chance rediscovered in the South Lopham area of east Norfolk, near the point where vice-counties 25, 26, 27 and 28 meet. Subsequent studies have shown it to be well established and common there but extremely local (Emmet, 1982a; 1982b). It is likely that other colonies await detection. Central and southern Europe.

PHYLLONORYCTER QUERCIFOLIELLA (Zeller)

Lithocolletis quercifoliella Zeller, 1839, *Isis, Leipzig* **1839**: 217.

Type locality: Germany; Glogau (now Poland; Glogów).

Description of imago (Pl.12, fig.6)

Wingspan 7–9mm. Head with vertical tuft of pale orange-brown scales, darker at tips; frons and labial palpus white; antenna whitish with light brownish band on distal edge of each segment, scape whitish. Thorax light golden brown; tegulae golden, anteriorly edged white; legs yellowish white banded dark fuscous, more strongly on foreleg, lightly on midleg and obscurely on hindleg; terminal tarsal segment darker on all legs. Forewing pale golden brown; four costal and three dorsal strigulae, and a long basal streak extending to three-fifths and slightly curved towards costa at outer extremity, all shining whitish; costal 1 and basal streak edged darker on both sides, other strigulae edged inwardly only; costal 1 at two-fifths, narrow and inclined at an acute angle; costal 2 triangular; costals 3 and 4 arc-shaped; dorsal 1 at two-fifths, small, often indistinct and strongly inclined; dorsal 2 more triangular and dorsal 3 narrow; apical spot black; fringe line strong from costal 4 to dorsal 3, a lilac sheen on apical scales; cilia yellowish white tipped black at apex, forming faint line. Hindwing light grey, cilia creamy white. Abdomen dark fuscous dorsally, yellowish white ventrally; caudal tuft whitish.

Similar species. *P. messaniella* (Zeller), in which the basal streak barely reaches middle of wing, terminating well before the apices of the first pair of strigulae. In *P. quercifoliella* the apical streak extends to almost two-thirds and terminates slightly beyond the apices of the first pair of strigulae.

Life history

Ovum. Laid on the underside of a leaf of deciduous oak (*Quercus* spp.).

Larva. Head light brown, slightly darker laterally. Body pale whitish green, gut showing through as a darker green dorsal line. July; September to early November.

Mine. Underside. Oval, elongate or irregular in shape, varying from 10 to 22mm in length, averaging 15mm; variously positioned in the leaf but often several together between two veins, contorting the leaf; when on the margin, causing the edge to fold over. On the upper surface a central patch of parenchyma is left uneaten, smaller in the summer generation; the lower surface has one strong central fold, which in the summer generation may diverge finely into a V-shape at one or both ends.

Pupa (figure 105r,w, p. 303). Light brown; setae long; abdominal segments with several rows of prominent

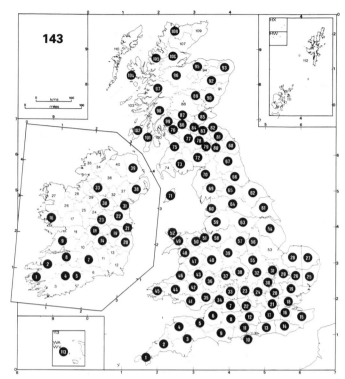

Phyllonorycter quercifoliella

spines on dorsal surface, segments 1–3 each with a single large spine on lateral margin; cremaster with outer pair of processes long, inner pair very small. In a strong white silken cocoon, this being smaller, attached solely to the green patch on the upperside and wholly frass-covered in the summer generation, but larger, attached to both surfaces and with the frass confined to the sides and posterior end in the autumn generation, when it shows as a dark U if held up against the light, often with a distinct waist in the middle part of the U. July to August; October to April.

Imago. Bivoltine. End of April to May; August to early September. It is one of the earlier species of *Phyllonorycter* to emerge in the spring.

Distribution (Map 143)

Found commonly throughout the British Isles except for the Hebrides, Orkney and Shetland. Throughout Europe.

PHYLLONORYCTER MESSANIELLA (Zeller)

Lithocolletis messaniella Zeller, 1846, *Linn.ent.* **1**: 221.
Type locality: Sicily; Messina.

Description of imago (Pl.12, fig.7)

Wingspan 7–9mm. Head with vertical tuft pale orange-
brown, some scales tipped dark brown; frons and labial
palpus shining white; antenna light fuscous banded slightly darker, scape pale orange-brown above, white below.
Thorax and tegulae light golden brown; legs slightly yellowish white, tarsi black-marked above on each segment,
more strongly on foreleg. Forewing shining pale golden
ochreous; four costal and four dorsal strigulae whitish,
more or less tinged with pale orange-brown and inwardly
edged dark brown, costal 1 and dorsal 2 also partly so
edged outwardly; a long, narrow, pale basal streak of same
colour to one-half, finely dark-edged on both sides; costal
1 at one-half, narrow, oblique and distinct; costal 2 arc-
shaped, less oblique; costals 3 and 4 more narrow, wedge-
shaped and at right angles to margin of wing; dorsal 1
small, obscure, often visible only as dark edging at one-
quarter; dorsal 2 opposite costal 1, very long, narrow and
oblique, extending as a fine line terminating between
apices of costal 2 and dorsal 3, appearing almost as extension of basal streak; dorsal 3 wide, more delta-shaped and
opposite costal 2; dorsal 4 obscure, its fine black edging
sometimes meeting that of costal 3 to form a slender,
slightly angled line across wing; apical spot black, isolated
and very small; fringe line from costal 4 to dorsal 4, an area
of purplish scales between it and apical spot; cilia yellowish white, dark-tipped from costal 4 to level of apical spot.
Hindwing greyish fuscous; costal cilia darker greyish,
dorsal cilia almost white. Abdomen fuscous above, white
below; caudal tuft light grey.

Similar species. *P. quercifoliella* (Zeller), *q.v.*

Life history

Ovum. Laid on the underside of a leaf of evergreen oak
(*Quercus ilex*) or deciduous oak (*Quercus* spp.); also less
often on hornbeam (*Carpinus betulus*), sweet chestnut
(*Castanea sativa*) or beech (*Fagus sylvatica*); more rarely
still on other trees including birch (*Betula* spp.), plum
(*Prunus* spp.) or apple (*Malus* spp.), although these are not
usual foodplants (Godfray, 1980).

Larva. Head brown, darker laterally. Body yellow, anterior segments more opaque whitish yellow. December to
March (only on evergreen oak); July; October.

Mine (Pl.2, fig.11). Underside. On evergreen oak the mine
is oval; the upper surface has a mottled appearance finally
turning brown; the lower epidermis has one strong central
crease which contracts the leaf, causing it to bend downwards; often two or three mines are found in one leaf. On

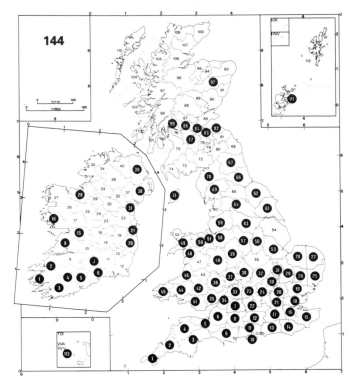

Phyllonorycter messaniella

deciduous oak the mine is small (9–14mm long), usually
with a single central crease on the underside; the upperside often has a strong fold which appears to narrow the
mine; all the parenchyma is usually consumed. On beech
and hornbeam the mine extends between two veins, the
length being effectively governed by the spacing of the
veins.

Pupa (figure 105A,v, p. 303). Dark brown; setae long;
abdominal segments 1–3 with single, massive, laterally-
directed spines; cremaster with two pairs of processes, the
inner pair smaller and tapering to acute tip. In a very
flimsy cocoon formed at one side of the mine, the edges
and the rear end being lined with a little of the frass. The
pupa is extruded through the underside before the emergence of the adult.

Imago. More or less continuous-brooded, the moths
occurring mainly from April to May, in August and in late
October to November, sometimes continuing until December (D. W. H. Ffennell, pers.comm.). It is of extreme
interest that such a common and cosmopolitan insect has a
life history that is still not fully understood. Overwintering is confirmed only in the larval stage, and then only on
evergreen oak; yet populations feeding on deciduous oak
persist even when no evergreen oaks are known within
tens of miles of the site. It is possible that this tiny moth

can regularly migrate these huge distances to recolonize but it is also possible that despite the attentions of entomologists for nearly 150 years the overwintering larva has some as yet undiscovered foodplant (Emmet, 1975b). However there are one or two recent reports of the adult being reared from mines collected in leaf-litter during winter (H. N. Michaelis and H. G. Heal, pers.comm.).

Distribution (Map 144)
Found throughout the British Isles as far north as Orkney, commonly in the southern half. Throughout central and southern Europe. In recent years a serious pest of oak and decorative trees in Australia, and New Zealand where it was accidentally introduced. In New Zealand up to 40 mincs per leaf have been reported (Swan, 1973) and in Canberra, Australia, mines reach an average of 15.3 per leaf (Common, 1976).

PHYLLONORYCTER MUELLERIELLA (Zeller)
Lithocolletis muelleriella Zeller, 1839, *Isis, Leipzig* **1839**: 217.
Elachista amyotella Duponchel, 1840, *in* Godart & Duponchel, *Hist.nat.Lépid.Fr.* **11**: 544.
Type locality: Germany; Glogau (now Poland; Głogów).

Description of imago (Pl.12, fig.8)
Wingspan 7.5–9.0mm. Head with vertical tuft dark brown mixed whitish anteriorly; frons shining white; labial palpus whitish; antenna light fuscous above, slightly darker before segmental joints, white below and with distal one-fifth wholly white. Thorax and tegulae shining golden brown; legs white, tarsal segments rather widely banded sooty brown. Forewing burnt umber with lighter golden sheen; three costal and three dorsal strigulae white, darker-edged inwardly, first pair and dorsal 2 also partly so edged outwardly; costal 1 at about two-fifths, wide at costa, sharply pointed and oblique towards apex; costal 2 at two-thirds, slightly less oblique and with outer edge blending smoothly into ground colour; costal 3 small but very distinct; dorsal 1 at about one-sixth, square or bluntly hook-shaped; dorsal 2 opposite costal 1, wide on dorsum but narrowing to a fine, curving apex reaching, but not quite blending with, the end of costal 1; dorsal 3 wide, triangular, opposite costal 2; the two pairs of opposite strigulae form two pointed chevrons, the dark edging from which occasionally forms fine lines extending towards the apex of the wing; apical spot indistinct; ground colour in apical third of wing sprinkled with darker scales, particularly along midline, sometimes forming a diffuse streak; fringe line from costal 3 to dorsal 3, blackest below apex, preceded by a white area with bluish sheen; cilia whitish fuscous with a pronounced darkened apical hook of type 3 extending from costal 3 to level of apex. Hindwing grey, cilia fuscous with brown sheen. Abdomen dark fuscous dorsally, light fuscous to white ventrally; caudal tuft brownish ochreous above, whitish below.

Variation occurs in the ground colour of the forewing; in dark specimens the dark edging of the strigulae merges into the ground colour.

Life history
Ovum. Laid on the underside of a leaf of deciduous oak (*Quercus* spp.).
Larva. Not described. July; September to October.
Mine. Underside. Rather long, usually between veins and adjacent to midrib, often several to one leaf; the upper epidermis often strongly arched upwards almost forming a tube and frequently with a small green central patch; lower epidermis with a strong central fold.
Pupa (figure 105q,u, p. 303). Golden brown; very similar

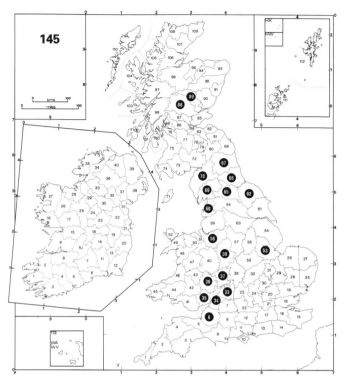

Phyllonorycter muelleriella

to that of *P. messaniella* (Zeller), but the laterally directed spines on abdominal segments 1–3 are smaller, more narrowly based and are curved rather than angled outwards. In a well-defined, frass-lined silken cocoon, attached more strongly to the upper epidermis. July to August; October to May.

Imago. Bivoltine. May; August.

Distribution (Map 145)

Found principally in ancient oak woodland along the Welsh border counties from Somerset to Cheshire; southern Lincolnshire; northern England, especially the Lake District; and Perthshire, Scotland (Shaw, 1982). Throughout Europe.

PHYLLONORYCTER OXYACANTHAE (Frey)

Lithocolletis oxyacanthae Frey, 1856, *Die Tineen und Pterophoren der Schweiz*: 336.
Type locality: Switzerland; Zürich.

Description of imago (Pl.12, fig.9)

Wingspan 6–8mm. Head with vertical tuft dark ochreous brown mixed white posteriorly; frons and labial palpus shining white; antenna whitish ochreous, segments marked fuscous above, scape greyish ochreous above, white below. Thorax golden brown, white anteriorly; tegulae golden brown, white posteriorly; legs whitish marked dark fuscous on segments above, more strongly on foreleg. Forewing tawny brown marked with scattered dark brown scales, particularly in distal half of wing; four costal and three dorsal strigulae brilliant white dark-bordered inwardly, sometimes very weakly; costal 1 small, narrow, strongly angled or delta-shaped; costals 2–4 small, arc-shaped; dorsal 1 long, narrow, strongly angled at apex and reaching beyond end of costal 1 but occasionally almost meeting it to form an acute-angled fascia, its fine dark edging continued around tip; dorsal 2 widely based, triangular, its dark edging sometimes meeting that of costal 2 to form a right-angled chevron; dorsal 3 nearly opposite costal 3, forming with it a curved white fascia broken in middle by apical streak; basal streak to one-third straight, brilliant white, dark-bordered above and for short distance round apex; a white dorsal patch below basal streak, sometimes weakly dark-bordered above; apical spot small, distinct, extending as a short streak below costal 4 to junction of third pair of strigulae, sometimes expanding to form a dark patch; fringe line from costal 4, strong round apex then fading to dorsal 2, enclosing dorsal 3 except in specimens where this strigula is well marked; cilia whitish fuscous. Hindwing grey, cilia whitish fuscous. Abdomen light fuscous dorsally, whitish ventrally; caudal tuft ochreous in male, fuscous in female.

A form occurs, more frequently in the north, in which the forewing ground colour is uniformly dark brown, so that the dark edges of the strigulae are not apparent.

Similar species. This species, *P. sorbi* (Frey) and *P. mespilella* (Hübner) cannot be separated on the basis of wing characteristics alone and the characters given by Meyrick (1928) are too variable within each species to be reliable; *P. sorbi*, however, is generally larger (wingspan *c*.8mm) than the other two (wingspan 7mm or less); most specimens of *P. mespilella* lack the white patch on the dorsum beneath the basal streak.

Determination using genitalic characters (figure 106, p. 316) may be made with the help of the following keys.

Male

1　Valvae strongly asymmetrical, right costa distinctly longer than left 2
–　Valvae hardly asymmetrical, right costa hardly longer than left; terminal spine straight .... *P. sorbi* (p. 315)
2(1)　Right costa with terminal spine strongly hooked ....
　　.. *P. oxyacanthae* (p. 314)
–　Right costa with terminal spine straight ....
　　.. *P. mespilella* (p. 317)

Female

1　Sterigma narrow with a small central ring of thickening which appears to form an ovoid disc on anterior side of sterigma .... *P. mespilella* (p. 317)
–　Sterigma without this character 2
2(1)　Sterigma abruptly angled caudally and extended to form a square-ended plate .... *P. sorbi* (p. 315)
–　Sterigma smoothly arched and not extended to form a plate .... *P. oxyacanthae* (p. 314)

Life history

Ovum. Laid on the underside of a leaf of hawthorn (*Crataegus* spp.), occasionally pear (*Pyrus* spp.) or quince (*Cydonia vulgaris*).

Larva. Head pale brown. Body pale greenish yellow, gut darker greenish brown. July; September to October.

Mine. Underside. Usually formed in the lobe of a leaf or near the petiole; lower surface with many creases and strongly contracted, causing the lobe or leaf-edge to fold over. For the difference between this mine and that made by *Parornix anglicella* (Stainton) see p. 280.

Pupa (figure 105c, p. 303). The pupae of this species and all those feeding on Rosaceae (*Phyllonorycter sorbi*, *P. mespilella*, *P. blancardella* (Fabricius), *P. cydoniella* ([Denis & Schiffermüller]), *P. spinicolella* (Zeller) and *P. cerasicolella* (Herrich-Schäffer)) are virtually indistinguishable; in all the setae are short, the dorsal surface of the abdominal segments has prominent spines, usually arranged in two distinct rows, and the cremaster consists of two well-developed processes of equal length, the outer pair stouter than the inner pair. That of *P. oxyacanthae* in the autumn generation is in a flimsy golden brown cocoon, in the summer generation almost without a cocoon, the frass being piled behind a silken pad at the extreme end of the lobe. July; October to May.

Imago. Bivoltine. May; August.

Distribution (Map 146)

Common wherever hawthorn is to be found throughout the British Isles as far north as Wester Ross. Throughout Europe.

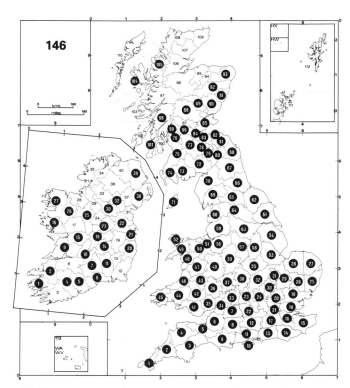

Phyllonorycter oxyacanthae

PHYLLONORYCTER SORBI (Frey)

Lithocolletis sorbi Frey, 1855, *Mitt.naturf.Ges.Zürich*: 608.
Type locality: Switzerland; Zürich.

Description of imago (Pl.12, fig.10)

Wingspan 7.5–8.5mm. Head with brownish vertical tuft, whitish posteriorly; frons and labial palpus shining white; antenna whitish, marked above with fuscous on distal edge of each segment, particularly in middle third, terminal segment grey, scape white. Thorax bright golden brown, broadly white anteriorly; tegulae golden brown edged white; legs whitish below, foreleg grey above, lightly marked white on tarsal segments, mid- and hindlegs tibiae light fuscous, tarsi whitish above. Forewing shining golden brown, more or less strongly suffused blackish brown in distal half of wing and sometimes near dorsum towards base; four costal and three dorsal strigulae shining white, sometimes with slight golden sheen, clearly bordered blackish brown inwardly, the first pair and dorsal 2 also around tip; costal 1 short, narrow and angled towards apex; costal 2 broad, arc-shaped, strongly bordered inwardly; costals 3 and 4 narrow, arc-shaped; dorsal 1 variable but normally long, narrow, sickle-shaped, terminating between costals 1 and 2; dorsal 2 wide, triangular, sometimes with inner edge curved inwards; dorsal 3

Figure 106

Phyllonorycter oxyacanthae (Frey)
(**a**) male genitalia (**b**) female genitalia

Phyllonorycter sorbi (Frey)
(**c**) male genitalia (**d**) female genitalia

Phyllonorycter mespilella (Hübner)
(**e**) male genitalia (**f**) female genitalia

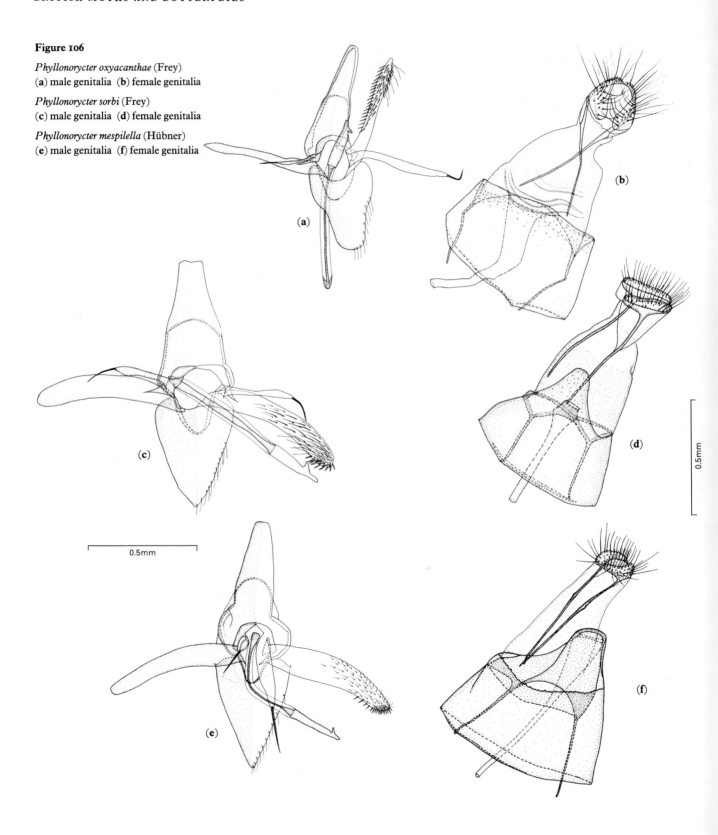

varying in width and positioned between costals 3 and 4, not usually cutting fringe line; basal streak broad, pointed, extending to one-third, dark-bordered above and for a short distance around apex; a small white patch on dorsum below basal streak; apical spot small, diffuse, with scattered blackish scales extending back towards junction of second pair of strigulae as an ill-defined streak; faintly purplish scales in fringe near apical spot; fringe line weak from costal 4 to dorsal 2, sometimes fading by dorsal 3; cilia whitish fuscous, slightly darker on dorsum. Hindwing grey, cilia light fuscous. Abdomen brownish fuscous dorsally, ochreous white ventrally; caudal tuft yellowish grey in male, ochreous white in female. Genitalia, see figure 106, and key, p. 315.

Similar species. P. oxyacanthae (Frey), *q.v.*

Life history

Ovum. Laid on the underside of a leaf of rowan (*Sorbus aucuparia*) or bird-cherry (*Prunus padus*); occasionally on common whitebeam (*Sorbus aria*).

Larva. Head pale brownish fuscous. Body very pale greenish yellow, abdominal segments slightly darker, gut dark greenish. Late June to July; September to October.

Mine. Underside. On rowan extending longitudinally beside the midrib or edge of the leaflet, sometimes fully occupying one complete half; the lower surface with a number of pronounced longitudinal creases, puckering the mine strongly and causing contortion of the leaf or completely folding over the leaf-edge; on bird-cherry more irregularly shaped and variably placed, either at the edge of the leaf or between veins; often the parenchyma within the mine is almost wholly consumed, leaving the upperside transparent and of a whitish grey colour.

Pupa. Similar to that of *P. oxyacanthae*, *q.v.* In a strong whitish cocoon with the frass piled near the middle of the mine; the pupa is extruded usually from the upper surface before the emergence of the adult. July to August; October to April.

Imago. Bivoltine. Late April to May; August.

Distribution (Map 147)

Found throughout the British Isles, being widespread and common in northern England and Scotland but rather more local and mainly confined to woodland in the south of England. Throughout Europe.

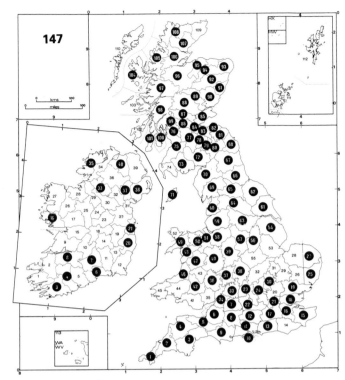

Phyllonorycter sorbi

PHYLLONORYCTER MESPILELLA (Hübner)

Tinea mespilella Hübner, [1805], *Samml.eur.Schmett.* **8**: pl.39, fig.272.
Lithocolletis pyrivorella Bankes, 1899, *Entomologist's mon. Mag.* **35**: 252.
Type locality: Europe.

Description of imago (Pl.12, fig.11)

Wingspan 6–8mm. Head with vertical tuft brownish, whitish posteriorly; frons and labial palpus white; antenna whitish, marked above with fuscous on distal edge of each segment, terminal segment grey, scape white below, greyish ochreous above. Thorax golden brown, white anteriorly; tegulae similar, inwardly edged white; legs all white beneath, on upperside foreleg dark grey lightly marked white, midleg similar but more whitish, hindleg whitish indistinctly banded with grey on tarsal segments. Forewing ranging from dark brown to golden brown with scattered dark brown scales; four costal and three dorsal strigulae, all white irregularly bordered inwardly with thin dark brown lines, the first pair also outwardly; costal 1 small, angled, wedge- or delta-shaped; costals 2–4 distinct, of similar size and usually wedge-shaped, slightly curved inwardly; dorsal 1 long, wider at base and slightly curved to reach just beyond costal 1; dorsal 2 triangular, opposite costal 2, often with many blackish scales between the two; dorsal 3 if present between costals 3 and 4, seen more as a small break in the region of blackish scales; basal streak to one-third white, dark-edged above and just round apex; a very faint line of unedged white scales along dorsum beneath basal streak in some specimens; apical spot distinct and extended inwards as a black streak or diffuse band of black scales to junction of second pair of strigulae; fringe line fairly strong from costal 4 to dorsal 2, enclosing dorsal 3 if present; cilia whitish fuscous. Hindwing grey, cilia light fuscous, darker towards apex. Abdomen dark fuscous dorsally, whitish ventrally; caudal tuft yellowish fuscous. Genitalia, see figure 106, p. 316.

Similar species. P. oxyacanthae (Frey), *q.v.*

Life history

Ovum. Laid on the underside of a leaf of wild service-tree (*Sorbus torminalis*) or wild pear (*Pyrus communis*), occasionally on rowan (*Sorbus aucuparia*), common whitebeam (*S. aria*) or some other rosaceous tree.

Larva. Head pale brown. Body whitish green, becoming yellowish posteriorly; orange patches under middle segments; gut green. July; September to October.

Mine. Underside. Long and narrow (20–30mm), between two lateral ribs, often near the petiole between the first two; lower epidermis contracted, causing the leaf to pucker strongly and form a tube usually not reaching either the midrib or the lobe of the leaf; underside with many longitudinal folds. In common whitebeam the mine does not discolour the underside of the leaf and the folds are not obvious through the thicker leaf-hairs. On all foodplants much of the parenchyma attached to the upper surface is consumed.

Pupa (figure 105g, p. 303). Similar to that of *P. oxyacanthae, q.v.* In a very pale brown, loose silken chamber or cocoon formed at one end of the mine, the frass being disposed in a long loose line behind the cocoon; the pupa is extruded through the lower surface of the mine before the emergence of the adult. July to August; October to May.

Imago. Bivoltine. May; August.

Distribution (Map 148)

Very local in England as far north as Durham and in south Wales, although it may be encountered in some numbers where it occurs in the larger, long-established woods containing wild service-tree; eastern Ireland (Beirne, 1941). Central Europe.

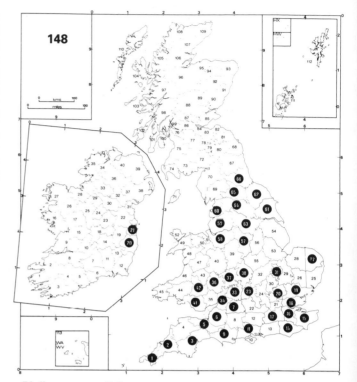

Phyllonorycter mespilella

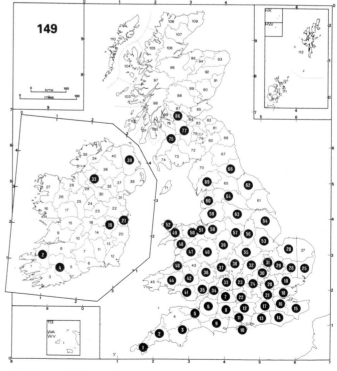

Phyllonorycter blancardella

PHYLLONORYCTER BLANCARDELLA (Fabricius)

Tinea blancardella Fabricius, 1781, *Spec.Ins.* 2: 305.
Lithocolletis concomitella Bankes, 1899, *Entomologist's mon.Mag.* **35**: 246.
Type locality: England.

Description of imago (Pl.12, figs 13–15)

Wingspan 6.0–8.5mm. Head with vertical tuft orange-brown, whitish posteriorly; frons shining white; labial palpus whitish; antenna whitish marked fuscous on segments above, white below, scape fuscous above, white below. Thorax brown, broadly white anteriorly; tegulae brown, white-edged inwardly; all legs whitish at proximal end of segments, fuscous mixed brown at distal end, this contrast more strongly expressed on foreleg. Forewing dark brown proximally with scattered blackish scales, more ochreous brown distally, and very variable; four costal and three dorsal strigulae, mostly pure white but outer strigulae sometimes slightly golden, all irregularly edged dark brown inwardly, first pair and dorsal 2 also round tips; costal 1 short, narrow and oblique or delta-shaped, often with inner border extending along costa, sometimes to base, as a dark line; costals 2–4 small, arc-shaped, each bordered outwardly with pale golden brown; dorsal 1 broad, oblique, sometimes with sinuate sides, and terminating near apex of costal 1, in some specimens being very broad and meeting costal 1 to form a fascia, or the basal streak to form a crescent, or dorsal 2 to form an inverted U-shape, or exhibiting a combination of these variations to form a complex area of white with the dark edging often absent (figs 14,15); dorsal 2 large, triangular, sometimes broadened to form a semicircle; dorsal 3 small, between costals 3 and 4 and not cutting fringe line; a broad white basal streak to one-third, usually dark-edged on costal side; an unedged white patch on dorsum beneath basal streak, sometimes rather large and extending as a fine white edge on each side to reach base and dorsal 1; apical spot small, surrounded by scattered black scales and extended as a streak towards junction of third pair, sometimes second pair, of strigulae; fringe line fine but distinct, from costal 4 to dorsal 2 enclosing dorsal 3; cilia whitish fuscous, darker from costal 4 to apex. Hindwing grey, cilia light fuscous. Abdomen brownish fuscous dorsally, yellowish white ventrally; caudal tuft ochreous.

Similar species. *P. cydoniella* ([Denis & Schiffermüller]), which is generally larger (wingspan 8–9mm), has the ground colour shining golden brown without the scattered blackish scales and the markings brighter silvery white; the dark edging to the strigulae is more sharply defined and the basal streak is narrower, with more distinct dark edging on its costal side and around the apex. Genitalia. *Male.* In *P. blancardella* the right costa is more than twice the length of the left and the terminal spine is straight; in *P. cydoniella* the right costa is less than one-third longer than the left and the terminal spine is sharply hooked. *Female.* In *P. blancardella* the sterigma extends in a straight line dorsally between the two apophyses anteriores and is extended ventrally as a large truncated flap; in *P. cydoniella* the dorsal side of the sterigma is strongly arched between the two apophyses but does not form a flap, and is only weakly visible on the ventral side. See figure 107, p. 321.

Life history

Ovum. Laid on the underside of a leaf of apple (*Malus* spp.), including cultivars.

Larva. Head pale brown. Body pale greenish orange, gut greenish. Late May to June; September to early November.

Mine. Underside. Between two veins, usually smaller than that of *P. cydoniella* and with more distinct creases than in that species; on crab-apple, the upper surface may be arched to form a tube; on cultivated apple where the leaf is much thicker, the mines are flatter and blend more into the leaf.

Pupa. Similar to that of *P. oxyacanthae* (Frey), *q.v.* In a large, white, silk-lined chamber with the frass heaped behind, but often with grains along the edge of or in front of the silken chamber and a few random grains distributed throughout the mine. The pupa is extruded through the underside of the mine before the emergence of the adult. July to August; October to May.

Imago. Bivoltine. May; August.

Distribution (Map 149)

Widespread and common wherever apple or crab-apple is to be found in southern and central England and Wales; more local in northern England, Scotland as far north as Stirlingshire and in Ireland. Throughout Europe and as an introduction in the apple-growing regions of north America (Pottinger & LeRoux, 1971).

PHYLLONORYCTER CYDONIELLA ([Denis & Schiffermüller])

Tinea cydoniella [Denis & Schiffermüller], 1775, *Schmett. Wien.*: 144.

Lithocolletis blancardella sensu Pierce & Metcalfe, 1935, *The genitalia of the tineid families of the Lepidoptera of the British Islands*: 72, pl.43.

Type locality: [Austria]; Vienna district.

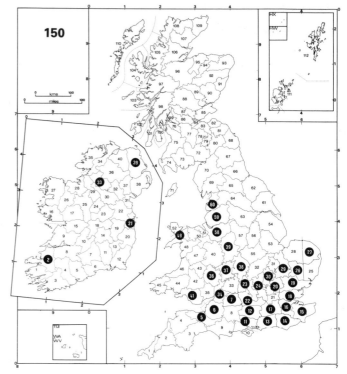

Phyllonorycter cydoniella

Description of imago (Pl.12, fig.12)

Wingspan 8–9mm. Head with vertical tuft orange-brown, lighter posteriorly; frons and labial palpus white; antenna greyish fuscous, segments banded dark fuscous, apical one or two segments dark grey, scape whitish. Thorax shining golden brown, white anteriorly; tegulae similar but white-edged inwardly; foreleg dark grey, segments banded white, midleg similar but more whitish, hindleg brownish on tibia, whitish on tarsus, banded fuscous. Forewing shining golden brown; four costal and three dorsal strigulae, all shining silvery white strongly black-edged inwardly, first pair and dorsal 2 also black-edged round tips; costal 1 short, delta-shaped, with dark edging extending a short distance along costa towards base; costal 2 almost triangular; costals 3 and 4 wedge-shaped; dorsal 1 long, broadly sickle-shaped, sometimes slightly sinuate, reaching midway between costals 1 and 2; dorsal 2 large, triangular, opposite costal 2; dorsal 3 between costals 3 and 4; basal streak to two-fifths, distinct, black-edged above and round apex; a white unedged patch on dorsum below dorsal streak; apical spot small but distinct, extending as a short streak to junction of third pair of strigulae; fringe line fine from costal 4 to dorsal 2, enclosing dorsal 3; cilia whitish fuscous. Hindwing grey, cilia fuscous. Abdomen dark fuscous dorsally, ventrally pale fuscous in male, yellowish white in female; caudal tuft dark fuscous above, ochreous fuscous below. Genitalia, see figure 107.

Similar species. P. blancardella (Fabricius), *q.v.*

Life history

Ovum. Laid on the underside of a leaf of quince (*Cydonia oblonga*), crab-apple (*Malus sylvestris*), wild service-tree (*Sorbus torminalis*) or occasionally some other species of Rosaceae.

Larva. Head light brownish. Body pale greenish orange, gut slightly darker green. June to July; September to October.

Mine. Underside. Larger than that of *P. blancardella* when on crab-apple, extending between two veins, often contorting the leaf, particularly if it contains several mines; the upperside is usually strongly arched upward forming a distinct tube, the underside with a strong crease or ridge, often with several other smaller creases alongside.

Pupa. Similar to that of *P. oxyacanthae* (Frey), *q.v.* In a large white to yellowish white silk-lined chamber, with the frass piled neatly behind. The pupa is extruded through the underside before the emergence of the adult. July to August; October to May.

Imago. Bivoltine. May; August.

Distribution (Map 150)

Very local in south-eastern England, along the Welsh border counties from Somerset as far north as Lancashire, and in Ireland; however, the species can be very common where it occurs. Throughout Europe.

Figure 107

Phyllonorycter blancardella (Fabricius)
(**a**) male genitalia (**b**) female genitalia

Phyllonorycter cydoniella ([Denis & Schiffermüller])
(**c**) male genitalia (**d**) female genitalia

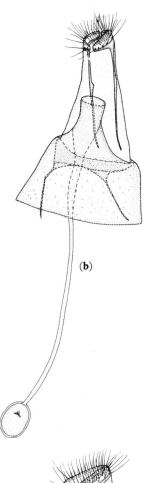

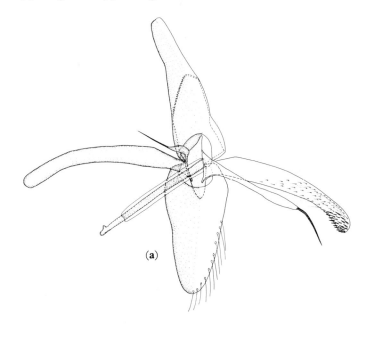

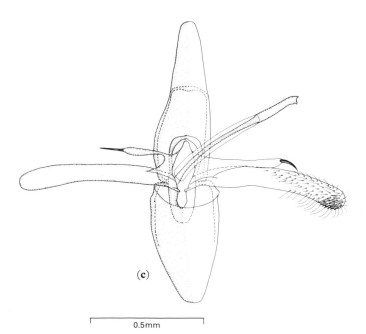

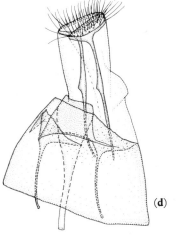

0.5mm

0.5mm

PHYLLONORYCTER JUNONIELLA (Zeller)

Lithocolletis junoniella Zeller, 1846, *Linn.ent.* **1**: 215.
Lithocolletis vacciniella Stainton, 1855, *Entomologist's Annu.* **1855**: 48.
Type locality: Germany; Glogau (now Poland; Głogów).

Description of imago (Pl.12, fig.16)
Wingspan 7–8mm. Head with vertical tuft orange-brown or brown, lighter posteriorly; frons white; labial palpus ochreous; antenna light fuscous banded dark fuscous, terminal segments dark grey, scape whitish. Thorax bright golden brown, finely edged white; tegulae similar but edged white inwardly; legs whitish banded fuscous, more prominently on foreleg. Forewing clear bright orange-brown; four costal and three dorsal strigulae, all clear white dark-edged inwardly, first pair and dorsal 2 also round tips; white basal streak to about one-third, distinct, dark-edged above and around pointed apex; costal 1 short, delta-shaped; costal 2 nearly triangular; costals 3 and 4 clear and arc-shaped; dorsal 1 large, smoothly sickle-shaped, nearly meeting costal 1 to form sharply acute angle; dorsal 2 large, triangular, opposite costal 2; dorsal 3 joining with costal 3 to form an almost parallel-sided, narrow white fascia curving smoothly inwards; apical spot strongly surrounded by black scales and extending to end of costal 4; fringe line from costal 4 to dorsal 2, enclosing dorsal 3; cilia whitish round apex to end of fringe line, thence fuscous mixed brownish. Hindwing brownish grey, cilia fuscous with brownish sheen. Abdomen greyish fuscous dorsally, whitish ventrally; caudal tuft ochreous.

Life history
Ovum. Laid on the underside of a leaf of cow-berry (*Vaccinium vitis-idaea*).
Larva. Head pale yellowish brown. Body pale amber yellow, gut yellowish brown. October to April, overwintering half-fed; sometimes also in July.
Mine. Underside. Occupies nearly the whole of a leaf, the lower surface contracting, drawing the edges of the leaf down and arching the upper surface; the parenchyma attached to the upper surface is incompletely eaten, giving a mottled appearance.
Pupa. Very pale brown; setae short; cremaster consisting of two pairs of short processes of equal length, the outer pair stouter than the inner pair. Extruded through the underside of the mine before the emergence of the adult.
Imago. Mainly univoltine, with the moth occurring in June and July. On low ground and in the southern part of its range it is sometimes bivoltine with the adult appearing again in August and September.

Distribution (Map 151)
Locally very common wherever its foodplant occurs on moorland in Wales, central and northern England and Scotland to Orkney. Throughout Europe in similar localities.

PHYLLONORYCTER SPINICOLELLA (Zeller)

Lithocolletis spinicolella Zeller, 1846, *Linn.ent.* **1**: 203.
Lithocolletis pomonella sensu Kloet & Hincks, 1972.
Type locality: Czechoslovakia; Nixdorf, Bohemia.
NOMENCLATURE. This species appears in recent lists as *Phyllonorycter pomonella* (Zeller). Even if *pomonella* and *spinicolella* were synonymous (which they are not, *pomonella* having been named from one of the apple-feeding species and *spinicolella* as its 'variety' on blackthorn (*Prunus spinosa*)), they were described in the same paper and neither name has priority. It is therefore unnecessary to make any change from the well-established name *spinicolella*.

Description of imago (Pl.12, fig.17)
Wingspan 6.5–7.5mm. Head with vertical tuft ochreous yellow, more whitish posteriorly; frons and labial palpus white; antenna whitish fuscous lightly banded grey, terminal segment dark grey, scape whitish. Thorax shining golden orange finely edged white and with white central line; tegulae golden brown, narrowly white inwardly; foreleg grey, marked white on tibia, midleg paler grey, hindleg whitish. Forewing golden orange; four costal and three dorsal strigulae and basal streak to past one-third all shining white, all strigulae except costal 4 dark-edged inwardly, costals 1 and 2 and dorsal 2 more strongly so; costal 1 short, sometimes extended minutely along costa towards base as a fine white line and occasionally with a few dark scales on distal edge; costals 2–4 usually distinct but small and arc-shaped; dorsal 1 long, narrow, often sinuate and curved towards apex, partly black-edged outwardly except near dorsum and often meeting costal 1 to form a sharply acute chevron, the apex of which may almost touch junction of black edging to second pair of strigulae; dorsal 2 large, triangular, with patch of black scales above which forms a bridge to the black edge of costal 2, thus often forming an obtusely angled blackish fascia; dorsal 3 faint; basal streak fine, often with distal third angled slightly towards costa, usually finely edged dark brown on costal side and around apex for a short distance on dorsal side; a narrow patch of white scales along dorsum below basal streak, sometimes finely edged brown; apical spot small and extended inwards as a narrow streak below costal 4 to junction of third pair of strigulae; fringe line fine but distinct from costal 4 to dorsal 2, enclosing dorsal 3; cilia light fuscous. Hindwing grey, cilia

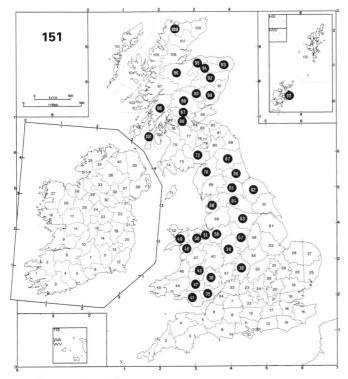

Phyllonorycter junoniella

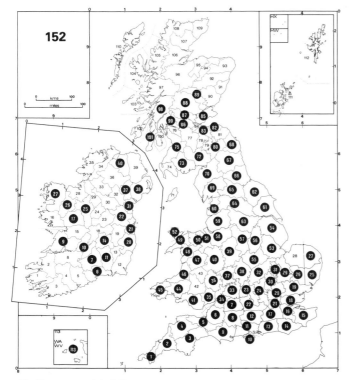

Phyllonorycter spinicolella

fuscous. Abdomen silvery grey dorsally, white ventrally; caudal tuft ochreous above, white below.

Similar species. *P. cerasicolella* (Herrich-Schäffer), which cannot be separated on the basis of wing-pattern alone; the characters given in the keys of Meyrick (1928), Jacobs (1945) and Bradley *et al.* (1969) do not always hold because the species are so variable. If long series are compared, the basal streak in *P. cerasicolella* tends to be slightly more irregular in outline and to have less dark edging on the costal side. Genitalia. *Male.* In *P. spinicolella* the valvae are slightly shorter and broader than in *P. cerasicolella*, the sacculus tends to be more rounded and has more black pigmentation, and the subapical spine is situated on a concave part of the valva and is not surrounded by short stiff bristles. *Female.* The ductus bursae extends beyond the anterior end of the seventh segment in *P. cerasicolella* but usually turns before this point in *P. spinicolella*. See figure 108, p. 324. For fuller details of the genitalia see Povolný (1949) and Gregor *et al.* (1963); for more general observations on the two species see Corbett (1898) and Tutt (1898).

Life history

Ovum. Laid on the underside of a leaf of blackthorn (*Prunus spinosa*) or wild plum (*P. domestica*). On the Continent it prefers cultivars, especially young trees, and populations on blackthorn make small mines from which the adults are very tiny, often with aberrant genitalia (Gregor *et al.*, *loc.cit.*; Povolný, 1977); no such tendency has been observed in the British Isles.

Larva. Head pale greenish ochreous. Body pale greenish white, anterior segments slightly amber; gut dark green anteriorly, purplish brown posteriorly; an orange spot on abdominal segment 6. July; September to April, the larva overwintering in the mine and pupating in the spring.

Mine. Underside, often between veins. The lower epidermis is strongly contracted by creases causing the upperside to arch. The mine is very similar to that of *Parornix finitimella* (Zeller), but that is smaller (*c*.5–10mm long), generally has the lower epidermis flecked with grey and occurs in association with the folded edges of leaves in which the larva feeds after vacating its mine (p. 286).

Pupa. Resembles that of *Phyllonorycter oxyacanthae* (Frey), *q.v.* In a white silk-lined chamber with the frass piled at one end. The pupa is extruded through the underside of the mine before the emergence of the adult. April; July to August.

Imago. Bivoltine. May; August.

Distribution (Map 152)

Widely distributed and generally common throughout the British Isles as far north as Perthshire. Throughout Europe.

Figure 108

Phyllonorycter spinicolella (Zeller)
(**a**) male genitalia (**b**) female genitalia

Phyllonorycter cerasicolella (Herrich-Schäffer)
(**c**) male genitalia (**d**) female genitalia

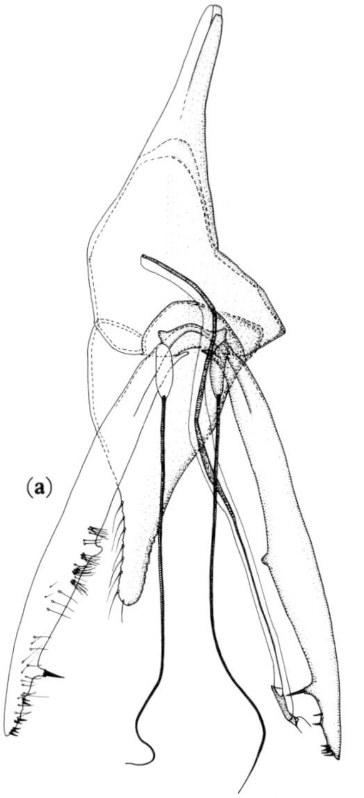

(**a**)

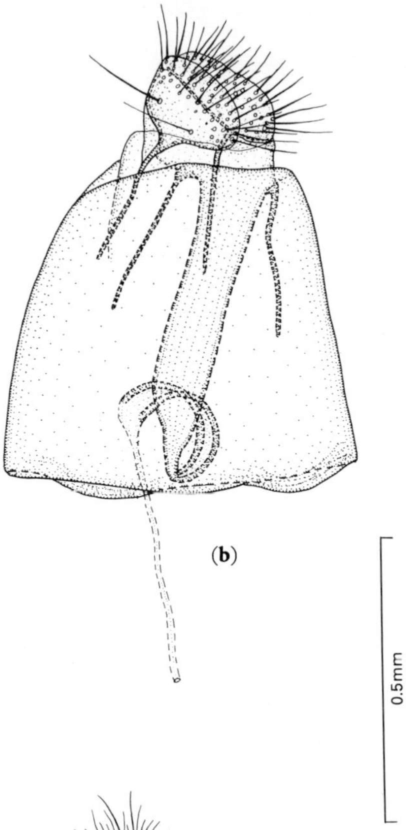

(**b**)

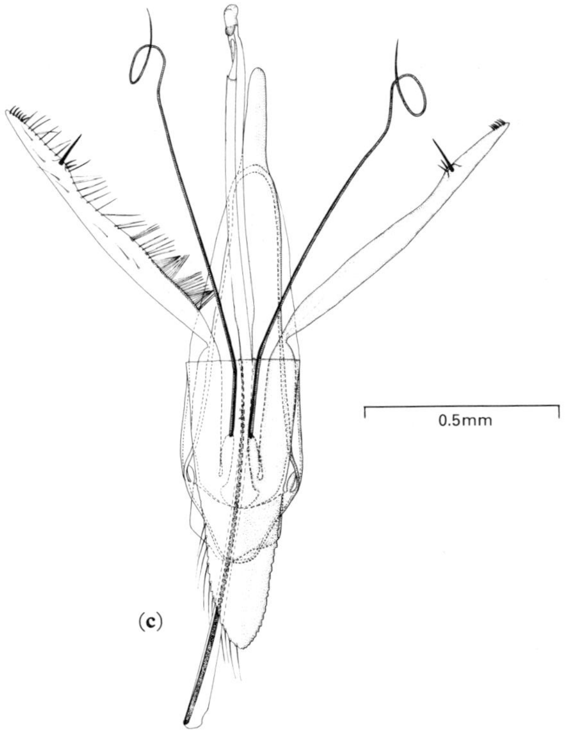

0.5mm

(**c**)

0.5mm

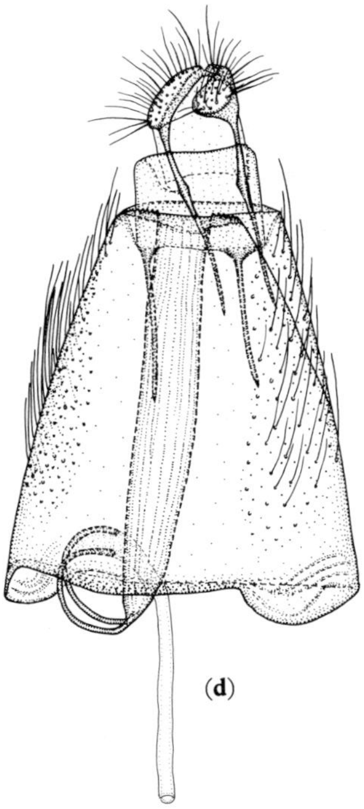

(**d**)

PHYLLONORYCTER CERASICOLELLA (Herrich-Schäffer)

Lithocolletis cerasicolella Herrich-Schäffer, 1855, *Syst. Bearb.Schmett.Eur.* **5**: 326.

Type locality: Germany; Frankfurt am Main.

Description of imago (Pl.12, figs 18,19)
Wingspan 7–8mm. Head with vertical tuft orange-brown, whitish posteriorly; frons and labial palpus shining white; antenna shining whitish grey, marked fuscous on segments, terminal segment dark fuscous, scape whitish ochreous. Thorax shining orange-brown, central line and anterior edge white; tegulae shining brownish orange, white-edged inwardly; foreleg grey marked white on tibia, midleg paler grey, hindleg whitish. Forewing bright golden brown; four costal and three dorsal strigulae and a basal streak to beyond one-third all shining white, all strigulae except costal 4 slightly dark-edged inwardly, second pair strongly so; costal 1 narrow, oblique and often very lightly edged outwardly, sometimes with the dark edging extending along costa to base; costals 2–4 generally small and slightly arc-shaped; dorsal 1 long, narrow, sinuate, oblique and usually dark-edged outwardly, often meeting costal 1 to form a sharply angulate fascia, the apex of which reaches nearly to two-thirds; dorsal 2 large, triangular, opposite to but not reaching costal 2, but with the dark edging expanding as a small patch of black scales and merging into the black edge of costal 2, so forming an obtusely angled dark fascia; dorsal 3 indistinct, often within fringe line; basal streak narrow, occasionally with fine dark edging on dorsal side, angled towards costa for the distal third; a small white dorsal patch below basal streak; apical spot not pronounced but preceded by an irregular line of dark scales to junction of second pair of strigulae; fringe line ill-defined from costal 4 to dorsal 2, enclosing dorsal 3; cilia light orange-brown, shading to light fuscous outwardly. Hindwing grey, cilia fuscous. Abdomen in male dark grey dorsally, yellowish white ventrally; in female ochreous grey dorsally, whitish ventrally; caudal tuft light ochreous. Genitalia, see figure 108.
Similar species. P. *spinicolella* (Zeller), *q.v.*

Life history

Ovum. Laid on the underside of a leaf of cultivated cherry (*Prunus* spp.), wild cherry (*P. avium*) or dwarf cherry (*P. cerasus*). Never on blackthorn (*P. spinosa*), wild plum (*P. domestica*) or cultivated plum (*P. insititia*) (Gregor *et al.*, 1963; Povolný, 1977).
Larva. Head dark brown. Body pale yellow, gut dark green; orange spot on abdominal segment 6. July; September to April, the larva overwintering in its mine and pupating in the spring.

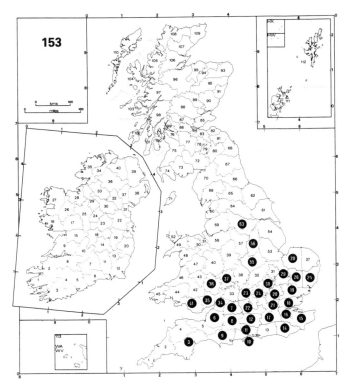

Phyllonorycter cerasicolella

Mine. Underside. Between two veins, often near the base of the leaf, strongly contorting the upperside to form a distinct pucker; often several mines are found in one leaf; occasionally the mine is formed along the leaf-edge, but it is never found along the midrib.
Pupa. Similar to that of P. *oxyacanthae* (Frey), *q.v.* In a large, tough, whitish cocoon formed from the sides of the mine, the frass being piled in a heap behind the cocoon. The pupa is extruded through the underside before the emergence of the adult. April; July to August.
Imago. Bivoltine. May; August.

Distribution (Map 153)

Widespread and locally common in south-east England and Wales, extending as far as south Devon, Glamorgan, Worcestershire and southern Yorkshire. The species occasionally occurs in large numbers on unsprayed cherries, especially domestic escapes or self-seeded plants. Throughout Europe.

PHYLLONORYCTER LANTANELLA (Schrank)

Tinea lantanella Schrank, 1802, *Fauna boica* 2(2): 138.
Type locality: Europe.

Description of imago (Pl.12, fig.20)

Wingspan 8–9mm. Head with vertical tuft white, mixed brown or pale orange-brown anteriorly; frons and labial palpus shining white; antenna whitish grey above, darker before intersegmental joints in middle third, terminal segment black, scape white, narrowly grey distally. Thorax golden brown with fine white central line and finely edged white; foreleg dark grey, lightly banded white on tarsus, mid- and hindlegs ochreous fuscous on tibiae, tarsi whitish marked with grey near joints, all legs whitish beneath. Forewing golden brown or orange-brown; four costal and three, sometimes four, dorsal strigulae, all brilliant silvery white, strongly dark-edged inwardly; a white basal streak to one-third, wider in middle and with dark edging above and around apex; costal 1 small, delta-shaped with the black edging often extending along costa to base; costals 2–4 all arc-shaped and distinct; dorsal 1 sickle-shaped, long, smoothly curved towards apex and passing end of costal 1, with black edging continuing around apex to halfway down outer edge; dorsal 2 large, triangular, with dark edging continuing along almost all of distal edge; dorsal 3, and 4 if present, small, finely edged and immediately opposite costals 3 and 4, sometimes almost forming interrupted fasciae; a small white dorsal mark below basal streak; apical spot small but distinct and often extended inwards as a black streak between pairs of strigulae, sometimes as far as end of dorsal 1; fringe line strong from costal 4 to dorsal 2, enclosing dorsals 3 and 4 if present and containing purple-reflecting scales at apex; cilia light brown round apex to end of fringe line, then fuscous. Hindwing light grey with slight brownish tinge; costal cilia light fuscous, darkening towards apex, dorsal cilia brownish. Abdomen in male greyish fuscous dorsally, light fuscous ventrally; in female light brown dorsally, ochreous white ventrally; caudal tuft dark ochreous fuscous in male, ochreous in female.

Life history

Ovum. Laid on the underside of a leaf of wayfaring-tree (*Viburnum lantana*), sometimes on a leaf of guelder-rose (*V. opulus*) or laurustinus (*V. tinus*) and very occasionally on rowan (*Sorbus aucuparia*). When present, seedling plants or leaves below overhead foliage appear to be very attractive for oviposition.

Larva. Head pale brown. Body pale amber with a faint greenish tinge, gut green from thoracic segment 3. July; September to April, the larva overwintering in the mine and pupating in the spring.

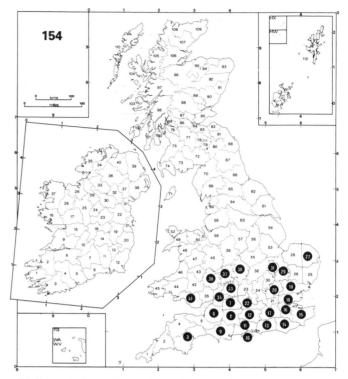

Phyllonorycter lantanella

Mine. Underside. Formed between two lateral veins causing a strong pucker in the upperside of the leaf, the lower surface initially white, both surfaces of the mine eventually turning brown. On guelder-rose the mine is longer and narrower.

Pupa (figure 105h, p. 303). Light brown; setae short; cremaster consisting of a pair of long processes. In a large, silk-lined chamber formed from the sides of the mine, without any real cocoon. The pupa is extruded through the upper surface of the mine near the midrib before the emergence of the adult. April; July to August.

Imago. Bivoltine. May; August.

Distribution (Map 154)

Widespread and very locally common on calcareous soils where its foodplants are found in south-eastern England and eastern Wales, extending north as far as Huntingdonshire and Warwickshire and west as far as south Devon and Glamorgan. A single record from south Tipperary, Ireland (Beirne, 1941), was based on a misidentification. Throughout Europe.

PHYLLONORYCTER CORYLIFOLIELLA (Hübner)

Tinea corylifoliella Hübner, 1796, *Samml.eur.Schmett.* **8**: pl.28, fig.194.

Lithocolletis betulae Zeller, 1839, *Isis, Leipzig* **1839**: 217.

Lithocolletis caledoniella Stainton, 1851, *Suppl.Cat.Br. Tineidae and Pterophoridae*: 12.

Type locality: Europe.

Description of imago (Pl.12, figs 21–23)

Wingspan 8–9mm. Head with vertical tuft bright chestnut-brown, whitish posteriorly; frons and labial palpus shining white; antenna white, segments strongly banded with fuscous above at apex, except in apical quarter, scape white. Thorax chestnut-brown finely edged white, with fine white central line; tegulae chestnut-brown, white-edged outwardly; foreleg grey banded white, midleg less strongly banded, hindleg with tibia dark ochreous and tarsal segments white ringed dark grey. Forewing rich matt chestnut-brown; one costal and one, occasionally two, dorsal strigulae whitish and very narrow; a long white basal streak to between two-fifths and one-half, angled obtusely upwards at about two-thirds of its length; costal 1 at about one-half, short, extending obliquely towards apex and almost meeting the slightly curved dorsal 1; dorsal 2 if present usually expressed by a few dark edging scales; apical spot usually obscure or absent, with an irregular band of dark brown scales varying in intensity and in some individuals extending almost to end of dorsal 1; a second patch of dark scales on costa on outer side of costal 1; fringe line fairly strong; a few blackish scales along dorsum; cilia light brownish fuscous. Hindwing light grey, cilia light brownish fuscous. Abdomen in male light fuscous dorsally, white ventrally, caudal tuft creamy white; in female dark grey dorsally, white ventrally, caudal tuft dark ochreous above, lighter ochreous below.

In f. *betulae* Zeller (fig.23) the forewing has whiter strigulae and is marked with striking dark brown patches constituting areas in basal third on costa and dorsum, large subtriangular marks preceding dorsal 1 and dorsal 2 and an area beyond costal 1 through disc to apex. This form grades into f. *caledoniella* Stainton in which the chestnut-brown ground colour of the typical form is almost completely replaced by dark brown. Both forms are found only on birch (*Betula* spp.) and replace the typical form in the north of England and Scotland; they are also occasionally found on birch in southern England.

Life history

Ovum. Laid on the upperside of a leaf of one of a number of rosaceous trees or shrubs or their cultivars, including apple (*Malus* spp.), wild pear (*Pyrus communis*), hawthorn (*Crataegus* spp.), rowan (*Sorbus aucuparia*), common

whitebeam (*S. aria*) and other *Sorbus* spp.; also on birch (*Betula* spp.). The statements by Jacobs (1945) and Ford (1949) that hazel (*Corylus avellana*) is a foodplant appear to have arisen through confusion of the early stages with those of *P. coryli* (Nicelli).

Larva. Head dark brownish. Body pale yellowish, greenish between segments, gut green; first thoracic and anal segments with darker green patch. July; September to October. The f. *betulae* and f. *caledoniella* appear to feed only in the autumn. (See also figure 94, p. 244).

Mine. Upperside. Soon after the formation of the epidermal blotch has begun, a differently coloured (generally paler) central patch becomes evident. This is caused by a mine within the mine, the larva, still in its second instar, having penetrated into the upper surface of the parenchyma. The mandibles are still adapted for sap-drinking and according to Hering (1951: 140–141) the larva has opened the upper part of the parenchymal cells in order to consume their sap; however, the presence of a patch of black, solid frass adhering to the lower surface of the inner mine suggests that some tissue is also ingested. The transition to the third, still sap-feeding, instar takes place in the inner mine. This has a hole in its upper wall through which the larva passes to and fro, alternately enlarging both mines. In the second instar the inner mine is almost coextensive with the outer, but in the third the epidermal mine is enlarged more rapidly until it has a diameter at least four times that of the inner mine. In the fourth instar, when the mouth parts are adapted for tissue-feeding, the larva eats the upper wall of the inner mine and eventually the only trace of it left is the patch of black frass. Hering (*loc.cit.*) states that this is the only known instance of a double mine. When on apple, the mine in its early stages resembles that of *Callisto denticulella* (Thunberg), *q.v.*, but can readily be distinguished by the differently coloured central patch caused by the inner mine.

The upper epidermis is at first silvery white but later characteristically more tinged with brown and speckled with rusty-coloured flecks. The completed mine is rather extensive and roughly fusiform; it is generally formed over the midrib or a major vein, causing the leaf to fold strongly upwards, the upper surface with many small creases, the parenchyma attached to the lower surface marbled but rarely all consumed.

Pupa. Entire dorsal surface covered with very small spines and with longer spines on anterior portion of abdominal segments 1–3; setae long; cremaster with two pairs of processes, the inner pair thin and hooked inwards, the outer pair wide-based, stouter and hooked outwards at tips. In an off-white silken cocoon formed from the sides of the mine and placed towards one end, the frass disposed along the central line behind the cocoon. The pupa is

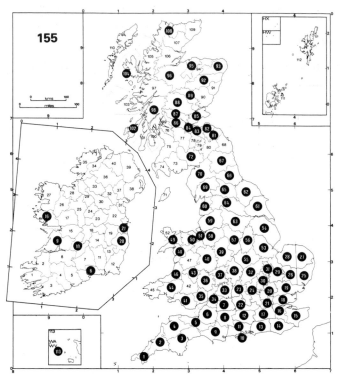

Phyllonorycter corylifoliella

extruded through the lower surface before the emergence of the adult. July to August; October to May.

Imago. Bivoltine in the south, with the moths occurring in May and June, and again in August. In the north f. *betulae* and f. *caledoniella* appear to have only one generation with the adults occurring in July and August.

Distribution (Map 155)

Widespread and generally common throughout the British Isles, unrecorded in the northern half of Ireland. Throughout Europe, but f. *betulae* is recorded only from central and northern Europe.

PHYLLONORYCTER VIMINIELLA (Sircom)

Argyromiges viminiella Sircom, 1848, *Zoologist* 6: 2271.
Type locality: England; [Brislington, Somerset].

Description of imago (Pl.12, fig.26)

Wingspan 7.5–8.0mm. Head with vertical tuft pale coppery brown, whitish posteriorly; frons and labial palpus white; antenna white, middle third sometimes banded greyish above, terminal segment black, scape white. Thorax ochreous brown, sometimes with fine central line and finely edged pale golden; tegulae light golden brown, paler-edged outwardly; foreleg white banded dark grey, midleg less conspicuously banded, hindleg with tibia yellowish white, tarsus white. Forewing pale coppery brown; four costal and three, sometimes two, dorsal strigulae varying from whitish to a slight lightening of the ground colour, edged inwardly with darker scales on the first two pairs, sometimes very indistinctly; a narrow, sinuous basal streak to slightly beyond one-third; costal 1 oblique, narrow and short; costal 2 arc-shaped, short; costals 3 and 4 often indistinct, sometimes joining to form a U-shaped strigula above apical streak; dorsal 1 long, narrow, oblique and curved towards apex, reaching beyond costal 1, nearly forming a sharply acute angle with it; dorsal 2 more triangular in shape; dorsal 3 diffuse; apical spot very indistinct, preceded by a narrow, somewhat irregular line of brown scales to junction of second pair of strigulae, often forming a distinct edge against costals 3 and 4; fringe line fairly strong from costal 4, fading dorsally to dorsal 2, usually enclosing dorsal 3; cilia pale ochreous brown. Hindwing grey, cilia pale ochreous brown, fuscous at apex. Abdomen light brown dorsally, white ventrally; caudal tuft ochreous fuscous above, brassy white below.

Life history

Ovum. Laid on the underside of a leaf of osier (*Salix viminalis*), crack willow (*S. fragilis*) or more rarely one of the other species of *Salix*.

Larva. Head pale brown, darker laterally. Body greenish yellow when young, pale amber when full-grown with the gut green, more conspicuous on abdominal segments 3 and 4, and an orange spot on abdominal segment 6. Late June to July; September to October or early November.

Mine. Underside. Usually at the extreme edge of a leaf, frequently near the petiole; often more than one in a leaf; the lower surface of the mine is strongly contracted, arching the upper epidermis and causing the leaf-edge to roll under, so forming a narrow tube.

Pupa. Varies from pale to dark brown; setae short; abdominal segments with anterior and posterior rows of short spines on dorsal surface; cremaster with two pairs of processes of equal length, the outer pair widely based and

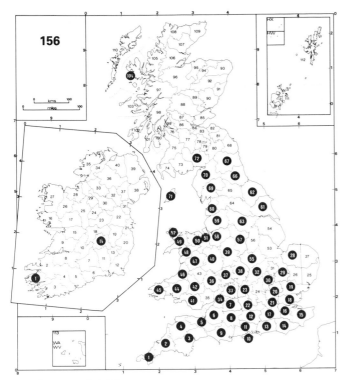

Phyllonorycter viminiella

stouter than the inner pair. Enclosed in a tough, very light brown cocoon formed from the sides of the mine; usually towards one end, with the frass disposed in a line throughout the rest of the mine's length. July to August; October to May.

Imago. Bivoltine. May and June; August.

Distribution (Map 156)

Most commonly found in fens, marshes and other wet areas where smooth-leaved *Salix* species are more frequently encountered. Locally common in England and Wales with isolated records from Scotland including the Inner Hebrides (P. H. Sterling, *in litt.*). In Ireland recorded only from south Kerry (Mere *et al.*, 1964) and Co. Leix (Beirne, 1945). Central and northern Europe.

PHYLLONORYCTER VIMINETORUM (Stainton)

Lithocolletis viminetorum Stainton, 1854, *Ins.Br.Lepid.* **3**: 272.

Type locality: England; Lewisham, Kent.

Description of imago (Pl.12, figs 27,28)

Wingspan 8–9mm. Head with vertical tuft brown, paler posteriorly; frons and labial palpus shining white, slightly brassy; antenna fuscous annulated darker, whitish at base, subterminal nine or ten segments white, terminal segment black, scape white. Thorax light golden brown, white anteriorly and laterally; tegulae light golden brown, widely edged white outwardly; foreleg white strongly annulated dark grey, midleg less strongly annulated, hindleg with tibia fuscous, tarsus white. Forewing light golden brown sprinkled with blackish brown scales; four costal and three dorsal strigulae, a sinuate basal streak almost reaching dorsal 1, and a strong dorsal spot reaching basal streak at about its middle and extending along dorsum to base, all white with only the faintest trace of darker edging, often reduced to a few isolated scales; costal 1 and dorsal 1 meeting slightly above middle line of wing and forming a right angle or occasionally a sinuate outward-curving fascia; costal 2 small; dorsal 2 larger, triangular and opposite, with a patch of darker scales in between; costals 3 and 4 and dorsal 3 indistinct; apical spot black, small, sometimes with a faint dark line extending back to junction of third pair of strigulae; a faint fringe line round apex from costal 4 nearly to dorsal 3, strongest at apex; cilia pale brownish fuscous, darker around tornus. Hindwing grey, cilia pale brownish fuscous. Abdomen light greyish brown dorsally, white ventrally; caudal tuft ochreous fuscous above, brassy white below.

Life history

Ovum. Laid on the underside of a leaf of osier (*Salix viminalis*).

Larva. Head pale brown. Body whitish yellow when small, pale amber when full-grown, gut greenish; an orange spot on abdominal segment 6. July; September to early November.

Mine. Underside. Usually formed near the petiole, invariably on the edge of the leaf, the underside being strongly contracted, arching the upper epidermis and causing the leaf-edge to roll completely over and form a very narrow tube. Very similar to the mine of *P. viminiella* (Sircom) but perhaps a little smaller and more tightly rolled. Often there is more than one mine in a leaf.

Pupa. Light yellow-brown; setae short; spines on dorsal surface of abdominal segments placed in distinct anterior and posterior rows; cremaster with one pair of processes, resembling *P. salicicolella* (Sircom) (figure 105m, p. 303).

In a chamber within the mine without a cocoon. The pupa is extruded through the lower epidermis before the emergence of the adult. July to August; October to May.

Imago. Bivoltine. May and June, very slightly later than *P. viminiella*; August.

Distribution (Map 157)

Restricted to osier beds and marshy areas where osier is well established. Very local and usually uncommon throughout England; in Ireland recorded only from Co. Wicklow (Beirne, 1941). Central Europe.

PHYLLONORYCTER SALICICOLELLA (Sircom)

Argyromiges salicicolella Sircom, 1848, *Zoologist* **6**: 2271.
Type locality: England; [Brislington, Somerset].

Description of imago (Pl.12, figs 30–33)

Wingspan 7.0–8.5mm. Head with vertical tuft golden brown or brown, whitish posteriorly; frons and labial palpus shining white, slightly brassy; antenna white, middle third annulated grey, terminal segment black, scape white. Thorax golden brown with faint white central line, sometimes absent; tegulae golden brown, white-edged inwardly; foreleg dark sooty fuscous annulated white, midleg lighter, hindleg with tibia ochreous brown, tarsus white. Forewing light sienna brown; four costal and three dorsal white strigulae, first two pairs usually but not always edged inwardly with blackish brown; first pair with dorsal larger than costal, the dark edges when present almost meeting to form an obtuse angle or curved line, the white forming a chevron or curved band; second pair nearly equal in size, again with the dark edging nearly meeting to form a slightly angulate line across wing, but with the white areas not meeting; costals 3 and 4 and dorsal 3 unedged and usually less distinct, often of a more yellowish white, dorsal between the two costals; a fine, unedged, sinuate basal streak to slightly beyond one-third; apical spot small, diffuse, directly adjacent to costal 4 and surrounded by a few scattered dark scales, sometimes forming a short streak; fringe line from costal 4 to dorsal 3, but only distinct around apex; cilia whitish fuscous. Hindwing grey, cilia pale fuscous, darker above. Abdomen dark grey dorsally, white ventrally; caudal tuft yellowish brown.

Similar species. P. dubitella (Herrich-Schäffer), in which the first pair of strigulae are elongate, curve strongly towards apex of wing and almost touch to form a sharply acute chevron. In *P. salicicolella* the first pair of strigulae almost meet to form a curved band, an obtusely angled chevron or a right angle (Ffennell, 1970). The fringe line on the termen is slightly concave in *P. dubitella*, slightly convex in *P. salicicolella*.

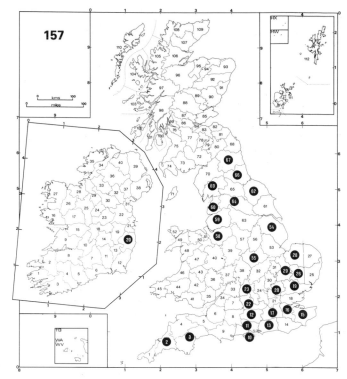

Phyllonorycter viminetorum

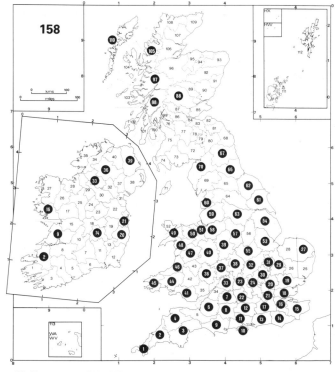

Phyllonorycter salicicolella

Life history

Ovum. Laid on the underside of a leaf of sallow (*Salix* spp.).

Larva. Head brown. Body whitish green, gut darker green; an orange spot on abdominal segment 6. July; September to October.

Mine. Underside. Positioned anywhere between the midrib and the edge of the leaf, when near the midrib usually between two veins, when near the margin usually in the middle or distal section of the leaf, crossing several veins and causing the edge of the leaf to curl over and making a noticeable pucker in the upper epidermis. The inside of the mine is lined with brown silk.

Pupa (figure 105m, p. 303). Pale brown; setae short; abdominal segments with distinct anterior and posterior rows of spines on dorsal surface; cremaster with one pair of processes. In a large, loosely woven, whitish or very pale yellow-brown cocoon, integral with the walls of the mine, the frass in a neat pile behind. The pupa is extruded through the lower surface of the mine before the emergence of the adult. July to August; October to May.

Imago. Bivoltine. May; late July and August.

Distribution (Map 158)

Generally widespread and common throughout England, Wales and Ireland, being most plentiful in the south of England; in Scotland recorded from as far north as the Outer Hebrides and appearing to have a mainly westerly distribution. Central and northern Europe.

PHYLLONORYCTER DUBITELLA
(Herrich-Schäffer)

Lithocolletis dubitella Herrich-Schäffer, 1855, *Syst.Bearb. Schmett.Eur.* 5: 325.

Type locality: Germany; Regensburg.

Description of imago (Pl.12, fig.29)

Wingspan 7–8mm. Head with vertical tuft orange-brown, whitish posteriorly; frons and labial palpus shining white; antenna white annulated light fuscous, two terminal segments black, scape white. Thorax golden ochreous with white central line and finely edged white; tegulae white, golden anteriorly; legs white beneath, above with femora and tibiae fuscous and tarsi whitish banded fuscous, more strongly on foreleg. Forewing golden brown or coppery brown; four costal and three dorsal strigulae, all white finely edged blackish inwardly; first pair narrow, oblique, elongate, curving strongly towards apex and nearly touching to form a sharply acute angle, dorsal slightly longer than costal; second pair of strigulae arc-shaped, with black edging meeting to form an obtuse-angled fas-

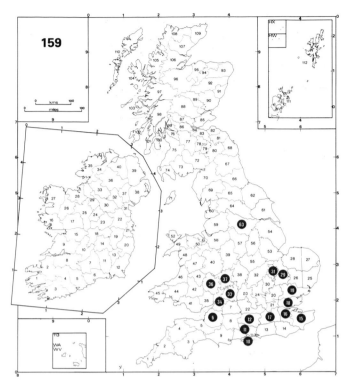

Phyllonorycter dubitella

cia; costals 3 and 4 and dorsal 3 more obscure with finer edging, third pair sometimes meeting to form a weak white fascia across wing; basal streak to two-fifths, slightly upturned for distal third, sometimes showing faint trace of dark border on lower edge at furthest point from base; a small white patch on dorsum between dorsal 1 and base; apical spot diffuse, integral with a line of black scales forming a short streak below costal 4 to dorsal 3; fringe line strong from costal 4 to dorsal 2, enclosing dorsal 3; cilia pale fuscous. Hindwing grey, cilia pale fuscous, darker towards apex. Abdomen fuscous dorsally, white ventrally; caudal tuft ochreous white above, whitish below.

Similar species. P. salicicolella (Sircom), *q.v.*; weakly marked specimens resemble *P. viminiella* (Sircom) in shape and colouring but the strigulae are more distinct.

Life history

Ovum. Laid on the underside of a leaf of goat-willow (*Salix caprea*).

Larva. Not described. July; September to October.

Mine. Underside. Fairly large, variously positioned on the leaf; the upper surface strongly puckered, the underside covered by small creases, these often being hidden by the leaf-hairs.

Pupa. Golden brown; setae long; other characters as for *P. salicicolella*. In a large, golden, silken cocoon, more resembling that of *P. spinolella* (Duponchel) than that of *P. salicicolella*, but slightly more thickly constructed and with the frass in a heap behind the cocoon. July to August; October to May.

Imago. Bivoltine. May and June; August.

Distribution (Map 159)

Locally common in southern England to south Yorkshire. It is a recent addition to the British list (Ffennell, 1970), and further studies may show it to be more widespread. Central and northern Europe.

PHYLLONORYCTER SPINOLELLA (Duponchel)

Elachista spinolella Duponchel, 1840, *in* Godart & Duponchel, *Hist.nat.Lépid.Fr.* **11**: 535.

Type locality: France.

Description of imago (Pl.12, figs 24,25)

Wingspan 7–9mm. Head with vertical tuft whitish, brown anteriorly; frons and labial palpus shining white; antenna grey with proximal third and subterminal 10 or 11 segments white, terminal segment black, scape white. Thorax brown, edged white; tegulae brown, edged white inwardly; legs with tibiae fuscous, tarsi white banded fuscous, more conspicuously on foreleg. Forewing golden brown, darkening towards apex; four costal and three, occasionally four, dorsal strigulae, all pure white dark-edged inwardly, particularly the first two pairs, the dark edge of costal 1 extending along costa, occasionally as far as base; costal 1 and dorsal 1 meeting to form a fairly wide outward-curving white fascia, sometimes angulated in middle; second pair of strigulae almost opposite, nearly meeting to form an obtuse angle; dorsal 3 between costals 3 and 4 and sometimes meeting one or other, rarely both, to form an angled fascia; a white basal streak to about one-third, broader towards apex, usually unedged but distal end occasionally with faint edging; a rather wide white dorsal spot between dorsal 1 and base; apical spot black, diffuse, edged by costal 4 and last dorsal; fringe line rather indistinct between costal 4 and dorsal 3; cilia fuscous with dark sheen, the white of costals 3 and 4 and dorsal 3 extending through to tips of cilia. Hindwing greyish fuscous, cilia fuscous with brownish sheen. Abdomen in male dark fuscous dorsally, light fuscous ventrally with caudal tuft ochreous brown; in female light fuscous dorsally, white ventrally, the caudal tuft ochreous white.

Life history

Ovum. Laid on the underside of a leaf of sallow (*Salix* spp.), usually goat-willow (*Salix caprea*), often on bushes

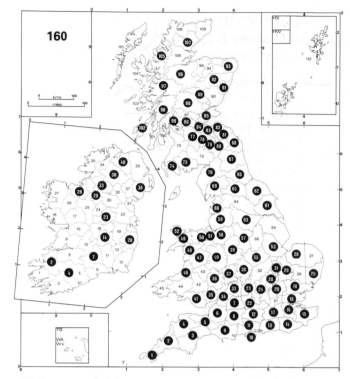

Phyllonorycter spinolella

or trees at the edge of woodland or woodland rides.

Larva. Head pale brown. Body lightish yellow. July; September to early November.

Mine. Underside. Fairly large and irregular, often in the middle part of the leaf between two veins. The upperside is strongly puckered, the underside covered in minute creases hidden by leaf-hairs.

Pupa (figure 105l, p. 303). Pale brown; cremaster with processes broader and their apical portion shorter, otherwise as for *P. salicicolella* (Sircom). In a pale golden brown silken cocoon in the pucker of the upper epidermis of the mine, the frass in a heap behind. The cocoon is tougher than that of *P. salicicolella*.

Imago. Bivoltine. The moths occur in late May and June, some two weeks later than *P. salicicolella*, and again in August.

Distribution (Map 160)

Widespread and common throughout the British Isles, being the most prevalent sallow-feeding *Phyllonorycter* in the north of England and Scotland. Throughout Europe.

PHYLLONORYCTER CAVELLA (Zeller)

Lithocolletis cavella Zeller, 1846, *Linn.ent.* **1**: 213.
Type locality: Germany; Glogau (now Poland; Glogów).

Description of imago (Pl.12, figs 34–36)

Wingspan 7–9mm. Head with vertical tuft whitish, brown or pale brown anteriorly; frons and labial palpus white; antenna white, marked above with dark brown on outer edge of each segment, less so distally, terminal segment blackish brown, scape white. Thorax shining golden brown with white central line; tegulae white, narrowly brown anteriorly; foreleg dark brownish fuscous marked white on inner edge of tarsal and tibial segments, mid- and hindlegs whitish, faintly marked light fuscous. Forewing shining golden brown, darker in central part of wing; four costal and three dorsal strigulae pearly white, rarely with a slight golden sheen on outer dorsals, all finely black-margined inwardly; a strong white basal streak to one-third, costal margin straight with black edge sometimes continuing slightly round apex, dorsal margin otherwise unedged and more sinuous; a fine white border along dorsum below basal streak but not usually reaching first dorsal strigula or base; first pair of strigulae opposite at two-fifths, in most specimens the costal short, almost delta-shaped, the dorsal long, smoothly curved, sickle-shaped, extending beyond middle line of wing and reaching just beyond end of costal, in other cases, however, the two strigulae meeting to form a broad fascia with their pronounced dark inside edge making an acute or right angle; the second pair opposite with the dorsal triangular and larger than the costal, the latter arc-shaped, not reaching the dorsal, but the dark edging almost meeting to form an obtusely angled fascia; third pair opposite, narrow, wedge-shaped, usually slightly curved so that the dark edging forms a faint, nearly continuous, inward-curving fascia across wing; costal 4 narrow with only a fine edging; apical spot small, black, diffuse, usually with a few white scales above and extending a short distance inwards; fringe line fairly strong from costal 4 to dorsal 2, interrupted by dorsal 3; cilia pale fuscous, lighter opposite dorsals 2 and 3. Hindwing pale grey, cilia pale fuscous. Abdomen dark fuscous dorsally, paler ventrally; caudal tuft ochreous fuscous above, whitish below.

Individuals with five costals, often four dorsals and with a slightly darker ground colour are referable to f. *milleri* Povolný & Gregor.

Life history

Ovum. Laid on the underside of a leaf of birch (*Betula* spp.), and, according to Povolný & Gregor (1950), occasionally on dwarf cherry (*Prunus cerasus*).
Larva. Head brownish. Body pale yellowish green, gut darker green; an orange patch on abdominal segment 6.

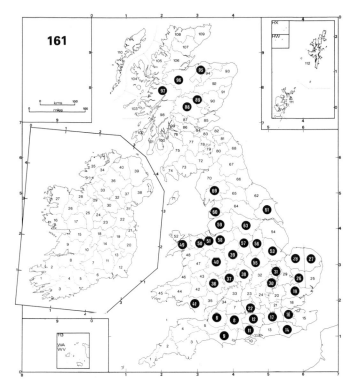

Phyllonorycter cavella

Late August to end of October.
Mine. Underside. Very large (15–20mm long), underside strongly contracted with some seven to twelve creases, causing the upperside to arch strongly and considerably contorting the leaf; very occasionally as a large silvery white mine on the upperside of the leaf, folding it strongly upwards.
Pupa (figure 105d, p. 303). Pale golden brown; dorsal surface of abdominal segments 1–3 with two distinct rows of large spines; setae short; cremaster with two pairs of processes, the middle pair less well developed. In a white cocoon in the mine. October to June.
Imago. Univoltine (Emmet, 1976). June and July. The claim by Stainton (1859) and subsequent authors that the species is bivoltine in Britain has not been substantiated.

Distribution (Map 161)

Found locally throughout England and Wales and in the Highlands of Scotland, perhaps most commonly in the north-western counties of England. Specimens of f. *milleri* determined as *P. blancardella* (Fabricius) in the S. Wakely collection were taken at Darenth, Kent, but extensive searching in that area subsequently has failed to reveal the presence of the species (Watkinson & Chalmers-Hunt, 1980). Central and northern Europe.

PHYLLONORYCTER ULICICOLELLA (Stainton)

Lithocolletis ulicicolella Stainton, 1851, *Suppl.Cat.Br. Tineidae and Pterophoridae*: 112.

Type locality: England; Durdham Down, Bristol.

Description of imago (Pl.12, fig.37)

Wingspan 6–7mm. Head with vertical tuft pale coppery brown, whitish posteriorly; frons and labial palpus whitish; antenna greyish above, white below, scape whitish. Thorax pale coppery brown with, or sometimes without, a white central line and white-edged laterally; tegulae pale coppery brown inwardly edged white; foreleg dark ochreous fuscous, midleg similar but banded paler on tarsus, hindleg whitish lightly banded fuscous. Forewing pale coppery or golden brown; four costal and three dorsal strigulae, and a slightly sinuate basal streak to about one-third, white without any strong edging; costal 1 short, narrow, obtuse, sometimes hook-shaped; costals 2–4 narrow, arc-shaped, not always clearly differentiated from ground colour; dorsal 1 long, narrow, obtuse, sometimes sinuate, reaching beyond midline of wing, reaching or passing end of costal 1 and sometimes extending along dorsum to base as fine white edging; dorsal 2 small, triangular, often with short obtuse extension of apex towards end of costal 2, sometimes with border slightly darker than ground colour on inner edge; dorsal 3 very faint, opposite to and occasionally blending with costal 3, sometimes not present and often not reaching fringe line; apical spot small, obscure, extending inwards as a short streak towards junction of third pair of strigulae and edged above with white; fringe line fairly strong from costal 4 to dorsal 2, enclosing dorsal 3; cilia pale fuscous with golden sheen. Hindwing light grey, cilia pale fuscous with light golden sheen. Abdomen dark fuscous dorsally, whitish fuscous ventrally; caudal tuft fuscous above, white below.

Life history

Ovum. Laid on the green bark near the tip of a shoot of gorse (*Ulex europaeus*).

Larva. Not described. September to May, feeding through the winter.

Mine. Somewhat long, formed in the epidermis of the green bark of a thin twig, the mine surface with fine longitudinal folds and strengthened inside with silk. The mine is almost invisible, even when its position is indicated by the extruded pupal exuviae.

Pupa. Brown; abdominal segments with small spines evenly distributed over dorsal surface; setae short; cremaster with two pairs of small processes, more or less equal in size. In a silk-lined chamber within the mine. May to June.

Imago. Univoltine. The moths occur in June and early

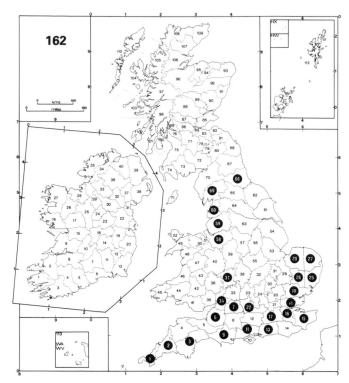

Phyllonorycter ulicicolella

July and may be obtained by beating gorse bushes.

Distribution (Map 162)

Very local, occurring mainly in the coastal counties of southern England from Norfolk to Cornwall and in the north-west. It may well be overlooked in many areas since, unlike other species of the genus, it is seldom encountered except as an adult. Abroad it is known only from northern France.

PHYLLONORYCTER SCOPARIELLA (Zeller)

Lithocolletis scopariella Zeller, 1846, *Linn.ent.* 1: 227.

Type locality: Germany; Glogau (now Poland; Glogów).

Description of imago (Pl.13, fig.1)

Wingspan 6–8mm.Head with vertical tuft brown, white posteriorly; frons and labial palpus whitish; antenna pale greyish, marked slightly darker grey on distal end of each segment, scape whitish. Thorax golden brown, usually with white central line; tegulae similar, white-edged inwardly; foreleg light fuscous marked darker on tarsal segments, midleg whitish very lightly banded fuscous on tibia and tarsal segments, hindleg whitish. Forewing golden brown, sometimes suffused with greyish brown; five costal and four dorsal strigulae variable in appearance, a basal streak to slightly beyond one-third, its distal end angled slightly towards costa, and a dorsal patch below basal streak between base and dorsal 1, all white and usually without any darker edging; costal 1 at two-fifths, narrow, very oblique; costals 2 and 3 arc-shaped, sometimes faintly dark-edged inwardly; costals 4 and 5 sometimes very faint and represented by a few white scales only; dorsal 1 long, narrow, obtuse and curved or sinuate, extending beyond costal 1 nearly to end of costal 2 but often connected with costal 1 to form an acute-angled fascia; dorsal 2 triangular or delta-shaped, opposite to but not reaching costal 2; dorsal 3 between costals 3 and 4 and sometimes connecting with one or other or both; dorsal 4 small, obscure, between costals 4 and 5; apical spot brown, small and obscure, sometimes extended inwards for a short distance as a faint diffuse line of brown scales; fringe line very weak from costal 4 to dorsal 2, enclosing dorsal 4 but not dorsal 3 or costal 5; cilia pale fuscous with golden sheen. Hindwing pale grey, cilia pale fuscous with golden sheen. Abdomen light fuscous dorsally, white ventrally; caudal tuft ochreous fuscous above, white below.

Life history

Ovum. Laid on the green bark near the tip of a shoot of broom (*Sarothamnus scoparius*).

Larva. Not described. September, feeding through to May of the following year.

Mine. Long (15–20mm), narrow and inflated, in the green bark of a thin twig between the small ridges, usually near the end of the twig and often on a sapling; the epidermis over the mine turns dark greenish grey. Difficult to find, though less so than *P. ulicicolella* (Stainton).

Pupa (figure 105s, p. 303). Light brown; abdominal segments with small spines evenly distributed over dorsal surface; setae short; cremaster with two pairs of short processes, outer pair distinctly smaller and gently curved outwards, inner pair curved upwards from dorsum almost at a right angle. In a silk-lined chamber within the mine. May to June.

Imago. Univoltine. The moths occur in June and early July and are best obtained by beating broom bushes.

Distribution (Map 163)

Widely scattered and locally common throughout England, Scotland and eastern Wales; in Ireland recorded from Cos Wicklow and Dublin (Beirne, 1941; 1945). The species may be overlooked in many areas since, unlike most other species of the genus, it is seldom encountered except as an adult. Central and western Europe.

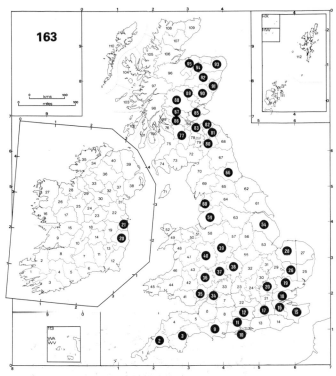

Phyllonorycter scopariella

PHYLLONORYCTER STAINTONIELLA (Nicelli)

Lithocolletis staintoniella Nicelli, 1853, *Proc.ent.Soc.Lond.*
2: 114.

Type locality: Germany; Frankfurt am Main.

Description of imago (Pl.11, fig.17)

Wingspan 4.5–6.0mm; British specimens are smaller than those from the Continent. Head with vertical tuft reddish brown mixed fuscous, frons and labial palpus white; antenna shining greyish white with darker annulations. Thorax reddish brown with white central line; tegulae edged white. Forewing reddish brown; basal streak to one-third, straight or slightly angled towards costa near apex; three or four costal and three or four dorsal strigulae whitish or pale ochreous without dark edging; costal 1 and dorsal 1 distinct, opposite and sometimes confluent to form an angled fascia; remaining strigulae indistinct; dark scaling between tips of costal and dorsal strigulae in apical area; fringe line absent; cilia pale fuscous. Hindwing rather dark fuscous, cilia paler.

Life history

Ovum. Laid on the upper surface of a leaf of hairy greenweed (*Genista pilosa*).

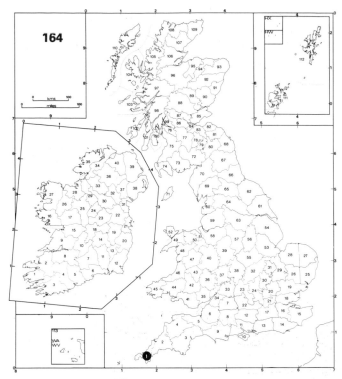

Phyllonorycter staintoniella

Larva. Head pale brown. Body deep yellow; gut dark green (Stainton, 1857). June to July. According to Stainton (*loc.cit.*) larvae also feed in early spring until May.

Mine. Upperside. In the sap-feeding phase, a small, greenish white, flat mine with a reddish brown mark in the vicinity of the egg-shell; in the tissue-eating phase, extensive spinning causes the leaf to fold upwards to resemble a pod almost concealing the mine. The frass is packed at the tip of the leaf.

Pupa. Dark brownish black. July.

Imago. Bivoltine on the Continent and probably also in Britain. In England adults have been reared in early August; there may also be an earlier generation.

Distribution (Map 164)

Discovered by Heckford (1984) near Perranporth on the north coast of Cornwall; its foodplant has a very restricted range in Britain and the moth may not, therefore, occur elsewhere. Denmark, France, western Germany, Austria and Czechoslovakia.

PHYLLONORYCTER MAESTINGELLA (Müller)

Phalaena (Tinea) maestingella Müller, 1764, *F.Inst.Frid.*: 58.
Lithocolletis faginella Zeller, 1846, *Linn.ent.* **1**: 204.
Type locality: Denmark; Copenhagen.

Description of imago (Pl.13, fig.2)

Wingspan 7.5–9.0mm. Head with vertical tuft brown or light brown, whitish posteriorly; frons shining white; labial palpus whitish; antenna white marked greyish fuscous on segments, more distinctly in apical third, scape white. Thorax light golden brown with white central line; tegulae white, narrowly brown anteriorly; legs white, foreleg with distal end of segments banded dark fuscous. Forewing varying from shining sienna brown to greyish brown; four costal and three dorsal white strigulae; basal streak to one-third white with apex pointed and only faintly dark-edged above; strigulae usually edged dark brown inwardly, the edging of costal 1 continuing a short distance along costa, sometimes to base; strigulae in pairs, first pair oblique, second pair less oblique, third pair almost vertically opposite; apical spot black with line of dark scales extending to end of costal 4; a fine line of white scales along dorsum near base; fringe line fairly strong from costal 4 round apex but fading gradually and not quite reaching dorsal 3; cilia light brown, becoming light fuscous on dorsum. Hindwing greyish fuscous, cilia fuscous with light brown sheen. Abdomen fuscous dorsally, white ventrally; caudal tuft brownish fuscous above, white below.

Life history

Ovum. Laid on the underside of a leaf of beech (*Fagus sylvatica*).

Larva. Head pale brownish, darker on edges. Body pale greenish yellow, gut darker green. July; late August to late October.

Mine. Underside. Between two lateral veins, often extending from midrib to leaf-edge, the lower surface contracted and the upper arched upwards to form a tube; the centre of the lower surface with many small folds, giving the impression of one larger fold; occasionally formed at the leaf-edge, causing the upper surface to curl under. Further details of the mining habits are given by Miller (1973).

Pupa. Pale brown; spines on dorsal surface of abdominal segments occupying reduced area, with larger spines in several rows towards anterior margin; setae long; cremaster with two pairs of processes of equal length, outer pair stouter than inner pair. In a loosely woven, white silken cocoon attached to both upper and lower surfaces, the frass in a neat pile in the middle of the mine. Late July to August; November to May.

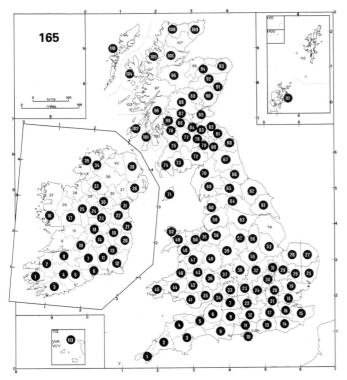

165

Phyllonorycter maestingella

Imago. Bivoltine. May and early June; August.

Distribution (Map 165)

Common throughout the British Isles. Throughout central and northern Europe.

PHYLLONORYCTER CORYLI (Nicelli)
Nut Leaf Blister Moth

Lithocolletis coryli Nicelli, 1851, *Stettin.ent.Ztg* **12**: 36.
Type locality: Germany; Stettin (now Poland; Szczecin).

Description of imago (Pl.13, fig.3)

Wingspan 7.0–8.5mm. Head with vertical tuft whitish, brown or light brown anteriorly; frons and labial palpus shining white; antenna white, segments on outer two-thirds lightly marked fuscous, scape white. Thorax pale orange-brown, anterior edge and central line white; tegulae pale orange-brown, white-edged inwardly; legs white lightly marked fuscous, more strongly on foreleg. Forewing light sienna brown; four costal and three dorsal strigulae white, dark-edged inwardly; costal 1 narrow, oblique and extending a short distance along costa towards base; costal 2 triangular; costals 3 and 4 arc-shaped; dorsal 1 long, narrow, oblique, both sides dark-edged on distal half, nearly meeting costal 1 to form a sharply acute angle, its white extending along dorsum for a short distance towards base; dorsal 2 distinct, triangular, nearly meeting costal 2 so that the dark edges form an obtuse angle; dorsal 3 small but distinct and within fringe line; apical spot small, black, extended as a short streak below costal 4 to meet junction of third pair of strigulae, a few dark scales below; fringe line strongly black from costal 4 to dorsal 2; cilia light brown tipped fuscous from costal 4 to apex to form very faint type 2 apical hook. Hindwing grey, cilia pale fuscous. Abdomen light brownish fuscous dorsally, white ventrally; caudal tuft whitish.

Life history

Ovum. Laid on the upperside of a leaf of hazel (*Corylus avellana*).

Larva. Head pale brown, paler at sides. Body with thoracic segments pale yellowish, abdominal segments 1–6 greenish, 7–10 yellowish, gut dark green and conspicuous on abdominal segments 7–9; a large orange patch on abdominal segment 6. July; September to October.

Mine. Upperside. In the sap-feeding instars an almost circular whitish blotch with particles of frass visible as brownish flecks; in the tissue-feeding instars the surface is markedly contracted with many small creases, causing strong folding of the leaf. If the mine is at the margin of the leaf the edge curls right over and resembles the fold (not a mine) made in its later instars by *Parornix devoniella* (Stainton). Often several mines are found in one leaf, causing it to be greatly contorted, though when this occurs some of the mines are aborted.

Pupa. Abdomen with central portion of segments covered with short spines; abdominal segments 1–3 also with anterior and posterior rows of longer spines; setae long;

cremaster with two pairs of processes, the inner pair thin and hooked inwards, the outer pair wide-based, stout and hooked outwards at tips. In a silk-lined chamber at one end of the mine, the frass stacked in a loose pile at the opposite end; the pupal site is often on an uneaten patch of parenchyma. July to August; October to May.

Imago. Bivoltine. May; August.

Distribution (Map 166)

Widespread and common throughout the British Isles except for Orkney, Shetland and the Outer Hebrides. Throughout central and northern Europe.

PHYLLONORYCTER QUINNATA (Geoffroy)

Tinea quinnata Geoffroy, 1785, *in* Fourcroy, *Ent.Paris* 2: 331.

Lithocolletis carpinicolella Stainton, 1851, *Suppl.Cat.Br. Tineidae and Pterophoridae*: 13.

Type locality: France; Paris.

Description of imago (Pl.13, fig.4)

Wingspan 7.5–9.0mm. Head with vertical tuft pale brown or pale golden brown anteriorly, white posteriorly; frons shining white; labial palpus pale golden; antenna shining white annulated fuscous, scape white. Thorax pale golden brown, anterior margin and central line white; tegulae pale golden brown, broadly white inwardly; legs white banded fuscous, more strongly on foreleg. Forewing pale golden brown; four costal and three dorsal white strigulae, narrowly black-edged inwardly; a narrow white basal streak to about two-fifths, slightly pointed at apex and sometimes slightly dark-edged above; costal 1 narrow and strongly angled, usually extended as a fine white edge along costa towards, but not reaching, base; costals 2–4 distinct and arc-shaped; dorsal 1 long, narrow, sinuate, at an acute angle to dorsum and extended along it to base; dorsal 2 triangular and opposite to costal 2; dorsal 3 triangular and within fringe line; first pair of strigulae nearly meeting at an acute angle, second pair nearly meeting at an obtuse angle, third pair sometimes meeting to form an angled fascia; apical spot black, extended to junction of third pair of strigulae as a short streak; fringe line from costal 4 to dorsal 2, enclosing dorsal 3, darkest from apex to dorsal 3 and preceded by a slight purple sheen; cilia pale fuscous, slightly dark-tipped from costal 4 to apex, forming faint type 2 hook. Hindwing pale grey, cilia light fuscous with brownish sheen. Abdomen golden fuscous above, white below; caudal tuft whitish.

Life history

Ovum. Laid on the upperside of a leaf of hornbeam (*Carpinus betulus*).

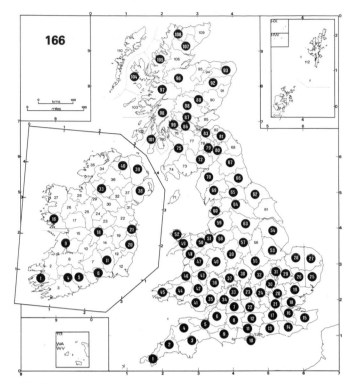

Phyllonorycter coryli

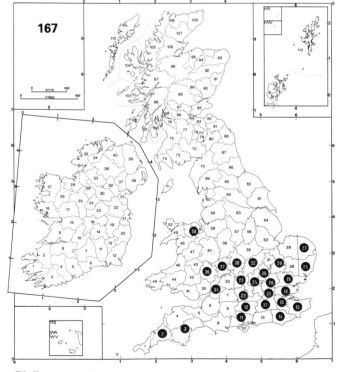

Phyllonorycter quinnata

Larva. Head pale brownish green, darker laterally. Body pale greenish white, gut darker green; a yellow spot on abdominal segment 6. July; September to October.

Mine. Upperside. In the sap-feeding phase a large, whitish, irregular-shaped blotch; in the tissue-feeding phase the upper surface contracts with many tiny creases strongly contorting the leaf, often completely folding it in half; usually the parenchyma is almost completely consumed.

Pupa (figure 105b, p. 303). Dorsal surface of abdominal segments 1–3 each with a prominent group of large spines placed centrally and towards anterior margin; setae short; cremaster with two pairs of well-developed processes of equal length, the outer pair stouter than the inner pair. In a small, tough, light brown cocoon formed at one extremity of the mine, the frass neatly hcapcd in anothcr part away from the cocoon. Late July to August; October to May.

Imago. Bivoltine. May; August.

Distribution (Map 167)

Locally common in south-eastern England to as far north as Northamptonshire, with isolated records outside this area from south Devon, the Severn valley and Denbighshire (H. N. Michaelis, pers.comm.). Central and northern Europe.

PHYLLONORYCTER STRIGULATELLA (Zeller)

Lithocolletis strigulatella Zeller, 1846, *Linn.ent.* **1**: 187.

Lithocolletis rajella Linnaeus, sensu auctt.

Lithocolletis alnifoliella Hübner, sensu auctt.

Type localities: Livonia (now Latvia S.S.R.) and Sweden.

Description of imago (Pl.13, fig.8)

Wingspan 7–9mm. Head with vertical tuft dark ochreous fuscous, some lighter scales posteriorly; frons and labial palpus shining white; antenna dark fuscous, paler beneath, subterminal eight segments white, terminal segment brown, scape white. Thorax and tegulae sooty brown, latter occasionally white-edged inwardly; legs dark brown, tarsi annulated paler. Forewing bright orange-brown, becoming sooty along costa and dorsum; four costal and three dorsal white strigulae, first two pairs thickly dark-edged inwardly and outwardly, third pair and costal 4 inwardly only; costal 1 long, narrow, truncate with parallel sides; costals 2–4 all distinct and arc-shaped; dorsals 1 and 2 triangular, 2 usually more pointed than 1; dorsal 3 smaller; narrow white basal streak to one-third, dark-edged on both sides; white dorsal patch between dorsal 1 and base, dark-edged, sometimes forming an extra strigula; apex of dorsal 1 under beginning of costal 1; apex of dorsal 2 between costals 2 and 3; apical spot

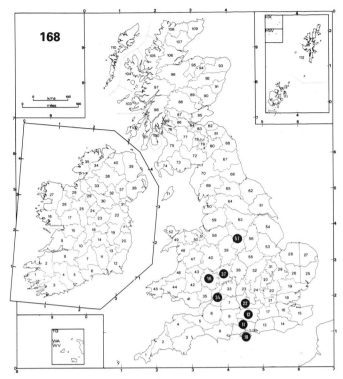

Phyllonorycter strigulatella

distinct; fringe line strong from costal 4 to dorsal 2, cut by dorsal 3; cilia fuscous, darker from costal 4 to apex. Hindwing grey, cilia fuscous with very slight brown sheen. Abdomen dark fuscous dorsally, light fuscous ventrally; caudal tuft dark ochreous fuscous.

Life history

Ovum. Laid on the underside of a leaf of grey alder (*Alnus incana*).

Larva. Head brownish, blackish laterally. Body pale yellowish. June to July; September to October.

Mine. Upperside. Long, narrow and sometimes irregular in outline, between two veins close to, but not quite reaching, the midrib; often several mines are adjacent; underside with barely visible small creases; parenchyma on upperside evenly marbled with feeding sites.

Pupa. Light brown; setae long; abdominal segments on dorsal surface each with one prominent anterior row of large spines, followed by rows of spines becoming progressively smaller towards posterior margin, on ventral surface with two rows of large, laterally directed spines; cremaster with two pairs of processes of equal length, the outer pair stouter than the inner pair. In a large, light yellow chamber in the middle of the mine, the frass having been cleared away from this area. July; October to May.

Imago. Bivoltine. Early May; late July and August.

Distribution (Map 168)

Unknown in the British Isles before 1928 when mines were collected by Waters (1929) at Cothill, Berkshire. Occurs in very local, isolated colonies in central southern England, the southern Welsh border counties and Derbyshire. Grey alder is not native in Britain but where *P. strigulatella* has been imported with the tree it can be very common. However, despite the increased planting of grey alder as a windbreak around fields, the species does not seem to be extending its range. Central and northern Europe.

PHYLLONORYCTER RAJELLA (Linnaeus)

Phalaena (*Tinea*) *rajella* Linnaeus, 1758, *Syst.Nat.* (Edn 10) **I**: 542.

Tinea alnifoliella Hübner, [1796], *Samml.eur.Schmett.* **8**: pl.28, fig.193.

Type locality: Europe.

Description of imago (Pl.13, figs 5–7)

Wingspan 7–9mm. A very variable species. Head with vertical tuft fuscous anteriorly, white posteriorly; frons shining white; labial palpus off-white; antenna pale fuscous, whitish at base and apex. Thorax and tegulae varying from dark fuscous to white; legs fuscous brown, tarsi dark fuscous narrowly banded paler before intersegmental joints. Forewing with ground colour ranging from light orange-brown to fuscous, the female tending to be paler than the male; four costal and three dorsal white strigulae narrowly dark-edged inwardly, although this is not always obvious on specimens with dark ground colour; broad white basal streak to two-fifths, with costal edge narrowly dark-bordered and arched towards costa, sometimes extensive enough to give basal half of wing a whitish appearance; costal 1 dark-lined on inner and outer edges, elongate and strongly angled towards apex of wing; costals 2–4 distinctly arc-shaped; dorsal 1 large with inner edge curving strongly towards apex until parallel with dorsum, outer edge sometimes meeting dorsum at a right angle; dorsal 2 triangular, usually with a fine extension of the apex of the triangle angled towards, but not meeting, apical spot; dorsal 3 small but distinct, cutting fringe line; sometimes there is an area of ochreous or even whitish scales above and below the basal streak, the dark edging of costal 1 being continued along costa to base; apical spot large and black; fringe line strong from above costal 4 to dorsal 2, cleanly cut by dorsal 3; cilia sooty between costal strigulae to apex, then pale fuscous. Hindwing grey, cilia fuscous with brown sheen darker towards apex. Abdomen

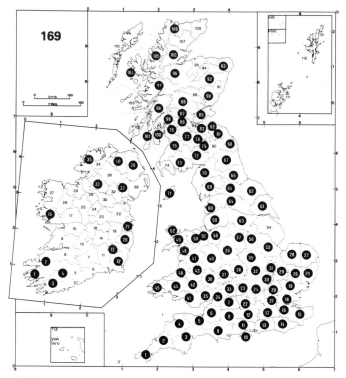

Phyllonorycter rajella

brownish fuscous dorsally, light fuscous ventrally; caudal tuft dark yellowish fuscous.

In the north, the dark areas of the wings may be extended to give a blackish fuscous ground colour.

Life history

Ovum. Laid on the underside of a leaf of alder (*Alnus* spp.).

Larva. Head pale brown, darker laterally. Body pale greenish, gut darker. July; September to October.

Mine. Underside. Small, usually against the midrib between two veins when it is somewhat triangular in shape, or slightly away from the midrib when it is more or less oval, only weakly contorting the upper epidermis; the lower surface is brownish and has one strong central fold with other weak folds alongside; the upper surface is mottled by irregular feeding in the parenchyma. Many mines of the autumn generation may be found in one leaf.

Pupa. Light brown; setae long; abdomen with dorsal surface of segments each with a prominent row of large spines, ventral surface of segment 6 with two rows of large laterally directed spines; setae long; cremaster consisting of two pairs of processes of equal length, the outer pair stouter than the inner pair. In a stout, oval, whitish ochreous cocoon with the frass incorporated along both sides and at the caudal end; the cocoon is attached to both surfaces of the mine. Late July; October to May.

Imago. Bivoltine. May; August. The adults, mines and life histories of this and other species found on alder are fully described and compared by Frankenhuyzen & Freriks (1976).

Distribution (Map 169)

Common throughout the British Isles wherever the food-plants occur. Central and northern Europe.

PHYLLONORYCTER DISTENTELLA (Zeller)

Lithocolletis distentella Zeller, 1846, *Linn.ent.* **1**: 181.
Type locality: Austria; Vienna.

Description of imago (Pl.13, fig.9)

Wingspan 8.0–9.5mm. Head with vertical tuft white, sometimes light orange-brown anteriorly; frons and labial palpus whitish; antenna light straw-coloured annulated very slightly darker, scape whitish. Thorax light golden with broad white median line; tegulae white, edged light golden; foreleg brownish fuscous above, mid- and hind-legs pale straw-coloured above, all with white bands on segments and white beneath. Forewing light golden brown, more or less sprinkled with blackish scales towards apex; four costal and two or three dorsal strigulae silvery white; costal 1 very oblique and extending along costa towards base, finely brown-edged inwardly; costals 2–4 wide, strongly black-edged inwardly, the edging extending into apical patch of black scales; dorsal 1 very oblique, finely edged dark brown inwardly and extending slightly beyond costal 1; dorsal 2 wide, black-edged, with edge continuing into apical patch; sometimes the faintest trace of dorsal 3 within fringe line; broad white basal streak to two-fifths, not usually dark-edged; fine white edging along dorsum between base and dorsal 1 sometimes continuous with dorsal 1; apical spot obscure within an area of black scales forming a diffuse streak; distinct apical hook of type 3, obliquely projecting from outer edge of costal 4 to tips of cilia; fringe line strong from costal 4 to dorsal 2; cilia whitish fuscous and noticeably shorter immediately below apical hook, lengthening again towards tornus. Hindwing whitish grey, cilia whitish with light yellowish brown tinge. Abdomen light golden brown anteriorly, ochreous posteriorly, white ventrally; caudal tuft white to light ochreous fuscous.

Life history

Ovum. Laid on the underside of a leaf of deciduous oak (*Quercus* spp.).

Larva. Not described. July; September to October.

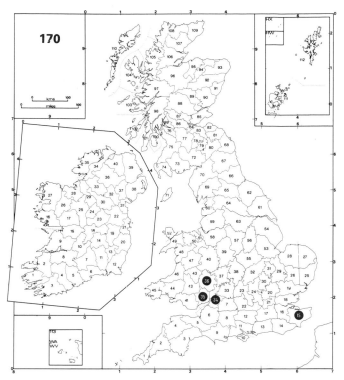

Phyllonorycter distentella

Mine. Upperside. Very large, averaging 25mm long, extending from midrib often to leaf-margin, frequently with several in one leaf; the lower surface is covered with minute creases, macroscopically giving the appearance of being smooth, although the contraction caused is so pronounced that the leaf is frequently completely folded over and if it contains several mines can be considerably distorted; the parenchyma on the upperside is all consumed with the exception of a large irregular central patch.

Pupa (figure 105 o, p. 303). Brown; setae long; abdominal segments with very short spines on dorsal surface; cremaster with two pairs of long, narrow processes. In a fine mesh of silken threads within the mine, these being generally confined to the region of the green patch; the frass is heaped and retained by a small pad of silk placed behind the pupal area. Late July; October to May.

Imago. Bivoltine. May; August.

Distribution (Map 170)

Very local and uncommon, the better-known areas being the old-established oak woodlands of Gloucestershire, Monmouthshire and Herefordshire, and the Blean Woods region near Canterbury, Kent, where it was first taken many years ago (Daltry, 1936) and was recently rediscovered (Heal, 1978). Central Europe.

PHYLLONORYCTER ANDERIDAE (Fletcher)

Lithocolletis anderidae Fletcher, 1885, *Entomologist's mon. Mag.* **22**: 40.

Type locality: England; Abbot's Wood, Sussex.

Description of imago (Pl.13, fig.10)

Wingspan 5.5–6.5mm. Head with vertical tuft dark brown or fuscous, whitish posteriorly; frons and labial palpus white; antenna dark brown, paler between segments, scape golden brown above, white below. Thorax golden brown with white central line; tegulae white, golden on anterior margin; foreleg brown, mid- and hindlegs light brown, all banded whiter on segments. Forewing shining golden brown; four costal and three dorsal strigulae shining silvery white and black-bordered inwardly, first two pairs thickly so; costal 1 and dorsal 1 joined to form a curved, or sometimes angulate, fascia, with a few dark scales on centre of outer edge; costals 2 and 3 curved towards apex; costal 4 above apical spot, often unedged but very distinct; dorsal 2 between costals 2 and 3, occasionally meeting former at a right angle or having its black edging extending to latter; dorsal 3 almost opposite costal 3, their black edges sometimes meeting to form a curved fascia; a white basal streak to one-third, finely dark-edged on costal side and outer third of dorsal side; apical spot small and black; short but distinct fringe line from costal 4 to dorsal 3, broken below; costal cilia fuscous, white on strigulae, a dark pencil extending from apical spot along outer edge of costal 4, dorsal cilia paler with golden brown sheen. Hindwing grey with golden sheen, cilia light fuscous with golden brown sheen. Abdomen brownish fuscous dorsally, white banded brown ventrally; caudal tuft dark brown.

Life history

Ovum. Laid on the underside of a leaf of birch (*Betula* spp.), the female having a marked preference for seedling plants growing on paths, in clearings or amongst heather and other low-growing plants.

Larva. Head dark brown. Body pale yellowish green, gut darker green. It always lacks the orange-yellow spot on abdominal segment 6 which is present in the other *Phyllonorycter* spp. feeding on birch. July; September to early November.

Mine. Underside. Small (*c.*10mm long), between two veins, the lower surface strongly contracted with several small folds, causing the upper surface to arch upwards and form a tiny tube. The larva of *Parornix betulae* (Stainton) makes a mine of similar size, also on seedlings, but this has the lower surface brown, whereas it remains green in that of *Phyllonorycter anderidae*.

Pupa. Light brown; setae short; spines on dorsal surface of abdominal segments placed in distinct anterior and posterior rows; cremaster with processes greatly reduced, inner pair absent and outer pair very small. Naked in the mine without a cocoon. Late July; November to May.

Imago. Bivoltine. May; August.

Distribution (Map 171)

A disjunct distribution, being recorded from the southern and south-eastern counties of England, a number of north Welsh and adjacent English counties to Westmorland (Cumbria), and from west Perthshire (E. C. Pelham-Clinton, pers.comm.). It is extremely local and changes its breeding sites from year to year to areas where its preferred seedling birches are growing. In the south it is found in heathy woods and in the north on mosses and moors. The adult is seldom encountered and the species may be overlooked in many areas. Belgium; Denmark.

PHYLLONORYCTER QUINQUEGUTTELLA (Stainton)

Lithocolletis quinqueguttella Stainton, 1851, *Suppl.Cat.Br. Tineidae and Pterophoridae*: 12.

Type locality: England; Cumberland (Cumbria).

Description of imago (Pl.13, fig.11)

Wingspan 6.0–7.5mm. Head with vertical tuft light golden brown, whitish posteriorly; frons and labial palpus white with coppery sheen; antenna rather dark fuscous, lighter below, scape white with coppery sheen. Thorax golden brown with white median line; tegulae golden brown, narrowly white inwardly; foreleg fuscous, mid- and hindlegs brown, all white-ringed, terminal segment blackish. Forewing shining golden brown, the scales narrowly tipped paler; five costal and three dorsal strigulae white, brown-edged inwardly; basal streak white thinly dark-edged above and partly below, bending upwards slightly towards costa; costal 1 at about one-third, extended as a fine line along costa to base; costal 2 reaching middle of wing, sides almost parallel, square-ended; costals 3–5 triangular; dorsal 1 between costals 1 and 2 and nearer to latter; dorsal 2 between costals 2 and 3, describing a strong arc; dorsal 3 almost meeting costal 4; a white dorsal mark under middle of basal streak, sometimes enlarged and often dark-edged; apical spot distinct and immediately below costal 5, the apex of which extends to inner edge of spot; fringe line weak, extending only from costal 5 to dorsal 3; apical cilia brownish fuscous above apex, whitish and fuscous-tipped below apex, dorsal cilia coppery. Hindwing light grey, cilia fuscous with light coppery brown sheen. Abdomen dark coppery fuscous; caudal tuft coppery fuscous, whitish below.

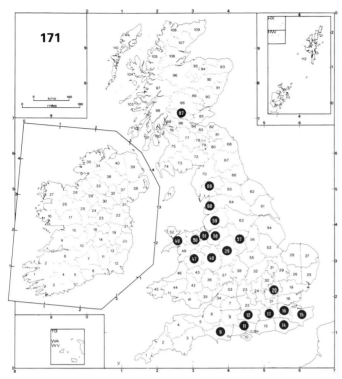

Phyllonorycter anderidae

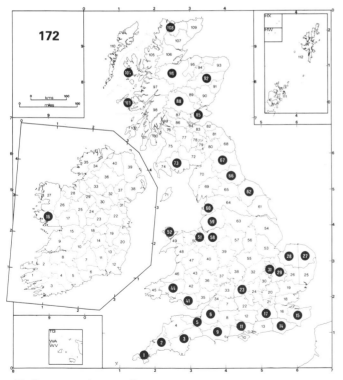

Phyllonorycter quinqueguttella

Life history

Ovum. Laid on the underside of a leaf of creeping willow (dwarf sallow) (*Salix repens*) or sandhill creeping willow (*S. arenaria*), often on plants well hidden in heather, grass or other low-growing plants.

Larva. Not described. July; September to October.

Mine. Underside. Occupies a large part of a leaf or the whole of a small leaf; the lower epidermis is strongly contracted with many small longitudinal creases causing the margins of the leaf to curl downwards and form a tube.

Pupa (figure 105k, p. 303). Golden brown; setae long; abdominal segments with dorsal surface clad in small spines not in definite rows; cremaster with two pairs of processes, inner pair short. In a brown, silk-lined chamber formed from the sides of the mine without any real cocoon, the frass being stacked in a heap at the opposite end of the mine. The pupa may be extruded through either surface prior to the emergence of the adult. Late July; October to May.

Imago. Bivoltine. May; August.

Distribution (Map 172)

Occurs mainly in coastal counties in Britain; in Ireland recorded only from west Galway (Emmet, 1971a). Very local, although it can be common on sandhills or in the marshy and heathy areas where its foodplant occurs. Palaearctic.

PHYLLONORYCTER NIGRESCENTELLA (Logan)

Lithocolletis nigrescentella Logan, 1851, *Trans.ent.Soc. Lond.* 1: 182.

Type locality: England; Morpeth, Northumberland.

Description of imago (Pl.13, figs 12,13)

Wingspan 7–9mm. Head with vertical tuft ochreous brown, paler posteriorly; frons and labial palpus leaden or silvery metallic; antenna greyish fuscous, paler distally, lightly annulated darker, scape greyish fuscous. Thorax and tegulae shining orange-brown, sometimes greyish; legs dark grey, banded white. Forewing shining golden orange-brown; one fascia, three, sometimes only two, costal and two dorsal strigulae, all silvery white, dark-edged, more strongly inwardly; basal streak to beyond one-quarter silvery white, dark-edged above and below; a white dorsal mark below basal streak; fascia at about two-fifths with an outwards-curved inner border and a more sinuate outer border, its dark border continued along dorsum to base, enclosing dorsal mark; costal 1 distinct, at right angles to costa and shaped as a truncated wedge or rectangle; costal 2 more arc-shaped; costal 3 small, narrow; dorsal 1 moderate in size, triangular, slightly beyond costal 1; dorsal 2 small, indistinct, sometimes represented only by a tiny patch of white scales, the white never extending into fringe or cilia; apical spot black, diffuse, sometimes extending inwards a short distance as an area of scattered blackish scales and occasionally with a small patch of white scales at inner margin; fringe line fairly strong from costal 3 to dorsal 1, broken by costal 3 if present and very occasionally interrupted by dorsal 2 if this is strong; cilia light brownish. Hindwing rather dark greyish fuscous, cilia shining brown with slight golden sheen. Abdomen greyish fuscous with some orange-brown scales dorsally, shining fuscous ventrally; caudal tuft fuscous mixed with orange-brown.

Similar species. P. *insignitella* (Zeller), *q.v.*

Life history

Ovum. Laid on the underside of a leaflet of bush vetch (*Vicia sepium*), reputedly on medick (*Medicago* spp.), often on plants buried in grass, under a hedge or beneath overhead tree-cover, rarely those in a fully exposed position.

Larva. Head pale brown. Body yellowish, thoracic segments 2 and 3 deeper yellow, gut strongly green. July; August to September.

Mine. Underside. Occupies the whole of a leaflet; the lower surface is contracted causing the edges to curl downwards, strongly contorting the leaflet; the upper surface becomes clear white and papery with the removal of the parenchyma and tenanted plants become very obvious, especially as several adjacent leaflets may be attacked, in extreme cases over 50 mines having been recorded on a single plant. Occasionally the tenanted leaves remain on the plant throughout the winter if the weather is mild and the plant is in a protected environment.

Pupa. Light brown; setae short; cremaster with a single pair of long processes. In a chamber loosely lined with silk, formed from the walls of the mine. Late July to August; October to April.

Imago. Bivoltine. April and May; late August and September.

Distribution (Map 173)

Very local in southern England where it occurs particularly in the south-east and south-west; more common in the north of England and throughout Ireland. Northern and central Europe.

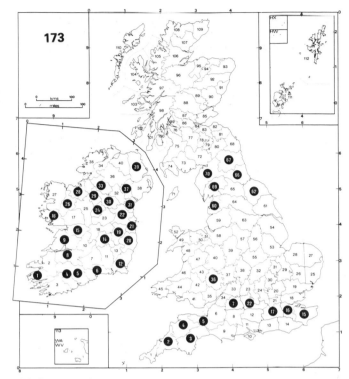

Phyllonorycter nigrescentella

PHYLLONORYCTER INSIGNITELLA (Zeller)

Lithocolletis insignitella Zeller, 1846, *Linn.ent.* 1: 193.
Type locality: Germany; Glogau (now Poland; Głogów).

Description of imago (Pl.13, figs 14,15)

Wingspan 6.5–8.5mm. Head with vertical tuft brown, whitish posteriorly; frons and labial palpus silvery white; antenna brownish, paler distally, scape brownish above, whitish below. Thorax orange-brown with fine white median line; tegulae orange-brown, edged white inwardly; legs dark grey, banded white. Forewing shining orange-brown, four costal and three dorsal strigulae all shining silvery white, finely edged black inwardly, occasionally also outwardly; a white basal streak to one-third, edged black above and below; a distinct black-edged white dorsal patch below basal streak; first pair of strigulae at two-fifths, either with costal small and triangular and the dorsal long, narrow, obtuse, slightly curved and reaching just beyond the end of the costal, or the two joined as a sinuate or right-angled fascia; second pair with costal small and broad and with dorsal larger, triangular and further from base; costals 3 and 4 narrow, wedge-shaped; dorsal 3 distinct, often broad, sometimes divided by a fine line of dark scales and with the white always extending through the fringe line into the cilia; apical spot small, diffuse and extended inwards as a narrow area of scattered black scales; fringe line fairly strong from costal 4 round apex but fading to dorsal 2 and strongly interrupted by dorsal 3; cilia dark brown above apex, whitish below to dorsal 3, then brownish bronze to tornus. Hindwing greyish fuscous, cilia brownish with golden sheen. Abdomen greyish fuscous dorsally, pale fuscous ventrally; caudal tuft fuscous.

Similar species. P. nigrescentella (Logan). In *P. insignitella* the white of the third dorsal strigula extends through the fringe line into the cilia; in *P. nigrescentella* the equivalent strigula is not always present and the fringe line is always entire in this position.

Life history

Ovum. Laid on the underside of a leaflet of clover (*Trifolium* spp.) or common restharrow (*Ononis repens*).

Larva. Not described. July; September to October.

Mine. Underside. Occupies almost half of a leaflet, being variously situated and extending across from one edge to the other; the lower surface is strongly contracted with several major folds visible only at the ends of the mine, the middle section being closed right up owing to the extreme contortion of the leaflet. All the upper parenchyma is consumed giving the surface a rather noticeable bleached appearance.

Pupa. Pale brown; setae short; abdominal segments on dorsal surface with several anterior and posterior rows of large spines with very small spines between; cremaster with two pairs of widely separated, very long, thin processes of approximately equal length, hooked at tips. In a very loosely woven white silken cocoon formed towards one end of the mine; the frass is heaped in a neat pile behind the cocoon without a silk pad; the pupa is extruded through the lower surface of the mine before the emergence of the adult. Late July; October to May.

Imago. Bivoltine. May; August.

Distribution (Map 174)

The earliest records were from Co. Durham (Robson, 1913) and the adjacent part of Yorkshire, where it was reported to be abundant. Since then it remained undetected for many years until it was recorded sparingly from the west of Ireland (Emmet, 1970; 1975a) and the Inner Hebrides (P. H. Sterling, *in litt.*). Records from elsewhere in the British Isles are unconfirmed. Central and northern Europe.

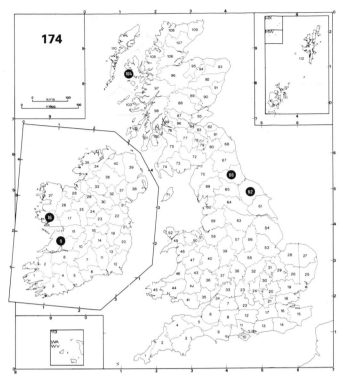

Phyllonorycter insignitella

PHYLLONORYCTER LAUTELLA (Zeller)

Lithocolletis lautella Zeller, 1846, *Linn.ent.* 1: 194.

Type localities: Austria; Vienna and Germany; Frankfurt am Main.

Description of imago (Pl.13, figs 16,17)

Wingspan 6–7mm. Head with vertical tuft blackish brown; frons and labial palpus metallic grey; antenna dark grey above, light greyish below, outer fifth white tipped brownish black, scape dark grey above, silvery grey below. Thorax and tegulae leaden metallic; legs shining brown above, white below. Forewing bright coppery brown; three costal and three dorsal strigulae silvery white, strongly black-edged inwardly and partly outwardly; a white basal streak to one-quarter black-edged above and below, narrowing inwardly with the white not quite reaching the base; a small white dorsal mark below basal streak, black-edged with the black extending along dorsum to dorsal 1; costal 1 small; dorsal 1 rather long and narrow with its black edging often reaching that of costal 1; costal 2 a little closer to base than dorsal 2, both smallish and each with dark edging round apex; costal 3 also black-edged round apex; dorsal 3 reduced to a round spot below costal 3 not reaching cilia; apical patch blackish; broad dark submetallic fringe line distinct from just beyond

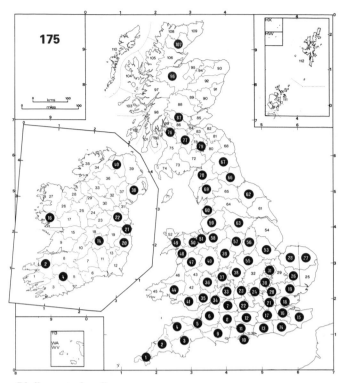

Phyllonorycter lautella

costal 3 to dorsal 2; cilia whitish at apex, becoming brown dorsally. Hindwing greyish fuscous, cilia fuscous with light brown sheen. Abdomen dark greyish fuscous dorsally, whitish ventrally; caudal tuft brownish fuscous above, silvery white below.

Ab. *irradiella* Scott (fig.17), in which the ground colour is sooty with a coppery sheen, occurs throughout the range but is more frequent in Wales and the north of Britain.

Life history

Ovum. Laid on the underside of a leaf of deciduous oak (*Quercus* spp.), often, but not exclusively, on seedlings, young saplings or stools round felled trees.

Larva. Head very pale greenish brown. Body pale greenish white, a slight yellowish tinge on thoracic segment 3 and abdominal segments 1 and 5, particularly in larger larvae; a blackish grey dorsal mark on abdominal segment 8; gut strongly visible as a blackish green dorsal line from thoracic segment 3 to abdominal segment 6. July; September to early November.

Mine. Underside. Long and slender (20–25mm long), usually extending from the midrib between two veins, rarely reaching the margin, often several in one leaf strongly contorting it; the upper surface is raised in a pointed rather than a rounded arch; the lower surface has

one strong central fold, sometimes divided at one end, flanked by numerous fine corrugations; much of the parenchyma is consumed but isolated uneaten cells are usually left at random over the upper surface.

Pupa. Golden brown; setae short; abdominal segments 1–3 with distinct anterior and posterior rows of spines on dorsal surface; cremaster with two pairs of short processes, outer pair wide-based and stouter than inner pair. In a whitish, papery, oval cocoon attached to the arched upper surface, the frass being piled behind the cocoon and retained by a small pad of silk. Late July; November to May.

Imago. Bivoltine. May; August. The species may be univoltine on high ground in the north. Papers on this and other oak-feeding *Phyllonorycter* are cited under *P. harrisella* (Linnaeus), p. 306.

Distribution (Map 175)

Locally common throughout the British Isles. Northern and central Europe.

PHYLLONORYCTER SCHREBERELLA (Fabricius)

Tinea schreberella Fabricius, 1781, *Spec.Ins.* **2**: 304.
Type locality: England.

Description of imago (Pl.13, fig.18)

Wingspan 6.5–7.5mm. Head with vertical tuft black; frons and labial palpus leaden metallic; antenna leaden black, the nine apical segments white, scape black above, leaden below. Thorax and tegulae leaden metallic; legs leaden black. Forewing shining orange-brown; three costal and three dorsal silvery white strigulae, all distinctly black-edged inwardly; first two pairs, just before one-quarter and one-half respectively, meeting to form two slightly sinuate, roughly parallel fasciae angled slightly outwards from costa to dorsum; costal 3 more whitish, at three-fifths, angled obliquely inwards; dorsal 3 strongly inclined, silvery at base, continuing as mirror-like scales; a dull black patch at base of costa; apical area metallic, obscuring apical spot; cilia dark fuscous with slight brownish sheen. Hindwing medium grey, cilia dark fuscous with brown sheen. Abdomen dark brownish fuscous; caudal tuft dark ochreous fuscous.

Life history

Ovum. Laid on the underside of a leaf of elm (*Ulmus* spp.), often one on a lower branch or a bush.

Larva. Head pale brown, darker laterally. Body pale amber, gut dark green. July; September to October.

Mine. Underside. Somewhat circular or oval, variously positioned on the leaf, often crossing a vein and strongly inflated, contorting a smaller leaf; the upperside with a small green patch of uneaten parenchyma, the underside with several long folds.

Pupa. Dark brown; dorsal surface of abdominal segments with a few, well-dispersed large spines; setae short; cremaster with two pairs of short processes. In a firm, cigar-shaped cocoon, more sharply pointed at one end, the colour varying between green and brown much as in *P. tristrigella* (Haworth). The cocoon is loosely attached to the uneaten patch of parenchyma on the upper epidermis and may be even more loosely attached to the lower surface of the mine. Late July; October to May.

Imago. Bivoltine. May; August.

Distribution (Map 176)

Locally common in England and Wales as far north as Lancashire and south Yorkshire; in Ireland, a single record from Co. Dublin (Beirne, 1941) lacks confirmation. Central and southern Europe.

PHYLLONORYCTER ULMIFOLIELLA (Hübner)

Tinea ulmifoliella Hübner, [1817], *Samml.eur.Schmett.* 8: pl.66, fig.444.

Type locality: Europe.

Description of imago (Pl.13, figs 19,20)

Wingspan 7–9mm. Head with vertical tuft ochreous brown; frons and labial palpus shining white; antenna fuscous, basally whitish below, terminal segment brownish fuscous, subterminal eight to ten segments white, scape fuscous above, white below. Thorax and tegulae golden brown; legs greyish fuscous banded white. Forewing golden brown; four costal and three dorsal strigulae shining white, all dark-edged inwardly; a white basal streak to one-third, dark-edged above, sometimes below; a pale, ill-defined white patch on dorsum between base and dorsal 1, sometimes dark-edged inwardly to form an extra strigula reaching end of basal streak; first pair meeting to form a slightly obtuse-angled, often curved, fascia; dorsals 2 and 3 lying between costals 2 and 3 and 3 and 4 respectively, often the dark edging of the second pair of strigulae meeting to form an angulate or sinuate dark line; third pair of strigulae occasionally similarly joined; apical spot small, black, lying immediately below costal 4; fringe line most obvious from costal 4 to dorsal 3, terminating between dorsals 2 and 3; a grey type 1 apical hook extending from apical spot into cilia; cilia grey to costal 4, thence whitish fuscous. Hindwing light grey in male, dark grey in female, contrasting strongly with cilia which are light fuscous in both sexes. Abdomen and caudal tuft in male dark grey; in female abdomen dark grey dorsally, pale brown ventrally, three terminal segments on dorsal sur-

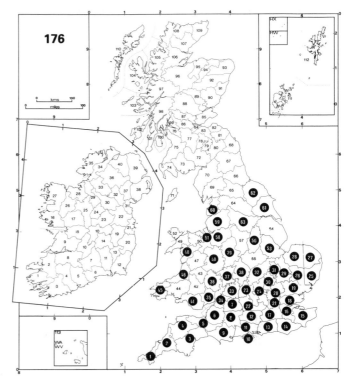

Phyllonorycter schreberella

face with white streak intensifying posteriorly; caudal tuft whitish.

Life history

Ovum. Laid on the underside of a leaf of birch (*Betula* spp.).

Larva. Head pale brown, darker laterally. Body pale yellowish green, gut dull greenish grey, a distinct yellow spot on abdominal segment 6. The larva turns completely yellow before pupation. July; September to early November.

Mine. Underside. Usually fairly small (8–14mm long); upperside brownish with a mottled, dried-up appearance; underside with several folds which coalesce in the centre of the mine to give the appearance of a larger central fold. Although small, the mine often contorts the leaf and causes the edges to curve downwards.

Pupa (figure 105f,z, p. 303). Dorsal surface of abdominal segments 1–3 clad in spines, long on anterior margin and shorter on posterior margin of each segment; ventral surface of abdominal segment 6 with from two to six large laterally directed spines; setae short; cremaster with two pairs of well-developed processes of equal length, outer pair stouter than inner pair. In a soft but tough white cocoon attached to the sides of the mine in the winter generation, but often extremely flimsy in the summer

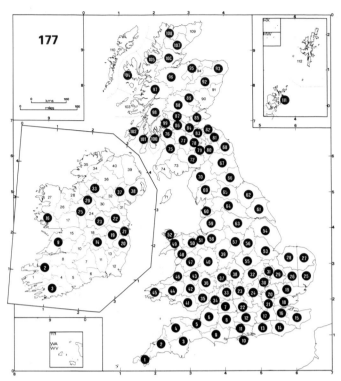

Phyllonorycter ulmifoliella

generation; the frass is not included in the cocoon and may be heaped immediately adjacent to or near one end. Late July; October to May.

Imago. Bivoltine. May; August.

Distribution (Map 177)

Widespread and generally common throughout the British Isles. Central and northern Europe.

PHYLLONORYCTER EMBERIZAEPENELLA (Bouché)

Ornix emberizaepenella Bouché, 1834, *Naturgesch. Insect.*: 132.

Type locality: not stated.

Description of imago (Pl.13, fig.21)

Wingspan 9–10mm. Head with vertical tuft pale ferruginous, whitish posteriorly; frons and labial palpus white; antenna fuscous above, banded white, scape white. Thorax light orange-brown, white anteriorly and laterally; tegulae light orange-brown anteriorly; legs white, banded brownish grey, spurs white. Forewing light orange-brown; four costal and five dorsal white strigulae; a short, narrow, sinuous white basal streak reaching costal 1; dorsum white-edged from base to dorsal 1; costal 2 and dor-

sals 2 and 3 narrowly dark-edged inwardly; first pair meeting to form a slightly curved fascia varying in width from costa to dorsum; second pair meeting to form an obtuse angle or a smooth, outward-curving fascia; third pair opposite and nearly meeting to form an acute angle; a small patch of dark brown scales between this and the fourth pair of strigulae; dorsal 5 immediately below apical patch; fringe line absent; cilia light brown apically, becoming fuscous on dorsum. Hindwing pale grey, cilia pale fuscous. Abdomen brownish grey marked white laterally at junction of segments, white ventrally; caudal tuft ochreous brown above, white below.

Life history

Ovum. Laid on the underside of a leaf of honeysuckle (*Lonicera* spp.) or snow-berry (*Symphoricarpos rivularis*); occasionally on Himalayan honeysuckle (*Leycesteria formosa*) (Sokoloff, 1979).

Larva. Head pale greenish, mouth parts and sides brown. Body pale yellowish white, gut dark green. July; September to October.

Mine. Underside. Very large, occupying a whole leaf of honeysuckle or a small leaf of snow-berry. The lower epidermis is gathered into strong folds greatly puckering the leaf, but since the mine is not folded diagonally, the leaf does not twist as with *P. trifasciella* (Haworth).

Pupa (figure 105j, p. 303). Setae short; cremaster greatly reduced, consisting of one pair of minute lateral processes. In a large, strong, firm, dull green or yellowish green cocoon, in shape very much like a rugby ball, placed at one end of the mine, the frass stacked at the other end. Mined leaves rot away during the winter and little remains other than the strong cocoons. Late July; October to May.

Imago. Bivoltine. May; August.

Distribution (Map 178)

Widely distributed, but local and rarely common, throughout Britain; in Ireland recorded only from Co. Clare (Bradley & Pelham-Clinton, 1967). Throughout Europe; north America.

PHYLLONORYCTER SCABIOSELLA (Douglas)

Lithocolletis scabiosella Douglas, 1853, *Trans.ent.Soc. Lond.* 2: 121.

Type locality: England; Croydon, Surrey.

Description of imago (Pl.13, fig.22)

Wingspan 6.5–8.5mm. Head with vertical tuft pale ferruginous, whitish posteriorly; frons white; labial palpus greyish white; antenna ochreous strongly banded dark fuscous above, whitish below, scape fuscous mixed ochreous above, white below. Thorax and tegulae orange-brown; legs dark grey, banded white. Forewing shining orange-brown ochreous; two transverse fasciae, two costal and one or two dorsal strigulae all shining silvery white sharply edged proximally with black; first fascia at nearly one-quarter, almost vertical and slightly curved outwards; second fascia at nearly one-half, slightly sinuate or obtusely angled with the angle above the midline of the wing and usually with the white forming a small extension distally from this angle, sometimes extending as a few whitish scales to tips of first pair of strigulae; first pair of strigulae opposite at nearly three-quarters, both narrow, arc-shaped and not usually meeting; second pair near to apex, both straight, at right angles to wing-edge proximal to a diffuse apical patch of dark-tipped scales; another suffusion of dark-tipped scales between dorsal strigulae; basal streak short, straight, unedged, reaching only to about one-ninth; small white dorsal patch below this streak, sometimes extending to base, sometimes almost absent; fringe line weak or absent; cilia brownish. Hindwing light grey, cilia light brownish with golden sheen. Abdomen fuscous dorsally, white ventrally; caudal tuft fuscous above, white below.

Life history

Ovum. Laid on the underside of a lower leaf of small scabious (*Scabiosa columbaria*), usually on seedling plants.

Larva. Head pale brown, mouth parts and sides darker. Body pale yellow, gut green. July to August; October, overwintering and feeding until April.

Mine. Underside. The mine usually occupies a considerable portion of the leaf; the lower surface has several longitudinal folds and is much contracted, drawing the upper surface downwards to form a strongly inflated blotch. The parenchyma attached to the upper surface is irregularly consumed, giving rise to a mottled appearance, and the epidermis is also tinged purple. A rosette of radical leaves often contains several mines.

Pupa. Dark brown; setae long; anal segment elongate, with cremaster consisting of a single pair of short processes; abdominal segment 7 with bulbous dorsolateral expansions. In a slender cocoon, the frass disposed in one or two heaps to one side. April to May; August.

Imago. Bivoltine. May; August and September.

Distribution (Map 179)
Found very locally along the North Downs in Surrey and Kent. Central Europe.

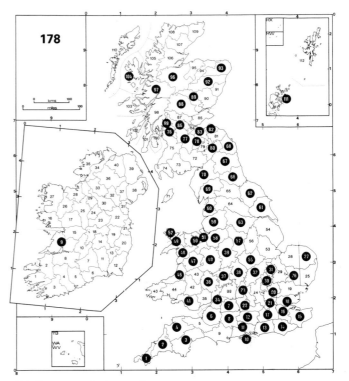

Phyllonorycter emberizaepenella

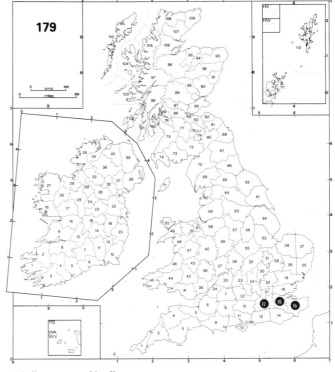

Phyllonorycter scabiosella

PHYLLONORYCTER TRISTRIGELLA (Haworth)

Tinea tristrigella Haworth, 1828, *Lepid.Br.*: 576.
Type locality: England; Coomb Wood, [Surrey].

Description of imago (Pl.13, fig.23)

Wingspan 7.5–8.5mm. Head with vertical tuft orange-brown, whitish posteriorly; frons and labial palpus shining white with a brassy sheen; antenna black banded white, apical eight to twelve segments white. Thorax shining orange-brown, whitish brown anteriorly; tegulae shining orange-brown, slightly white-edged posteriorly; legs with tibiae dull brown, tarsi white banded black. Forewing shining orange-brown; basal streak absent, but sometimes a small whitish basal spot; two shining white fasciae one at one-quarter and one at one-half, both usually curved distad but first sometimes straight or sinuate; beyond fasciae two costal and one dorsal strigulae; costal 1 and dorsal 1 opposite, strongly curved and nearly meeting in a chevron pointing towards apex; costal 2 a crescent, curved back proximad and often meeting costal 1 to form a semicircle; both major fasciae and to a lesser extent first strigulae finely dark-edged, the edging shading off uniformly inwards through brown into the orange-brown ground colour, more obviously towards dorsum; lower half of wing from dorsal strigula to apex shaded dark brown; apical spot small, often not present; fringe line absent; cilia brown round apex, becoming fuscous on dorsum. Hindwing shining pale grey, cilia brown on costa and round apex to dorsum, then fuscous. Abdomen dark greyish fuscous, caudal tuft pale brown above, shining whitish below.

Similar species. P. stettinensis (Nicelli), *P. froelichiella* (Zeller), *P. nicellii* (Stainton) and *P. kleemannella* (Fabricius), *q.v. P. tristrigella* differs from these in having the white pattern much less metallic and its black edging fading more evenly into the ground colour.

Life history

Ovum. Laid on the underside of a leaf of elm (*Ulmus* spp.).
Larva. Head brown, darker laterally. Body very pale whitish green, gut forming a darker green dorsal line from thoracic segment 3 to abdominal segment 4, reappearing on abdominal segment 7; yellow dorsal patches on abdominal segments 5 and 6. July; September to October.
Mine. Underside. On small-leaved elm (*Ulmus carpinifolia*) and wych-elm (*U. glabra*) between two veins with the underside strongly contracted by several long folds so that the veins almost meet to form a long, narrow, tube-like mine sometimes extending from the midrib almost to the leaf-edge, the upper surface with much of the parenchyma consumed except towards one end; on English elm (*U. procera*) forming more of an oval mine. Often there are several mines in a leaf.

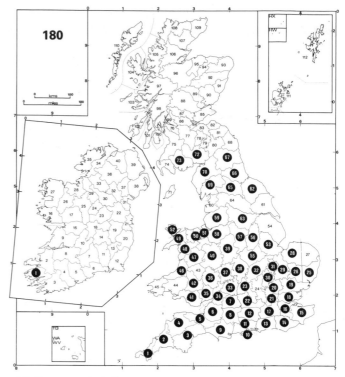

Phyllonorycter tristrigella

Pupa. Light brown; dorsal surface of abdomen with spines smaller, more numerous and more scattered than in *P. schreberella* (Fabricius), *q.v.*; setae short; cremaster consisting of two pairs of small processes. In a strong, greenish brown cocoon attached firmly with silk to the lower epidermis and unattached or only loosely attached to the upper epidermis; usually situated on the uneaten area of parenchyma near to one end of the mine; the cocoon is slightly broader and shorter than that of *P. schreberella* and is more rounded at one end. Late July; October to May.

Imago. Bivoltine. May; August.

Distribution (Map 180)

Widespread and generally common in England and Wales, just reaching the Scottish border counties; in Ireland recorded only from south Kerry (Mere *et al.*, 1964). Central and northern Europe.

PHYLLONORYCTER STETTINENSIS (Nicelli)

Lithocolletis stettinensis Nicelli, 1852, *Stettin.ent.Ztg* **13**: 219.

Type locality: Germany; Stettin (now Poland; Szczecin).

Description of imago (Pl.13, fig.24)

Wingspan 6.5–7.5mm. Head with vertical tuft blackish brown; frons and labial palpus shining white; antenna black with the nine apical segments whitish in male, white in female, scape grey; antenna of male noticeably thicker than that of female. Thorax and tegulae dark leaden or brassy brown; legs fuscous mixed coppery brown, foreleg in female with the two apical tarsal segments white. Forewing coppery brown; two shining silvery white fasciae, one at one-quarter and one at one-half, both slightly curved distad; beyond fasciae, three costal and two dorsal silvery white strigulae; dorsal 1 between costals 1 and 2, dorsal 2 between costals 2 and 3, costal 3 and dorsal 2 often obscure; all fasciae and strigulae inwardly with strong dark brown edge blending smoothly into ground colour and extended along dorsum at base; a short, silvery basal streak not reaching first fascia, edged above with dark brown, sometimes almost obsolete so that only the edging remains; apical patch black, diffuse; fringe line distinct from costal 3 to dorsal 1, not broken by dorsal 2; cilia fuscous. Hindwing grey, cilia fuscous with slight brown sheen. Abdomen fuscous, silver laterally, whitish ventrally; caudal tuft dull dark brown.

Sometimes the dark brown edging of the fasciae and strigulae is extended so that almost the whole ground colour is dark brown.

Similar species. *P. tristrigella* (Haworth), *P. froelichiella* (Zeller), *P. nicellii* (Stainton) and *P. kleemannella* (Fabricius), *q.v.* *P. stettinensis* is the only one of the group to have a blackish brown vertical tuft on the head. It may also be distinguished by its short basal streak; *P. tristrigella* usually has a pale basal spot, but the other species no indication even of this.

Life history

Ovum. Laid on the upperside of a leaf of alder (*Alnus glutinosa*), usually near the midrib or a major vein.

Larva. Head pale brown, darker laterally. Body pale greenish white, gut darker green. July; late September to October.

Mine. Upperside. Small, oval and normally situated over the midrib or a major vein; at first green but later becoming brownish; upper surface of a flimsy appearance and with a strong central crease, but remaining flat and usually causing very little distortion of the leaf. Hymenopterous upper surface mines on alder have the upper epidermis translucent and uncreased.

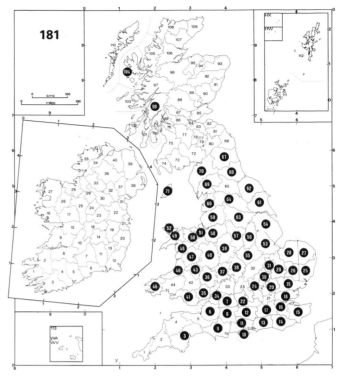

Phyllonorycter stettinensis

Pupa. Light brown; dorsal surface of abdominal segments with large, prominent spines; setae long; cremaster with two pairs of processes of equal length, outer pair stouter than inner pair. In a tough cocoon. July; October to May.

Imago. Bivoltine. May; August.

Distribution (Map 181)

Widespread and locally common in England and Wales; in Scotland recorded only from the west, in Glen Nant, Argyll (J. R. Langmaid and D. J. L. Agassiz, pers. comm.), and the Inner Hebrides (P. H. Sterling, *in litt.*). Central Europe.

PHYLLONORYCTER FROELICHIELLA (Zeller)

Lithocolletis froelichiella Zeller, 1839, *Isis, Leipzig* **1839**: 218.

Type localities: Germany; Frankfurt am Main and Glogau (now Poland; Glogów).

Description of imago (Pl.13, fig.25)

Wingspan 9–10mm. Head with vertical tuft light orange-brown; frons and labial palpus shining white; antenna fuscous banded whitish at inner end of segments, the terminal segment brown, the nine subterminal segments white, scape white. Thorax and tegulae golden brown; legs sooty brown, banded white. Forewing clear copper-brown, brassy at extreme base; two shining silvery white fasciae, one at one-quarter and one at one-half, the first almost straight, the second very slightly curved distad; beyond fasciae, two costal and three dorsal strigulae all shining silvery white; first pair curved towards apex, sometimes meeting to form angulated third fascia; dorsal 2 weak, commencing between costals 2 and 3, but with its apex sometimes joining costal 2 to form a sharp-angled fascia; costal 3 barely visible, often absent; all fasciae and larger strigulae finely edged blackish brown inwardly; apical patch black, formed by scattered black scales; fringe line weak, black; cilia light brown, becoming fuscous dorsally. Hindwing grey, cilia fuscous brown. Abdomen sooty brown, caudal tuft light brown.

Similar species. *P. tristrigella* (Haworth), *P. stettinensis* (Nicelli), *P. nicellii* (Stainton) and *P. kleemannella* (Fabricius), *q.v. P. froelichiella* is by far the largest species in the group, its copper-brown ground colour is characteristic, it has the narrowest and most sharply defined black edging to the white markings and the first pair of strigulae are smoothly curved distad with their apices almost touching to form an acute- or right-angled chevron.

Life history

Ovum. Laid on the underside of a leaf of alder (*Alnus glutinosa*), often on a higher branch of a young bush.

Larva. Head brown. Body grey, anal segment whitish; gut dark green. September to October.

Mine. Underside. Very large (*c.*25mm long), often several together between veins and extending from midrib almost to leaf-edge, the angle between veins and midrib being fully excavated and sometimes tinged reddish; the upperside with a green patch of uneaten parenchyma slightly offset from the centre, the underside with several tiny folds but macroscopically giving the impression of being smooth.

Pupa. Light brown; setae short; cremaster with two pairs of processes, the outer pair large, the inner greatly reduced. In a large off-white to light brown silken chamber

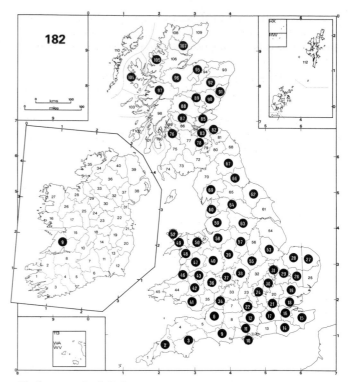

Phyllonorycter froelichiella

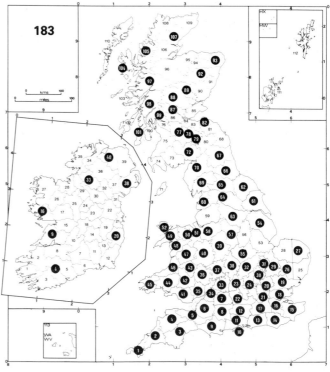

Phyllonorycter nicellii

or cocoon attached to the central green patch and clearly visible from outside as a bulge in the underside of the mine, the frass being neatly heaped usually at the inner end of the cocoon. October to May.

Imago. Although described as bivoltine by Stainton (1859), his opinion being followed by subsequent authors (Meyrick, 1928; Ford, 1949; Emmet, [1979]c), it is probably univoltine throughout the British Isles and certainly so in Scotland (E. C. Pelham-Clinton, pers.comm.), the moths occurring in July and August.

Distribution (Map 182)

Locally common throughout England, Wales and Scotland; in Ireland recorded only from the Burren, Co. Clare (Bradley & Pelham-Clinton, 1967). Central and northern Europe.

PHYLLONORYCTER NICELLII (Stainton)

Lithocolletis nicellii Stainton, 1851, *Zoologist* **9** Appendix: clxxii.

Type locality: not stated.

Description of imago (Pl.13, fig.26)

Wingspan 7–8mm. Head with orange-brown vertical tuft; frons shining white with slight brownish sheen; labial palpus whitish; antenna sooty fuscous banded whitish at inner end of segments, the nine to twelve apical segments whitish, scape glossy pale brown. Thorax and tegulae brown; legs dark brown, banded white. Forewing pale glossy orange-brown; two fasciae, one at one-quarter and one at one-half, the first sometimes, the second always, curved outwards; three, sometimes two, costal and two dorsal strigulae beyond fasciae; first pair opposite, curving outwards, sometimes meeting to form a third fascia; dorsal 2 very diffuse between costals 2 and 3; all fasciae and strigulae shining white, the white outwardly blending smoothly into ground colour, inwardly sharply edged dark brown and preceded by a band of brown irroration; apical spot indistinct, set in an area of blackish scaling; fringe line strong from costal 3 to dorsal 1, enclosing dorsal 2; cilia light brownish. Hindwing brownish fuscous, cilia brownish, slightly darker towards apex. Abdomen dark shining brown above, off-white below; caudal tuft yellowish brown.

The shining white fasciae and strigulae may be brassy-tinged and in some specimens are quite dull.

Similar species. *P. tristrigella* (Haworth), *P. stettinensis* (Nicelli) and *P. froelichiella* (Zeller), *q.v.* *P. kleemannella* (Fabricius) and *P. nicellii* are two of the more difficult of the group to distinguish. *P. kleemannella* has a deeper orange ground colour, the fasciae are more shining and often bordered inwardly by a broad band of deep coppery

brown rather than the brown irroration of *P. nicellii*; the first pair of strigulae rarely meet. In *P. nicellii* the first pair of strigulae often touch, forming a third angulated fascia.

Life history

Ovum. Laid on the underside of a leaf of hazel (*Corylus avellana*).

Larva. Head and prothoracic plate blackish brown. Body pale greenish white. July; September to October.

Mine. Underside. Elongate, of moderate size (15–20mm long), generally situated between veins. The underside is contracted by spinning and the upperside strongly arched; the upper epidermis has a central green patch of uneaten parenchyma, the surrounding area being mottled by smaller patches. The similar mine of *Parornix devoniella* (Stainton) is smaller, subrectangular, less strongly arched and has the lower epidermis brown, not green as in *Phyllonorycter nicellii*.

Pupa. Setae short; cremaster with one pair of processes. In a white cocoon attached to both the upper and lower epidermis; the frass is not incorporated but is disposed in a heap behind the cocoon. July; October to May.

Imago. Bivoltine. May; August.

Distribution (Map 183)

Widely distributed and generally common throughout the British Isles, but more plentiful in the south. Central and northern Europe.

PHYLLONORYCTER KLEEMANNELLA (Fabricius)

Tinea kleemannella Fabricius, 1781, *Spec.Ins.* **2**: 509.

Type locality: Germany; Hamburg.

Description of imago (Pl.13, fig.27)

Wingspan 8mm. Head with vertical tuft orange-brown; frons and labial palpus whitish with brassy sheen; antenna uniform fuscous except for the nine apical segments which are white, scape fuscous. Thorax and tegulae brown with brassy sheen; legs mottled brown above, shining whitish with brassy sheen below. Forewing shining orange-brown; two fasciae, one at one-quarter and one at one-half, the first slightly inward-curved at costa and dorsum, meeting dorsum slightly nearer to apex than its costal origin, the second smoothly curved outwards and meeting dorsum opposite its costal origin; beyond fasciae two costal and two dorsal strigulae in opposite pairs, not meeting, the two costals with their inner margins curved and of equal size; dorsal 1 larger than costals and strongly curved inwardly and dorsal 2 represented by a small spot or obsolete; all strigulae and fasciae shining silvery white, often becoming brassy and shading smoothly into ground colour outwardly and bordered dark brown inwardly; apical spot in an area of black scales extending to fringe line which is well defined from costal 2 to dorsal 1, enclosing dorsal 2; cilia light brownish. Hindwing pale brownish buff, cilia light brownish, darker towards apex. Abdomen fuscous with brassy sheen, posterior segments whitish below; caudal tuft dark brown.

Paler specimens have the ground colour clear orange-brown, the fasciae clear silvery white without a brassy outer margin and with a narrow dark brown inner border. More frequently the fasciae become brassy on their outer edge and shade into a dark coppery brown ground colour. Intermediate forms occur in which each fascia is preceded by a dark coppery brown band.

Similar species. P. tristrigella (Haworth), *P. stettinensis* (Nicelli), *P. froelichiella* (Zeller) and especially *P. nicellii* (Stainton), *q.v.*

Life history

Ovum. Laid on the underside of a leaf of alder (*Alnus glutinosa*) or, occasionally, on *A. cordata* or grey alder (*A. incana*).

Larva. Head pale brown, darker laterally. Body yellowish white, gut greyish green or reddish. The yellowish tinge distinguishes this larva from that of *P. rajella* (Linnaeus). September to October.

Mine. Underside. The smallest of the alder-feeding *Phyllonorycter* mines, often several together, between veins and variously positioned on the leaf; much smaller than that of *P. froelichiella* and more rounded in appearance; when the mine reaches the midrib, the parenchyma in the acute angle formed by the side vein is seldom entirely mined and this part is never reddish-tinged; the underside of the mine is slightly transparent in appearance and has several indefinite longitudinal creases; the upperside has a patch of uneaten parenchyma, usually towards the distal end.

Pupa. Light brown; setae short; abdominal segments 3–5 each with two distinct rows of large spines on dorsal surface, ventral surface without large spines; cremaster with two pairs of processes of equal length, outer pair stouter than inner pair. In a stout white cocoon attached to the green patch of parenchyma; the frass is not incorporated into the structure of the cocoon but is piled neatly at the opposite end of the mine. October to May.

Imago. Like *P. froelichiella*, *q.v.*, probably univoltine throughout the British Isles and certainly so in Scotland (E. C. Pelham-Clinton, pers.comm.), the moths occurring in August.

Distribution (Map 184)

Widely distributed and locally common throughout Britain, except southern Scotland; in Ireland recorded only from Cos Kerry and Wicklow (Beirne, 1941). Central and northern Europe.

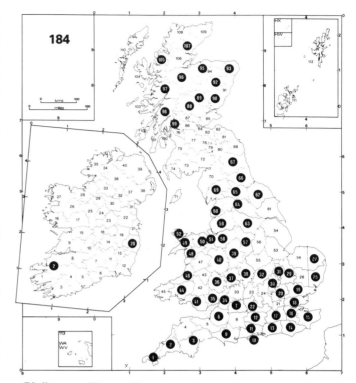

Phyllonorycter kleemannella

PHYLLONORYCTER TRIFASCIELLA (Haworth)
Tinea trifasciella Haworth, 1828, *Lepid.Br.*: 576.
Type locality: England; Coomb Wood, [Surrey].

Description of imago (Pl.13, figs 28,29)
Wingspan 7–8mm. Head with vertical tuft pale orange-brown becoming whitish posteriorly; frons and labial palpus whitish; antenna above sooty fuscous banded off-white, more whitish on under surface, scape whitish. Thorax and tegulae light pinkish or orange-brown, latter with variable amount of dark brown anteriorly; legs whitish broadly marked dark brown on segments above and on middle third of spurs. Forewing light pinkish or orange-brown; two fasciae, one at about one-fifth and one at one-half, the first sinuate outwards in costal half, the second angulate outwards in costal half, less prominently so in dorsal half; second fascia broadly edged blackish inwardly in upper and lower portions, less so in middle; two costal and one dorsal strigulae beyond fasciae; costal 1 and dorsal 1 opposite at about three-quarters, both curving outwards and preceded by dark brown patches widest on costa and dorsum; a band of blackish scales extending from junction of strigulae obliquely towards tornus; costal 2 indistinct, preceded by a small group of blackish scales on costa; apical spot black, surrounded by a few blackish scales; cilia light brownish. Hindwing light fuscous, cilia light brown. Abdomen light fuscous brown; caudal tuft light sooty brown.

Life history
Ovum. Laid on the underside or very occasionally the upperside of a leaf of honeysuckle (*Lonicera* spp.), occasionally on snow-berry (*Symphoricarpos rivularis*) or Himalayan honeysuckle (*Leycesteria formosa*) (Sokoloff, 1979).
Larva. Head pale brown, mouth parts slightly darker. Body pale yellowish green, gut darker green. March (February in Isles of Scilly) to April; July to August; October.
Mine. Underside; very occasionally upperside (Emmet, 1972; Gregory, 1972). Initially flat, extending from the midrib to the leaf-edge; later the lower epidermis is gradually contracted diagonally across the mine, the edge of the leaf being twisted often so as to form a cone and cause great distortion, leaves thus affected being very conspicuous. Often the parenchyma nearest to the apex of the cone is all consumed, leaving the epidermis clear white; the lower epidermis has long, flange-like creases.
Pupa (figure 105t, p. 303). Light brown; setae long; abdominal segment 7 with lateral bulbous expansions; terminal segment elongate; cremaster reduced, with two pairs of small processes. In a chamber within the mine, the frass being heaped to one side. April; July; October.

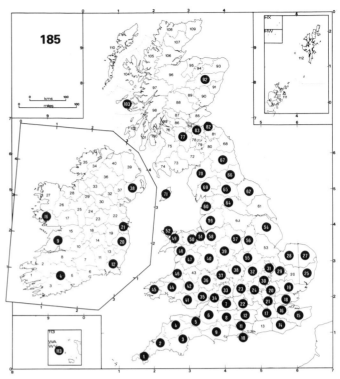

Phyllonorycter trifasciella

Imago. Usually in three generations, the moths occurring in May, August and November. It is not known in which stage the winter months are passed, the early spring larvae being found mining leaves of the new growth which were not present the previous autumn.

Distribution (Map 185)
Widespread and common throughout England and Wales; more local in Scotland; in Ireland recorded only from coastal counties. Throughout Europe.

PHYLLONORYCTER SYLVELLA (Haworth)

Tinea sylvella Haworth, 1828, *Lepid.Br.*: 579.
Type locality: England.

Description of imago (Pl.13, figs 30,31)

Wingspan 8mm. Head with vertical tuft white, usually dark brown anteriorly; frons and labial palpus white with slight brassy sheen; antenna white, segments narrowly banded fuscous, scape white. Thorax and tegulae white, latter with a few dark fuscous scales anteriorly; foreleg with tarsus white banded fuscous, femur and tibia mainly fuscous; midleg white lightly banded fuscous; hindleg with a wide fuscous band on tarsal segment 1 and narrower bands on other segments. Forewing white; three fasciae, two costal and one dorsal strigulae all pale yellow-brown, edged and sometimes irrorate black; first fascia very weakly expressed, consisting of two short dark costal strigulae at one-tenth and one-fifth, and one dorsal strigula at one-sixth almost touching second costal and not always quite reaching dorsum, the brownish suffusion between these strigulae very faint or more often absent; second fascia at one-third forming a right-angled chevron (figure 109a), the apex of the angle usually constricted and occasionally extended to third fascia by a thin line of brownish scales; third fascia at two-thirds, acutely angled; costal strigula 1 arc-shaped, dark-edged distally, often touching apex of third fascia; costal 2 small, hardly dark-edged and often merging with apical patch; dorsal 1 strongly black-edged, reaching apex of third fascia; apical spot small, set in area of orange-brown scales and extended as a dark streak to dorsal 1; fringe line distinct from costal 2 to dorsal 1; cilia pale yellowish brown. Hindwing and cilia greyish white. Abdomen light fuscous; caudal tuft white.

Similar species. *P. platanoidella* (Joannis), in which the dorsal strigula forming part of the rudimentary first fascia does not nearly touch and make a straight fascia with the second costal strigula as in *P. sylvella*; the second fascia forms an obtuse angle or a smooth curve, whereas in *P. sylvella* it forms an acute or right angle; the yellow-brown coloration of the pattern is deeper than on *P. sylvella*. *P. geniculella* (Ragonot), *q.v.*

Life history

Ovum. Laid on the underside of a leaf of field maple (*Acer campestre*), usually near the margin.

Larva. Head pale yellowish brown. Body yellowish, gut dark green, a yellow spot on abdominal segment 6. July; September to October.

Mine. Underside. Small, usually at the edge of the leaf, very often at the apex of a lobe, the contraction of the lower epidermis causing it to turn completely under. Often the upper surface shows dappled markings because of the removal of the parenchyma beneath.

Pupa (figure 105p, p. 303). Rather dark brown; setae long; anterior rows of widely separated large spines on dorsal surface of abdominal segments 1–3; cremaster with two pairs of long processes. In a delicate greyish ochreous cocoon. Late July; October to May.

Imago. Bivoltine. May; August.

Distribution (Map 186)

Coextensive with the native range of the foodplant; locally common in south Wales and England as far north as Derbyshire with isolated records reaching Northumberland and Westmorland (Cumbria). Throughout Europe.

PHYLLONORYCTER PLATANOIDELLA (Joannis)

Lithocolletis platanoidella Joannis, 1920, *Annls Soc.ent.Fr.* **89**: 411.
Type locality: France; Bois de Boulogne, near Paris.

Description of imago (Pl.13, figs 32,33)

Wingspan 8mm. Head with vertical tuft white usually slightly mixed brown especially anteriorly; frons and labial palpus shining white; antenna white narrowly banded fuscous on distal edge of segments, scape white. Thorax and tegulae shining white, latter with a small patch of dark fuscous scales on anterior margin; foreleg with tarsus white banded fuscous, tibia and femur mainly fuscous; midleg white, lightly banded fuscous; hindleg white with a fuscous band on tarsal segment 1. Forewing white; three fasciae, two costal and one dorsal strigulae all yellow-brown, edged blackish brown; first fascia very weakly expressed, consisting of two short dark costal strigulae at one-tenth and one-fifth and one dorsal strigula at about one-seventh, the latter terminating midway between the two costals opposite, some trace of brownish scaling between the strigulae usually being present; the second fascia at one-third, the inner edge obtusely angled or occasionally curved, the outside edge smoothly curved (figure 109b); the third fascia at about two-thirds, obtusely angled, right-angled or occasionally acutely angled, the dark edging extended from apex of angle to apical spot;

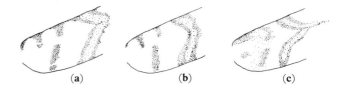

Figure 109 Forewing base of *Acer* feeders to show species differences (a) *Phyllonorycter sylvella* (b) *P. platanoidella* (c) *P. geniculella*

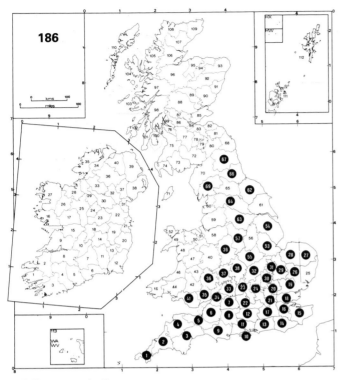

Phyllonorycter sylvella

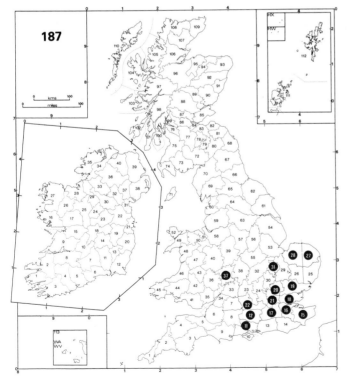

Phyllonorycter platanoidella

costal strigula 1 generally arc-shaped, dark-edged distally, joining apex of third fascia; costal 2 smaller; dorsal 1 strongly black-edged reaching apex of third fascia; apical spot small, set in area of orange-brown scales; fringe line distinct from costal 2 to dorsal 1; cilia greyish white. Hindwing and cilia greyish white. Abdomen light fuscous dorsally, shining white ventrally; caudal tuft white.

Similar species. P. *sylvella* (Haworth) and P. *geniculella* (Ragonot), *q.v.*

Life history

Ovum. Laid on the underside of a leaf of Norway maple (*Acer platanoides*).

Larva. Not described. July; September to October.

Mine. Underside. Usually fairly large and rounded, between two veins but not necessarily touching both and seldom at the edge of the leaf; the lower surface has several small creases and becomes whitish; frequently there are several mines in a leaf, causing distortion.

Pupa. Not described. Late July; October to May.

Imago. Bivoltine. May; August. Although this species and P. *sylvella* have slightly different wing-patterns and to some extent different types of mine and mining habits on their respective foodplants, it is by no means certain that they are distinct species. Gregor & Povolný (1950) could show no difference in wing-pattern or genitalia between individuals reared from field maple (*Acer campestre*) and Norway maple in Czechoslovakia. However, when P. *platanoidella* on Norway maple and P. *sylvella* on field maple occur close together in Britain, the populations apparently do not intermix, this being assessed on the basis of wing-pattern.

Distribution (Map 187)

Locally common in the south-eastern counties of England and in Worcestershire (Chalmers-Hunt, 1965). Central Europe.

PHYLLONORYCTER GENICULELLA (Ragonot)

Lithocolletis geniculella Ragonot, 1874, *Petites Nouv.Ent.* 5: 346.

Type locality: France; Raincy.

Description of imago (Pl.13, fig.34)

Wingspan 8mm. Head with vertical tuft white, sometimes mixed with brown scales, especially anteriorly; frons and labial palpus shining white; antenna and scape white. Thorax white marked brown posteriorly; tegulae white, outwardly marked pale yellow-brown; legs white, banded fuscous. Forewing white; three fasciae, three costal and three dorsal strigulae pale yellow-brown, edged brownish black; first fascia weakly expressed, consisting of two short dark costal strigulae, one at about one-twelfth and one at one-fifth, and a similar dorsal strigula at about one-seventh, usually some yellow-brown between these strigulae, especially near costa; second fascia at about one-third, strongly angled to form an acute-angled chevron (figure 109c, p. 356), its apex, including both brown and yellow-brown scaling, often extended to reach third fascia; third fascia strongly angulate and dark-edged on both sides, apex extended as a dark streak to small costal spot; costal and dorsal strigulae generally arc-shaped, dark-edged distally and reaching dark subapical streak; fringe line from costal 2 to dorsal 3, sometimes rather weak; cilia yellowish white. Hindwing pale brownish fuscous, cilia pale brownish. Abdomen light fuscous dorsally, whitish ventrally; caudal tuft yellowish fuscous.

Similar species. P. sylvella (Haworth) and *P. platanoidella* (Joannis). *P. geniculella* differs in having the second chevron sharply angled with its apex extended as a dark-bordered yellow line to the third chevron; the rudimentary first chevron at the base of the wing contains more yellowish brown scales than either of the other two species. See figure 109, p. 356.

Life history

Ovum. Laid on the underside of a leaf of sycamore (*Acer pseudoplatanus*).

Larva. Not described. July; September to October.

Mine. Underside. Rather small, round or irregular in outline, when on the edge of the leaf causing the upper surface to curl under, but when away from the margin causing only slight distortion.

Pupa. Very dark brown, darker than that of *P. sylvella*; abdominal segments 1–3 each with an anterior row of small, closely packed spines on the dorsal surface; otherwise similar to that of *P. sylvella*. Late July; October to May.

Imago. Bivoltine. May; August.

Distribution (Map 188)

Widespread and locally common throughout Britain as far north as Perthshire. Central Europe.

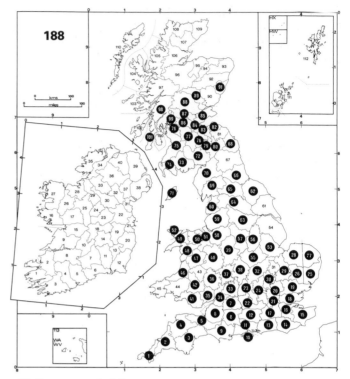

Phyllonorycter geniculella

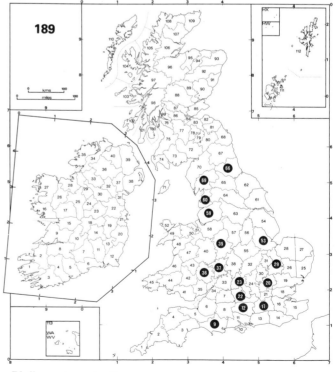

Phyllonorycter comparella

PHYLLONORYCTER COMPARELLA (Duponchel)

Elachista comparella Duponchel, 1843, *in* Godart & Duponchel, *Hist.nat.Lépid.Fr.* Suppl. 4: 318.

Type locality: France.

Description of imago (Pl.13, figs 35,36)
Wingspan 8mm. Head with vertical tuft white, brown anteriorly; frons and labial palpus shining white; antenna white, segments narrowly ringed fuscous, less obviously towards apex, scape white. Thorax and tegulae white sprinkled fuscous, more densely anteriorly; legs distinctly banded fuscous. Forewing white, sprinkled with light fuscous scales; four or five costal and three or four dorsal darker-edged fuscous strigulae, the costals roughly at base, one-quarter, one-half, three-quarters and midway thence to apex, the first three dorsals slightly distad of the first three costals, costals 2–4 and dorsals 2 and 3 in the form of parallelograms inclining towards apex; in some individuals the second pair of strigulae meet to form an angled fascia and dorsal 3 curves so as just to meet costal 4; in other specimens costals 1 and 2 are combined to give an area of fuscous scales extending from base to one-quarter; costal 4 extends towards apex with the outer dark edging becoming a distinct apical streak, widening to become a triangular patch before a rather weak fringe line (figure 110a); costal 5 forms an inverted triangle with its apex barely touching the apical streak; dorsal 4, if present, is represented by a small patch between inner end of apical streak and tornus; cilia pale fuscous, becoming whiter towards apex. Hindwing pale fuscous, cilia pale fuscous becoming slightly brownish dorsally. Abdomen light fuscous dorsally, white ventrally; caudal tuft ochreous white.
Similar species. *P. sagitella* (Bjerkander), *q.v.*

Life history
Ovum. Laid on the underside of a leaf of white poplar (*Populus alba*) or grey poplar (*P. canescens*).
Larva. Head pale brown, darker near mouth parts. Body pale green, abdominal segments 3–5 more whitish, 7–10 yellowish, gut dark green. July; September to early October.
Mine. Underside. The upper surface only slightly or not discoloured and having a mottled appearance due to patches of uneaten parenchyma; the lower epidermis smooth or with very small creases, so causing little distortion to the leaf.
Pupa (figure 105i, p. 303). Setae long; ventral surface of abdominal segment 6 with a group of large, laterally directed spines; cremaster much reduced with processes minute. In a centrally placed, oval, flimsy cocoon, the frass in a heap at one end of the mine. Late July; October.
Imago. Bivoltine. The first generation occurs in August

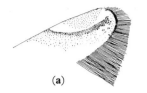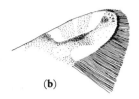

(a)   (b)

Figure 110 Forewing tips of *Populus* feeders to show species differences
(**a**) *Phyllonorycter comparella* (**b**) *P. sagitella*

and the second emerges in October and overwinters, surviving until the following May.

Distribution (Map 189)
Very local and generally uncommon in England as far north as Westmorland (Cumbria) and Durham. Central Europe eastwards to Turkestan.

PHYLLONORYCTER SAGITELLA (Bjerkander)

Phalaena (*Tinea*) *sagitella* Bjerkander, 1790, *K.Vet.Ac. Nya Handl.* 11: 132.

Type locality: not stated.

Description of imago (Pl.13, fig.37)
Wingspan 8.5–9.0mm. Head with vertical tuft white, darkening to brown anteriorly; frons and labial palpus white; antenna white, banded grey before joints on upper surface. Thorax white with sparsely scattered fuscous scales; tegulae fuscous, white posteriorly; legs ochreous white banded dark fuscous above, two terminal tarsal segments white. Forewing with ground colour white, generally sprinkled with dark fuscous scales; five or six costal and three or four dorsal strigulae marked by rather loosely placed fuscous scales; costal 1 from base to about one-fifth; costal 2 just before one-half, sometimes closely mixed with costal 1; costal 3 beyond one-half, curving towards apex and joining dorsal 3, the two extending together as an elongate streak to small apical spot (figure 110b); costals 4 and 5 close together, small and outwardly curved, each with lower edge barely touching apical streak; costal 6 if present represented by only one or two dark scales close to apex; dorsal 1 a diffuse zigzag line extending from about one-eighth to costal 1; dorsal 2 at about one-quarter, small, represented by a few dark scales and separated from costal 1 by a diffuse line of ground colour; dorsal 3 at about one-half, triangular, fading apically but joining costal 3; dorsal 4 if present represented only by a few dark scales near to tornus; apical spot black with a few dark fuscous scales continuing line to apex; fringe line strong but narrow from costal 6 to dorsal 4,

with a few dark scales projecting from above apex into cilia; cilia pale ochreous fuscous. Hindwing grey, cilia pale fuscous at base becoming paler at tips. Abdomen shining white ventrally, slightly ochreous dorsally; caudal tuft pale ochreous.

Similar species. P. comparella (Duponchel), in which the fifth costal strigula is to be found in the area of white scales bordered by the apical streak and the dark edge to costal 4, forming a fuscous inverted triangle or semicircle on the wing-edge, being completely surrounded on its inner periphery by white scales. In *P. sagitella* the same area of white is crossed by two distinct patches of fuscous scales, finely separated by white. Generally the white areas on *P. sagitella* are more heavily irrorate with scattered fuscous scales than on *P. comparella*. See figure 110, p. 359.

Life history

Ovum. Laid on the underside of a leaf of aspen (*Populus tremula*), usually on saplings (Hering, 1957).

Larva. Head brownish yellow. Body yellowish, paler at extremities. June; August to October.

Mine. Underside. Irregular or circular, between two veins but variously positioned on the leaf; the upper surface strongly arched and the lower surface usually contracted with small creases but sometimes smooth; often reddish-coloured.

Pupa. Not described. In the mine with a few threads of silk to secure it, without any real cocoon; the frass is neatly piled at the opposite end of the mine.

Imago. Bivoltine. May; July and August.

Distribution (Map 190)

A recent addition to the British list and so far recorded only from Gloucestershire (Price, 1977), Worcestershire (Simpson, 1978) and Denbighshire (Michaelis, 1979). However, a hitherto misidentified specimen in the City Museum, Worcester dating from the mid-nineteenth century shows that it is not a recent colonist (A. N. B. Simpson, pers.comm.). Central and northern Europe to western U.S.S.R.

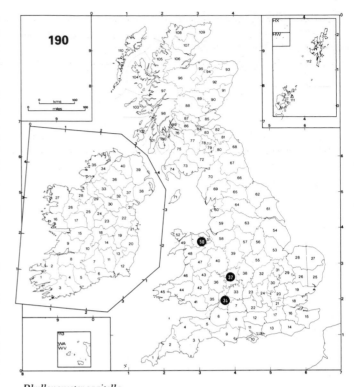

Phyllonorycter sagitella

References

Beavis, I. C., 1973. Notes on the pupation and emergence of *Zygaena lonicerae* Scheven. *Entomologist's Rec. J. Var.* **85**: 267.

Beirne, B. P., 1941. A list of the Microlepidoptera of Ireland. *Proc. R. Ir. Acad.* **47**(B): 53–147.

——, 1945. Irish Lepidoptera collecting in 1944. *Entomologist's Rec. J. Var.* **57**: 51–53, 63–66, 86–89.

Bolam, G. M., 1931. The Lepidoptera of Northumberland and the eastern Borders. *Hist. Berwicksh. Nat. Club* **27**: 221–265.

Bradley, J. D., Jacobs, S. N. A. & Tremewan, W. G., 1969. A key to the British and French species of *Phyllonorycter* Hübner (*Lithocolletis* Hübner) (Lep., Gracillariidae). *Entomologist's Gaz.* **20**: 3–33.

—— & Pelham-Clinton, E. C., 1967. The Lepidoptera of the Burren, Co. Clare, W. Ireland. *Ibid.* **18**: 115–153.

Brown, S. C. S., 1947. *Caloptilia* Hübn., a genus of Tineina. *Proc. Trans. S. Lond. ent. nat. Hist. Soc.* **1946–47**: 157–167, 1 col. pl.; reprinted in [Agassiz, D. J. L.] (Ed.), 1978. *Illustrated papers on British Microlepidoptera*: 38–48.

Chalmers-Hunt, J. M., 1965. Observations on *Lithocolletis acerifoliella* Zeller (*sylvella* Haworth) ssp. *joannisi* Le Marchand (*platanoidella* Joannis *nom. preoc.*) and a note on its occurrence in Britain. *Entomologist's Rec. J. Var.* **77**: 247–248, 1 pl.

——, 1982. On some interesting Lepidoptera in the National Museum of Ireland. *Ir. Nat. J.* **20**: 531–537.

Chapman, T. A., 1902. The classification of *Gracilaria* and allied genera. *Entomologist* **35**: 81–88, 138–142, 159–164.

Chrétien, P., 1908. Microlépidoptères nouveaux pour la faune française: 5. *Gracilaria pyrenaeella* n.sp. *Naturaliste* **30**: 246.

Common, I. F. B., 1976. The oak leaf miner, *Phyllonorycter messaniella* (Lepidoptera: Gracillariidae), established in Australia. *J. Aust. ent. Soc.* **15**: 471–473.

Corbett, H. H., 1898. The '*pomifoliella–spinicolella*' group of the Lithocolletidae. *Entomologist's Rec. J. Var.* **10**: 168–170.

Daltry, H. W., 1936. *Eupista* (*Coleophora*) *sylvaticella* Wood, and *Lithocolletis distentella* Zell., in East Kent. *Entomologist* **69**: 114.

Emmet, A. M., 1970. Exhibits – annual exhibition 1969. *Proc. Trans. Br. ent. nat. Hist. Soc.* **3**: 21.

——, 1971a. More Lepidoptera in West Galway. *Entomologist's Gaz.* **22**: 3–18.

——, 1971b. *Caloptilia rufipennella* Hübner (Lep., Gracillariidae), a species new to Britain. *Entomologist's Rec. J. Var.* **83**: 291–295.

——, 1972. *Phyllonorycter trifasciella* Haw. (Lep., Gracillariidae), an upper surface mine. *Ibid.* **84**: 205.

——, 1975a. Exhibits – annual exhibition 1974. *Proc. Trans. Br. ent. nat. Hist. Soc.* **8**: 17.

——, 1975b. Notes on the oak-feeding species of *Phyllonorycter* Hübner (Lep.: Gracillariidae). *Entomologist's Rec. J. Var.* **87**: 240–245.

——, 1976. The voltinism of *Phyllonorycter roboris* (Zeller) and *P. cavella* (Zeller). *Ibid.* **88**: 158–160.

——, 1977. The influence of the hot summer of 1976 on leaf-mining lepidoptera. *Ibid.* **89**: 123–124.

——, 1979a. Microlepidoptera in Scotland, 1978. *Ibid.* **91**: 92–96, 122–125.

——, 1979b. The voltinism of *Phyllonorycter roboris* (Zeller). *Ibid.* **91**: 174.

——, (Ed.), [1979]c. *A field guide to the smaller British Lepidoptera*, 271 pp. London.

——, 1981. The distribution of *Caloptilia rufipennella* (Hübner) (Lep.: Gracillariidae) in northern England. *Entomologist's Rec. J. Var.* **93**: 233.

——, 1982a. *Phyllonorycter saportella* (Duponchel) (*hortella* Fabricius) in East Norfolk. *Ibid.* **94**: 119–120.

——, 1982b. Further notes on *Phyllonorycter saportella* (Duponchel) in East Anglia. *Ibid.* **94**: 244.

——, 1983. Exhibits – meeting 11 November 1982. *Proc. Trans. Br. ent. nat. Hist. Soc.* **16**: 58.

Ffennell, D. W. H., 1975. Further notes on the oak-feeding species of *Phyllonorycter* Hübner (Lep.: Gracillariidae). *Entomologist's Rec. J. Var.* **87**: 245–247.

Fletcher, T. Bainbrigge, 1942. *Caloptilia* (*Gracillaria*) *phasianipennella* Hb. *Ibid.* **54**: 125–126.

Ford, L. T., 1933. *Gracilaria pyrenaeella*, Chrét., an addition to the British list. *Entomologist* **66**: 230.

——, 1949. *A guide to the smaller British Lepidoptera*, 230 pp. London.

Frankenhuyzen, A. van & Freriks, J. M., 1976. De in Nederland voorkomende *Phyllonorycter*–soorten op *Alnus*. *Levende Natuur* **76**: 264–273. [In Dutch].

Goater, B., 1974. *The butterflies and moths of Hampshire and the Isle of Wight*, xiv, 439 pp. Faringdon.

Godfray, H. C. J., 1980. Foodplants of *Phyllonorycter messaniella* (Zell.) (Lep.: Gracillariidae). *Entomologist's Rec. J. Var.* **92**: 204.

Gregor, F., 1952. The quercicolous *Lithocolletis* Hb. in ČSR. *Zool. ent. Listy*: 24–56, 3 col. pls [summaries in English].

Gregor, F. & Povolný, D., 1950. The members of *Lithocolletis* Hb. mining *Acer* and *Alnus*. *Ent. Listy (Folia Entom.)* **13**: 129–151 [extensive summaries in English].

——, —— & Řezáč, M., 1963. Systematic oligophagy of the Central European species of the genus *Lithocolletis* Hbn. and *Argyresthia* Hbn. (Lep.) on Prunoideae. *Cas. čsl. Spol. ent.* **60**: 81–93.

Gregory, J. L., 1972. *Phyllonorycter trifasciella* Haw. (Lep. Gracillariidae), an upper surface mine. *Entomologist's Rec. J. Var.* **84**: 78.

——, 1973. *Caloptilia stigmatella* Fabr. (Lep. Gracillariidae). *Ibid.* **85**: 31.

Harper, M. W. & Langmaid, J. R., 1978. Observations on the mines of oak-feeding species of *Phyllonorycter* (Lep.: Gracillariidae). *Ibid.* **90**: 162–166.

Heal, N. F., 1978. *Phyllonorycter distentella* (Zeller) in East Kent. *Ibid.* **90**: 275.

Heckford, R. J., 1984. *Phyllonorycter staintoniella* (Nicelli) (Lepidoptera: Gracillariidae) new to the British Isles. *Entomologist's Gaz.* **35**: 73–75, pl.4, figs 1, 2.

Hering, E. M., 1951. *Biology of the leaf miners*, iv, 420 pp., 180 text figs. 's-Gravenhage.

——, 1957. *Bestimmungstabellen der Blattminen von Europa*, **1**–**3**, 1185, 221 pp., 725 figs. 's-Gravenhage.

Hodgkinson, J. B., 1891. Notes on collecting, *etc.*: Ashton-on-Ribble. *Entomologist's Rec. J. Var.* **2**: 111.

Jacobs, S. N. A., 1945. On the British species of the genus *Lithocolletis*, Hb. *Proc. Trans. S. Lond. ent. nat. Hist. Soc.* **1944-45**: 34–59, text figs, 1 col. pl.; reprinted *in* [Agassiz, D. J. L.] (Ed.), 1978. *Illustrated papers on British Microlepidoptera*: 49–76.

Kloet, G. S. & Hincks, W. D., 1972. A check list of British insects: Lepidoptera (Edn 2). *Handbk Ident. Br. Insects* **11** (2), viii, 153 pp.

Le Marchand, S., 1936. Clé ou table analytique pour la determination des espèces françaises de *Lithocolletis* (Famille des Gracillariidae). *Amat. Papillons* **8**: 83–118.

Leraut, P., 1980. Liste systématique et synonymique des Lépidoptères de France, Belgique et Corse. *Alexanor*, suppl. **1980**, 334 pp.

Mere, R. M., Bradley, J. D. & Pelham-Clinton, E. C., 1964. Lepidoptera in Ireland, May–June, 1962. *Entomologist's Gaz.* **15**: 66–92.

Meyrick, E., 1928. *A revised handbook of British Lepidoptera*, vi, 914 pp. London.

Michaelis, H. N., 1979. Exhibits – annual exhibition 1978. *Proc. Trans. Br. ent. nat. Hist. Soc.* **12**: 9.

Miller, P. F., 1973. The biology of some *Phyllonorycter* species (Lepidoptera: Gracillariidae) mining leaves of oak and beech. *J. nat. Hist.* **7**: 391–409.

Newton, J., 1959. *Caloptilia sulphurella* Haw. var. *aurantiella* Peyer. in Gloucestershire. *Entomologist's Rec. J. Var.* **71**: 89.

——, 1981. *Caloptilia hemidactylella* D. & S. (Lep.: Gracillariidae) in Gloucestershire. *Ibid.* **93**: 85.

Palmer, R. M., Pelham-Clinton, E. C. & Young, M. R., 1984. *Callisto coffeella* (Zetterstedt) (Lep., Gracillariidae): a species new to Britain. *Ibid.* **96**: 41–42, 1 fig.

Pelham-Clinton, E. C., 1967. *Parornix alpicola* (Wocke) (Gracillariidae) and other Lepidoptera feeding on *Dryas octopetala* L. in Scotland. *Entomologist's Gaz.* **18**: 69–72.

Pierce, F. N., 1917. Occurrence in England of *Parornix finitimella* Z., a species of Gracilariidae new to the British list. *Entomologist's mon. Mag.* **53**: 9–10.

—— & Metcalfe, J. W., 1935. *The genitalia of the tineid family of the Lepidoptera of the British Isles*, xxii, 116 pp., 34 pls. Oundle.

Porritt, G. T., 1907. Lepidoptera. *Victoria County History of Yorkshire* **1**: 245–276.

Pottinger, R. P. & LeRoux, E. J., 1971. The biology and dynamics of *Lithocolletis blancardella* (Lepidoptera: Gracillariidae) on apple in Quebec. *Mem. ent. Soc. Can.* **77**: 1–437.

Povolný, D., 1949. The members of the genus *Lithocolletis* Hb. mining Prunoidea and Pomoidea. *Acta Univ. agric. et silvic.*, *Brno* Sign. C **45**: 1–57.

——, 1977. Die systematische Oligophagie mancher Blattminiermotten und ihre Beziehung zur taxonomischen Stellung und Pflege von Aprikose Pfirsich im Rahmen von Prunoideae und Pomoideae. *Mitt. Klosterneuburg* **27**(1): 23–26.

—— & Gregor, F., 1950. Contributions to the knowledge of genus *Lithocolletis* Hb. *Ent. Listy (Folia Entom.)* **13**: 33–36, 1 col. pl. [partly in English].

Price, L., 1977. *Phyllonorycter sagitella* (Bjerkander, 1790) = *tremulae* Zeller, 1846 (Lep., Gracillariidae). A species new to Britain. *Entomologist's Rec. J. Var.* **89**: 106–107.

Robson, J., 1913. A catalogue of the Lepidoptera of Northumberland, Durham and Newcastle-upon-Tyne. *Nat. Hist. Trans. Northumb.* **15**(1912): 107–289.

St Edmundsbury and Ipswich [Whittingham, W. G.], 1937. Unusual larval habit of an *Ornix*. *Entomologist* **70**: 31.

Shaw, M. R., 1981. The northerly distribution of *Caloptilia rufipennella* (Hübner) (Lepidoptera: Gracillariidae) in Britain. *Entomologist's Rec. J. Var.* **93**: 148–149.

——, 1982. *Phyllonorycter muelleriella* (Zeller) (Lepidoptera: Gracillariidae) in Scotland. *Entomologist's Gaz.* **33**: 77–78.

Sich, A., 1911. *Gracilaria syringella*, F., mining in *Phillyrea media*. *Entomologist's mon. Mag.* **47**: 227–228.

Simpson, A. N. B., 1978. *Phyllonorycter sagitella* (Bjerkander) (Lep.: Gracillariidae) in Worcestershire. *Entomologist's Rec. J. Var.* **90**: 86.

Sokoloff, P. A., 1979. Foodplants of *Phyllonorycter trifasciella* Haworth. *Ibid.* **91**: 130.

Stainton, H. T., 1857. *The natural history of the Tineina*, **2**, vii, 317 pp., 8 col. pls. London.

——, 1859. *A manual of British butterflies and moths*, **2**, xi, 480 pp. London.

——, 1864. *The natural history of the Tineina*, **8**, vii, 315 pp., 8 col. pls. London.

——, 1868. Observations on the Tineina – *Gracilaria imperialella* Mann. *Entomologist's Annu.* **1868**: 147–149.

——, 1870. Observations on the Tineina – *Gracilaria imperialella* Mann. *Ibid.* **1870**: 12.

Swan, D. I., 1973. Evaluation of biological control of the oak leaf-miner *Phyllonorycter messaniella* (Zell.) (Lep., Gracillariidae) in New Zealand. *Bull. ent. Res.* **63**: 49–55.

Trägårdh, I., 1913. Contributions towards the comparative morphology of the trophi of the lepidopterous leaf-miners. *Ark. Zool.* **8**: No. 9.

Tutt, J. W., 1898. On the British species of *Lithocolletis* of the *spinicolella* group. *Entomologist's Rec. J. Var.* **10**: 164–168.

Vári, I., 1961. *South African Lepidoptera*, **1**, 238 pp., 23 col. pls, 499 text figs. Pretoria.

Wakely, S., 1946. Collecting notes from Warnham, Sussex. *Entomologist's Rec. J. Var.* **58**: 100–101.

——, 1961. Notes on *Acrocercops imperialella* Mann and its appearance at Wood Walten [*sic*] Fen, Hunts. *Ibid.* **73**: 83–86.

——, 1966. Notes on *Caloptilia* (*Gracillaria*) *falconipennella* Hübner (Lepidoptera). *Ibid.* **78**: 49–50.

Waters, E. G. R., 1929. Notes on some species of *Lithocolletis*, with an addition to the British list. *Entomologist's mon. Mag.* **65**: 163–170.

Watkinson, I. A. & Chalmers-Hunt, J. M., 1980. Field meeting – 28 October 1979. *Proc. Trans. Br. ent. nat. Hist. Soc.* **13**: 135.

Wood, J. H., 1890. Notes on *Gracilaria populetorum, elongella* and *falconipennella*. *Entomologist's mon. Mag.* **26**: 133–137.

PHYLLOCNISTIDAE
A. M. Emmet

A family with world-wide distribution but absent from New Zealand. It includes some of the smallest Lepidoptera known. The wing-pattern of the imagines is very similar to that of *Leucoptera* Hübner (Lyonetiidae) and accordingly *Phyllocnistis* Zeller was included in that family by early systematists. However, Chapman (1902) pointed out that the early stages showed affinity with the Gracillariidae; the larva has similar adaptations as a sap-feeder and likewise is dimorphic, although in *Phyllocnistis* the second morph occurs only in the final instar; it consumes no food and exists wholly for cocoon-spinning, the mouth parts and most organs other than the spinneret being atrophied. Accordingly he transferred *Phyllocnistis* to the Gracillariidae with subfamily status. Modern systematists have followed Chapman's arrangement or, as here, accord full family status.

PHYLLOCNISTIS Zeller

Phyllocnistis Zeller, 1848, *Linn. ent.* **3**: 264.

A large genus with world-wide distribution. There are three British species, all of which are associated with Salicaceae.

Imago. Head smooth; antenna about four-fifths length of forewing; scape somewhat dilated and concave beneath to form a small eyecap; labial palpus of moderate length, porrect, filiform, pointed; maxillary palpus very short. Tibia of hindleg with series of long bristles above, tarsus bristly above towards base (figure 23, p. 65). Forewing lanceolate and weakly caudate in some species; veins 3 (Cu_1) and 4 (M_3) absent, 6 (M_1) and 7 (R_5) stalked, 8 (R_4) generally absent, 11 (R_1) from beyond middle of cell. Hindwing narrow lanceolate with long cilia; veins 3 (Cu_1) and 4 (M_3) absent, 5 (M_2) and 6 (M_1) stalked (figure 111).

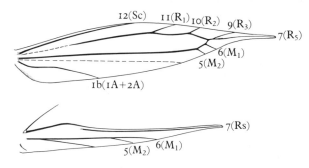

Figure 111 *Phyllocnistis unipunctella* (Stephens), wing venation

Ovum. Oval; chorion at first smooth but becoming wrinkled before hatching (Sundby, 1953: 104). Soon afterwards it collapses entirely, leaving no recognizable character. Laid on a leaf of the foodplant.

Larva (figure 112a). Throughout the whole stage a sap-feeder, mining epidermal cells of leaves or, in the case of *P. saligna* (Zeller) during part of its life, of twigs. It hereby differs from a gracillariid larva which feeds thus only in its earlier instars. The morphology is adapted accordingly. The body is flattened dorsoventrally but is relatively broad, especially the thoracic segments, and has the intersegmental divisions deeply incised. The three posterior segments, however, differ from those of gracillariids. Abdominal segment 8 is narrower and has four rather large outgrowths, two projecting in a dorsolateral and two in a ventrolateral direction; the former pair are often pressed against the cuticle of the leaf and may serve the purpose of legs. Segment 9 is still narrower and barrel-shaped; it can be withdrawn into segment 8. Segment 10 is elongate and tubular with its distal part forked (Sundby, *loc. cit.*: 107). Thoracic legs and prolegs absent. Important modifications occur also in the head-capsule, which is relatively broad, triangular and exceedingly thin. The stemmata are reduced to a single functional pair (Hering, 1951: 120, fig.94, 125) and there is no spinneret. The head is prognathous; each mandible consists of a flattened horizontal disc with a serrate edge (Trägårdh, 1913: 28, fig. 45; Hering, *loc. cit.*: 131, fig. 105) which operates by sawing, not chewing. The labrum is even more enlarged than in the Gracillariidae, but lacks the serration of that family; its function appears to be solely to shield the cuticle from the action of the mandibles and it does not also shave off the fibrous remnants of the epidermal cells (Hering, *loc. cit.*: 126–127).

The following observations of the feeding of *P. unipunctella* (Stephens) were made under a microscope with light transmitted from below. The mandibles sweep rapidly through a wide arc, severing the walls of the epidermal cells; during this process the sap can clearly be seen passing backwards into the alimentary canal, the head-capsule being transparent; no solid matter is absorbed. After 10 to 20 'chews' the mandibles become stationary, but imbibing continues until all the liberated sap has been swallowed. The process is then repeated.

Defecation takes place seldom. When it does, a surprisingly large drop of colourless liquid is suddenly voided. For two or three seconds the droplet remains intact; then its walls collapse and the liquid diffuses over the breadth of the mine, drawing down the cuticle which adheres to the surface of the parenchyma. The leaf-tissues soon absorb the liquid and the cuticle and the parenchyma separate once more. During this process the larva clearly benefits from its elongate and telescopic anal segments which enable it to defecate at a distance and avoid wetting itself. One larva swung its head round and apparently reimbibed its excrement, but this habit was not observed amongst the others which were under observation and possibly is not obligate. Herein a phyllocnistid differs from a gracillariid larva which cannot escape reconsuming the liquid frass voided in the blotch-making phase, for this is absorbed by the area of parenchyma which will be eaten after the change in mouth parts. The quick dispersal of the liquid frass in the phyllocnistid mine is evidence that the haste with which the larva consumes the sap it has liberated from the cells is essential to forestall its loss by absorption.

As in the Gracillariidae, the larva undergoes hypermetamorphosis, but in a different manner and for a different purpose. In gracillariid larvae structural changes accompany the transition from sap-feeding to tissue-feeding, but a phyllocnistid completes its growth as a sap-feeder and is not subjected to changes until its third and final moult which takes place solely for the purpose of cocoon-spinning. The head-capsule is discarded at ecdysis and with it mandibles and stemmata; the antennae and labial palpi, however, are retained in a modified form. The spinneret, which was absent in the feeding instars, is now strongly developed. The larva no longer looks lepidopterous and Sich (1903: 39) on first seeing it mistook it for a parasite (figure 112b,c). The cocoon is spun within the mine in a chamber prepared by the larva before the last ecdysis, generally at the margin of the leaf. Contracting silk causes the leaf-edge to curl, rendering the cocoon more conspicuous than the mine itself.

Pupa (figure 112d,e). Head with a curved, horn-like 'cocoon-piercer', serrate on its inner surface (Hering, *loc. cit.*: 69, fig.56). Appendages shorter than abdomen and not fused; abdominal segments 4–6 in female and 4–7 in male free; abdominal segments 1–8 with lateral projections (Hering, *loc. cit.*: 41, fig.29); segment 10 with cremaster differing between species (Sundby, *loc. cit.*: 112, fig.10). The pupa projects from the cocoon before eclosion.

Imago. The remarkable similarity of the forewing pattern to that of certain species of Lyonetiidae and of *Leucoptera* in particular has already been noted (p. 363), yet what has been written above shows the wide phylogenetic separation between the Phyllocnistidae and the Lyonetiidae. How then does it come about that they are superficially so similar?

Tropical butterflies of different families sometimes resemble each other because one which is palatable mimics another which is distasteful, or because both mimic the same distasteful model. Mimicry is more apparent in tropical regions, and both the Lyonetiidae and the Phyllocnistidae are tropical families with Palaearctic members. When a moth of either family is at rest with its costa touching the substrate, its wing-pattern mimics a small black and yellow hymenopteron facing away from the head of the moth. The pattern is discussed in detail in the introduction to the Cemiostominae (p. 213). It is interesting that in the Cemiostominae the apical cilial pencil, which represents the antenna of the hymenopteron, is made more convincing by gaps in the cilia on either side; in *Phyllocnistis*, although there is no complete break, the adjacent cilia are shorter and more sparse, showing a tendency towards the same adaptation.

The three British Phyllocnistidae overwinter as adults. Most species of other families doing this have the exposed surfaces of dull and cryptic coloration. By contrast, the forewing pattern in *Phyllocnistis* is strongly emphasized against the shining white ground colour. It is as if the moth 'wished' to be seen. Protection is achieved because the pattern is mimetic of an aposematic model and also deflectionary. If the moth is found by a predator, it may be mistaken for a dangerous or distasteful hymenopteron and shunned in consequence; should an attack nonetheless be made, there is a good chance that it will be directed against the less vulnerable wing-tip. As far as is known, the explanation here given for the wing-pattern of the Cemiostominae and the Phyllocnistidae has not previously been proposed.

Key to species (imagines) of the genus *Phyllocnistis*

I — Forewing with two parallel median longitudinal dark fuscous lines .... *saligna*(p. 366)
– Forewing without such longitudinal lines 2
2(I) Forewing with fuscous suffusion on dorsum near base and in disc anterior to transverse fascia; four costal strigulae .... *unipunctella*(p. 367)
– Forewing without fuscous suffusion in these areas but with median dorsal fuscous spot; usually five costal strigulae .... *xenia*(p. 368)

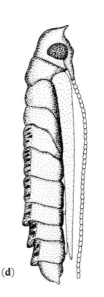
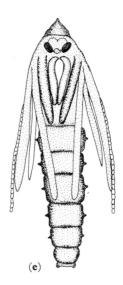

Figure 112 *Phyllocnistis* spp., larva and pupa
(**a**) larva, first to third instars. After Hering
(**b**,**c**) *Phyllocnistis unipunctella* (Stephens), head in final instar showing spinneret and modified palpi and antennae, greatly enlarged. After Trägårdh
(**d**,**e**) phyllocnistid pupa, viewed laterally and ventrally. After Hering

PHYLLOCNISTIS SALIGNA (Zeller)

Opostega saligna Zeller, 1839, *Isis, Leipzig* **1839**: 214.

Type localities: Germany; Berlin, Glogau (now Poland; Glogów) and Frankfurt am Main.

Description of imago (Pl.11, fig.18)

Wingspan 6–7mm. Head white; antenna grey with darker annulations, obsolescent beneath. Forewing white, apical one-third suffused pale ochreous fuscous; two parallel median longitudinal dark fuscous lines from base of costa to beyond one-half, enclosed space usually ochreous-tinged; a slightly curved transverse fascia at three-quarters, preceded by an outward-oblique bar from costa which sometimes extends almost or wholly across wing as an anterior parallel fascia, the intervening space then being suffused ochreous; cilia white, tips in termen with fuscous reflections; costal cilia with two dark bars; a black dot at base of apical cilia whence three divergent dark bars extend into terminal cilia and below which a dark fuscous line extends near base of cilia to tornus. Hindwing pale purplish grey; cilia white.

Life history

Ovum. Laid on either side of a leaf of purple willow (*Salix purpurea*), more commonly on the underside. Crack willow (*S. fragilis*) and other smooth-leaved willows are less often selected.

Larva. Head transparent and almost colourless, mouth parts pale brown; body whitish yellow; gut yellow. In later instars, head pale brown, body whitish green.

Mine (Pl.1, figs 10,11). Feeding starts in a sinuous epidermal mine on the side of the leaf on which the egg was laid, the frass forming rather a broad, brownish central line. The larva then mines via the petiole into the twig; upper surface mines are nearly always aborted because the larva mines down the upper surface of the petiole and on reaching the base finds itself trapped in the sharp angle between the petiole and the twig. The under surface miner reaches the twig successfully and then mines upwards or downwards in the epidermis in a pale buff-coloured gallery with a fine central line of frass adhering to the cuticle. The main direction of movement is upwards and when nearly full-grown the larva mines via the petiole back into a leaf which is almost always distal to the one in which feeding began. In this second leaf the mine is still epidermal and similar to, though broader than, the mine of the young larva. The mine usually ends at the margin of the leaf. June to July; August to September.

Pupa. The cocoon is spun under the epidermis at the end of the mine, usually at the margin of the leaf, which curls over to conceal it. Late June to July and late August to September.

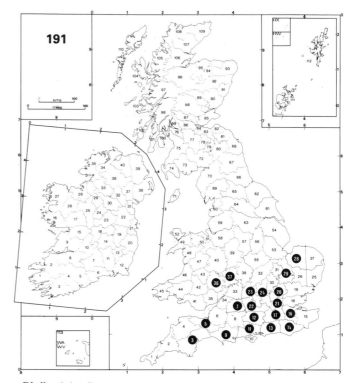

Phyllocnistis saligna

Imago. Bivoltine, flying in July and from September until April; the winter generation often overwinters in thatch.

Distribution (Map 191)

Local and uncommon south of a line from the Wash to Herefordshire. Throughout Europe, where it is more common than in Great Britain.

PHYLLOCNISTIS UNIPUNCTELLA (Stephens)

Argyromiges unipunctella Stephens, 1834, *Ill.Br.Ent.* (Haust.) **4**: 260.
Opostega suffusella Zeller, 1847, *Isis, Leipzig* **1847**: 894.
Type locality: England; Coomb Wood [Surrey].

Description of imago (Pl.11, fig.19)

Wingspan 7–8 mm. Head creamy white; antenna greyish white with obscure darker annulations, obsolescent below. Forewing white, apical one-third suffused ochreous; a fuscous suffusion on dorsum near base and a similar suffusion in disc beyond middle; an almost straight fuscous fascia at three-quarters with posterior white edging expanded into triangular white spots on costa and dorsum, preceded by an outward-oblique fuscous bar from costa; cilia white, tips on termen with fuscous reflections; costal cilia with two fuscous bars; a prominent black dot at base of apical cilia whence three divergent fuscous bars extend into terminal cilia and below which a fuscous line extends near base of cilia to tornus. Hindwing grey; cilia white with fuscous reflections, especially on costa.

Occasionally the forewing is suffused yellow and July adults often lack the fuscous suffusion near base of dorsum; Waters (1925: 88) mistook this form for *P. tremulella* [Fischer von Röslerstamm] (*labyrinthella* (Bjerkander)), but later corrected his mistake (Waters, 1928: 176).

Life history

Ovum. Oval, milk-white, with inconspicuous, irregular sculpturing (Sich, 1903: 36). Laid on either side of a leaf of black poplar (*Populus nigra*) or Lombardy poplar (*P. nigra* var. *italica*), but more often on the upperside.

Larva. Head colourless. Body pale greenish white; prothoracic plate more or less semicircular, blackish.

Mine (Pl.1, fig.9). Epidermal on the side of the leaf on which the egg was laid, taking the form of a long, irregular gallery. The mine is hard to see except from an angle, when the raised epidermis appears silvery, as if a snail had crawled over the leaf. The larva voids clear liquid 'frass' and there is no dark frass-line as in other British *Phyllocnistis*. Feeding and defecation are described above in the generic introduction (p. 364). The mine nearly always ends at the margin of the leaf. June and August.

Pupa. Blackish. Head equipped with a prominent, horn-like 'cocoon-piercer', its inner margin serrate (Hering, 1951: 68, fig.56, in which the left side of the figure represents the dorsal surface); abdominal segments 1–8 on each side with a projection bearing a long seta, the posterior of segments 5–8 also with a prominent, transverse, lateral ridge; abdominal segments 4–9 each with a small area of strong, posteriorly directed, thorn-like spikes on dorsum; cremaster consisting of two widely divergent spines. In a

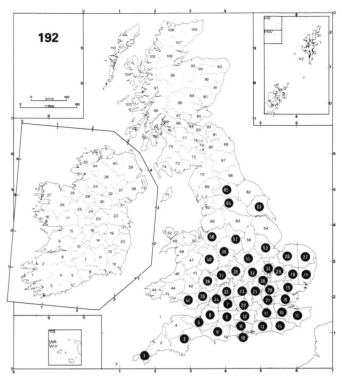

Phyllocnistis unipunctella

relatively stout silken cocoon in a chamber at the end of the mine, generally in the leaf-margin which curls over conspicuously. June to July; August to September.

Imago. Bivoltine. July; September to April; the second generation often overwinters in thatch or hayricks.

Distribution (Map 192)

Frequents open country, woodland, gardens and all situations where the foodplant grows. Widespread and common in England as far north as Cheshire; south Wales. Europe, including Fennoscandia; Asia Minor.

PHYLLOCNISTIS XENIA Hering

Phyllocnistis xenia Hering, 1936, *Eos, Madr.* **11**: 369.
Type locality: Spain; Torre del Mar, Andalusia.

Description of imago (Pl.11, fig.20)

Wingspan 6–7mm. Head shining white; antenna white with obscure darker annulations. Forewing shining white, apical one-third suffused pale ochreous and greyish fuscous; a fuscous median dorsal spot; a transverse fuscous fascia at three-quarters, often interrupted in disc, preceded by one outward-oblique and followed by two or three direct bars from costa; cilia white, tips on termen with fuscous reflections; a prominent black dot at base of apical cilia, from which emanate three divergent dark bars and below which there is a dark line near base of cilia extending to tornus. Hindwing purplish grey; cilia shining white, on costa with fuscous reflections.

Life history

Ovum. Laid on the upperside of a leaf of grey poplar (*Populus canescens*), usually a tender terminal leaf of a sapling.

Larva. Head transparent and colourless, but sutures obscurely darkened. Body pale yellow, prothoracic plate colourless.

Mine (Pl.1, fig.12). A silvery gallery in the upper epidermis with a dark central line of frass showing up vividly against the silver background; the course is gracefully sinuous and the colour and pattern combine to produce a very beautiful mine. The mine ends at the margin of the leaf. June, July; August, September.

Pupa. In a cocoon at the end of the mine in the leaf-margin, which folds over it. July; August, September.

Imago. Bivoltine, flying in July and August and again in September, when it presumably overwinters until April or May. Some authorities regard this species as synonymous with *P. labyrinthella* (Bjerkander), the larva of which feeds in the leaves of aspen (*Populus tremula*), generally on the underside. Differences between the imagines are very slight, but the mines show clear distinctive features even where both occupy the same territory in central Europe (Pelham-Clinton, 1976: 163).

Distribution (Map 193)

Known in Britain only in eastern Kent where it was discovered in 1974 by E. C. Pelham-Clinton (*loc.cit.*). It was evidently a recent arrival as the grey poplars had been searched a year or two previously by D. W. H. Ffennell for *Phyllonorycter comparella* (Duponchel) (Gracillariidae), but he had seen no sign of the conspicuous *Phyllocnistis* mines. The finding of a second colony 27km (17 miles) from the first (Heal, 1984) may be evidence that it is

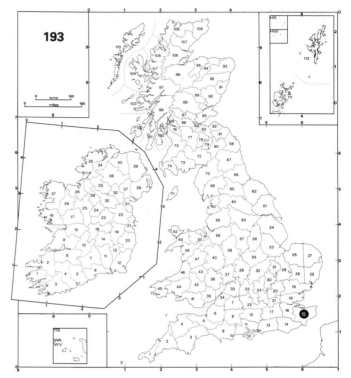

Phyllocnistis xenia

now extending its range. Europe, from Poland to Spain, but local and uncommon.

References

Chapman, T. A., 1902. The classification of *Gracillaria* and allied genera. *Entomologist* **35**: 81–88, 138–142, 159–164.

Heal, N. F., 1984. A second British locality for *Phyllocnistis xenia* Hering. *Entomologist's Rec. J. Var.* **96**: 98.

Hering, E. M., 1951. *Biology of the leaf miners*, iv, 420 pp., 180 text figs. 's-Gravenhage.

Pelham-Clinton, E. C., 1976. *Phyllocnistis xenia* Hering, 1936, a recent addition to the British list of Lepidoptera. *Entomologist's Rec. J. Var.* **88**: 161–164, 1 fig.

Sich, A., 1903. Observations on the early stages of *Phyllocnistis suffusella* Zell. *Proc. S. Lond. ent. nat. Hist. Soc.* **12**: 30–48.

Sundby, R., 1953. Studies on the leaf-mining moth, *Phyllocnistis labyrinthella* Bjerk. 1. *Nytt Mag. Zool.* **1**: 98–128, 15 figs.

Trägårdh, I., 1913. Contributions towards the comparative morphology of the trophi of the lepidopterous leaf-miners. *Ark. Zool.* **8**(9): 1–48, 67 text figs.

Waters, E. G. R., 1925. Three additions to the British list of Tineina. *Entomologist's mon. Mag.* **61**: 82–89.

———, 1928. Tineina in the Oxford district (part). *Ibid.* **64**: 172–178.

SESIIDAE
B. R. Baker

Clearwing moths are almost world-wide in distribution with about 1000 species now described (Naumann, 1971). The imagines are diurnal, swift of flight and seldom observed unless special search is undertaken. The larvae are all internal feeders within the wood of certain trees or in the stems and roots of particular herbaceous plants. Only 15 species have so far been recorded in the British Isles, 13 of which were already known to Edward Newman as early as 1832 (Newman, 1832), but when we read of *Synanthedon andrenaeformis* (Laspeyres) being an 'extremely rare species' (Bankes, 1906) it is an indication of the imperfect state of knowledge existing at that time regarding a now generally recognized common species. It is also a reminder that patient field-work could well result in extending the known distributions of many of our species or in discovering species new to the British Isles. When clearwing life histories are fully understood and the tell-tale signs of occupation appreciated it is not a difficult task to come by the moths. Few real rarities exist and it is more accurate to speak of extreme localisation when referring to such species as *S. scoliaeformis* (Borkhausen) or *Bembecia chrysidiformis* (Esper).

Imago. Medium-sized to small. Haustellum reduced (*Sesia* Fabricius) or well developed (*Synanthedon* Hübner and *Bembecia* Hübner); antennae sexually dimorphic, males exhibiting a variety of surface-increasing specializations; labial palpus moderately long, colour of scaling of some specific significance. Forewing extremely narrow due to reduction of anal area. Wing-coupling unique, the forewing posterior margin being folded downwards and the hindwing anterior margin folded upwards, both folds interlocking by a series of recurved spines (Common, 1970). Additionally present are the more normal frenulum and retinaculum. Scales generally absent from the greater part of both pairs of wings. Abdomen with scale-covering giving well-defined, often brightly coloured, specific patterning; prominent expansible anal tuft.

Ovum. Conforming to a generalized pattern of a flat ovoid, flattened at one pole, and bearing a fine reticulation which is only visible under high magnification. Colour variable, mostly through shades of brown to black. Inserted in bark crevices, axils of branches, on leaves or in the ground close to the roots of the foodplant. Very small in *Sesia*.

Larva. Well adapted for tunnelling within plant tissue. Mandibles prominent; head-capsule sclerotized and capable of being withdrawn within the prothorax which itself bears varying degrees of sclerotization. Prolegs well developed. Integument appearing wrinkled, generally of a whitish or whitish yellow transparency through which the dorsal vessel and gut contents are frequently visible; anal plate often lightly sclerotized.

Pupa. Head heavily sclerotized and bearing a frontal process with which the pupa ruptures the cocoon (if present) and tunnel cap. Most abdominal segments with one or two transverse rows of backwardly directed spines; segment 7 bears a double row of spines in the male but only a single row in the female (Kemner, 1922).

Biology. All the clearwings are colonial in habit and colonies may be concentrated upon one or two trees, leaving others unaffected. Much time can be saved if an infected tree can be quickly recognized and the following signs should be looked for.

(1) Presence of old emergence holes. These are often very conspicuous in the case of the larger species. Sizes of holes are given in the descriptions of species.

(2) Presence of frass in bark crevices or at the bases of trunks.

(3) Presence of pupae extruded from the host plants.

Old emergence holes are evident throughout the year, frass is more readily observed in early spring and the presence of extruded pupae will coincide with the emergence of the particular species. These extruded pupae are not as delicate as one might think and though best sought for at the beginning of an emergence the exuviae can remain in position on a trunk for up to a year provided they extrude in a fissure of a much-wrinkled bark surface.

Once a host tree is found, adult moths can be expected on the trunk provided it is practicable to visit the tree regularly for several days and one is prepared to keep watch there from about 08.00 to 11.00 hrs on each occasion! Rarely is this possible and in practice one must have recourse to breeding clearwings through to the adult by carefully taking sections of trunks, stems or stump-tops. These sections or stem-lengths should be stood in a 150mm depth of damp sand with a topping of sphagnum and the container kept in a sheltered position outside until emergences begin. It is important to keep the moss and stems regularly sprayed with water and helpful, though not essential, for the container to receive an hour or so of early morning sunshine. Aquarium tanks make very suitable containers and these should be fitted with framed perforated-zinc tops, each top bearing two lift-up doors. This enables one to reach any position in the tank when wishing to remove newly emerged specimens, while at the same time minimizing the risk of losing one of these skittish insects.

Breeding clearwings is a fascinating and useful exercise, for the mapping of distributions in Great Britain has only been possible through the information received from many field-workers whose records are largely based upon the results of breeding experiments. Much valuable information on collecting clearwings is given by Classey *et al.* (1946).

Breeding clearwings from egg to adult would be difficult to achieve, but this would seem to be the only sure way to determine accurately the length of the life-cycle of the various species. The only published account known to the writer concerns *Sesia apiformis* (Clerck) (Bretherton, 1946). In the following species descriptions therefore one constantly has to use the qualifying word 'probably' when referring to one-, two- or three-year cycles, and only in a few instances has it been found possible to be precise on this point.

The generic descriptions which follow are based on Fibiger & Kristensen (1974).

Figure 113

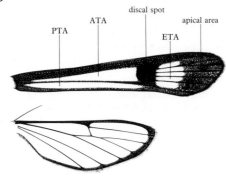

(**a**) *Synanthedon tipuliformis* (Clerck)
Wing areas: PTA – posterior transparent area; ATA – anterior transparent area; ETA – exterior transparent area. After Fibiger & Kristensen (*loc.cit.*)

Key to species (imagines) of the Sesiidae

Based on Meyrick (1928); terminology of wing areas after Fibiger & Kristensen (*loc.cit.*).

1   Forewing almost entirely covered with dark fuscous scales; ATA and PTA only apparent as small patches at wing-bases; hindwing with veins 3 (Cu_1) and 4 (M_3) separate .... *Paranthrene tabaniformis* (p. 373)
 –   Forewing without large area of dark fuscous scales; ATA and PTA easily visible; hindwing with veins 3 (Cu_1) and 4 (M_3) stalked (figure 113a) 2

2(1)   Forewing with vein 7 (R_5) to termen (figure 113b); large robust insects 3
 –   Forewing with vein 7 (R_5) to apex (figure 113c); more delicate, thinly bodied insects 4

3(2)   Head yellow; anterior portion of tegulae yellow; patagia dark brown .... *Sesia apiformis* (p. 371)
 –   Head blackish; thorax blackish brown; patagia bright yellow .... *S. bembeciformis* (p. 372)

4(2)   Frons wholly black 5
 –   Frons white at least on sides 6

5(4)   Antenna with white subapical band ....
.... *Synanthedon spheciformis* (p. 377)
 –   Antenna without white band ....
.... *S. andrenaeformis* (p. 380)

6(4)   Forewing with PTA reaching discal spot 7
 –   Forewing with PTA absent, partially obliterated or not reaching discal spot 13

7(6)   Forewing with discal spot distally orange ....
.... *S. vespiformis* (p. 376)
 –   Forewing with discal spot not distally orange 8

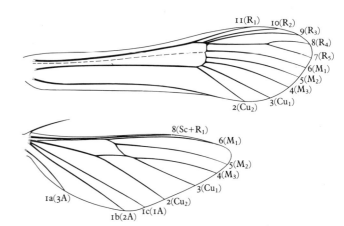

(**b**) *Sesia bembeciformis* (Hübner), wing venation

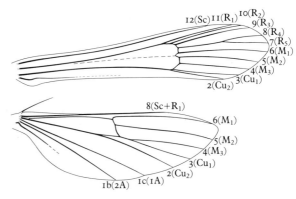

(**c**) *Synanthedon tipuliformis* (Clerck), wing venation

8(7) Forewing with terminal fascia red ....
.... *S. formicaeformis* (p. 382)
–   Forewing with terminal fascia not red 9
9(8) Forewing with terminal fascia suffused yellow 10
–   Forewing with terminal fascia not suffused yellow 11
10(9) Head with yellow posterior ring; tegulae edged yellow
.... *S. tipuliformis* (p. 375)
–   Head without yellow posterior ring; tegulae black ....
.... *S. flaviventris* (p. 379)
11(9) Abdomen with two yellow rings ....
.... *S. scoliaeformis* (p. 378)
–   Abdomen with one red (seldom yellow) ring 12
12(11) Forewing base suffused red; labial palpus reddish
orange beneath .... *S. culiciformis* (p. 383)
–   Forewing base not suffused red; labial palpus not orange
beneath .... *S. myopaeformis* (p. 381)
13(6) Forewing marked orange-red 14
–   Forewing without red scales ....
.... *Bembecia muscaeformis* (p. 386)
14(13) Anal tuft reddish orange .... *B. chrysidiformis* (p. 387)
–   Anal tuft not orange .... *B. scopigera* (p. 385)

Sesiinae

SESIA Fabricius

Sesia Fabricius, 1775, *Syst.Ent.*: 547.
Sphecia Hübner, [1819], *Verz.bekannt.Schmett.* : 127.

A small genus with two species occurring in Great Britain
and Ireland.

Imago. Characterized by having the forewing with vein 7
(R_5) to the termen (figure 113b) and the presence of thickening of processes on the valvae (Naumann, 1971).

SESIA APIFORMIS (Clerck)
Hornet Moth

Sphinx apiformis Clerck, 1759, *Icones Insect.rar.* **1**: pl.9,
fig.2.
Type locality: [Sweden].

Description of imago (Pl.14, figs 1,2)

Wingspan 33–46mm. Head bright yellow; antenna lamellate in male. Thorax ferruginous brown except for anterior
portion of tegulae which is bright yellow; posterior edge of
mesothoracic crest bears traces of yellow. Legs ferruginous brown, mid- and hindfemora streaked yellow. Forewing costa and discal spot with dense ferruginous brown
scaling; veins ferruginous brown; patches of ferruginous
brown scales, often very slight, may occur in any of the
three transparent areas of forewing (ATA, PTA and ETA)
(figure 113a). Hindwing hyaline; veins ferruginous. Cilia
of both wings dark fuscous in male, ferruginous brown in
female. Abdomen blackish brown on segment 3 and at
intersegmental divisions, otherwise bright yellow. Some
variability of the blackish clouding on the yellow.

Life history

Ovum. Dark red, elliptical; long axis 1.1mm. Laid low
down in bark crevices or in old emergence holes on black
poplar (*Populus nigra*) and other species of *Populus*. The
egg potential of each female is about 1400. Eggs hatch in
13 days at 20°C.

Larva. Full-fed *c.*30mm long. Head chestnut-brown;
mandibles prominent, black-tipped; head can retract into
prothorax which is lightly sclerotized. Body yellowish
white, wrinkled in appearance with dorsal vessel faintly
visible; spiracles edged brown.

The larva makes extensive tunnels between bark and
wood in the lower parts of the trunk and into the roots,
living at least two years. Some evidence indicates a three-
year life-cycle (Cockayne, 1933). The larva is sometimes
responsible for causing the death of large poplars (Chrystal, 1937).

Pupa. 22–30mm long. Dark reddish brown with rows of
backwardly directed spines on dorsum of abdominal segments. In a cocoon, 25–35mm long, elongate with rounded ends, the surface roughened with pieces of gnawed-off
wood fragments arranged criss-cross in the cocoon matrix.
The cocoons are formed close to the end of the capped
tunnel and remain visible within old tunnels for several
years. Formation of cocoons takes place in autumn and
these may be detected throughout the winter months by
carefully rasping away the bark of trees which show at
their bases numerous holes of previous emergences. The
larvae overwinter within the cocoons and pupate in April.

Imago. The moths emerge from mid-June to mid-July,
rarely in August. They may be found on trunks of black

poplar and other species of *Populus* from 07.00 until 11.00 hrs by which time pairing has often taken place. On 4 July 1983 a female which had emerged at 08.45 hrs attracted a male at 10.25 hrs and the pair remained *in copula* for just over an hour. In very wet weather the moths may remain on the trunks throughout the day, but under sunny or at least dry conditions, dispersal will have taken place by late morning. Old emergence holes, 8mm in diameter, around trunk-bases indicate infected trees.

Distribution (Map 194)

Predominantly in the central, southern and south-eastern parts of England, though also recorded from Durham and north and south Wales. Local in Ireland with scattered records from a few eastern and southern counties. Widely distributed in the western and central parts of the Palacarctic region and accidentally introduced into north America.

SESIA BEMBECIFORMIS (Hübner)
Lunar Hornet Moth

Sphinx bembeciformis Hübner, 1797, *Beitr.Gesch. Schmett.*: 92.

Sphinx crabroniformis Lewin, 1797, *Trans.Linn.Soc.Lond.* **3**: 1.

Type locality: not stated.

Description of imago (Pl.14, figs 3,4)

Wingspan 32–42mm. Head blackish; antenna dilate sub-apically, lamellate in male. Thorax blackish brown with bright yellow patagial collar. Legs predominantly yellow, but tibiae markedly reddish yellow. Forewing with little trace of scaling; costa and discal spot blackish brown in male, ferruginous brown in female; base of dorsum orange-brown; veins of ETA blackish or ferruginous brown. Hindwing with veins blackish in male, ferruginous brown in female; wing-borders narrow in male. Abdomen with segment 1 blackish, and variable black or brown clouding on segments 3 and 4, segment 2 and terminal segment bright yellow.

Life history

Ovum. Reddish brown, elliptical. Laid low down on the trunks of various species of *Salix*. The eggs occur in diffuse patches on the bark and are fairly conspicuous when searched for.

Larva. Full-fed 27–30mm long. General appearance yellowish white and much wrinkled. Head chestnut-brown, mandibles black-tipped. Pronotum lightly sclerotized and showing two faint brownish diagonal streaks; mesonotum widest segment, reaching 7mm in width; spiracles ringed brown.

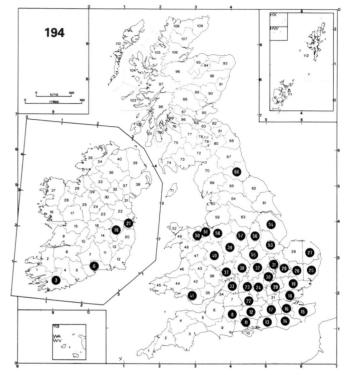

Sesia apiformis

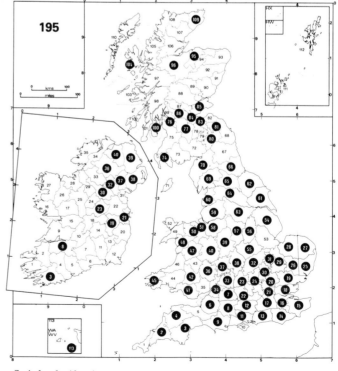

Sesia bembeciformis

During the first year haphazard tunnelling takes place below the bark at and below ground level, and fine granular frass is apparent around trunk-bases. In the second year the larvae bore deeper into the wood and excavate vertical tunnels 25–50cms above ground level. At this stage the presence of *S. bembeciformis* in the trunks may be indicated by gashes made by woodpeckers which make unsightly scars as they search for larvae and pupae. Old emergence holes are generally 10–20cms above ground level and have an average diameter of 8mm. Capped tunnels are prepared during the second autumn and require careful searching for their detection.

Pupa. 21–29mm long. Yellowish brown with rows of backwardly directed spines on the dorsum of abdominal segments. The cocoon is situated at the upper end of the larval boring and attached to the extremity of the silkenlined tunnel. The cocoon is soft and delicate around the point of attachment, but toughened with wood scrapings at the downward-facing head end which is free of the tunnel wall; internally with a dense white silk pad at head end.

Imago. The moths emerge from the first to third week of July and, like *S. apiformis* (Clerck), may be found on the host trunks in early morning. Females intent on ovipositing may at times be seen flying low around sallows from about 15.00 to 17.00 hrs.

Distribution (Map 195)

Generally distributed throughout the British Isles, with most records in central and southern England. Probably under-recorded in Ireland. Throughout north-western and central Europe.

Paranthreninae

PARANTHRENE Hübner

Paranthrene Hübner, [1819], *Verz.bekannt.Schmett.* : 128.
Sciapteron Staudinger, 1854, *Diss.ent.Sesiis agri berol.*: 39,43.

This genus is represented in Great Britain by a single species.

Imago. Characterized by having the hindwing with veins 3 (Cu_I) and 4 (M_3) separate.

PARANTHRENE TABANIFORMIS (Rottemburg)
Dusky Clearwing

Sphinx tabaniformis Rottemburg, 1775, *Naturforscher, Halle* 7: 110.

Type locality: Germany; Landsberg an der Warthe.

Description of imago (Pl.14, fig.5)

Wingspan 26–34mm. Head black with white vertical bar anterior to each eye and a yellow posterior ring; antenna black, strongly bipectinate in male. Thorax black, small yellow dots at base near the outer edge of patagium. Legs with mid- and hindtibiae orange, black-banded. Forewing almost entirely covered with brownish black scales; costal streak violet-blackish; only towards wing-bases are hyaline areas of the ATA and PTA evident. Hindwing hyaline, veins and termen dark fuscous. Abdomen black with dorsal yellow bands on segments 2, 4, 6 and (males) 7.

Life history

As there is no account of this species having been bred in this country, continental authors have been quoted below.

Ovum. Placed on the bases of leaves on twigs or suckers, or in crevices near ground level on trunks of aspen (*Populus tremula*), black poplar (*P. nigra*), Italian poplar (*P.* × *canadensis*) and very occasionally sallow (*Salix* spp.) (Urbahn & Urbahn, 1939).

Larva. Head and prothoracic plate black-brown. Body yellowish white bearing single dark setae; dorsal line dark (Bartel, 1912).

Larval development may follow one of two patterns.

(1) The newly emerged larva lives in a web outside the bark, and overwintering takes place inside a gall made by the beetle *Saperda populnea*. In the spring the larva bores a tunnel, thereby producing a gall which is distinguishable from the *Saperda* gall in being pyriform or ellipsoidal instead of globular. The second winter is spent in the gall and pupation takes place in May.

(2) The larva bores in the roots or inside the bark. After the first hibernation it makes tunnels in the wood or between bark and wood, and after the second hibernation it pupates in a chamber below the trunk surface covered only by a thin layer of bark. The first mentioned developmental type is probably the more usual (Fibiger & Kristensen, 1974).

Larvae and pupae were discovered in France at Uvernet, Basses Alpes, in the unusual foodplant sea-buckthorn (*Hippophae rhamnoides*) (Curtis, 1957) and B. Goater, when in France in 1982, discovered *P. tabaniformis* in the same foodplant (pers.comm.).

Pupa. Yellowish brown, situated in a tunnel beneath the bark which is gnawed very thin. No cocoon is made (Bartel, *loc.cit.*).

Imago. Adults occur from the end of May until mid-July. The moths rest on twigs or foliage of the foodplants.

Distribution (Map 196)

There are few British records, with only two in this century: Cosham, Hampshire, 1909 and Tubney, Berkshire, 1924. In 1839 Doubleday found three in his garden at Epping, Essex, on an aspen trunk brought in from Epping Forest as a support, and afterwards he found pupal exuviae protruding from a living trunk. This species is probably our least-known resident (Bretherton, 1951). Europe to eastern Siberia and China, and north Africa (Meyrick, 1928). Unknown in the New World until 1952 when larvae were found in Newfoundland in *Populus balsamifera* (Jacobs, 1957).

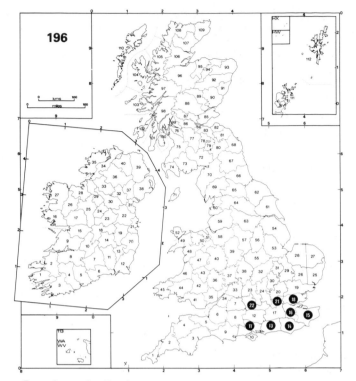

Paranthrene tabaniformis

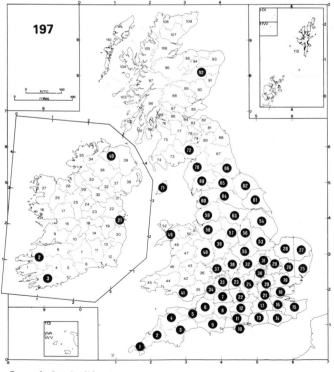

Synanthedon tipuliformis

SYNANTHEDON Hübner

Synanthedon Hübner, [1819], *Verz.bekannt.Schmett.*: 129.
Conopia Hübner, [1819], *Verz.bekannt.Schmett.*: 129.

Nine species of this genus occur in Great Britain, three of which also occur in Ireland.

Imago. Characterized by the forewing having PTA reaching the discal spot (figure 113a, p. 370).

SYNANTHEDON TIPULIFORMIS (Clerck)
Currant Clearwing

Sphinx tipuliformis Clerck, 1759, *Icones Insect.rar.* 1: pl.9, fig.1.
Sphinx salmachus Linnaeus, 1758, *Syst.Nat.* (Edn 10) 1: 493.

NOMENCLATURE. The name *tipuliformis* Clerck has been protected against the *nomen oblitum salmachus* Linnaeus by a Ruling of the International Commission of Zoological Nomenclature (Opinion 1288, in press.).

Type locality: [Sweden].

Description of imago (Pl.14, figs 9,10)

Wingspan 17–20mm. Head black with vertical bar of white scales anterior to each eye, and a posterior ring of yellow scales; antenna black; labial palpus yellow-scaled below. Thorax black, tegulae dorsally edged yellow; prominent yellow patch ventral to forewing insertion. Foreleg with femur black with lateral yellow streaks distally; proximal joint of tarsus with tuft of yellow hairs; mid- and hindlegs predominantly black, tibial spurs yellow. Forewing hyaline; costa and dorsum violet-black with yellow speckling; terminal fascia black-rayed with yellow infilling, narrowing to tornus; veins and discal spot violet-black. Hindwing hyaline; a black triangular mark on upper part of transverse vein; remaining veins and termen violet-black. Abdomen in male with yellow rings on segments 2, 4, 6 and 7; in female with yellow rings only on segments 2, 4 and 6; anal brush black with very slight admixture of yellow ventrally in male.

Life history

Ovum. 0.75mm long, ovoid, one end flattened; dull, finely reticulate, brown (nearly black (Tonge, 1933)). Laid on exposed pith of recently pruned currant bushes or in the leaf axils. Bushes of black currant (*Ribes nigrum*) are said to be preferred to those of red currant (*R. rubrum*), but on 3 July 1983 a female was observed ovipositing on both species.

Larva. Head warm brown, retractile into prothorax, which is weakly sclerotized bearing two diagonally opposed streaks; mandibles black-tipped. Body dull white with dark mid-dorsal line; anal segment faintly speckled brown and bearing fine whitish hairs; thoracic legs brown; spiracles brown-ringed. General appearance dull white with dark mid-dorsal line.

The larva bores in both lateral and main shoots but remains inactive in the winter months. In December 1974 larvae of two distinct sizes, 9mm and 12mm long, were discovered in burrows of 2mm and 3mm diameter. These larvae were torpid and each was shrouded in filmy white silken threads. Perret & Mussillon (1975) describe the construction of hibernation chambers composed of chewed wood and silken threads, the larva being isolated in its chamber by an anterior and posterior plug (bouchon). They state that these hibernation chambers are always found in the main gallery and that the size of each larva in its chamber varied from 7 to 16mm. Larvae become active in spring and then dark brown frass appears on pruned stem ends and from cracks in the thin bark. The exit holes, covered by a thin skin and often situated near a knot, are formed by early May. The thin covering of skin will frequently split but requires careful detection, as do exit holes whose edges are often concealed by bark fragments.

Pupa. 11–12mm long. Slender, pale yellow; frontal process terminating in two minute black points. No cocoon is made.

Imago. The moths emerge from late May until early July, peak time being the third week of June. The flight is rapid and difficult to follow and the moths are best observed by standing still in one's fruit cage and waiting for the insects to settle on the leaves of the bushes. This they do before and after midday but are more apparent between 15.00 and 16.00 hrs when the females are ovipositing. In very hot weather adult life seems to last only a few days. Confirmation of a one-year life-cycle has been provided by B. Skinner (pers.comm.) who carefully opened larval workings after the main emergence of moths and found no signs of larvae.

Distribution (Map 197)

Referred to as 'common' in earlier works – certainly not the case at the present time. Widely distributed from southern England northwards to Aberdeenshire; north and south Wales. Scattered records from Ireland. Throughout Europe to Siberia. Introduced with the food-plant into north America, Australia and New Zealand.

SYNANTHEDON VESPIFORMIS (Linnaeus)
Yellow-legged Clearwing

Sphinx vespiformis Linnaeus, 1761, *Fauna Suecica* (Edn 2): 289.

Sphinx oestriformis Rottemburg, 1775, *Naturforscher, Halle* 7: 109.

Sphinx cynipiformis Esper, 1783, *Schmett.* 2 (Fortsetzung): 214.

Sphinx chrysorrhoea Donovan, 1795, *Br.Ins.* 4: 21.

Type locality: Sweden.

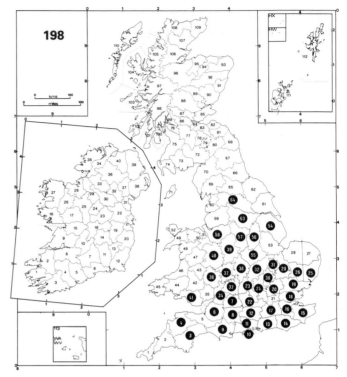

Synanthedon vespiformis

Description of imago (Pl.14, figs 11–13)
Wingspan 17–24mm. Head black with a prominent white bar of vertical scaling anterior to each eye and a posterior yellow ring; labial palpus black with yellow ventral scaling. Thorax black, tegulae thinly edged with yellow; yellow bar on metanotum, sometimes divided medially with black. Legs with femora blue-black, that of foreleg with external lateral yellow stripe; tibiae bright yellow, mid- and hindtibiae with distal and proximal black bands; tarsi yellow. Forewing hyaline; costal and dorsal streaks black with orange speckling; discal spot black, bright orange distally; veins violet-black. Hindwing hyaline; a triangular black mark on upper part of transverse vein. Abdomen black with yellow rings on segments 2, 4 and 6; faintly yellow on 5, in male also yellow on 7; anal brush, in male black with central yellow admixture ventrally; in female, yellow medially, black ventrally.

Ab. *rufomarginata* Spuler (fig.13) has forewing costal and dorsal streaks reddish orange.

Life history
Ovum. Laid usually along the edges or within bark crevices of fresh stumps of oak (*Quercus* spp.), although stumps two and three years old will still attract ovipositing females. Also laid in crevices of excrescences on trunks. Other host plants are sweet chestnut (*Castanea sativa*), beech (*Fagus sylvatica*), walnut (*Juglans regia*), excrescences on wych-elm (*Ulmus glabra*) (Allen, 1975) and birch (*Betula* spp.) stumps (Pratt, 1977).

Larva. Full-fed 20mm long. Dull white and transparent, showing blue-grey, sometimes greenish inclusions. Larvae extracted from their cocoons are bone-white and without transparency. Head reddish brown. Body with prothorax lightly sclerotized, with two outwardly directed bars matching colour of head; anal segment weakly sclerotized; spiracles brown-ringed; thin whitish hairs overall; some minute irregular dotting on the integument; thoracic legs pale brown.

A one-year life-cycle seems likely, for although very small larvae, fully grown ones and cocoons may all be found in close proximity, the emergence extends over several weeks. Tenanted stumps have frass exuding in patches over the face of the bark, appearing as aggregations of coarse 'sawdust'.

Pupa. *c.*11mm long. Pale yellowish brown, abdominal spine rows and edges of wings darker brown. Cocoon ovoid 13mm long, hard to the touch with brown granular exterior and lining of whitish silk. Situated on the inside of the bark layer.

Imago. Adults emerge in early morning from 08.00 hrs onwards, and there is a long emergence period from late May to early August. On hot sunny afternoons in early July females have been noted buzzing up and down the crevices of oak stumps. In flight the tail fan is particularly evident as a whitish 'blob'. This species will rapidly colonize cleared oak woodland.

Distribution (Map 198)
Widely distributed over central southern England northwards to Yorkshire. No record from the most western counties, or from Scotland or Ireland. Western Palaearctic.

SYNANTHEDON SPHECIFORMIS ([Denis & Schiffermüller])
White-barred Clearwing

Sphinx spheciformis [Denis & Schiffermüller], 1775, *Schmett.Wien.*: 306.

Type locality: [Austria]; Vienna district.

Description of imago (Pl.14, fig.6)

Wingspan 26–31mm. Head entirely black; antenna in both sexes with a yellowish white subapical band; labial palpus with yellowish white scaling ventrally. Thorax black except for yellowish white edging of tegulae. Legs predominantly black; fore- and hindtibiae streaked yellowish white ventrally; tarsi yellowish white. Forewing hyaline; costa, dorsum and termen violet-black, narrowing along termen to tornus; prominent spot on cross-vein and remaining veins violet-black. Hindwing hyaline; triangular mark on cross-vein; remaining veins and termen violet-black; underside, marked with yellow scaling on costa. Abdominal segment 2 with a narrow yellowish white band dorsally, sometimes continued forward laterally to segment 1; ventral surface with posterior margin of segment 4 strongly, of segments 5 and 6 weakly, coloured yellowish white; anal brush black with slight yellow scaling ventrally.

Life history

Ovum. Length 0.8mm, ovoid, one end flattened; dull, pitted all over, dark brown (Tonge, 1933). Laid in ground near bark crevices (Fibiger & Kristensen, 1974).

Larva. First year. Head shining red-brown, retractile into prothorax, mandibles black-tipped. Body with prothorax broad, with two diagonal light brown streaks posteriorly; remaining segments pinkish white excepting anal plate which is faintly dotted brown; spiracles brown-ringed. Less pink in second year.

The larvae feed in alder (*Alnus* spp.) and birch (*Betula* spp.) and in winter may go deep into the roots, tunnelling upwards again in spring. Second-year larvae eject frass which collects in conspicuous little heaps at stem-bases. This frass bleaches to a pale brown colour, contrasting strongly with the small-sized and continuously produced darker accumulations from first-year feeding. On heathland where birch-cutting has taken place, the stumps in spring will readily show whether *S. spheciformis* larvae are present. Round 5mm diameter tunnels indicate second-year larvae, oval 2–3mm tunnels those of first year.

There are differing opinions regarding length of larval life. Classey *et al.* (1946) state that *S. spheciformis* has a life-cycle extending over two years, South (1961) says 'full grown in third year after hatching from the egg' and Newman & Leeds (1913) also indicate a three-year larval life. Tugwell (1891) states that the larvae feed for at least

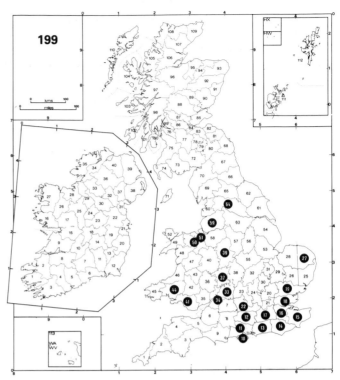

Synanthedon spheciformis

three years and that they are in the larval state for part of the fourth year. None of these accounts appears to be based on evidence from larval measurements, but recent field work has shown that by April two distinct sizes of larvae, 12mm and 19–20mm, can be found in infected trunks or stumps. This is suggestive of a two-year life-cycle, but on the evidence of the discovery of a larva measuring 32mm by October, it would seem that length of larval life does not always conform to a rigid pattern.

Pupa. Averaging 16mm in length. Yellowish brown, more distinctly yellow at intersegmental divisions; a prominent black frontal process, sharp-pointed with raised lateral margins. The pupa is situated at the top of the burrow which is lined with chewed wood and silk. No cocoon is constructed. The emergence holes, of 5mm diameter, are generally situated low down on trunks.

Imago. The moths emerge from mid-May until mid-June. Eclosion from the pupa is from 07.00 to 10.00 hrs according to weather.

Distribution (Map 199)

A local species, but widespread over much of southern and central England; north and south Wales; extends northwards to Yorkshire with only a few records from eastern England. Widely distributed across central, northern and eastern Europe.

SYNANTHEDON SCOLIAEFORMIS (Borkhausen)
Welsh Clearwing

Sphinx scoliaeformis Borkhausen, 1789, *Naturgesch.eur. Schmett.* **2**: 173.

Type locality: Europe.

Description of imago (Pl.14, figs 7,8)

Wingspan 24–35mm. Head black with a silvery white band of scales anterior to each eye and an indefinite posterior yellow ring; antenna black, whitish yellow towards apex, this being more evident in the female; labial palpus with yellowish orange scaling and black tip. Thorax blue-black, tegulae edged yellow. Legs with femora black, that of foreleg with external yellow scaling; mid- and hind-tibiae dark with dull yellowish suffusion and a ring of blackish hairs at apex; tarsi mottled black and dull yellow. Forewing hyaline; costal and dorsal streaks deep black, mottled with yellow; discal spot very prominent, angled towards base, deep black; veins and broad terminal fascia narrowing to tornus, deep black. Hindwing hyaline, with an almost triangular black mark on uppermost part of cross-vein; remaining veins black. Terminal black fascia broader in female. Abdomen with a narrow yellow ring on posterior edge of segment 2 and a broader one on segment 4; ventrally the ring on segment 4 is broader with a diffuse anterior edge; anal tuft of male orange-brown interspersed with black, ventrally with a central area of yellow; the tuft of the female is a much brighter orange with less extent of black and no yellow ventrally.

Life history

Ovum. Ovoid, 0.7×0.4mm; brown with white reticulations discernible only under high magnification. Laid in old emergence holes or in bark crevices in the lower part (below 2m) of old birch trunks (Fibiger & Kristensen, 1974).

Larva. Head reddish brown, mandibles black-tipped. Prothorax yellow-brown with traces of two darker brown streaks outwardly directed from posterior edge of the segment; rest of body off-white with minute traces of brown speckling; thoracic legs pale brown with terminal claws blackish brown; spiracles brown-ringed.

The larvae make very tortuous tunnels below the bark of well-grown birches and have a three-year life-cycle (Classey *et al.*, 1946). Full-grown larvae and newly formed pupae are to be found about the third week of May. Emergence holes are usually 1–2m above ground level but have been noted as high as 6m. Old emergence holes, *c.*5mm in diameter, are readily seen, but capped holes require careful searching for. They may be revealed by brushing the trunks with a wire brush, thereby removing the lichen cover and exposing the small amount of dry frass at the mouth of the capped or ruptured emergence hole.

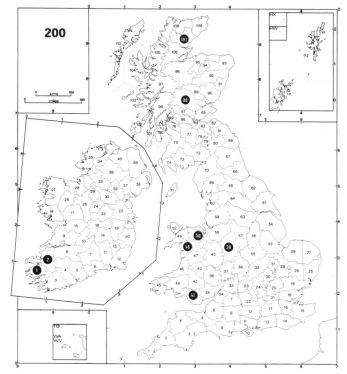

Synanthedon scoliaeformis

Pupa. 15–18mm long. Pale yellowish brown with darker brown abdominal spine-rows. The cocoon, 16–20mm in length, is elongate with rounded ends; a compact construction of granular wood particles bonded with silk. Dark reddish brown, situated below the bark layer.

Imago. The moths emerge from mid-June to mid-July and may be found on the trunks of well-grown birches in early morning in favoured localities. Females have been noted ovipositing in mid-afternoon upon trees bearing the extruded pupae of recent emergences, supporting the view that within birch woodland certain trees are preferred to the total exclusion of others.

History and distribution (Map 200)

As recent records relate only to Scotland or southern Ireland it is of interest to consider the early history of this species. It was discovered near Llangollen, north Wales, about 1854 (South, 1961) and Stainton (1857) says 'taken by Mr Ashworth at Bryn Hyfryd, near Llangollen'. In 1862 seven specimens, four bred and three captured, were exhibited at a meeting of the Northern Entomological Society (Cooke, 1862). The following year Gregson was able to figure various stages of the life history by observing infected trunks taken to his residence from the Llangollen locality (Gregson, 1863). Newman (1869) was still referring to Ashworth's specimens as the only ones in existence though the Rannoch locality is said to have been known by 1867 (South, *loc.cit.*). How long *S. scoliaeformis* persisted at Llangollen is uncertain but Day (1903) says 'the insect does not appear to have been taken there in recent years'. Regarding other localities South (*loc.cit.*) mentions Delamere Forest, Cheshire, two, 1901 and 1905; and says 'recorded from Hereford' and, 'a pupa obtained and moth captured at Cannock Chase, Staffordshire in 1913'. The Cannock specimens are fully recorded by their captor H. C. Hayward (1913), who also speaks of seeing other sesiid pupae at Cannock. He doubted whether these were *S. scoliaeformis* because they were rather small compared with the older Welsh specimens he had seen in collections.

A very local species even in its known Scottish and Irish localities. There is no recent record from the north of England or from Wales, but systematic field-work should lead to the rediscovery of this species in the country whose name it bears. Widely distributed in the western Palaearctic region.

SYNANTHEDON FLAVIVENTRIS (Staudinger)
Sallow Clearwing

Sesia flaviventris Staudinger, 1883, *Stettin.ent.Ztg* **38**: 177.
Type locality: Germany; Mecklenburg.

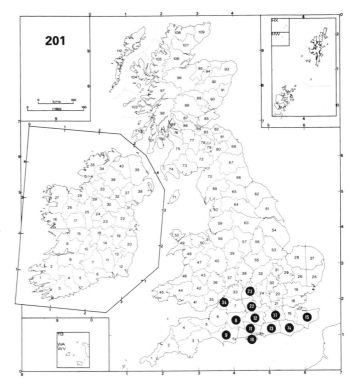

Synanthedon flaviventris

Description of imago (Pl.14, fig.15)
Wingspan 17–20mm. Head black with vertical bar of white scales anterior to each eye, posterior yellow ring absent; antenna black with pale ventral streak; labial palpus yellow-scaled below. Thorax uniformly black dorsally, laterally with a patch of yellow scales below insertion of forewing. Femora black, that of foreleg with yellow streaking inwardly; tibiae black with some admixture of yellow below, spurs yellow; tarsi predominantly yellow but with some black streaking below. Forewing hyaline; costa and dorsum violet-black, lightly speckled yellow; terminal fascia black-rayed, infilled with yellow, more evident on underside; ETA higher than broad with oblique apical extension; discal spot black, streaked with yellow distally. Hindwing hyaline; a black triangular mark on upper part of transverse vein; other veins black. Abdomen black with yellow bands on segments 2, 4 and 6 dorsally; a yellow patch ventrally from segments 2–6; in male, segment 7 occasionally with a few yellow scales dorsally, a more frequent suffusion ventrally; anal brush entirely black in male, black with some lateral yellow suffusion in female.

Life history
Ovum. Laid in axils on slender (diameter less than 1cm) branches of *Salix* spp. (Fibiger & Kristensen, 1974).

Larva. Reaches a length of 17–18mm by the spring of its second year. Head pale yellowish brown except for mandibles which are deeper brown and black-tipped. The head can retract into prothorax which is faintly yellowish brown but with two more-clearly defined brown diagonal streaks; remainder of larva dull white; thoracic legs yellowish brown, tarsal claws black; prolegs fleshy white, crochets brown; spiracles brown-ringed.

The first-year larva excavates a peripheral tunnel around a sallow stem which shows no outward sign of its tenant. In the spring of the following year the larva bores deeper into the tissues of the stem and then excavates a vertical tunnel of some 3mm diameter and 50–75mm length. The characteristic, pear-shaped gall associated with this species, and especially noticeable in the second autumn of larval life, is attributable to frass being pressed into cavities between bark and wood; leaves of the branch distal to the swelling are of an abnormal pale green colour (Fibiger & Kristensen, *loc.cit.*). The galls are most evident in the second winter when the sallows are free of leaves. There is some variation in size, one in a freshly cut branch being 25mm long and 16mm across at its broadest region. All records of the occurrence of galls refer to odd-even winters, future favourable periods being 1985/86, 1987/88 and so on.

Pupa. 11–13mm long. Pale yellowish; pupal frontal process a low transverse ridge (Fibiger & Kristensen, *loc.cit.*). The pupae, which are formed head downward within a chamber above the gall, invariably extrude from the side of the swelling by rupturing the specially prepared exit cap. No cocoon is made.

Imago. The moths emerge from about the third week of June until mid-July.

Distribution (Map 201)

Not discovered in the British Isles until 1926 when it was found near Southampton by Fassnidge (1926). So far recorded from only a few vice-counties in central southern and south-eastern England. Abroad probably also under-recorded; known at present from Denmark, Finland and the lowlands of northern Europe eastwards to the U.S.S.R. and Rumania.

SYNANTHEDON ANDRENAEFORMIS (Laspeyres)
Orange-tailed Clearwing

Sesia andrenaeformis Laspeyres, 1801, *Sesiae eur.*: 20.

Sphinx anthraciniformis Esper, 1798, *Schmett.* Suppl. (Abschnitt 2): 29.

Trochilium allantiforme Newman, 1832, *Ent.Mag.* 1: 79.

NOMENCLATURE. The name *andrenaeformis* Laspeyres has been protected against the *nomen oblitum anthraciniformis* Esper by a Ruling of the International Commission of Zoological Nomenclature (Opinion 1287, 1984).

Type locality: [Europe].

Description of imago (Pl.14, fig.14).

Wingspan 18–22mm. Head entirely black; antenna black; labial palpus with yellowish white scaling ventrally. Thorax dorsally black with some violet iridescence; a prominent yellow patch below insertion of forewing. Legs having femora and tibiae black with violet iridescence; an apical ring of yellow hairs on mid- and hindtibiae, spurs yellow; tarsi yellow with an admixture of black scaling. Forewing hyaline; costa, dorsum and discal spot black with violet iridescence, the last with the proximal face angled into ATA; terminal fascia broad, narrowing abruptly to tornus; the areas between the black veins violet with few golden specks. Hindwing hyaline; all veins black with upper part of cross-vein having a black triangular mark. Abdomen violet-black with a thin yellow band on segment 2 and a wider one on segment 4; laterally the yellowish white scaling on segment 2 extends forward to segment 1; ventrally segment 4 with a broad patch of white scales which in the male extends to segment 6, in the female only to part of segment 5; the expanded anal brush of the male black with broad orange-yellow scaling from middle to tip, that of the female only strongly orange-yellow towards the tip; abdomen of the male laterally compressed.

Life history

Ovum. 0.75mm long, ovoid, one end flattened; dull, finely reticulate, nearly black (Tonge, 1933). Laid in bark crevices of the wayfaring-tree (*Viburnum lantana*), less often guelder-rose (*V. opulus*).

Larva. Head yellowish brown, mandibles darker brown with black tips. Prothorax shining white with weak sclerotization; rest of body ochreous white, semitransparent with few scattered whitish hairs; legs ochreous white with brown claws; spiracles small, brown-ringed; the segmental divisions of the larva are well marked.

The larvae live at least two years and occur in stems from pencil thickness up to 25mm or more in diameter. The larval tunnel, 100mm long in a thin stem, turns at right angles and is terminated by a circular bark disc cut by the larva in its last year of growth, discs being about 7mm in diameter. The tunnel may also extend downwards 20–25mm below the emergence disc. Disc may fall off and then a recess with a small central hole which may be plugged with yellowish frass is revealed. A greater exudation of frass often indicates a parasitized larva within. Diameter of emergence hole below outer cap 3mm. Height of emergence holes above ground level varies from 0.3 to 6.0m (Britton, 1977). Infected stems can be cut throughout the winter months until early spring when leaf-growth makes their detection more difficult.

Pupa. Length 14–16mm. Pale yellowish brown except the dorsal spine-rows, which are darker brown, and the weak frontal process which is black-tipped. No cocoon is formed.

Imago. The moths emerge from mid-May until the end of June and favour slopes and edges of woods where the wayfaring-tree flourishes. In such areas it is but a matter of patient search to discover the stems with larvae or pupae; adults however, except when newly emerged in early morning, are seemingly non-existent. Sweeping low herbage or beating branches in dull weather does not reveal them: possibly they rest high up on the host plant.

Distribution (Map 202)

Widespread on the downland of southern England to Worcester and Huntingdon. Through central Europe to Asia Minor and central U.S.S.R.

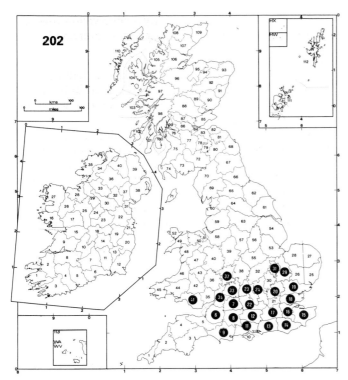

Synanthedon andrenaeformis

SYNANTHEDON MYOPAEFORMIS (Borkhausen)
Red-belted Clearwing

Sphinx myopaeformis Borkhausen, 1789, *Naturgesch. eur.Schmett.* **2**: 169.
Sphinx zonata Donovan, 1797, *Br.Ins.* **6**: 35.
Sphinx mutillaeformis Laspeyres, 1801, *Sesiae eur.*: 26.
Type locality: Europe.

Description of imago (Pl.14, fig.17)
Wingspan 15–22mm. Head black with a vertical bar of white scales anterior to each eye; antenna uniformly black; labial palpus of male black with whitish ventral suffusion, uniformly black in female. Thorax dull black with a patch of bright orange below wing insertions. Legs, femora and tibiae black with bluish iridescence; a patch of white scales overlapping top of forefemur; spurs dull black; tarsi pale with darker speckling, whitish ventrally. Forewing hyaline; costa, dorsum and discal spot black, shot with bluish iridescence; broad terminal fascia narrowing to tornus, violet-black; ventrally, costa, dorsum, discal spot and spaces between veins across terminal fascia marked with orange scaling. Hindwing hyaline; veins violet-black; a narrow black triangular mark extending entire length of cross-vein. Abdomen black with a prominent orange-red belt on segment 4; in the male this belt is broken ventrally by a patch of white scales which extends complete to segment 6, but becomes obsolescent on segment 7; in the female the red belt is broken mid-ventrally by a patch of black scales with minimal white scaling on the edge of the segment; anal brush bluish black, that of male with orange and white scaling mid-ventrally.

Life history
Ovum. Ovoid, 0.5×0.3mm. Brownish green with white reticulations, laid in bark crevices of apple (*Malus* spp. including cultivars) and, less usually, pear (*Pyrus communis*), almond (*Prunus amygdalus*), peach (*P. persica*), rowan (*Sorbus aucuparia*) (B. Skinner, pers.comm.) and hawthorn (*Crataegus* spp.).
Larva. Full-fed 16mm long. Head flattened, reddish brown, mandibles black-tipped. Pronotum broadly quadrate with slight light brown sclerotization, otherwise waxy white as are successive segments; thin yellowish hairs especially laterally on head, prothorax and around anal plate; spiracles ringed blackish brown; thoracic legs whitish with brown claws.

Signs of larval activity are given by small quantities of reddish frass which accumulate in, or hang from, bark crevices during the winter until early spring.
Pupa. 13–15mm long. Pale yellowish brown, darker brown on head-capsule and on terminal segment; frontal process centred on a black transverse ridge. Empty pupae,

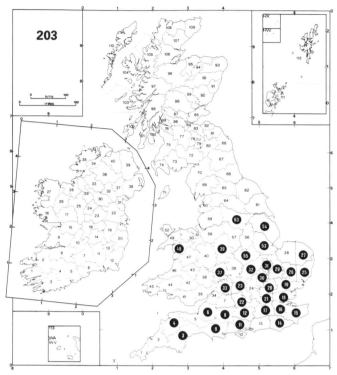

Synanthedon myopaeformis

protected by bark crevices, will remain in position for up to a year. Cocoon 15–16mm in length, 5mm broad, with rounded ends; made of small bark particles bonded with silk, internally lined with whitish silk; colour variable, light to dark brown. Situated on the inner surface of the bark. Diameter of emergence hole 4mm.

Imago. The moths emerge about the third week of June to the end of July. Eclosion from the pupa is later than with some other clearwings and occurs from 10.00 to 12.00 hrs. An apple-tree bearing a colony of this clearwing may continue to do so for many years; one tree under observation for more than a decade produced annually 30–40 imagines. On sunny days males could be seen flying rapidly over the topmost branches, and females returned to the tree to oviposit in early afternoon. After the main emergence of moths had taken place from another tree, the sectioned trunk was excavated carefully and revealed no larvae, indicating a probable one-year life-cycle.

Distribution (Map 203)

Southern England northwards to Yorkshire; north Wales. Central and southern Europe to Asia Minor and south-eastern European U.S.S.R.

SYNANTHEDON FORMICAEFORMIS (Esper)
Red-tipped Clearwing

Sphinx formicaeformis Esper, 1779, *Schmett.* 2: 216.

Type locality: [Europe].

Description of imago (Pl.14, fig.16)

Wingspan 17–19mm. Head black with vertical bar of white scales anterior to each eye; antenna uniformly black; labial palpus black with bright orange scales ventrally. Thorax black. Legs bluish black; midtibia with a tuft of white hairs at apex; hindtibia with a white median ring, suffused on inner surface, and a second white tuft at apex; tarsi whitish yellow with black suffusion. Forewing hyaline; costa and dorsum black streaked with red; discal spot black with some red scaling distally; apical area, narrowing to tornus, bright red crossed by black veins; cilia dark fuscous. Hindwing hyaline; veins and a spot on upper part of transverse vein black; cilia dark fuscous. Abdomen bluish black, segments 2 and 3 with dusting of whitish yellow scales; a prominent red belt on segment 4 extending partially to segment 5 dorsally and to segments 5 and 6 ventrally; anal tuft bluish black, in male with yellowish white scaling lateroventrally, in female with white streaks dorsolaterally.

Life history

Ovum. Ovoid with one end flattened, 0.7×0.3mm; dark brown with whitish reticulations visible only under high magnification. Laid in bark crevices, axils and stump edges of *Salix* spp.

Larva. Head yellowish brown, mandibles darker brown with three 'teeth' in a curve to apex. Prothorax transparent, whitish, with two outwardly directed brownish streaks; remainder of body whitish yellow, dorsal vessel most visible on mid-abdominal segments; thoracic legs pale brown with darker claws; prolegs whitish, crochets brown; spiracles brown-ringed.

In the British Isles osier (*Salix viminalis*) is a favoured host plant but the larva gives little external sign of its presence within other than a slight browning, sometimes accompanied by frass, at the edge of a broken stem or stump edge. The emergence holes are *c.*2.5mm in diameter but frequently hidden by a thin cap which often remains *in situ*. The larvae bore between the wood and the bark for a few centimetres when they may be revealed in their shallow tunnels by peeling back the bark. Some larvae may tunnel slightly deeper into the wood, but all terminate their boring in a shallow, silk-lined, pupal chamber, the tunnel leading to this being packed with wood-chips and frass. There are occasional records in this country of *S. formicaeformis* having been bred from galls collected ostensibly as those of *S. flaviventris* (Staudinger) but Fassnidge (1926) points out that whereas some galls

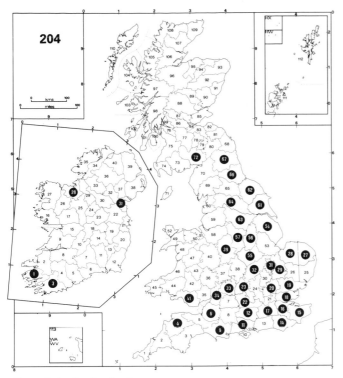

Synanthedon formicaeformis

made by the two species are similar others made by *S. formicaeformis* have a peculiar shape, for the diameter of the stem *above* the gall is greater than that below and continues so for some distance. The galls found by Fassnidge were on goat-willow (*Salix caprea*) and in nearly all cases a characteristic circular scar, made by the young larva in its passage round the stem just under the bark, could be plainly seen. In Scandinavia the swellings made by *S. formicaeformis* on creeping willow (*Salix repens*) are reputed to occur low down (up to 30cm above ground level) whereas those on goat-willow occur fairly high (1 to 3m) (Fibiger & Kristensen, 1974). The galls found by Fassnidge, however occurred from ground level to the top of the tree. In April, 1984 larvae were discovered in goat-willow as secondary tenants of the large excrescences caused by mites. Fresh 'sawdust' on these excrescences indicate presence of *S. formicaeformis* (pers.obs.). Life-cycle said to be one year.

Pupa. 12–14mm long. Slender, pale yellowish with darker spine-rows and anal tubercles; prominent frontal process, black-edged and bifid centrally. No cocoon is constructed.

Imago. The moths emerge over a lengthy period from the end of May until late July with a peak in mid-June. Occasionally found low on osiers in early morning or feeding at a variety of flowers later in the day.

Distribution (Map 204)
Widely distributed from southern England northwards to Dumfriesshire, but not recorded elsewhere in Scotland. Also recorded from south Wales and a few well-separated localities in Ireland. Palaearctic.

SYNANTHEDON CULICIFORMIS (Linnaeus)
Large Red-belted Clearwing
Sphinx culiciformis Linnaeus, 1758, *Syst.Nat.* (Edn 10) 1: 493.
Type locality: Europe.

Description of imago (Pl.14, figs 18,19)
Wingspan 22–25mm. Head black with a vertical bar of white scales anterior to each eye; antenna uniformly black; labial palpus black dorsally, orange-red ventrally. Thorax dull black with prominent orange patch below wing insertions. Legs blue-black, tibiae yellowish on inner surfaces; tarsi with first segments yellow, following segments yellowish speckled with black. Forewing with costa and dorsum black with reddish anterior streaks; a red suffusion at base extends on to a small area of the ATA and PTA; this suffusion is absent in the otherwise similar species *S. myopaeformis* (Borkhausen); discal spot and veins blue-black; terminal fascia blue-black with some orange suffusion on underside. Hindwing hyaline; veins blue-black with a black triangular mark on upper portion of cross-vein. Abdomen blue-black with a complete red belt on segment 4; occasionally this belt is yellow (ab. *flavocingulata* Spuler (fig.19)); sometimes a small red-yellow stripe laterally on segments 1 and 2; anal tuft black.

Life history
Ovum. Ovoid, 0.7×0.5mm, of a pale honey colour with white reticulations. Well hidden within the cracks and crevices of birch (*Betula* spp.) stumps and stems, also on alder (*Alnus* spp.) (Newman & Leeds, 1913).

Larva. Full-fed *c*.18mm long. Head red-brown, mandibles darker brown with black tips. Prothorax pale yellow with two darker brown, diagonal, outwardly directed streaks; remaining segments dull white with short brown hairs; thoracic legs brown with black claws; prolegs white with black-brown ring of crochets; spiracles brown-ringed.

The larvae feed steadily during the summer and by October are about 12mm in length. They are to be found 25–75mm down in superficial tunnels in birch stumps which have been created about a year previously. Thin particles of wood fill the lower part of the tunnel and from these the spindle-shaped cocoon is constructed during the autumn and winter. By October there are still little excrescences of brown frass issuing at the top or from lateral crevices of tenanted birch stumps, but by early April most

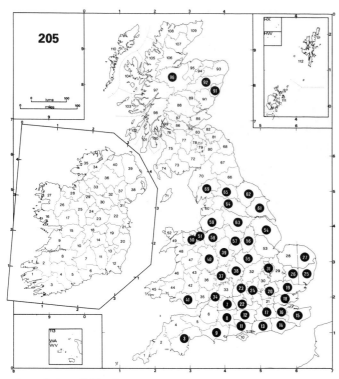

Synanthedon culiciformis

BEMBECIA Hübner

Bembecia Hübner, [1819], *Verz.bekannt.Schmett.*: 128.
Pyropteron Newman, 1832, *Ent.Mag.* **1**: 75.
Dipsosphecia Pungeler, 1910, *in* Spuler, *Schmett.Eur.* **2**: 316.

Only three species of this genus occur in Great Britain, one of which also occurs in Ireland.

Imago. Characterized by the forewing having PTA absent, partially obliterated or not reaching discal spot.

of this will have disappeared and the only outward sign will be a small hole 2–3mm in diameter situated at the stump edge between wood and bark, or rarely centrally.
Pupa. 16–18mm long. Slender, pale yellowish with darker spine-rows. The spindle-shaped cocoon, 21–25mm long, is made from thin particles of wood, each 4–10mm long, and bonded with silk. It is situated in the larval boring 25–75mm below the small emergence hole. Cocoons are best sought in late April. To effect least disturbance it is advisable to saw off completely those stumps which bear the small tell-tale emergence holes.
Imago. The earliest of our clearwings to emerge, appearing usually during the first two weeks of May, but sometimes continuing into June. The moths emerge between 11.00 and 13.00 hrs, that is, later in the day than most other clearwings. On warm sunny afternoons females can be seen flying rapidly around recently cut birch stumps, settling periodically to oviposit. This species has a one-year life-cycle (Baker, 1970).

Distribution (Map 205)
Widely distributed over heathy areas and open woodland with an abundance of birch. Under-recording may be responsible for the discontinuity shown on the map of mainland Britain. Not so far known from Ireland. Holarctic.

BEMBECIA SCOPIGERA (Scopoli)
Six-belted Clearwing

Sphinx scopigera Scopoli, 1763, *Ent.Carn.*: 188.
Sphinx ichneumoniformis [Denis & Schiffermüller], 1775, *Schmett.Wien.*: 44.
Type locality: Carniola (Yugoslavia; Slovenija).

Description of imago (Pl.14, figs 20,21)
Wingspan 15–21mm. Head dark fuscous dorsally, with a variable amount of whitish yellow scaling anterior and ventral to eye; antenna dorsally unicolorous black in male, with a paler subapical patch in female, ventrally with orange suffusion in both sexes; labial palpus black- and yellow-scaled dorsally, with a pronounced tuft of black hairs on ventral surface of second segment in male. Thorax bluish black variably speckled with yellow; tegulae edged yellow; yellow transverse band on metanotum. Legs with foretibiae dark with yellow hairs at apex; mid- and hindtibiae yellowish orange with black external streak and apical black band; tarsi yellowish. Forewing hyaline; costa, dorsum and veins dark fuscous, inner portion of dorsum suffused with orange; discal spot dark fuscous with orange distal patch; variable orange suffusion between dark veins of apical area. Hindwing hyaline; veins, a triangular mark on upper part of cross-vein and termen dark fuscous. Abdomen blue-black with yellow belts on segments 2–7 in male, 2–6 in female; anal tuft black with lateral areas yellow and a few yellow hairs centrally in the male.

Life history
Ovum. A black ovoid with longest axis *c.*0.6mm, broadest *c.*0.4mm, depth *c.*0.3mm. Laid on the leaves of bird's-foot trefoil (*Lotus* spp.) or kidney-vetch (*Anthyllis vulneraria*); also on horse-shoe vetch (*Hippocrepis comosa*) (W. G. Tremewan, pers.comm.), occasionally on stems or exposed roots. The empty shells appear as black dots which remain visible on the leaves until these wither after about three months.

Larva. Full-fed *c.*13mm long prior to pupation in June. Head pale brown mottled with lighter areas; mouth parts darker brown. Prothorax broad, transparent whitish; rest of body bone white, with a few thin whitish hairs; thoracic legs whitish with brown claws.

Finding larvae can be a lengthy process for leaves of the host plant give no indication of larval presence in the roots. One method which has proved successful is to dig up the plants at the edges of patches of bird's-foot trefoil, or any which have been isolated by disturbance of the ground. The larvae leave obvious signs of feeding by producing a granular 'sawdust' covering 50mm or more in extent along the side of the primary root, this being often of a lighter colour than the root.

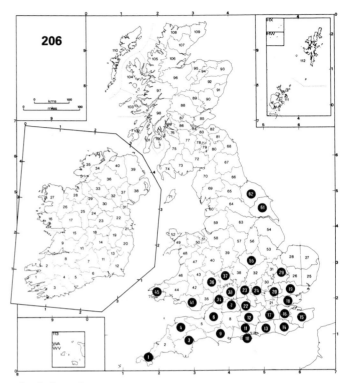

Bembecia scopigera

*Pupa. c.*13mm long. Dark brown but lighter brown at intersegmental divisions; frontal process broad, black-ridged, with slight central projection. Pupation may take place either in the silk-lined tunnel within the root, or in a separate silken tube extending upward to the ground surface. In either event a specially prepared lid is constructed through which the pupa can extrude. No cocoon is made.

Imago. The moths occur from the end of June until mid-August. They should be looked for in chalk and limestone areas where there is an abundance of bird's-foot trefoil. Cliffs, edges of downland trackways, pits, quarries and rough upland fields are typical habitats. This species is exceptional in that adults are more easily obtained than the immature stages. Systematic sweeping should produce one or two adults if a colony is present. If this evidence is found a careful search of grasses and other plant stems should reveal the imagines at rest. Evenings are a better time to search than earlier in the day when the insects would be more active.

Distribution (Map 206)
Recorded from all southern and south-western English counties, extending north to Cambridgeshire and Herefordshire; east Yorkshire; south Wales. Widely distributed in the western and central Palaearctic region.

BEMBECIA MUSCAEFORMIS (Esper)
Thrift Clearwing
Sphinx muscaeformis Esper, 1783, *Schmett.* 2: 217.
Sphinx philanthiformis Laspeyres, 1801, *Sesiae eur.*: 31.
Type locality: [Europe].

Description of imago (Pl.14, figs 24,25)
Wingspan 15–18mm. Head with a tuft of black hairs
mid-dorsally, whitish scales anterior to eyes and a ring of
brownish yellow hairs posteriorly; antenna uniformly
black in male, a subapical patch of white scales in female;
labial palpus white-scaled dorsally and inwardly, laterally
with an admixture of black and whitish scales. Thorax dull
black with a variable amount of yellowish white scales,
often in the form of two dorsolateral streaks but occasionally these defined only thinly with an additional thin
mid-dorsal streak; laterally a patch of yellowish white
scales extending below wing insertions. Foreleg with
femur white-scaled externally; tibiae and tarsi predominantly dark fuscous; mid- and hindtibiae and tarsi with
variable white and dark fuscous scaling. Forewing with
costa and dorsum densely black-scaled; ATA only hyaline
to discal spot, PTA obliterated by blackish scales; terminal fascia slightly narrowing to tornus, black overlaid with a
speckling of yellowish white scales. Hindwing hyaline; a
triangular mark on transverse vein; veins dark fuscous.
Abdomen blackish with variable speckling of white scales;
frequently a mid-dorsal line of white spots; white transverse bands on segments 2 (sometimes indistinct), 4 and 6,
and in the male sometimes on the posterior edge of 7; anal
tuft black with varying admixture of whitish yellow scales,
these generally central in the male.

Life history
Ovum. Ovoid, black with white reticulations visible only
under high magnification, longest axis *c*.0.7mm. Laid on
thrift (*Armeria maritima*).
Larva. Full-fed *c*.13mm long by early June. Head reddish
brown with a darker triangular mark on frons. Prothorax
and anal plate pale brown; rest of body ochreous white.
   The larvae sometimes make little red patches of frass on
the cushions of thrift and should be looked for in May.
Even without any sign externally the larvae can be found
by breaking the dead portions of thrift cushions, it being a
matter of practice to recognize likely looking patches: the
larvae occur just below the surface of the cushions by late
May. Sometimes the larva constructs a silk tube covered
externally with grains of frass and root fragments which
extends vertically or laterally 25–30mm outward from the
tuft of thrift. Towards the extremity the tube is generally
rounded off abruptly. The larval mine in the roots leads
upwards in the form of a silk-lined tunnel which extends
to the outer part of the root at surface level.

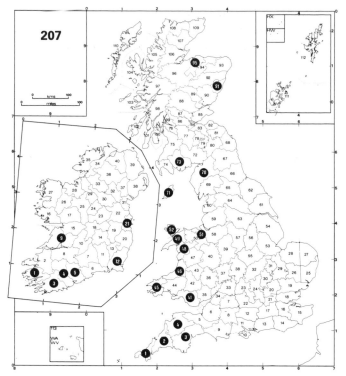

Bembecia muscaeformis

Pupa. 12–14mm long, slender and tapering; light reddish
brown with thorax, eyes and areas of wings of darker
brown; dorsal spine-rows black. Frontal process in the
form of a broad curved ridge and lacking a central projection. No cocoon is constructed.
Imago. The moths appear from the end of June and
throughout July, favouring rocky coastal areas where
there is an abundance of thrift, but in hot summers they
are over well before the end of July. Emergence takes
place between 10.30 and 11.30 hrs. On 9 July 1983 a
female emerged at 10.43 hrs. The fully inflated wings were
raised within seven minutes and lowered again after a
further nine minutes when the moth rested quietly among
the thrift for almost half an hour. The wings bear a loose
yellowish white scale covering which disappears on the
first flight but which enhances the insect's protective resemblance to the dry thrift stems during the initial resting
period. The moths, which are very rapid in flight, are
particularly attracted to patches of thyme upon which they
may be seen feeding. They may be found not only on
exposed cliff faces but also along the high banks of sunken
lanes a little way in from the sea provided there are patches
of thyme and an abundance of the foodplant.

Distribution (Map 207)
Coastal; south-west England to Cumbria and the Isle of Man; north-east Scotland. Southern Ireland. Could well await rediscovery on coasts of western Scotland. Denmark, Sweden, central and eastern Europe to southern U.S.S.R.

BEMBECIA CHRYSIDIFORMIS (Esper)
Fiery Clearwing
Sphinx chrysidiformis Esper, 1782, *Schmett.* **2**: 210.
Type locality: [Europe].

Description of imago (Pl.14, figs 22,23)
Wingspan 15–23mm. Head with fronto-clypeus white-scaled, a tuft of black hairs (sometimes orange in female) between antennal bases, and an inconspicuous ring of orange hairs posteriorly; antenna blue-black with a subapical patch of white scales dorsally in female; labial palpus with long black hairs basally and terminal segment yellowish. Thorax blue-black with white shoulder spot and scattered red hairs dorsally. Foreleg with femur white-scaled externally, mid- and hindfemora blue-black; tibiae of all legs densely covered with orange scales, tarsi orange basally, otherwise whitish yellow. Forewing with costal streak black; discal spot black with broad orange-red distal margin; terminal suffusion black-purple; extensive orange-red scaling obliterating PTA and most of ETA. Hindwing hyaline; veins, a spot on upper part of transverse vein and termen black. Abdomen blue-black with whitish yellow bands on segments 4 and 6 and, in male, an ill-defined band on 7; anal tuft mediodorsally orange-red, black laterally.

Life history
Ovum. A flattened ovoid depressed dorsally and ventrally, measuring *c.*0.7mm×*c.*0.5mm×*c.*0.3mm; violet-black with white reticulation discernible only under high magnification.
Larva. Full-fed *c.*19.5mm long in late April. Head broad, slightly flattened, chestnut-brown with two paler lateral streaks; mandibles black, markedly toothed. Prothorax pale yellow and broader than head; anal flap rounded, yellowish with dark brown hairs; rest of body dull white, much wrinkled; thoracic legs fleshy white, tarsal claws brown; prolegs fleshy white with dark brown crochets.
   The larvae mine the thick roots of curled dock (*Rumex crispus*), water-dock (*R. hydrolapathum*) and common sorrel (*R. acetosa*). Old peripheral galleries contrast markedly with fresh ones, being much blackened. Larvae of 19mm length have been discovered in December from roots taken the previous April, and Ullyett (1871) bred moths the year after roots were collected early in the previous

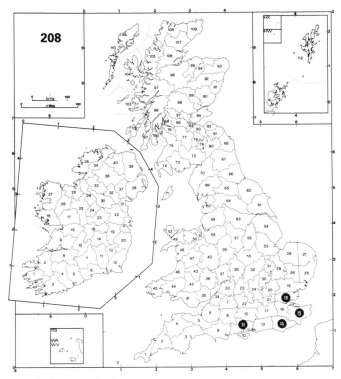

Bembecia chrysidiformis

year, indicating a two-year life-cycle.
Pupa. 17–19mm long. Warm brown, with darker abdominal spine-rows and eight prominent black-tipped anal spines; frontal process a broad, slightly curved blackish ridge. No cocoon is made. The pupae may extrude from low down on dock and sorrel stems or from specially prepared tough silken tubes covered with granular root scrapings. These tubes extend outwards from the root at ground surface for 25–30mm.
Imago. The moths appear during the last week of June and continue into July. They fly over dock and sorrel plants under sunny conditions and have been noted resting on bare chalk or feeding at the flowers of various Compositae.

Distribution (Map 208)
Very local on sea-cliffs in Kent, possibly still existing in similar habitats in Sussex and Hampshire. Occurred near Southend-on-Sea, Essex, until about 1860. Central and southern Europe.

Conservation
This is an endangered species which occurs only in a few restricted localities.

References

Allen, A. A., 1975. Notes on a colony of *Synanthedon vespiformis* L. in S.E. London with special reference to the breeding site. *Entomologist's Rec. J. Var.* **87**: 47–49.

Baker, B. R., 1970. Notes on clearwing moths associated with birch. *Entomologist's Gaz.* **21**: 161–166, 1 pl.

Bankes, E. R., 1906. Notes on the larva of *Trochilium andrenaeforme*. *Trans. ent. Soc. Lond.* **1906**: 474.

Bartel, M., 1912. Family: Aegeriidae (Sesiidae) *In* Seitz, A. (Ed.), *Macrolepidoptera of the World*, **2**: 375–416, Stuttgart.

Bretherton, R. F., 1946. Some clearwings in the Oxford District. *Entomologist's mon. Mag.* **82**: 213–217.

——, 1951. Our lost butterflies and moths. *Entomologist's Gaz.* **2**: 236–237.

Britton, M. R., 1977. The scarcity of the orange-tailed clearwing (*Aegeria andrenaeformis* Laspeyres). *Entomologist's Rec. J. Var.* **89**: 192–194, 2 pls.

Buckler, W., 1886. *The larvae of the British butterflies and moths*, **2**. London.

Chrystal, R. N., 1937. *Insects of the British woodlands*. London.

Classey, E. W., Cockayne, E. A., Fassnidge, W., Fletcher, J. B. & Parfitt, R. W., 1946. Collecting clearwings. *Leafl. Amat. Ent. Soc.* **18**, 12 pp.

Cockayne, E. A., 1933. The larval period of *Aegeria apiformis* (*Trochilium apiforme*, Cl.). *Entomologist's Rec. J. Var.* **45**: 97.

Common, I. F. B., 1970. Lepidoptera. *In* Mackerras, I. M., *The Insects of Australia*: 765–866. Melbourne.

Cooke, N., 1862. [Note]. *Zoologist* **20**: 8249.

Curtis, W. P., 1957. *Sciapteron tabaniformis* Rott. *Entomologist's Rec. J. Var.* **69**: 218–219.

Day, G. O., 1903. A list of Lepidoptera found in the counties of Cheshire, Flintshire, Denbighshire, Caernarvonshire, and Anglesey. *Proc. Chester Soc. nat. Sci.* **5**.

Fassnidge, W., 1926. A new British aegeriid:– *Synanthedon flaviventris* Stgr. *Entomologist's Rec. J. Var.* **38**: 113–115, pl. 3.

——, 1927. Notes on *Synanthedon formicaeformis*, Esp., in South Hampshire. *Ibid.* **39**: 67–70.

Fibiger, M. & Kristensen, N. P., 1974. *The Sesiidae (Lepidoptera) of Fennoscandia and Denmark*, 91 pp., 144 figs. Gadstrup, Denmark.

Gregson, C. S., 1863. Notes on insects injurious to fruit and forest trees. *Trans. Hist. Soc. Lancs & Chesh.* **15**: 11–12.

Hayward, H. C., 1913. [Note]. *Entomologist* **46**: 246–247.

Jacobs, S. N. A., 1957. Current notes. *Entomologist's Rec. J. Var.* **69**: 170.

Kemner, N. A., 1922. Zur Kenntnis der Entwicklungsstadien einiger Sesiiden. *Ent. Tidskr.* **1922**: 41–57.

Meyrick, E., 1928. *A revised handbook of British Lepidoptera*, vi, 914 pp. London.

Naumann, C., 1971. Untersuchungen zur Systematik und Phylogenese der holarktischen Sesiiden (Insecta, Lepidoptera). *Bonn. zool. Monogr.* **1**, 190 pp.

Newman, E., 1832. Monographia Aegeriorum Angliae. *Ent. Mag.* **1**: 66–84.

——, 1869. *The natural history of British moths*, vii, 486 pp. London.

Newman, L. W. & Leeds, H. A., 1913. *Textbook of British Butterflies and Moths*, 216 pp. St Albans.

Perret, J. & Mussillon, P., 1975. Contribution à la connaissance de *Synanthedon tipuliformis* Clerck, la sèsie du groseillier ct du cassissier. *Alexanor* **9**: 137-144, 163-171, 14 figs.

Pratt, C., 1977. Another pabulum for *Aegeria vespiformis* L. in Sussex. *Entomologist's Rec. J. Var.* **89**: 30.

South, R, 1961. *The Moths of the British Isles* (Edn 4), **2**, 379 pp., 141 pls. London.

Stainton, H. T., 1857. *A manual of British butterflies and moths*, **2**, 480 pp.

Tonge, A. E., 1933. The ova of British Lepidoptera. Pt.III. Geometridae, etc. *Proc. Trans. S. Lond. ent. nat. Hist. Soc.* **1932–33**: 11–25, pls I–IV.

Tugwell, W. H., 1891. Note on *Sesia sphegiformis* Fabr. *Entomologist* **24**: 204-206.

Ullyett, H., 1871. Notes on *Sesia chrysidiformis*. *Entomologist's mon. Mag.* **8**: 88.

Urbahn, E. & Urbahn, H., 1939. Die Schmetterlinge Pommerns mit einem vergleichenden Uberblick über den Ostseeraum. *Stettin. ent. Ztg* **100**: 185–826.

CHOREUTIDAE
E. C. Pelham-Clinton

This family has until recently been included in the composite family Glyphipterigidae following the classification of Meyrick (1928 and earlier works). Now Brock (1968) and Heppner (1977) have shown convincingly that the genus *Glyphipterix* Hübner is not closely related to the four genera here included in the Choreutidae, and this split has found general acceptance. New ideas on classification which would place the segregated families far apart are not yet stabilized and, to preserve the scheme of classification of Kloet & Hincks (1972) as far as possible, the Choreutidae and Glyphipterigidae are kept together in this volume in the superfamily Yponomeutoidea, maintaining the arrangement given in *MBGBI* 1: 148.

Imago. Sexes similar. Head smooth-scaled, lateral vertical scale-tufts directed inwards to cover occiput and meet on top of head. Antenna half to two-thirds length of forewing, ciliate in male, simple or nearly so in female, clearly banded blackish brown and white on upper surface in all British species; scape smooth-scaled except in *Tebenna* Billberg, without pecten. Ocelli present. Labial palpus porrect or ascending, length up to twice diameter of eye; vestiture providing good generic characters (figure 114). Maxillary palpus greatly reduced one to two segmented. Haustellum well developed, scaled at base (in this respect showing an apparent relationship to Gelechioidea). Legs rather short and stout, sometimes with whorls of scales, usually with sparse long hair (figure 31, p. 66); in all species tarsal segments with apical whitish rings. Fore- and hindwings with all veins present and separate except 7 (R_5) and 8 (R_4) which are stalked in *Choreutis diana* (Hübner), and 3 (Cu_1) and 4 (M_3) of hindwing which may be stalked or coincident (figures 115,116). Hindwing broad, length at most two and a half times greatest width, cilia short. Abdominal segments margined white or silvery.

A very distinct family in general appearance. The broad wings, and particularly the very broad hindwings, are of tortricoid proportions, and indeed when Brock (*loc.cit.*) first recognized the distinction of this family from the Glyphipterigidae he transferred it on other characters to the Tortricoidea (later to a position near Sesiidae). Most nineteenth-century authors included it amongst the Tortricidae. However, the moths have resting attitudes quite different from any tortricid; usually the wings are held slightly apart in a horizontal position and more or less corrugated longitudinally, but in *Tebenna* the wings may be tightly closed with apices bent down and pressed together. The moths fly mostly in sunshine and sometimes visit flowers. When settled and active they walk with short jerky movements.

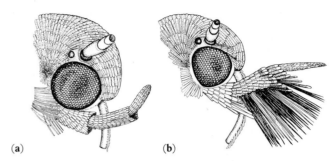

(a)   (b)

Figure 114 Heads (**a**) *Anthopila* (**b**) *Tebenna*

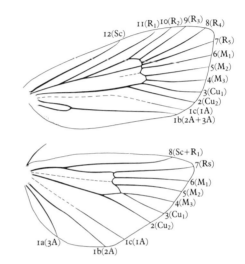

Figure 115 *Anthopila fabriciana* (Linnaeus), wing venation

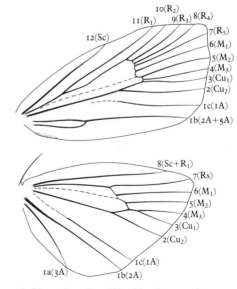

Figure 116 *Choreutis pariana* (Clerck), wing venation

The British species were previously illustrated in colour by Ford (1954).

Ovum. Described only for *Anthophila fabriciana* (Linnaeus), *q.v.*

Larva. In general appearance resembling the larvae of some Tortricidae, but the pinacula are prominent and the prothoracic plate weakly developed. L group of prothorax trisetose (unlike those of Glyphipterigidae). The larvae feed exposed in a web on a variety of flowering plant families and except for the first stage larvae of species of *Prochoreutis* Diakonoff & Heppner they do not mine.

Pupa. Abdominal segments with single row of spines; segments 3–7 free. In or near the larval spinning; protruded on emergence of the adult.

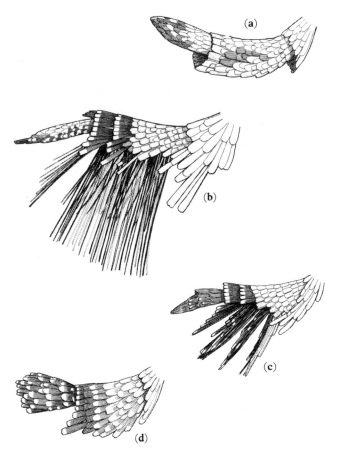

Figure 117 Palps (**a**) *Anthopila* (**b**) *Tebenna* (**c**) *Prochoreutis* (**d**) *Choreutis*

Key to species (imagines) of the Choreutidae

1   Labial palpus segment 2 tufted with long scales beneath, the scales almost or quite as long as segment 2; forewing with silvery spots 2

–   Labial palpus segment 2 at most much thickened with scales and rough beneath; forewing without silvery spots 4

2(1)   Scape of antenna with ventral scale-tuft; labial palpus segment 2 with ventral scales forming a dense brush (figure 117b) .... *Tebenna bjerkandrella* (p. 393)

–   Scape of antenna not tufted; ventral scales of labial palpus segment 2 grouped into bundles to give a pectinate effect (figure 117c) (*Prochoreutis*) 3

3(2)   Forewing with postmedian line of silver spots followed by a fascia of scales paler than ground colour. Genitalia figure 119a,b, p. 396 .. *Prochoreutis sehestediana* (p. 394)

–   Forewing without paler fascia. Genitalia figure 119c,d, p. 396 .... *P. myllerana* (p. 395)

4(1)   Labial palpus segments 2 and 3 smooth-scaled and cylindrical (figure 117a); hindwing with short whitish line near anal angle .... *Anthophila fabriciana* (p. 391)

–   Labial palpus segments 2 and 3 thickened with scales, segment 2 rough beneath, segment 3 short and obtuse (figure 117d); hindwing unmarked (*Choreutis*) 5

5(4)   Forewing mainly olive or greenish grey with fuscous fringe .... *Choreutis diana* (p. 398)

–   Forewing mainly brown or reddish brown, usually with reddish fringe .... *C. pariana* (p. 397)

ANTHOPHILA Haworth

Anthophila Haworth, 1811, *Lepid.Br.*: 471.

A small genus which includes nine Palaearctic species of which two are European. At least one closely related species occurs in north America. Other species described in *Anthophila* (or in *Simaethis* Leach, a junior objective synonym) may properly belong in other genera. Meyrick (1928) and Ford (1954) included in this genus those species now in *Choreutis* Hübner.

Imago. Antenna in male with fine cilia, their length up to twice diameter of shaft, in female simple; labial palpus segments 2 and 3 cylindrical, smooth-scaled (figure 114a, p. 389).

Larva. Feeding mostly on Urticaceae.

ANTHOPHILA FABRICIANA (Linnaeus)

Phalaena (Tortrix) fabriciana Linnaeus, 1767, *Syst.Nat.* (Edn 12) **1**: 880.

Type locality: Sweden; Hammerby.

Description of imago (Pl.11, fig.25)

Wingspan 11–15mm. Labial palpus ochreous variably marked with brown on segments 2 and 3. Head, thorax and forewing dark brown irrorate with whitish. Forewing irroration variable, most dense in basal area, absent from certain areas to form dark bands of ground colour, antemedian and median bands most distinct; ante- and postmedian whitish spots on costa, the latter continued as a fine irregular line to tornus; cilia beyond a dark brown basal band fuscous, alternating with whitish patches below apex and above tornus. Hindwing fuscous brown; a narrow whitish postmedian line from vein 2 (Cu$_2$) to anal angle; cilia repeating forewing pattern but whitish areas more extensive except on anal margin.

Similar species. Prochoreutis spp. are marked with silvery spots. *Choreutis pariana* (Clerck) is without whitish markings except along costa.

Life history

Ovum. Truncated conical, flat-topped, *c.*0.3mm diameter at base and a little less in height; translucent whitish becoming tinged with orange, sculptured with a reticulate pattern of vertical ridges on sides, and with ridges on top radiating from the micropyle which is surrounded with a foliate pattern (figure 118). Laid on a stem or on either surface of the leaf of the foodplant.

Larva. Head pale brown with prominent dark brown markings; at back of head a pair of dorsal and a pair of lateral blotches, in front of these a smaller dorsolateral pair. Prothoracic plate translucent brown with a central

Figure 118 Ova of *Anthophila fabriciana* (Linnaeus)
(**a**) as laid on leaf (×60 approx.) (**b**) showing patterns surrounding micropyle (×130 approx.)
Scanning electron micrographs. D. Claugher, Electron Microscope Unit, BMNH

pale stripe. Body pale yellowish green with large convex dark brown pinacula bearing rather long whitish hairs.

Feeds on nettle (*Urtica* spp.) and occasionally on pellitory-of-the-wall (*Parietaria judaica*); continental authors also give comfrey (*Symphytum* spp.) as a foodplant. The larva usually starts spinning on the upperside at the base of a leaf and develops an extensive loose web which is enclosed in the leaf folded upwards. The life-cycle is variable, but full-grown larvae are normally found in April and May and of a second generation in July. Larvae may overwinter in first, second or third instars (of a total of four – see Shaw, 1981).

Pupa. Enclosed in a dense whitish spindle-shaped cocoon, *c.*8–9mm in length, spun in or near the larval web.

Imago. At least bivoltine, moths appearing in May and June and from July onwards. Sometimes moths may be

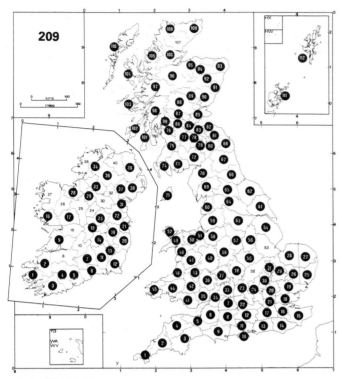

Anthophila fabriciana

TEBENNA Billberg

Tebenna Billberg, 1820, *Enum.Ins.Mus.Blbg*: 90.

A small genus of world-wide distribution, with most species found in the New World and six Palaearctic species.

Imago. Antenna in male with fine cilia, length up to twice diameter of shaft, in female simple; scape with scale-tuft arising from middle of ventral surface; labial palpus segment 2 with brush of long scales beneath whole length of segment; segment 3 slender, acute, more or less drooping (figure 114b, p. 389). Hindwing with veins 3 (Cu$_1$) and 4 (M$_3$) coincident.

Larva. Feeding mostly on Compositae.

seen on the wing as late as October, perhaps resulting from a third generation, and it is possible that, as suggested by Tremewan (1981), some adults may overwinter. The moths fly rapidly at any time of day, not always near nettles, sometimes in such numbers over hedgerow vegetation that it may become difficult to recognize and capture other species.

Distribution (Map 209)

One of the most abundant and widely distributed of British Lepidoptera, found almost wherever nettles grow in all parts of the British Isles including the Outer Hebrides, Orkney and Shetland Islands. Widely distributed in the Palaearctic region from Spain to Kashmir and east Siberia. Formerly recorded from north America, but this population is now recognized as a distinct species, *Anthophila alpinella* (Busck).

TEBENNA BJERKANDRELLA (Thunberg)

Tinea bjerkandrella Thunberg, 1784, *Diss.ent.sistens Insecta Suecica* **1**: 36, pl.3, figs 23,24.

Type locality: Sweden; Mt Kirmekulle.

Description of imago (Pl.11, fig.26)

Wingspan 11–14mm. Head brown, the scales paler-tipped; labial palpus dark brown mixed with whitish (figure 117b, p. 390). Thorax pale ferruginous with three silver lines, one median and a pair on inner edges of tegulae. Forewing brown, heavily irrorate with white-tipped scales forming two fasciae, antemedian and at three-quarters wing length, the latter outwardly angulated, the fasciae linked with whitish irroration in disc; two pale ferruginous patches extending from base, separated and bounded by silvery lines; slightly convex irregular silver spots surrounded with black marks, the largest in centre of wing, one nearer dorsum and one inwardly black-edged on termen; other smaller black and silver spots in median and subterminal areas; cilia fuscous, white-tipped. Hindwing fuscous with short white postmedian line from about vein 5 (M$_2$) to 1c (1A); cilia fuscous, darker at base, with subbasal and apical white lines.

Life history

Larva. Head very pale brown, translucent; mouth parts and some small posterior patches pale reddish brown; ocellar patch blackish. Body pale green, paler and more ochreous below; prothoracic plate not developed; setae on conspicuous small black pinacula.

The foodplant in Britain appears to be common fleabane (*Pulicaria dysenterica*) exclusively, though continental authors give *Carlina*, *Carduus* and *Eryngium*. On common fleabane the larval spinning is extensive and may include several leaves; some frass is usually visible, caught in the loose web. Usually the lower green leaves are selected and the larva feeds on the underside, making blotches which appear brown above; it does not make holes in the leaf or eat the leaf margin as does the small larva of *Ebulea crocealis* (Hübner) (Pyralidae) which may feed at the same time. *Coleophora* species (Coleophoridae) may make similar blotches, but no webbing. Several larvae may occupy the same web.

When numerous larvae were found in 1982 the feeding period extended throughout August and September. These larvae probably resulted from an immigration earlier in the year but there is no evidence of the number of generations produced here. The species is probably continuously brooded in southern Europe and it is unlikely to be able to survive our winter in any stage.

Pupa. In a narrow white cocoon, 10–12mm long, tapered to both ends, placed on a leaf surface under the feeding web or under a separate outer spinning.

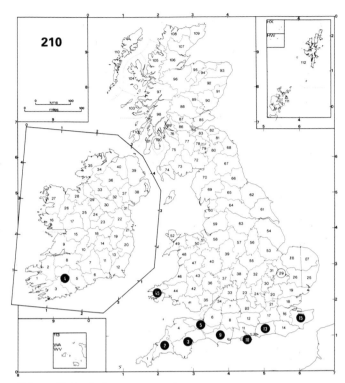

Tebenna bjerkandrella

Imago. A rare species in the British Isles, but found commonly in one year, 1982. Imagines have been found here mainly in August and September and have been recorded visiting flowers of *Pulicaria*, *Achillea* and Umbelliferae.

Distribution (Map 210)

The records of this species in the British Isles have been listed by Heckford (1984) who was the first to find larvae in 1982 on the south coast of Devon. In that year it was found for the first time in Ireland (Bond, 1983). Up till 1983 it was uncertain whether resident populations could maintain themselves in mild areas, but the complete absence of the species in 1983 in localities in which it had been common in 1982 showed that its sporadic occurrence in the British Isles is best explained by immigration.

An Old World species, widely distributed in the Palaearctic region from southern Europe to Japan. Records from the southern hemisphere should probably apply to the closely related *Tebenna bradleyi* Clarke, which is distributed from India to New Zealand and the south Pacific.

PROCHOREUTIS Diakonoff & Heppner

Prochoreutis Diakonoff & Heppner, 1980, *Ent.Ber., Amst.*
40: 196.

Choreutis sensu auctt.

A cosmopolitan genus of which four rather closely related
species are found in Europe. The two British species were
separated by Haworth (1811), but subsequently the distinction was overlooked until Pierce (1939) examined the
genitalia.

Imago. Antenna in male with fine cilia, length up to two
and a half times diameter of shaft, shortly ciliate in female;
labial palpus segment 2 with long scales beneath, gathered
into four or five more or less separate tufts; segment 3
slender and acute (figure 117c, p. 390).

Larva. Feeding on Labiatae.

PROCHOREUTIS SEHESTEDIANA (Fabricius)

Pyralis sehestediana Fabricius, 1777, *Gen.Insect.*: 293.
Anthophila punctosa Haworth, 1811, *Lepid.Br.*: 472.
Choreutis myllerana sensu Pierce & Metcalfe, 1935, *Genit.
Tineid Families Lepid.Br.Islands*: pl.23 (♀ genitalia).
Type locality: Germany; Kiel.

Description of imago (Pl.11, fig.27)

Wingspan 9–12mm. Head brown; labial palpus segment 1
white, 2 and 3 brown mixed with white. Thorax brown
with a pair of small white postero-lateral spots; tegulae
usually edged inwardly with silver. Forewing brown, variably mixed with darker brown; three small white spots on
costa, one antemedian, one postmedian and a larger
beyond this; three other white spots in disc, one antemedian and two in median area; a subrectangular area of
white irroration extending from the discal spots to dorsum; scattered bluish silver spots, the most definite forming a postmedian line beyond which is a broad fascia
slightly paler than ground colour; cilia white with brown
basal and median lines. Hindwing brown, darker towards
termen; a short submarginal white line from about vein 3
(Cu₁) to 1c (1A); cilia as in forewing. Genitalia, see figure
119, p. 396.

Similar species. *P. myllerana* (Fabricius), *q.v.*

Life history

Larva. Not distinguished from that of *P. myllerana*, *q.v.*

Imago. Owing to the confusion of identities there is little
that can be stated about this species that might distinguish
it in biology or habits from those of *P. myllerana.* However, it does appear that *P. sehestediana* is more often taken
flying well away from its known foodplants than *P. myller-*

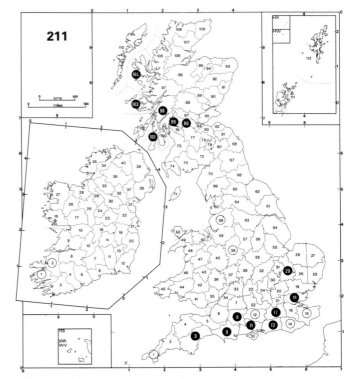

Prochoreutis sehestediana

ana, sometimes around trees or over open moorland;
Pierce (1939) made a point of this.

Distribution (Map 211)

Pierce (*loc.cit.*) showed that Haworth's (1811) specimens
came from east Sussex and added some Dorset records.
Since then the distribution has been found to extend to
Scotland and Ireland, but although the two species are
often found together *P. sehestediana* is decidedly less common than *P. myllerana.* Widely distributed in northern
Europe from France to Finland and western U.S.S.R.

For the significance of open circles on the map, see *P.
myllerana* (p. 395).

PROCHOREUTIS MYLLERANA (Fabricius)

Pyralis myllerana Fabricius, 1794, *Ent.syst.* **3**(2): 277.
Type locality: Sweden.

Description of imago (Pl.11, fig.28)

Wingspan 10–14mm. Differs from *P. sehestediana* (Fabricius) only as follows: forewing with white spots on average larger and more rounded; patch of white irroration extending from dorsum at most halfway across wing; no pale fascia beyond postmedian line of silvery spots; termen slightly more curved and apex less acute. Genitalia, see figure 119, p. 396. Males may be identified quite easily without dissection by the position of the costal process of the valva.

Life history

Larva. Head pale brown, sutures darker; back of head blotched with darker brown; a blackish patch covering ocelli. Body grey-green, paler below; prothoracic plate transparent pale brown with darker pinacula; pinacula prominent but small, blackish; abdominal segment 8 with small blackish dorsal plate; anal plate blackish.

Feeds on skullcap (*Scutellaria galericulata*) and lesser skullcap (*S. minor*), at first mining, making a very small full-depth mine on the lowest leaf of a plant (Hering, 1957), later making a loose spinning in a shoot or between two leaves. According to Hering (*loc.cit.*) and other continental authors it feeds also on *Lamium*, and Toll (1956) mentions white and red dead-nettle (*Lamium album* and *L. purpureum*). Fully grown larvae are found in June and from July to August.

Pupa. In a dense white spindle-shaped cocoon *c.*10mm long spun in a fold of a leaf.

Imago. At least two generations a year, moths being found in late May, June and July, and from August to early September. It is not known how the winter is passed, but most probably in the adult stage. During winter the usual foodplant, skullcap, is very often flooded and would not produce sufficient growth early in the year to support larvae. The larvae of this species and *P. sehestediana* may be taken off the same plant at the same time, but usually *P. myllerana* is a little earlier in all its stages.

Moths are usually taken at rest on or flying over the foodplant, but South (1894) recorded that they were common at honey-dew on sallow.

Distribution (Map 212)

The records of this species and *P. sehestediana* are confused, not only because up to 1939 the two were mostly considered as a single species, but also because Pierce (1939) gave a character (white submarginal spot on the underside of the forewing in *P. myllerana*) which is not at

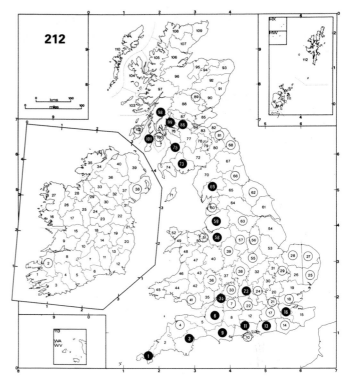

Prochoreutis myllerana

all reliable. The map therefore shows unconfirmed records as open circles. The species has a wide distribution in mainland Britain and occurs in Ireland. The European distribution appears to be much the same as that of *P. sehestediana*, from France to Scandinavia.

Figure 119

Prochoreutis sehestediana (Fabricius)
(**a**) male genitalia (**b**) female genitalia

Prochoreutis myllerana (Fabricius)
(**c**) male genitalia (**d**) female genitalia

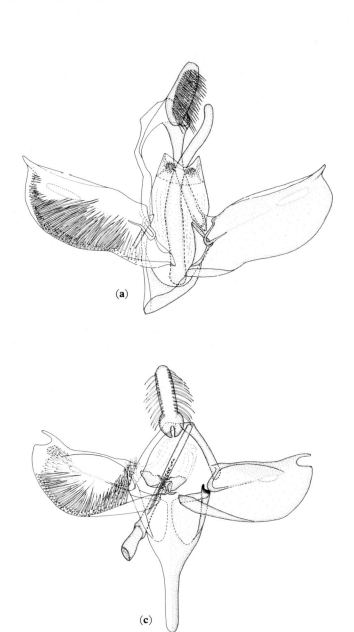

(**a**)

(**c**)

0.5mm

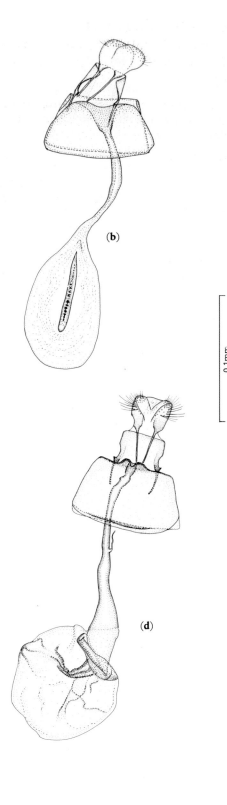

(**b**)

(**d**)

0.1mm

CHOREUTIS Hübner

Choreutis Hübner, [1825], *Verz.bekannt.Schmett.*: 373.

Eutromula Frölich, 1828, *Enum.Tortricum Würtembergiae*: 11.

An extensive cosmopolitan genus with three European species of which one, *C. nemorana* (Hübner), is found on fig (*Ficus*) in southern Europe.

Imago. Antenna in male with fine cilia, up to one and a half times diameter of shaft, simple in female; labial palpus thickened with scales; segment 2 rough beneath; 3 short, concealed in a blunt-tipped brush of scales (figure 117d, p. 390). Mid- and hindtibiae much thickened with scales and with prominent whorls of projecting scales in middle and at apex.

Larva. Usually in a web on upper surface of a leaf, mainly on trees of various families, the British species on Betulaceae and Rosaceae.

CHOREUTIS PARIANA (Clerck)

Apple Leaf Skeletonizer

Phalaena pariana Clerck, 1759, *Icones Insect.rar.* 1: pl.10, fig.9.

Type locality: not stated.

Description of imago (Pl.11, figs 29–31)

Wingspan 11–15mm. Head dark brown; labial palpus dark brown mixed with white. Forewing very variable, usually reddish brown, sometimes grey-brown or ochreous brown; indistinct whitish costal spots, antemedian, postmedian and at least one beyond; basal area sometimes bounded by an irregular blackish line, rarely almost completely dark grey; antemedian area usually of ground colour; median area often with irroration of whitish-margined scales; sometimes a partial irregular postmedian line of blackish or red-brown scales, most distinct on costa; a subapical fascia usually present, either red-brown or blackish or a mixture of these; cilia reddish or purplish brown with dark basal line and whitish-tipped patches below apex and below middle. Hindwing dark brownish fuscous; cilia partly reddish brown, at least at apex and in middle, with dark brown basal line and whitish margin.

In spite of its great variability this species is easily recognized by its brown colouring, usually with bright red-brown cilia, and absence of silvery markings.

Life history

Larva. Head pale brown with blackish markings on vertex and around ocelli. Prothorax with black pinacula linked by dark markings to form a pair of roughly U-shaped marks. Body dull greenish ochreous; pinacula black, prominent; a pale yellow broken dorsal stripe and a subdorsal series of yellow spots.

Feeds on the upperside of a leaf under a fine web, eating all but the lower cuticle which remains as a brown blotch. Most often on crab-apple (*Malus sylvestris*), both wild and cultivated, occasionally on hawthorn (*Crataegus* spp.), wild pear (*Pyrus communis*), rowan (*Sorbus aucuparia*) or other rosaceous trees: recorded from birch (*Betula* spp.) by continental authors. Larvae are found in May and June and again in August and are sometimes a minor pest of apple-trees.

Pupa. In a whitish cocoon 15–20mm long, consisting of a dense inner spindle-shaped part and an outer covering of less dense silk. Usually on the underside of a leaf, not always near the larval feeding place, often lower on the tree.

Imago. Bivoltine. July; September, the second generation overwintering in the adult stage.

Distribution (Map 213)

Widely distributed in Britain but very local. Recorded

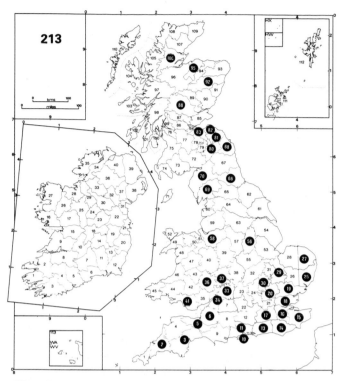

213

Choreutis pariana

from two vice-counties in Ireland but I have seen no Irish specimens. Sometimes occurs on particular trees in great abundance, but fluctuates greatly in numbers. In the Palaearctic region it extends from France to Japan and has been introduced during this century into North America where it now has an extensive distribution (Heppner, 1978).

CHOREUTIS DIANA (Hübner)

Tortrix diana Hübner, [1822], *Samml.eur.Schmett.* **7**: pl.44, fig.274.
Type locality: Europe.

Description of imago (Pl.11, fig.32)

Wingspan 14–18mm. Head pale fuscous variably mixed with greenish grey or whitish. Forewing pale olive, variably mixed with pale greenish blue and white; a white antemedian costal spot, usually continued as a fine irregular line to dorsum; two or three white costal spots in apical third, the first two usually continued as a broad irregular and diffuse whitish fascia to near tornus; blackish markings variable but usually including a crescentic postmedian mark near dorsum and a subapical fascia of spots; cilia shining fuscous with dark brown line at base and variable whitish patches below apex and below middle. Hindwing dark fuscous; cilia dark brown at base, outwardly whitish with brownish central line.

A variable but very distinct species, the largest of the British Choreutidae. The mottled pattern of olive, pale blue and white scales usually gives a greenish effect.

Life history

Larva. Not yet discovered in the British Isles. Described by Forbes (1923) from north American larvae as follows: 'Translucent greenish yellow. Head pale reddish with a black line on sides behind; tubercles black; body with a clear white dorsal, and blurred lateral lines.' Toll (1956) on the other hand, describes it as 'yellowish green in colour, with pale yellow dorsal bands and a black head and prothoracic plate . . .'. Toll gives the foodplant as grey alder (*Alnus incana*) whereas in Britain and north America birch (*Betula* spp.) is certainly the foodplant, so that perhaps he described another species: it would be unusual for a choreutid larva to have a black head and prothoracic plate. Unfortunately Forbes included a closely related north American species, *Choreutis betuliperda* (Dyar), in his description so that here is another possibility of error. The larva feeds on the upper surface of a leaf under a slight web, eating the upper surface of the leaf only.

Pupa. Beneath a leaf in a dense spindle-shaped cocoon enclosed in a looser outer spinning, some distance from the larval feeding position. In Britain cocoons have been found on birch but neither larvae nor larval spinnings have been detected.

Imago. Probably univoltine; in Britain throughout August.

History and distribution (Map 214)

First found in Scotland by Mackworth-Praed (see Durrant, 1920) in Glen Affric, Inverness-shire, flying amongst birches in sunshine and settling on bracken. It was found

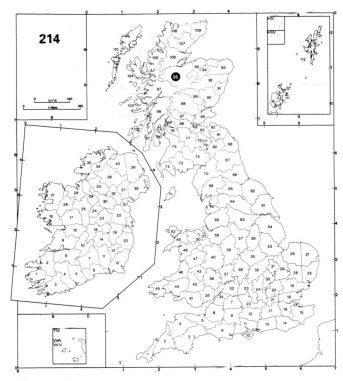

Choreutis diana

by the author in the same area in 1954 and 1977, in the latter year visiting flowers of creeping thistle (*Cirsium arvense*). In 1978 J. M. Chalmers-Hunt found under birch leaves two cocoons from which moths were bred.

Only known to occur in the Glen Affric locality in the British Isles, but it is doubtful whether an adequate search has been made for it in other areas. In Europe it has been found locally from France to central Europe and Finland, in north America from Nova Scotia to New Hampshire, British Columbia and Utah.

References

Bond, K. G. M., 1983. *Tebenna bjerkandrella* Thunberg (Lep.: Glyphipterigidae) in Ireland. *Entomologist's Rec. J. Var.* **95**: 28.

Brock, J. P., 1968. The systematic position of the Choreutinae (Lep., Glyphipterygidae). *Entomologist's mon. Mag.* **103**: 245–246, 2 figs.

Durrant, J. H., 1920. *Allononyma diana*, Hb. – a genus and species new to the British list (Lep.-Tin.). *Proc. ent. Soc. Lond.* **1919**: xliv–xlv.

Forbes, W. T. M., 1923. The Lepidoptera of New York and neighboring states. Primitive forms, Microlepidoptera, Pyraloids, Bombyces. *Mem. Cornell Univ. agric. Exp. Stn*, **68**, 729 pp., 439 figs.

Ford, L. T., 1954. The Glyphipterygidae and allied families. *Proc. Trans. S. Lond. ent. nat. Hist. Soc.* **1952–53**: 90–99, 1 col. pl.; reprinted [Agassiz, D. J. L. (Ed.)], 1978, in *Illustrated papers on British Microlepidoptera*: 77–86.

Haworth, A. H., 1811. *Lepidoptera Britannica* (part 3): 377–512. London.

Heckford, R. J., 1984. Notes on *Tebenna bjerkandrella* (Thunberg). *Entomologist's Rec. J. Var.* **96**: 58–63.

Heppner, J. B., 1977. The status of the Glyphipterigidae and a reassessment of relationships in yponomeutoid families and ditrysian superfamilies. *J. Lepid. Soc.* **31**: 124–134, 1 fig.

——, 1978. *Eutromula pariana* (Clerck) (Lepidoptera: Choreutidae), the correct name of the apple-and-thorn skeletonizer. *J. ent. Soc. Br. Columb.* **75**: 40–41.

Hering, E. M., 1957. *Bestimmungstabellen der Blattminen von Europa*, **1** and **2**, 1185 pp., **3**, 221 pp., 725 figs. 's-Gravenhage.

Kloet, G. S. & Hincks, W. D., 1972. A check list of British insects: Lepidoptera (Edn 2). *Handbk Ident. Br. Insects* **11**(2), viii, 153 pp.

Meyrick, E., 1928. *A revised handbook of British Lepidoptera*, vi, 914 pp. London.

Pierce, F. N., 1939. *Choreutis punctosa* Haworth a good species. *Entomologist* **72**: 257–260, 1 fig.

Shaw, M. R., 1981. Overwintering of *Anthophila fabriciana* (L.) (Lepidoptera: Choreutidae) in England, with notes on some of its parasites (Hymenoptera). *Entomologist's Gaz.* **32**: 203–204.

South, R., 1894. Tortrices at Northwood, Middlesex. *Entomologist* **27**: 323.

Toll, S., 1956. Glyphipterygidae. *Klucze Oznacz. Owad. Pol.* **27** (39), 36 pp., 88 figs.

Tremewan, W. G., 1981. Does *Anthophila fabriciana* (L.) (Lepidoptera: Choreutidae) overwinter in the adult stage? *Entomologist's Gaz.* **32**: 2–3.

GLYPHIPTERIGIDAE
E. C. Pelham-Clinton

As stated in the introduction to the Choreutidae, this family is restricted here to the genus *Glyphipterix* Hübner, but the arrangement of families given in *MBGBI* 1: 148 is preserved so that the family remains in the Yponomeutoidea.

The family is of world-wide distribution, 326 species being listed by Heppner (1982).

Imago. Sexes similar, but female often larger with longer abdomen. Head smooth-scaled; lateral vertical scale-tufts directed inwards and meeting on top of head (figure 120). Antenna just over half length of forewing, simple or shortly ciliate in both sexes; scape smooth-scaled, without pecten. Ocelli present. Labial palpus more or less upcurved, somewhat flattened dorsoventrally, often rough-scaled beneath; segment 3 acute. Maxillary palpus vestigial. Haustellum well developed, bare. Legs smooth-scaled. Both wings with all veins present and separate except that veins 7 (R_5) and 8 (R_4) of forewing and 3 (Cu_1) and 4 (M_3) of hindwing are sometimes stalked. Forewing with accessory cell usually well developed; membrane between veins 10 (R_2) and 12 (Sc) sometimes thickened along costa to form a pterostigma (figure 121). Hindwing narrow oval or suboblong, maximum cilia width about equal to width of hindwing; frenulum in female usually of two bristles.

Glyphipterigids are small day-flying moths which may be recognized in the field by the curious habit of raising and lowering the wings when at rest. Apart from *Glyphipterix simpliciella* (Stephens), which commonly visits flowers such as those of buttercup (*Ranunculus* spp.) and hawthorn (*Crataegus* spp.), moths are not often found on flowers except those of their foodplant. When settled for a short time they often rest transversely on stems or grass blades. Cabinet specimens of some of the larger species might be mistaken for Tortricidae, the wing-pattern being similar to some species of *Cydia* Hübner, but the labial palpi are more slender and upcurved and the hindwings much narrower.

The British species were previously figured in colour by Ford (1954).

Larva. Rather stout, with well-developed prothoracic and anal plates and small pinacula. Prothoracic dorsal plate divided medially, prespiracular plate separated from it. L group of setae on prothorax bisetose. Feeding internally on seeds or mining in stems and leaves. In most of these respects differing from the larvae of Choreutidae.

Pupa. In a thin but tough silken cocoon. Unspined and not protruded on emergence of adult.

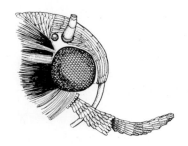

Figure 120 *Glyphipterix* sp., head

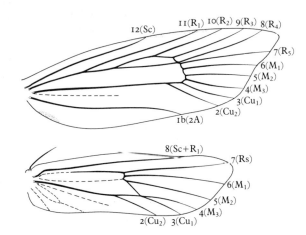

Figure 121 *Glyphipterix thrasonella* (Scopoli), wing venation

Key to species (imagines) of the genus *Glyphipterix*

1   Forewing with white or silver spots or strigulae 2
–   Forewing almost or quite unmarked 7

2(1)   Forewing with blackish apical spot enclosing a silvery dot .... *forsterella* (p. 404)
–   Forewing without silvery dot in blackish apical spot ... 3

3(2)   Forewing with a white strigula, outwardly oblique from before or at middle of dorsum, usually curved and reaching middle of wing (sometimes obsolete on dorsum in *schoenicolella*) 4
–   Forewing without white strigula from dorsum; usually two or three silvery dots enclosed in an oblong blackish mark along termen near tornus ....
                 .... *thrasonella* (part) (p. 406)

4(3)   Wingspan greater than 11mm .... *haworthana* (p. 405)
–   Wingspan less than 11mm 5

5(4)   Forewing with two distinct silvery dots in apical area, one in disc and one near tornus; base of wing pale grey; dorsal strigula very strong .... *equitella* (p. 403)
–   Forewing lacking distinct discal pre-apical silvery spot; usually much darker at base; dorsal strigula weaker, sometimes obsolescent 6

6(5)   Hindwing in basal half with dorsal cilia white ....
                 .... *schoenicolella* (p. 403)
–   Hindwing with dorsal cilia entirely pale fuscous ....
                 .... *simpliciella* (p. 401)

7(1)   Forewing termen sinuate below apex ....
                 .... *thrasonella* (part) (p. 406)
–   Forewing termen almost straight; costa very narrowly whitish in apical half .... *fuscoviridella* (p. 406)

GLYPHIPTERIX Hübner

Glyphipterix Hübner, [1825], *Verz.bekannt.Schmett.*: 421.

NOMENCLATURE. It is unfortunate that the generic name has been wrongly spelt in the past as '*Glyphipteryx*', this usage predominating over '*Glyphipterix*'. Since *Glyphipteryx* Curtis, 1827, is at present a valid name used for a genus in the Momphidae (Kloet & Hincks, 1972) application is being made to the I.C.Z.N. for its suppression (case 2115, Diakonoff & Heppner, 1977).

The genus is the largest of the Glyphipterigidae, 253 species being listed by Heppner (1982).

Imago and early stages. As described for family. The following markings of the imago are common to all species unless stated otherwise: antenna white beneath near base; frons and face lined whitish above antenna and eye; forewing with 'apical hook' of dark-tipped cilia from costa to apex; tarsal segments and hindtibia with pale apical rings; abdominal segments margined paler. The forewing markings of the first five species are similar; they are described fully for *G. simpliciella* (Stephens) and the following four by reference to this species.

Of the seven British species the larva of one mines Crassulaceae, two mine stems of Juncaceae and the remainder feed in seeds of Gramineae or Cyperaceae.

The species are most often encountered in the adult stage and the larvae of most are relatively difficult to find, though then it is usually a simple matter to breed the moths. Progress in the knowledge of life histories has been slow and the larvae of only three British species were known to Stainton (1870).

GLYPHIPTERIX SIMPLICIELLA (Stephens)
Cocksfoot Moth

Heribeia simpliciella Stephens, 1834, *Ill.Br.Ent.* (Haust.) **4**: 263.
Aechmia fischeriella Zeller, 1839, *Isis, Leipzig* **1839**: 204.
Type localities: England; Hertford and Coombe Wood, [Surrey].

Description of imago (Pl.11, fig.21)
Wingspan 6–9mm. Head and thorax greyish fuscous; labial palpus white banded with dark fuscous, more strongly below, segment 3 a little shorter than segment 2. Forewing fuscous, darker and slightly coppery in apical third; five silvery white costal strigulae in apical half, the first outwardly oblique, those following progressively less so; an oblique silvery white strigula from middle of dorsum almost meeting first costal at an acute angle; an erect silvery white strigula from tornus meeting second costal at an obtuse angle; an oval or elongate silver spot near termen

above tornus; an apical blackish spot; cilia silvery at base with dark median line broken below apex by a silvery white strigula leading to a silver dot, outer part pale fuscous. Hindwing fuscous.

Similar species. G. schoenicolella (Boyd) with hindwing dorsal cilia white in basal half; *G. forsterella* (Fabricius) with silver dot in blackish apical spot; *G. equitella* (Scopoli) with additional silver dot in disc before apex.

Life history

Larva. Head shining black. Prothoracic plate brown, tapered to sides and sinuate posteriorly. Body greenish or yellowish white with small dark pinacula; a narrow transverse dorsal plate on abdominal segment 9 and anal plate dark brown. Feeds in seeds of cock's-foot (*Dactylis glomerata*) or tall fescue (*Festuca arundinacea*) (Pelham-Clinton, 1978). When the larva is fully grown in July or August it leaves the seed-head and enters the stem, making a small hole usually beneath one of the upper leaf-sheaths. Several larvae may enter the stem by the same hole. Inside the stem a whitish cocoon is spun incorporating particles of pith, placed on one side of the stem or filling a narrow stem. The larva overwinters in this cocoon, pupating in April or May. Sometimes a pest of cock's-foot grown for seed.

Imago. Univoltine, appearing from May to early July, sometimes in very large numbers. A second generation may occur in southern France where it has been recorded as late as September. Often frequents flowers.

Distribution (Map 215)

A widely distributed and usually abundant species. It has not been recorded from the north mainland of Scotland or from Shetland, perhaps because it has not been sought in those areas. Most easily recorded during the winter months by searching for cocoons in cock's-foot stems. Throughout most of Europe, including Iceland, extending to north Africa and Asia Minor.

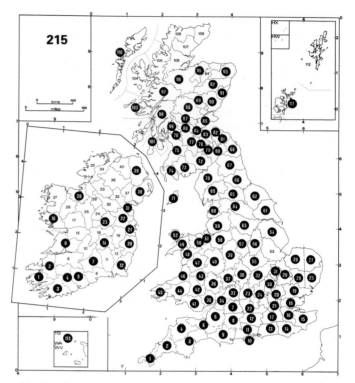

Glyphipterix simpliciella

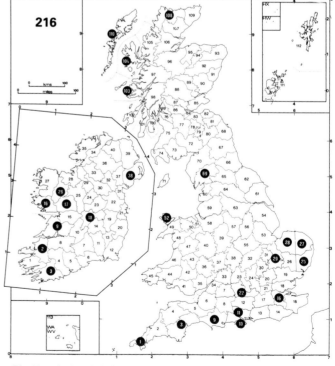

Glyphipterix shoenicolella

GLYPHIPTERIX SCHOENICOLELLA (Boyd)

Glyphipteryx schoenicolella Boyd, 1858, *Entomologist's wkly Intell.* **4**: 144.
Type locality: England; Lizard peninsula, Cornwall.

Description of imago (Pl.11, fig.22)
Wingspan 6–9mm. Very similar to *G. simpliciella* (Stephens), differing as follows: forewing more evenly coloured, hardly tinged with copper in apical third; costal strigulae less evenly spaced, usually a larger gap between third and fourth from base: hindwing dorsal cilia whitish-tipped, becoming entirely white towards base.
Similar species. As for *G. simpliciella*, *q.v.*

Life history
Described by Waters (1928).
Larva. Head and prothoracic plate shining black. Body pale yellowish green with small dark pinacula. Ninth segment dorsal plate and anal plate black. Feeds inside seeds of black bog-rush (*Schoenus nigricans*) from May till August. In spite of this long feeding period Waters (*loc.cit.*) concluded that there was only one extended generation, particularly as larvae were not to be found in autumn. Larvae and pupae collected at any time during summer produce moths the same year.
Pupa. In a flimsy cocoon spun inside a spikelet of the foodplant.
Imago. Moths may be found from May to September, but probably represent only one generation. They are often very abundant amongst the foodplant.

Distribution (Map 216)
Most common from Dorset to Cornwall, in the fens of East Anglia, in the west and north of Scotland and the west of Ireland, this distribution corresponding to the areas of greatest abundance of the foodplant. There are records from areas in which the foodplant does not occur (see, for instance, Poole (1909) for an Isle of Wight record) indicating the possibility of an alternative foodplant. Distributed abroad in northern Europe from France to Sweden.

GLYPHIPTERIX EQUITELLA (Scopoli)

Phalaena equitella Scopoli, 1763, *Ent.Carn.*: 254.
Glyphipteryx minorella Snellen, 1882, *Vlinders Nederland*: 753.
NOMENCLATURE. The correct name for this species depends on whether or not *G. minorella* (Snellen) is considered specifically distinct from *G. equitella* (Scopoli) (= *majorella* (Heinemann & Wocke, 1877)). Diakonoff (1976) treated them as distinct species and Bradley & Fletcher (1979) accordingly used *G. minorella* (Snellen) for the segregate occurring in Britain. More recently Diakonoff (*in litt.*) has concluded that the two are not distinct and that the oldest name for the complex, *G. equitella* (Scopoli), should be used for the British population.
Type locality: Carniola (Yugoslavia; Slovenija).

Description of imago (Pl.11, fig.23)
Wingspan 9–10mm. Head and thorax light grey or light greyish fuscous; labial palpus white banded with dark fuscous, segment 2 rough-scaled below, segment 3 about equal in length to 2. Forewing light greyish fuscous at base shading to fuscous with coppery tinge in apical third; forewing markings very similar to those of *G. simpliciella* (Stephens) and *G. schoenicolella* (Boyd), differing from both as follows: strigulae bolder, especially the first dorsal, often faintly dark-margined, posterior more silver in disc; sometimes six or seven costal strigulae; an additional silver dot in disc before blackish apical spot. Hindwing light fuscous or fuscous.
Similar species. The differences in forewing markings between this and the two foregoing species are not conspicuous and *G. equitella* is more easily recognized at a glance by its paler head, thorax and forewing base and by its slightly greater size.

Life history
Larva. Head black. Prothoracic plate blackish brown. Body pale yellow with small dark pinacula; a small transverse dorsal plate on abdominal segment 9 and anal plate dark brown. Thoracic legs brown-ringed. Mines in stems and leaves of biting stonecrop (*Sedum acre*), probably also English stonecrop (*Sedum anglicum*), and on the Continent recorded from other species of *Sedum*; affected parts of the plant turn whitish and wither. The larvae are fully grown at the end of May.
Pupa. In a fairly thick cocoon, usually between leaves of the foodplant but sometimes spun on a rock.
Imago. Univoltine, the moths flying in June and July.

Distribution (Map 217)
Widely distributed in England; Wales. In Scotland there is an old record from Perthshire (not on map) and a recent

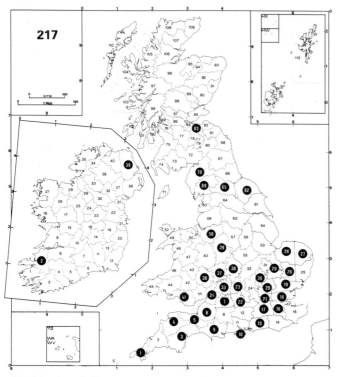

Glyphipterix equitella

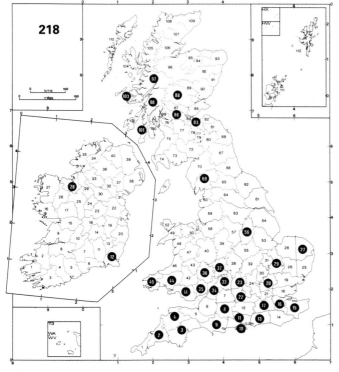

Glyphipterix forsterella

one from Midlothian. In Ireland recorded from Cos Kerry and Antrim. It is very local and perhaps scarcer than formerly, but common where it occurs. Distributed throughout Europe into Asia Minor.

GLYPHIPTERIX FORSTERELLA (Fabricius)

Tinea forsterella Fabricius, 1781, *Spec.Ins.* **2**: 509.

Type locality: [Germany]; Hamburg.

Description of imago (Pl.11, fig.24)

Wingspan 8–11mm. Head and thorax fuscous; labial palpus white, lined beneath with dark fuscous, segment 3 a little longer than segment 2. Forewing relatively a little broader than in the preceding three species, fuscous, more or less mixed ochreous especially in apical half; markings similar to those of *G. simpliciella* (Stephens) except first dorsal strigula is relatively stout, second costal ends well apicad of second dorsal, all except first dorsal become silvery in disc, the elongate subterminal silver spot is replaced by two silver dots and there is a distinct silver dot in the apical blackish spot.

Similar species. G. forsterella is distinguished from the three preceding species and from *G. haworthana* (Stephens) by the clear silver dot in the blackish apical spot. It is a very variable species in size but is usually larger than the first three and smaller than *G. haworthana*.

Life history

Early stages apparently undescribed. The larvae feed on seeds of various species of sedge (*Carex*), both large species such as true fox-sedge (*Carex vulpina*) and small-flowered species such as remote sedge (*Carex remota*). They feed from August onwards and according to Fletcher (1885) remain as larvae in the female spikes until the following April.

Imago. Univoltine. Moths are on the wing in May and June and can be found on flowers of the foodplant.

Distribution (Map 218)

Locally common throughout mainland Britain and recorded from two vice-counties in Ireland. It is strange that in the past this species seems to have been overlooked in some areas, but perhaps it is increasing its range. At Rannoch, Perthshire, for instance, it is common in an area which was formerly popular with microlepidopterists and yet it was not even recorded from Scotland before 1959. It is found over the greater part of Europe, from France to Scandinavia and eastwards as far as Asia Minor.

GLYPHIPTERIX HAWORTHANA (Stephens)

Heribeia haworthana Stephens, 1834, *Ill.Br.Ent.* (Haust.) 4: 262.

Type locality: England; Birmingham district.

Description of imago (Pl.11, fig.33)

Wingspan 11–15mm. Head and thorax dark fuscous; labial palpus white, variably mixed dark fuscous, rough-scaled, segment 3 slightly longer than segment 2. Forewing broad, dark brownish or ochreous fuscous, apical third pale brownish ochreous; pattern of the same type as that of *G. simpliciella* (Stephens) but first costal strigula almost connected to second dorsal by a brown mark, second costal continued as a silvery line usually connected to silver dot on termen near tornus and two other subterminal silvery dots, the upper in disc before apical blackish spot; most markings in apical third narrowly margined with brown; cilia white outside median line except for blackish apical hook and brown patch near tornus. Hindwing fuscous; cilia pale fuscous, whitish outwardly and near base.

The species varies very much in size and in the details of the forewing markings.

Similar species. Distinguished from the four previous species by its greater size. This is the *Glyphipterix* which among British species most resembles some of the Tortricidae; see note in introduction (p. 400).

Life history

Larva. Head and prothoracic plate brown. Body whitish with small inconspicuous brown pinacula; abdominal segment 8 with a narrow transverse posterior dorsal plate, segments 9 and 10 with entire plates, all light brown. Larvae feed from July onwards on the seeds of cotton-grass (*Eriophorum* spp.), remaining unchanged in the seed-heads till April. Each seed-head normally contains only one larva, but sometimes two. The seed-heads are spun onto adjoining vegetation so that they are not blown away during the winter, and in spring the great majority of seed-heads that may still be found contain larvae.

Pupa. In a flimsy cocoon in the seed-head.

Imago. Probably univoltine in Britain. The moths fly in May, sometimes as early as late April. According to Spuler (1910), Lhomme (1946) and Toll (1956), there is a second generation on the Continent in July and August.

Distribution (Map 219)

Widely distributed in bogs in mainland Britain. Recorded from the east of Ireland. Probably the species has a wider range than the map indicates since it is easily overlooked, both larvae and adults appearing rather early in the year. It has a wide distribution in northern Europe, from France

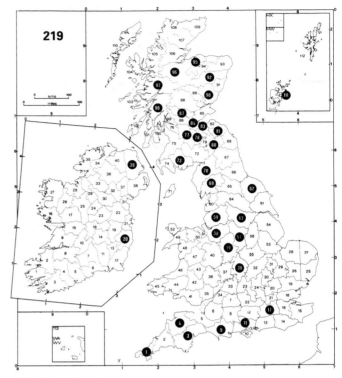

Glyphipterix haworthana

to Scandinavia, and has been recorded from north America.

GLYPHIPTERIX FUSCOVIRIDELLA (Haworth)

Tinea fuscoviridella Haworth, 1828, *Lepid.Br.*: 569.
Type locality: Great Britain.

Description of imago (Pl.11, fig.34)

Wingspan 10–16mm. Head and thorax brassy fuscous. Head and antenna without whitish lines; labial palpus segment 3 acute, about same length as segment 2. Forewing brassy fuscous, browner at base, sometimes irrorate ochreous in apical third; costa from middle to before apex and dorsum before tornus narrowly edged whitish; apex sometimes weakly margined darker; termen almost straight; cilia with basal half concolorous with wing, outer half obscurely whitish; no apical hook. Hindwing fuscous.
Similar species. Quite distinct amongst *Glyphipterix* species by reason of its plain brassy forewing and straight termen, the latter distinguishing it from plain forms of *G. thrasonella* (Scopoli).

Life history

Larva. Discovered and described by Sich (1900). Head pale shining amber. Prothoracic plate pale ochreous. Body very pale ochreous, pinkish brown on dorsum, pinacula small. On abdominal segment 8 three small dorsal plates, the central one elongate, segment 9 with anterior transverse plate and anal plate, all brownish ochreous. Feeds in stems of field wood-rush (*Luzula campestris*) and perhaps other species of *Luzula*, mining down to the roots; tenanted plants may be recognized by browned young leaves. The larva is fully grown in April and then pupates at the bottom of the mine.
Pupa. In an open network cocoon among the roots of the foodplant.
Imago. Univoltine, the moths flying in May and June. Females are probably relatively inactive as they are scarce in collections.

Distribution (Map 220)

Common in dry uncultivated grasslands throughout England and Wales and in south-east Scotland. On the Continent it is known only from Spain, France and northern Italy.

GLYPHIPTERIX THRASONELLA (Scopoli)

Phalaena thrasonella Scopoli, 1763, *Ent.Carn.*: 253.
Type locality: Carniola (Yugoslavia; Slovenija).

Description of imago (Pl.11, figs 35,36)

Wingspan 10–15mm. Head and thorax brassy fuscous. Head and antenna without whitish lines; labial palpus white, variably mixed with fuscous, segment 3 about same length as segment 2. Forewing brassy fuscous, sometimes

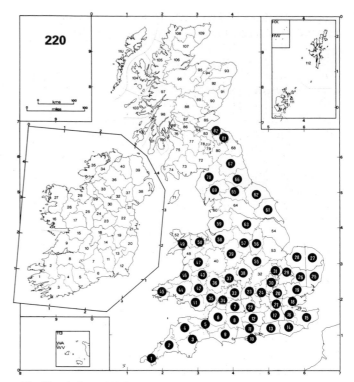

Glyphipterix fuscoviridella

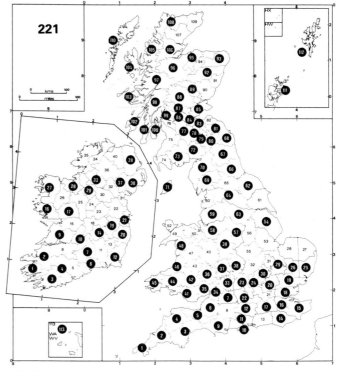

Glyphipterix thrasonella

copper-tinted, variably irrorate with ochreous; sometimes an obscure, pale, very oblique strigula from base of dorsum to fold; six evenly spaced silvery costal strigulae finely brown-margined, the first at one-third from base, short and oblique, second and third curved and continued to dorsum, fourth to sixth short, second to sixth ending on costa in whitish spots; a silver dot in disc beyond third strigula, which is often broken; an elongate coppery black mark along lower part of termen, including three silver dots; a silver dot on termen below apex; a coppery black spot at apex; cilia silvery fuscous at base with darker median line indented and broken below apex by a whitish spot, beyond this pale fuscous; small whitish spots on dorsum at ends of second and third strigulae. Hindwing fuscous or dark fuscous, cilia paler.

There is much variation in size, colour and extent of the forewing markings. The markings may be much reduced and in extreme forms are reduced to a few blackish scales: such a form was described as a separate species, *G. cladiella*, by Stainton ([1857]) from Wicken Fen specimens. A form of the opposite extreme, ab. *nitens* Bankes, has the outer part of the wing almost completely filled by a large silver blotch.

Similar species. The form *cladiella* Stainton (fig.36) is similar to *G. fuscoviridella* (Haworth), *q.v.* Other *Glyphipterix* species have the forewing strigulae mainly white, with a conspicuous oblique mid-dorsal strigula.

Life history
Little known, and the larva apparently undescribed. According to Toll (1956) the larvae feed from autumn to spring in the seeds of rush (*Juncus* spp.) and also on sun-dew (*Drosera* spp.). The latter is a most unlikely foodplant, and if the larvae really fed in the seeds of *Juncus* they would surely be better known. Sich (1915) recorded the females ovipositing in the heads of *Juncus*, but the larvae all left the rush heads on hatching. Several authors including Meyrick (1928) state that the larva feeds in rush stems in April and May and pupates in this situation.

Imago. Probably univoltine, with an extended flight period from May till August.

Distribution (Map 221)
Usually abundant amongst rushes throughout the British Isles, the distribution extending to the Outer Hebrides, St Kilda, and the Orkney and Shetland Islands. Abroad the species occurs throughout most of Europe, from Spain and Italy to European U.S.S.R. and Scandinavia.

References

Bradley, J. D. & Fletcher, D. S., 1979. *A recorder's log book or label list of British butterflies and moths*, [vi], 136 pp., London.

Diakonoff, A., 1976. Aantekeningen over de Nederlandse Microlepidoptera 3 (Glyphipterigidae). *Ent. Ber., Amst.* **36**: 82–84.

—— & Heppner, J. B., 1977. Proposed use of the plenary powers to designate a type-species for the nominal genus *Glyphipterix* Hubner [sic], [1825] (Lepidoptera, Glyphipterygidae) Z. N. (S) 2115. *Bull. zool. Nom.* **34**: 81–84.

Fletcher, W. H. B., 1885. *Glyphipteryx oculatella* bred. *Entomologist's mon. Mag.* **22**: 42.

Ford, L. T., 1954. The Glyphipterygidae and allied families. *Proc. Trans. S. Lond. ent. nat. Hist. Soc.* **1952–53**: 90–99, 1 col. pl.; reprinted [Agassiz, D. J. L. (Ed.)], 1978, in *Illustrated papers on British Microlepidoptera*: 77–86.

Heppner, J. B., 1982. Synopsis of the Glyphipterigidae (Lepidoptera: Copromorphoidea) of the world. *Proc. ent. Soc. Wash.* **84**: 38–66.

Kloet, G. S. & Hincks, W. D., 1972. A check list of British insects: Lepidoptera. (Edn 2). *Handbk Ident. Br. Insects* **11**(2), viii, 153 pp.

Lhomme, L., 1946. *Catalogue des lépidoptères de France et de Belgique* 2 (part, pp. 489–520). Le Carriol.

Meyrick, E., 1928. *A revised handbook of the British Lepidoptera*, 914 pp. London.

Pelham-Clinton, E. C., 1978. A previously unrecorded foodplant of *Glyphipterix simpliciella* (Stephens). *Entomologist's Rec. J. Var.* **90**: 166.

Poole, H. F., 1909. Lepidoptera. *In* Morey, F., (ed.), *A guide to the natural history of the Isle of Wight*: 394–438. Newport & London.

Sich, A., 1900. Note on the larva of *Glyphipteryx fuscoviridella*. *Entomologist's Rec. J. Var.* **12**: 192–193.

——, 1915. Notes on the micro-lepidoptera of south-west London (part). *Ibid.* **27**: 150–151.

Spuler, A., 1903–10. *Die Schmetterlinge Europas* 2, 523 pp. Stuttgart.

Stainton, H. T., [1857]. Lepidoptera. New British species in 1858. *Entomologist's Annu.* **1858**: 145–157.

——, 1870. *The natural history of the Tineina* 11, xiii, 330 pp., 8 col. pls. London.

Toll, S, 1956. Glyphipterygidae. *Klucze Oznacz. Owad. Pol.* **27** (39), 36 pp., 88 figs.

Waters, E. G. R., 1928. Observations on *Glyphipteryx schoenicolella* Boyd. *Entomologist's mon. Mag.* **64**: 252–253.

DOUGLASIIDAE
D. J. L. Agassiz

A family of more than 20 species, placed in two genera, which occur in Europe, the Middle East, Asia, Australia and north America. Gaedike (1974) has revised the Palaearctic species.

Imago. Head smooth-scaled; haustellum developed, bare; antenna scape without pecten; labial palpus short, straight and porrect, segment 3 rather blunt. Forewing lanceolate with scales bicolorous, pale basally, dark-tipped; venation (figure 122) veins R with four or five branches to costa, 7 (R_5) and 6 (M_1) stalked, 2 (Cu_2) weak or absent; female with subcostal retinaculum. Hindwing lanceolate; frenulum in female of one or two bristles; venation reduced, veins 7 (Rs) at or near long axis of wing, 3 (Cu_1) free. Female genitalia with a signum consisting of a stellate cluster of long spines; ostium sclerotized.

Similar in appearance to Elachistidae (*MBGBI* **3**) from which it differs in its resting attitude – anterior part raised in Douglasiidae, appressed to surface in Elachistidae.

Larva. Fusiform; prolegs small, crochets absent. Mines in leaves, tunnels in stems or feeds on flowers.

TINAGMA Zeller

Tinagma Zeller, 1839, *Isis, Leipzig* **1839**: 204.
Douglasia Stainton, 1854, *Ins.Br.Lepid.*: 179–180, pl.6, figs 5a–5c.

There are ten species of which two are found in Britain.

In the male genitalia of *Klimeschia* Amsel, the only other genus in the family, the uncus is present and the valvae are asymmetrical; in *Tinagma* the uncus is absent and the valvae are symmetrical (Gaedike, 1974).

Both British species feed on *Echium* in the larval stage; the other species of which the life history is known feed on other genera of Boraginaceae or on Rosaceae. All mine in leaves or stems or feed concealed in flower-heads. Adults can often be found flying over the foodplant.

Key to species (imagines) of the genus *Tinagma*

1　Forewing dark grey, without markings, finely irrorate whitish .... *ocnerostomella* (p. 409)
–　Forewing blackish grey, with a white tornal spot or a white fascia .... *balteolella* (p. 408)

TINAGMA BALTEOLELLA (Fischer von Rösler-stamm)

Aechmia balteolella Fischer von Röslerstamm, 1840, *Abb.Ber.Erg.Schmetterlingsk., Centurie* 1: 247.
Tinagma borkhauseniella Herrich-Schäffer, 1855, *Syst. Bearb.Schmett.Eur.* **5**: 259.
Type localities: Germany; Mark Brandenburg and Austria; Vienna.

Description of imago (Pl.11, figs 38,39)

Wingspan 8–9mm. Head dark grey; antenna and labial palpus black. Thorax black, tegulae dark grey. Forewing blackish grey, in male with a creamy white tornal spot; in female with a straight creamy white fascia to tornus; cilia in both sexes powdered with black and whitish scales. Hindwing grey, darker towards apex. Abdomen dark grey, each segment edged with black scales.

Life history

Ovum. Laid on flower-heads of viper's bugloss (*Echium vulgare*).
Larva. Head pale brown. Body pale yellowish with sparse long pale hairs; six rows of pale orange spots on each side one dorsal submedian, one dorsolateral, two subspiracular and two ventral, the dorsal spots more rounded and others more elongate; body colour paler in earlier instars. Feeds in July and early August among the flowers of viper's bugloss, spinning them together and causing some brownish discoloration. A cocoon is spun in August but pupation probably does not take place until the following spring.
Pupa. In a cocoon in the dead flowers.
Imago. Univoltine. Late May and June, when it can be disturbed from the foodplant around which it flies in sunshine.

Distribution (Map 222)

Hitherto found only on the coastal sand-dunes of east Kent where it was first discovered in 1975 (Agassiz, 1976) and is common. In Europe from France to Poland and the Balkans and also in Lebanon.

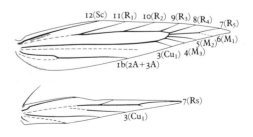

Figure 122 *Tinagma ocnerostomella* (Stainton), wing venation

TINAGMA OCNEROSTOMELLA (Stainton)

Gracilaria ? ocnerostomella Stainton, 1850, *Proc.ent.Soc. Lond.* **1**: 6.

Douglasia echii Herrich-Schäffer, 1855, *Syst.Bearb. Schmett.Eur.* **5**: 259; 1854, *ibid.* **5**: pl.118, fig.961 (legend non-binominal).

Type locality: England; Mickleham, Surrey.

Description of imago (Pl.11, fig.37)

Wingspan 8–9mm. Head grey; antenna dark greyish fuscous. Thorax dark grey. Forewing dark grey, finely irrorate whitish; cilia with darker lines. Hindwing grey. Abdomen grey.

Life history

Ovum. Laid amongst flowers of viper's bugloss (*Echium vulgare*) (Meyrick, 1928).

Larva. Full-fed 4mm long. Head brown. Body whitish, anal plate dark brown. Feeds on the pith within a stem of viper's bugloss, giving no external evidence. In the autumn when full-fed it spins a tough cocoon in the feeding place, about twice the length of the larva and slightly tapered at each end. Overwinters in its cocoon as a prepupa with legs hardly discernible, a black anal plate with a strong cremaster, and a blackish plate on the penultimate segment (Agassiz, pers.obs.).

Pupa. Within the cocoon in the stem. Pupation takes place in the spring.

Imago. Univoltine. Mid-June to mid-July. It can often be disturbed from the foodplant by day.

Distribution (Map 223)

Often common where the foodplant occurs, it has been recorded in south and south-east England from Devon to Lincolnshire. A record from Dublin, Ireland, requires confirmation (Beirne, 1941). In western Europe from France to Denmark and Fennoscandia; U.S.S.R.; the Middle East.

References

Agassiz, D. J. L., 1976. *Tinagma balteolellum* (Fischer von Röslerstamm) (Lep., Douglasiidae), a species new to the British Isles. *Entomologist's Gaz.* **26**: 291–293, 3 figs.

Beirne, B. P., 1941. A list of the Microlepidoptera of Ireland. *Proc. R. Ir. Acad.* **47**(B)4: 53–147.

Gaedike, R., 1974. Revision der paläarktischen Douglasiidae (Lepidoptera). *Acta faun. ent. Mus. natn. Pragae* **15**: 79–102, 59 figs.

Meyrick, E., 1928. *A revised handbook of British Lepidoptera*, vi, 914 pp. London.

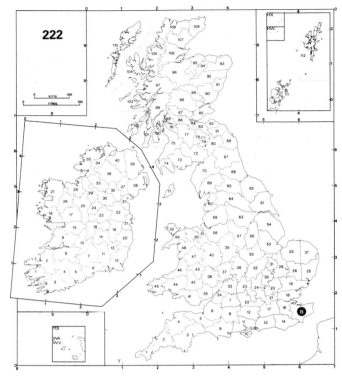

Tinagma balteolella

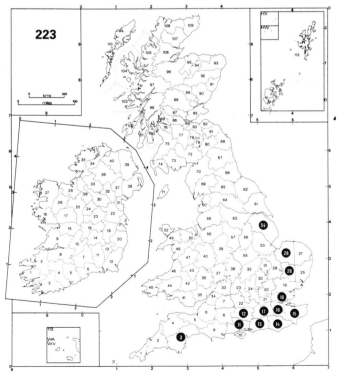

Tinagma ocnerostomella

HELIODINIDAE
A. M. Emmet

A family variously interpreted by systematists. Meyrick (1928) included *Schreckensteinia* Hübner and *Stathmopoda* Herrich-Schäffer and the former is still retained by some taxonomists (Common, 1970; Hodges *et al.*, 1983). The arrangement adopted in this work is that of Kloet & Hincks (1972) and Leraut (1980).

Characteristics are described under the single genus occurring in Europe.

HELIODINES Stainton

Heliodines Stainton, 1854, *Ins.Br.Lepid.*: 243.

A genus of world-wide distribution but poorly represented in the Palaearctic region. One species found in central and southern Europe used to occur in southern England.

Imago. Head smooth-scaled; antenna slightly shorter than forewing; haustellum naked, of moderate length; maxillary palpus very small; labial palpus short, drooping or recurved. Hindleg with whorls of stiff bristles at bases of spurs and apices of tarsi. Wings lanceolate with long cilia; forewing in most species brightly coloured, often with metallic spots. Wing venation, figure 123.

Larva. Prolegs well developed. Feeds externally in a web on foliage or fruits.

Pupa. Abdominal segments 3–7 in male, 3–6 in female, movable. In an open-mesh cocoon on the foodplant.

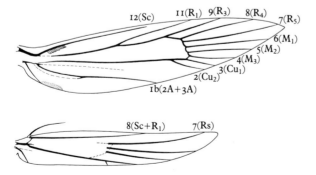

Figure 123 *Heliodines* spp., wing venation. After Wedgbrow

HELIODINES ROESELLA (Linnaeus)
Phalaena (*Tinea*) *roesella* Linnaeus, 1758, *Syst.Nat.* (Edn 10) **1**: 541.
Type locality: not stated.

Description of imago (Pl.11, fig.40)
Wingspan 10–11mm. Head smooth-scaled, metallic orange-fuscous; antenna slightly shorter than forewing, fuscous, apex whitish. Thorax and tegulae glossy brownish fuscous. Forewing deep orange with costa and dorsum irregularly edged black; a short black basal streak bearing two metallic silver spots; four costal and two dorsal black-edged metallic silver spots, first costal and first dorsal often united to form a fascia, third costal and second dorsal occasionally similarly joined; terminal area with ground colour strongly mixed with metallic silver scales; cilia brownish fuscous, glossy. Hindwing rough-scaled, reddish fuscous, paler basally. Abdomen fuscous, variably mixed with metallic silver scales.

Life history
Ovum. Laid on a leaf of a species of Chenopodiaceae, especially common orache (*Atriplex patula*) and good-King-Henry (*Chenopodium bonus-henricus*).
Larva. Head black. Body pale green, dorsal line darker; prothoracic plate consisting of two black spots. Feeds, often gregariously, in spun leaves of the foodplant. June.
Pupa. In an open network cocoon.
Imago. Emerges in July, overwinters and continues on the wing until May.

Distribution
Was recorded sparingly in the London district (Haworth, 1828), Darenth Wood, Kent, and Devon (Stephens, 1834–35), but has not been taken in Britain since about 1820. Central and southern Europe to Turkestan; extinct in Belgium (Lhomme, 1963).

References

Common, I. F. B., 1970. *In* Mackerras, I. M., *The insects of Australia*, 765–778. Melbourne.

Haworth, A. H., 1828. *Lepidoptera Britannica*, xxxvi, 609 pp. London.

Hodges, R. W. *et al.*, 1983. *Check list of the Lepidoptera of America north of Mexico*, xxiv, 284 pp. London.

Kloet, G. S. & Hincks, W. D., 1972. A check list of British insects: Lepidoptera (Edn 2). *Handbk Ident. Br. Insects* **11**(2), viii, 153 pp.

Leraut, P., 1980. Liste systématique et synonymique des Lépidoptères de France, Belgique et Corse. *Alexanor*, suppl., 334 pp.

Lhomme, L., 1963. *Catalogue des Lépidoptères de France et de Belgique* 2, 1253 pp. Le Carriol.

Meyrick, E., 1928. *A revised handbook of British Lepidoptera*, vi, 914 pp. London.

Stephens, J. F., 1834–35. *Illustrations of British entomology* (Haustellata), **4**, 436 pp, col. pls. London.

Addenda and Notes

Addenda and Notes

Addenda and Notes

Addenda and Notes

Addenda and Notes

Addenda and Notes

Addenda and Notes

Addenda and Notes

Addenda and Notes

Addenda and Notes

Addenda and Notes

THE PLATES

Plate 1: Lyonetiidae, Phyllocnistidae

Figs 1–12, × 1

1 *Leucoptera laburnella* (Stainton): larval mine in *Laburnum anagyroides* (laburnum). *Page 215*

2 *Leucoptera lathyrifoliella* (Stainton): larval mine in *Lathyrus sylvestris* (narrow-leaved everlasting pea). *Page 219*

3 *Leucoptera malifoliella* (O. G. Costa): larval mines in *Crataegus monogyna* (hawthorn). *Page 221*

4 *Leucoptera orobi* (Stainton): larval mines in *Lathyrus montanus* (bitter vetch). *Page 218*

5 *Leucoptera wailesella* (Stainton): larval mine in *Genista tinctoria* (dyer's greenweed). *Page 217*

6 *Leucoptera spartifoliella* (Hübner): larval mine in *Sarothamnus scoparius* (broom). *Page 217*

7 *Lyonetia clerkella* (Linnaeus): larval mine in *Betula pendula* (silver birch). *Page 224*

8 *Leucoptera lotella* (Stainton): larval mine in *Lotus uliginosus* (greater bird's-foot trefoil). *Page 220*

9 *Phyllocnistis unipunctella* (Stephens): larval mine in *Populus nigra* (black poplar). *Page 367*

10 *Phyllocnistis saligna* (Zeller): larval mine in *Salix* sp. (sallow), after Hering. *Page 366*

11 *Phyllocnistis saligna* (Zeller): larval mine in *Salix purpurea* (purple willow). *Page 366*

12 *Phyllocnistis xenia* Hering: larval mine in *Populus canescens* (grey poplar). *Page 368*

Plate 1: Lyonetiidae, Phyllocnistidae

Figs 1–12, × 1

Plate 2: Lyonetiidae, Gracillariidae

Figs 1–9, 11, × 1; 10, × 9

1 *Bedellia somnulentella* (Zeller): larval mine in *Convolvulus arvensis* (field bindweed). *Page 226*

2 *Bucculatrix nigricomella* Zeller: larval mine in *Leucanthemum vulgare* (oxeye daisy). *Page 230*

3 *Bucculatrix thoracella* (Thunberg): larval mines in *Tilia cordata* (small-leaved lime). *Page 236*

4 *Bucculatrix frangulella* (Goeze): larval mine in *Frangula alnus* (alder buckthorn). *Page 233*

5 *Bucculatrix cidarella* Zeller: larval mines in *Alnus glutinosa* (alder). *Page 234*

6 *Bucculatrix crataegi* Zeller: larval mine in *Crataegus monogyna* (hawthorn). *Page 237*

7 *Bucculatrix albedinella* Zeller: larval mine in *Ulmus carpinifolia* (small-leaved elm). *Page 234*

8 *Bucculatrix demaryella* (Duponchel): larval mines in *Betula pendula* (silver birch). *Page 238*

9 *Bucculatrix ulmella* Zeller: larval mines in *Quercus robur* (pedunculate oak). *Page 237*

10 *Bucculatrix* sp.: cocoon, after Jäckh. *Page 228*

11 *Phyllonorycter messaniella* (Zeller): larval mine in *Quercus ilex* (evergreen oak). *Page 312*

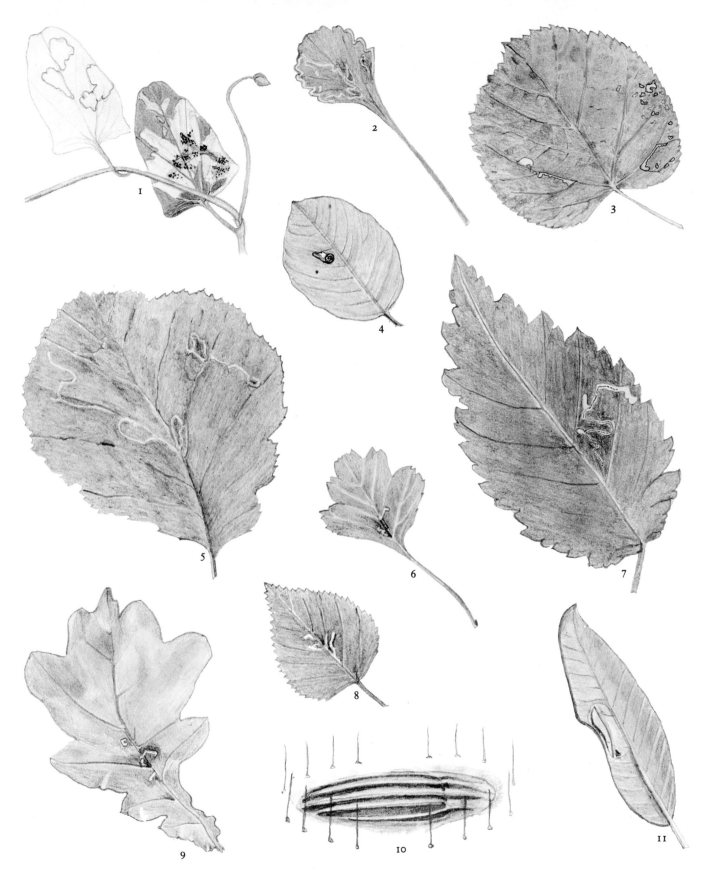

Plate 2: Lyonetiidae, Gracillariidae

Figs 1–9, 11, × 1; 10, × 9

Plate 3: Cossidae, Limacodidae

Figs 1–14, × 1

1 *Zeuzera pyrina* (Linnaeus) ♂ Leopard Moth. *Page 71*

2 *Zeuzera pyrina* (Linnaeus) ♀ Leopard Moth. *Page 71*

3 *Zeuzera pyrina* (Linnaeus) ab. *confluens* Cockayne ♀ Leopard Moth. *Page 71*

4 *Phragmataecia castaneae* (Hübner) ♂ Reed Leopard. *Page 70*

5 *Phragmataecia castaneae* (Hübner) ♀ Reed Leopard. *Page 70*

6 *Cossus cossus* (Linnaeus) ♂ Goat Moth. *Page 72*

7 *Cossus cossus* (Linnaeus) ♀ Goat Moth. *Page 72*

8 *Heterogenea asella* ([Denis & Schiffermüller]) ♂ Triangle. *Page 126*

9 *Heterogenea asella* ([Denis & Schiffermüller]) ab. *nigra* Tutt ♂ Triangle. *Page 126*

10 *Heterogenea asella* ([Denis & Schiffermüller]) ♀ Triangle. *Page 126*

11 *Apoda limacodes* (Hufnagel) ♂ Festoon. *Page 125*

12 *Apoda limacodes* (Hufnagel) ab. *suffusa* Seitz ♂ Festoon. *Page 125*

13 *Apoda limacodes* (Hufnagel) ab. *assella* Esper ♂ Festoon. *Page 125*

14 *Apoda limacodes* (Hufnagel) Esper ♀ Festoon. *Page 125*

Plate 4: Zygaenidae

Figs 1–64, × 1

1 *Adscita (Jordanita) globulariae* (Hübner) ♂ Scarce Forester. *Page 82*

2 *Adscita (Jordanita) globulariae* (Hübner) ♀ Scarce Forester. *Page 82*

3 *Adscita (Adscita) geryon* (Hübner) ♂ Cistus Forester. *Page 83*

4 *Adscita (Adscita) geryon* (Hübner) ♀ Cistus Forester. *Page 83*

5 *Adscita (Adscita) statices* (Linnaeus) ♂ Forester. *Page 84*

6 *Adscita (Adscita) statices* (Linnaeus) ♀ Forester. *Page 84*

7 *Zygaena (Zygaena) exulans subochracea* White ♂ Scotch or Mountain Burnet. *Page 104*

8 *Zygaena (Zygaena) exulans subochracea* White ♀ Scotch or Mountain Burnet. *Page 104*

9 *Zygaena (Zygaena) loti scotica* (Rowland-Brown) ♂ Slender Scotch Burnet. *Page 106*

10 *Zygaena (Zygaena) loti scotica* (Rowland-Brown) ♀ Slender Scotch Burnet. *Page 106*

11 *Zygaena (Zygaena) viciae ytenensis* Briggs ♂ New Forest Burnet. *Page 106*

12 *Zygaena (Zygaena) viciae ytenensis* Briggs ♀ New Forest Burnet. *Page 106*

13 *Zygaena (Zygaena) viciae argyllensis* Tremewan ♂ New Forest Burnet. *Page 106*

14 *Zygaena (Zygaena) viciae argyllensis* Tremewan ♀ New Forest Burnet. *Page 106*

15 *Zygaena (Zygaena) filipendulae stephensi* Dupont ♂ Six-spot Burnet. *Page 108*

16 *Zygaena (Zygaena) filipendulae stephensi* Dupont f. *conjuncta* Tutt ♂ Six-spot Burnet. *Page 108*

17 *Zygaena (Zygaena) filipendulae stephensi* Dupont ♂ (suffused confluent form) Six-spot Burnet. *Page 109*

18 *Zygaena (Zygaena) filipendulae stephensi* Dupont f. *grisescens* Oberthür ♂ Six-spot Burnet. *Page 109*

19 *Zygaena (Zygaena) filipendulae stephensi* Dupont f. *flava* Robson ♂ Six-spot Burnet. *Page 109*

20 *Zygaena (Zygaena) filipendulae stephensi* Dupont f. *aurantia* Tutt ♂ Six-spot Burnet. *Page 109*

21 *Zygaena (Zygaena) filipendulae stephensi* Dupont f. *intermedia* Tutt ♂ Six-spot Burnet. *Page 109*

22 *Zygaena (Zygaena) filipendulae stephensi* Dupont ♀ Six-spot Burnet. *Page 108*

23 *Zygaena (Zygaena) filipendulae stephensi* Dupont f. *nigrolimbata* Cockayne ♀ Six-spot Burnet. *Page 109*

24 *Zygaena (Zygaena) filipendulae stephensi* Dupont ♀ (sooty black form) Six-spot Burnet. *Page 109*

25 *Zygaena (Zygaena) trifolii decreta* Verity ♂ Five-spot Burnet. *Page 110*

26 *Zygaena (Zygaena) trifolii decreta* Verity ♂ (spots 3 and 4 separate) Five-spot Burnet. *Page 110*

27 *Zygaena (Zygaena) trifolii decreta* Verity ♂ (confluent form) Five-spot Burnet. *Page 110*

28 *Zygaena (Zygaena) trifolii decreta* Verity f. *daimon* Porritt ♂ Five-spot Burnet. *Page 110*

29 *Zygaena (Zygaena) trifolii decreta* Verity ♀ Five-spot Burnet. *Page 110*

30 *Zygaena (Zygaena) trifolii decreta* Verity ♀ (six-spotted form) Five-spot Burnet. *Page 110*

31 *Zygaena (Zygaena) trifolii palustrella* Verity ♂ Five-spot Burnet. *Page 110*

32 *Zygaena (Zygaena) trifolii palustrella* Verity f. *obsoleta* Tutt ♂ Five-spot Burnet. *Page 110*

33 *Zygaena (Zygaena) trifolii palustrella* Verity ♂ (broad hindwing border) Five-spot Burnet. *Page 110*

34 *Zygaena (Zygaena) trifolii palustrella* Verity f. *lutescens* Cockerell ♂ Five-spot Burnet. *Page 110*

35 *Zygaena (Zygaena) trifolii palustrella* Verity ♀ Five-spot Burnet. *Page 110*

36 *Zygaena (Zygaena) trifolii palustrella* Verity ♀ (confluent form) Five-spot Burnet. *Page 110*

37 *Zygaena (Zygaena) trifolii palustrella* Verity f. *extrema* Tutt ♀ Five-spot Burnet. *Page 110*

38 *Zygaena (Zygaena) trifolii palustrella* Verity ♀ (brown form) Five-spot Burnet. *Page 110*

39 *Zygaena (Zygaena) lonicerae latomarginata* (Tutt) ♂ (Filey, Yorkshire) Narrow-bordered Five-spot Burnet. *Page 112*

40 *Zygaena (Zygaena) lonicerae latomarginata* (Tutt) ♂ (Brookwood, Surrey) Narrow-bordered Five-spot Burnet. *Page 112*

41 *Zygaena (Zygaena) lonicerae latomarginata* (Tutt) f. *centripuncta* Tutt ♂ Narrow-bordered Five-spot Burnet. *Page 112*

42 *Zygaena (Zygaena) lonicerae latomarginata* (Tutt) ♂ (confluent form) Narrow-bordered Five-spot Burnet. *Page 112*

43 *Zygaena (Zygaena) lonicerae latomarginata* (Tutt) f. *grisescens* Cockayne ♂ Narrow-bordered Five-spot Burnet. *Page 112*

44 *Zygaena (Zygaena) lonicerae latomarginata* (Tutt) ♂ (yellow form) Narrow-bordered Five-spot Burnet. *Page 112*

45 *Zygaena (Zygaena) lonicerae latomarginata* (Tutt) ♂ (orange form) Narrow-bordered Five-spot Burnet. *Page 112*

46 *Zygaena (Zygaena) lonicerae latomarginata* (Tutt) ♂ (pseudo-orange form) Narrow-bordered Five-spot Burnet. *Page 112*

47 *Zygaena (Zygaena) lonicerae latomarginata* (Tutt) ♀ (Filey, Yorkshire) Narrow-bordered Five-spot Burnet. *Page 112*

48 *Zygaena (Zygaena) lonicerae latomarginata* (Tutt) ♀ (Ranmore Common, Surrey) Narrow-bordered Five-spot Burnet. *Page 112*

49 *Zygaena (Zygaena) lonicerae jocelynae* Tremewan ♂ Narrow-bordered Five-spot Burnet. *Page 112*

50 *Zygaena (Zygaena) lonicerae jocelynae* Tremewan ♀ Narrow-bordered Five-spot Burnet. *Page 112*

51 *Zygaena (Zygaena) lonicerae jocelynae* Tremewan ♀ (confluent form) Narrow-bordered Five-spot Burnet. *Page 112*

52 *Zygaena (Zygaena) lonicerae insularis* Tremewan ♂ Narrow-bordered Five-spot Burnet. *Page 112*

53 *Zygaena (Zygaena) lonicerae insularis* Tremewan ♀ Narrow-bordered Five-spot Burnet. *Page 112*

54 *Zygaena (Mesembrynus) purpuralis segontii* Tremewan ♂ Transparent Burnet. *Page 114*

55 *Zygaena (Mesembrynus) purpuralis segontii* Tremewan f. *obscura* Tutt ♂ Transparent Burnet. *Page 114*

56 *Zygaena (Mesembrynus) purpuralis segontii* Tremewan ♀ Transparent Burnet. *Page 114*

57 *Zygaena (Mesembrynus) purpuralis caledonensis* Reiss ♂ Transparent Burnet. *Page 114*

58 *Zygaena (Mesembrynus) purpuralis caledonensis* Reiss ♂ (brownish red form) Transparent Burnet. *Page 114*

59 *Zygaena (Mesembrynus) purpuralis caledonensis* Reiss ♂ (orange form) Transparent Burnet. *Page 114*

60 *Zygaena (Mesembrynus) purpuralis caledonensis* Reiss ♂ (yellow form) Transparent Burnet. *Page 114*

61 *Zygaena (Mesembrynus) purpuralis caledonensis* Reiss ♂ (suffused confluent form) Transparent Burnet. *Page 114*

62 *Zygaena (Mesembrynus) purpuralis caledonensis* Reiss ♀ Transparent Burnet. *Page 114*

63 *Zygaena (Mesembrynus) purpuralis sabulosa* Tremewan ♂ Transparent Burnet. *Page 114*

64 *Zygaena (Mesembrynus) purpuralis sabulosa* Tremewan ♀ Transparent Burnet. *Page 114*

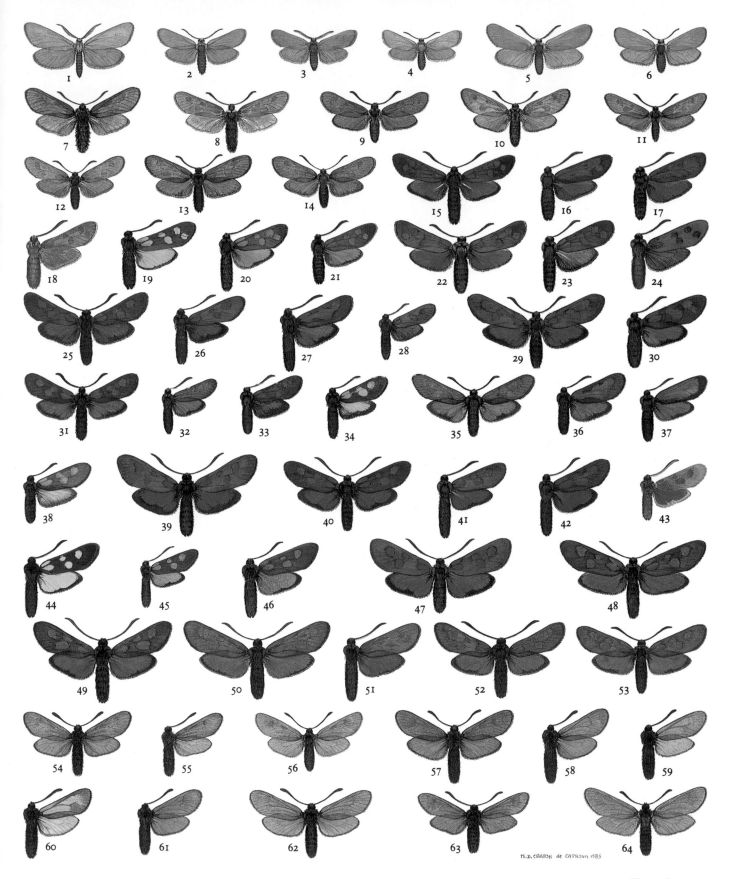

M.D. CRATON de CATROUN 1983

Plate 4: Zygaenidae

Figs 1–64, × 1

Plate 5: Zygaenidae

Figs 1–22, final instar larvae of *Zygaena* species

Figs 1, 2 × 2½; 3, 4 × 2¾; 5, 6 × 3¼; 7–12 × 2⅓; 13, 14 × 2½; 15–20 × 2⅓; 21, 22 × 2¾

See also Key to species (larvae) of the genus *Zygaena* (p. 102).

1, 2 *Zygaena (Zygaena) exulans* (Hohenwarth). *Page 105*

3, 4 *Zygaena (Zygaena) loti* ([Denis & Schiffermüller]). *Page 106*

5, 6 *Zygaena (Zygaena) viciae* ([Denis & Schiffermüller]). *Page 107*

7, 8 *Zygaena (Zygaena) filipendulae* (Linnaeus), normal form. *Page 109*

9, 10 *Zygaena (Zygaena) filipendulae* (Linnaeus), heavily marked form. *Page 109*

11, 12 *Zygaena (Zygaena) filipendulae* (Linnaeus), lightly marked form. *Pages 91, 109*

13, 14 *Zygaena (Zygaena) trifolii* (Esper). *Page 110*

15, 16 *Zygaena (Zygaena) lonicerae* (Scheven). *Page 112*

17, 18 *Zygaena* hybrid ex *Z. lonicerae* ♂ × *Z. filipendulae* ♀. *Page 98*

19, 20 *Zygaena* hybrid ex *Z. filipendulae* ♂ × *Z. lonicerae* ♀. *Page 98*

21, 22 *Zygaena (Mesembrynus) purpuralis* (Brünnich). *Page 114*

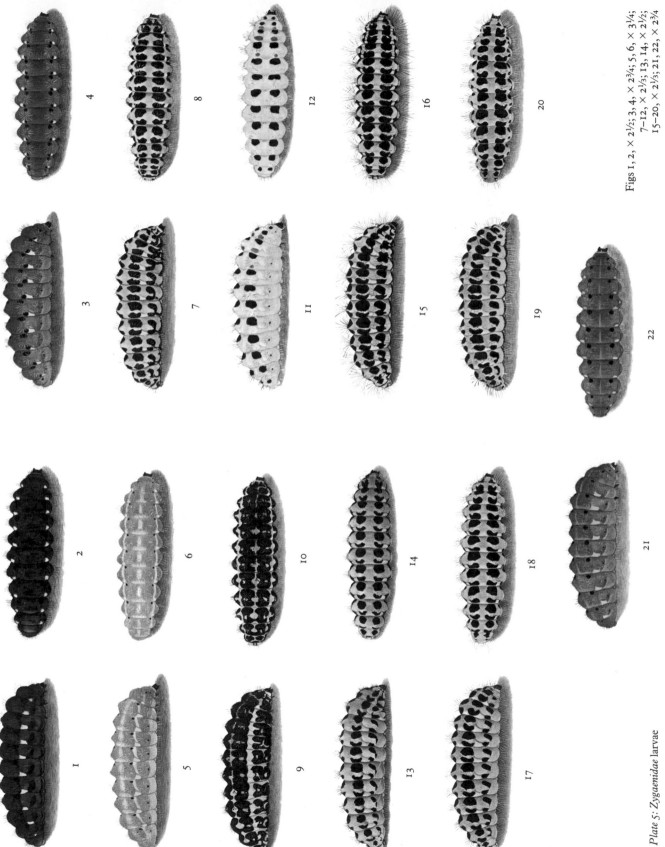

Figs 1, 2, × 2½; 3, 4, × 2¾; 5, 6, × 3¼;
7–12, × 2⅓; 13, 14, × 2½;
15–20, × 2⅓; 21, 22, × 2¾

Plate 5: *Zygaenidae* larvae

Plate 6: Zygaenidae

Figs 1–11

All photographs except fig. 3 were taken in the wild.
See also Key to species (cocoons) of the genus *Zygaena* (p. 103).

Plate 6: Zygaenidae cocoons

Figs 1–11

Plate 7: Psychidae

Figs 1–24, × 3; 25–32, × 1½; 33–52, × 3; 53–62, × 1½

1 *Narycia monilifera* (Geoffroy) ♂. *Page 132*

2 *Narycia monilifera* (Geoffroy) ♀. *Page 132*

3 *Dahlica inconspicuella* (Stainton) ♂ Lesser Lichen Case-bearer. *Page 134*

4 *Dahlica inconspicuella* (Stainton) ♂ Lesser Lichen Case-bearer. *Page 134*

5 *Dahlica inconspicuella* (Stainton) ♀ Lesser Lichen Case-bearer. *Page 134*

6 *Bankesia douglasii* (Stainton) ♂. *Page 137*

7 *Dahlica triquetrella* (Hübner) ♀. *Page 134*

8 *Dahlica lichenella* (Linnaeus) ♀ Lichen Case-bearer. *Page 135*

9 *Diplodoma herminata* (Geoffroy) ♂. *Page 131*

10 *Diplodoma herminata* (Geoffroy) ♀. *Page 131*

11 *Taleporia tubulosa* (Retzius) ♂. *Page 136*

12 *Taleporia tubulosa* (Retzius) ♀. *Page 136*

13 *Bacotia sepium* (Speyer) ♂. *Page 141*

14 *Bacotia sepium* (Speyer) ♀. *Page 141*

15 *Luffia lapidella* (Goeze) ♂. *Page 139*

16 *Psyche casta* (Pallas) ♂. *Page 143*

17 *Psyche casta* (Pallas) ♀. *Page 143*

18 *Whittleia retiella* (Newman) ♂. *Page 146*

19 *Psyche crassiorella* Bruand ♂. *Page 144*

20 *Psyche crassiorella* Bruand ♀. *Page 144*

21 *Proutia betulina* (Zeller) ♂. *Page 142*

22 *Proutia betulina* (Zeller) ♀. *Page 142*

23 *Epichnopterix plumella* ([Denis & Schiffermüller]) ♂. *Page 145*

24 *Epichnopterix plumella* ([Denis & Schiffermüller]) ♀. *Page 145*

25 *Acanthopsyche atra* (Linnaeus) ♂. *Page 147*

26 *Acanthopsyche atra* (Linnaeus) ♀. *Page 147*

27 *Pachythelia villosella* (Ochsenheimer) ♂. *Page 148*

28 *Pachythelia villosella* (Ochsenheimer) ♀. *Page 148*

29 *Lepidopsyche unicolor* (Hufnagel) ♂. *Page 149*

30 *Lepidopsyche unicolor* (Hufnagel) ♀. *Page 149*

31 *Sterrhopterix fusca* (Haworth) ♂. *Page 150*

32 *Sterrhopterix fusca* (Haworth) ♀. *Page 150*

33 *Narycia monilifera* (Geoffroy): larval case. *Page 132*

34 *Dahlica triquetrella* (Hübner): larval case. *Page 134*

35 *Dahlica inconspicuella* (Stainton) Lesser Lichen Case-bearer: larval case. *Page 134*

36 *Bankesia douglasii* (Stainton): larval case. *Page 137*

37 *Dahlica lichenella* (Linnaeus) Lichen Case-bearer: larval case. *Page 135*

38 *Diplodoma herminata* (Geoffroy): larval case. *Page 131*

39 *Taleporia tubulosa* (Retzius): larval case. *Page 136*

40 *Bacotia sepium* (Speyer) ♂: larval case. *Page 141*

41 *Bacotia sepium* (Speyer) ♀: larval case. *Page 141*

42 *Luffia lapidella* (Goeze) ♂: larval case. *Page 139*

43 *Luffia lapidella* (Goeze) ♀: larval case. *Page 139*

44 *Luffia ferchaultella* (Stephens): larval case. *Page 140*

45 *Psyche casta* (Pallas) ♂: larval case. *Page 143*

46 *Psyche casta* (Pallas) ♀: larval case. *Page 143*

47 *Psyche crassiorella* Bruand ♂: larval case. *Page 144*

48 *Psyche crassiorella* Bruand ♀: larval case. *Page 144*

49 *Proutia betulina* (Zeller) ♂: larval case. *Page 142*

50 *Proutia betulina* (Zeller) ♀: larval case. *Page 142*

51 *Whittleia retiella* (Newman) ♂: larval case. *Page 146*

52 *Whittleia retiella* (Newman) ♀: larval case. *Page 146*

53 *Epichnopterix plumella* ([Denis & Schiffermüller]) ♂: larval case. *Page 145*

54 *Epichnopterix plumella* ([Denis & Schiffermüller]) ♀: larval case. *Page 145*

55 *Acanthopsyche atra* (Linnaeus) ♂: larval case. *Page 147*

56 *Acanthopsyche atra* (Linnaeus) ♀: larval case. *Page 147*

57 *Pachythelia villosella* (Ochsenheimer) ♂: larval case. *Page 148*

58 *Pachythelia villosella* (Ochsenheimer) ♀: larval case. *Page 148*

59 *Lepidopsyche unicolor* (Hufnagel) ♂: larval case. *Page 149*

60 *Lepidopsyche unicolor* (Hufnagel) ♀: larval case. *Page 149*

61 *Sterrhopterix fusca* (Haworth) ♂: larval case. *Page 150*

62 *Sterrhopterix fusca* (Haworth) ♀: larval case. *Page 150*

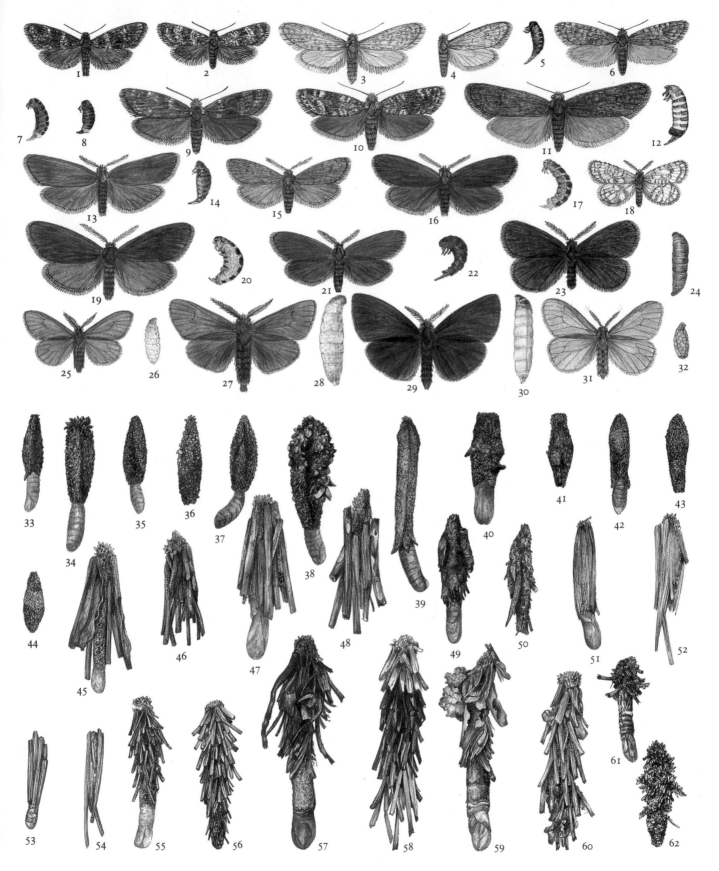

Plate 7: Psychidae

Figs 1–24, × 3; 25–32, × 1½; 33–52, × 3; 53–62, × 1½

Plate 8: Tineidae

Figs 1–4, × 4; 5–8, × 5; 9, × 2½; 10–37, × 3

Plate 8: Tineidae

Figs 1–4, × 4; 5–8, × 5; 9, × 2½; 10–37, × 3

Plate 9: Tineidae, Ochsenheimeriidae, Lyonetiidae

Figs 1–18, × 3; 19–38, × 5

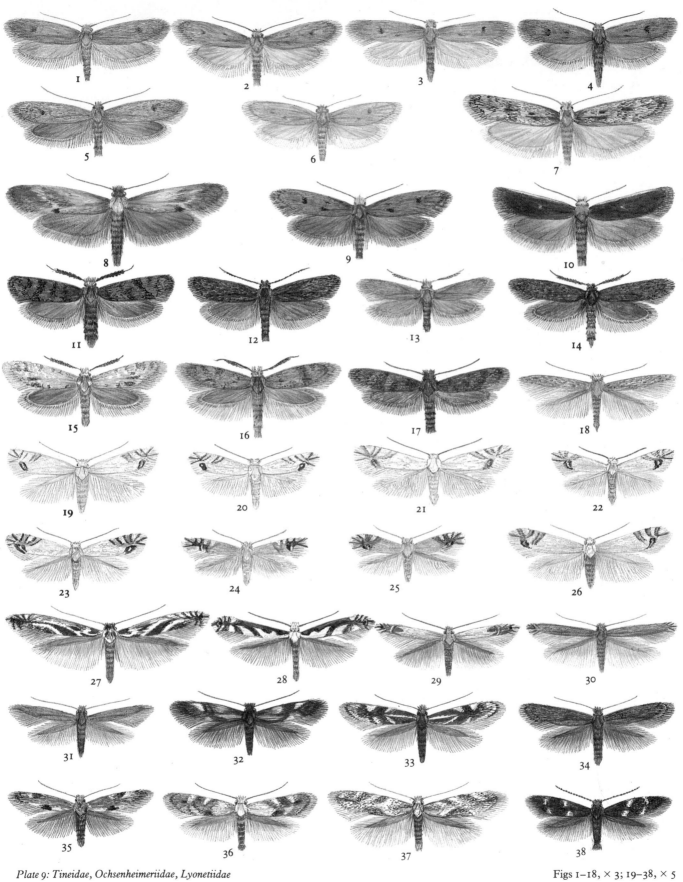

Plate 9: Tineidae, Ochsenheimeriidae, Lyonetiidae

Figs 1–18, × 3; 19–38, × 5

Plate 10: Lyonetiidae, Hieroxestidae, Gracillariidae

Figs 1–4, × 5; 5–33, × 3; 34–37, × 4

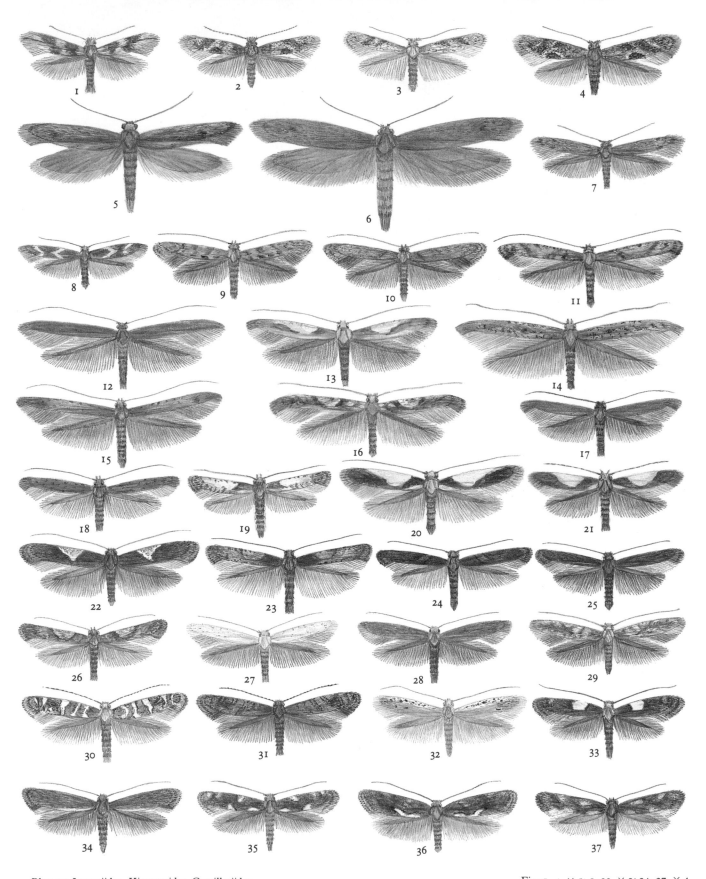

Plate 10: Lyonetiidae, Hieroxestidae, Gracillariidae

Figs 1–4, × 5; 5–33, × 3; 34–37, × 4

Plate 11: Gracillariidae, Phyllocnistidae, Choreutidae, Glyphipterigidae, Douglasiidae, Heliodinidae

Figs 1–12, × 4; 13–24, × 5; 25–36, × 3; 37–40, × 4

1   2   3   4

5   6   7   8

9   10   11   12

13   14   15   16

17   18   19   20

21   22   23   24

25   26   27   28

29   30   31   32

33   34   35   36

37   38   39   40

Plate 11: Gracillariidae, Phyllocnistidae, Choreutidae,
Glyphipterigidae, Douglasiidae, Heliodinidae

Figs 1–12, × 4; 13–24, × 5; 25–36, × 3; 37–40, × 4

Plate 12: Gracillariidae Lithocolletinae

Figs 1–37, × 4½

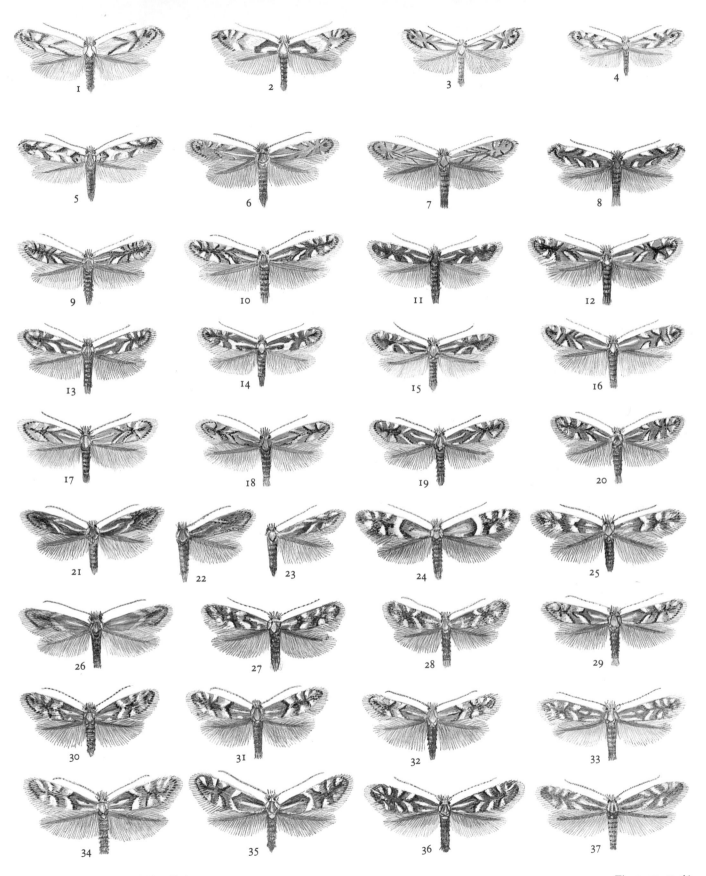

Plate 12: *Gracillariidae Lithocolletinae*

Figs 1–37, × 4½

Plate 13: Gracillariidae Lithocolletinae

Figs 1–37, × 4½

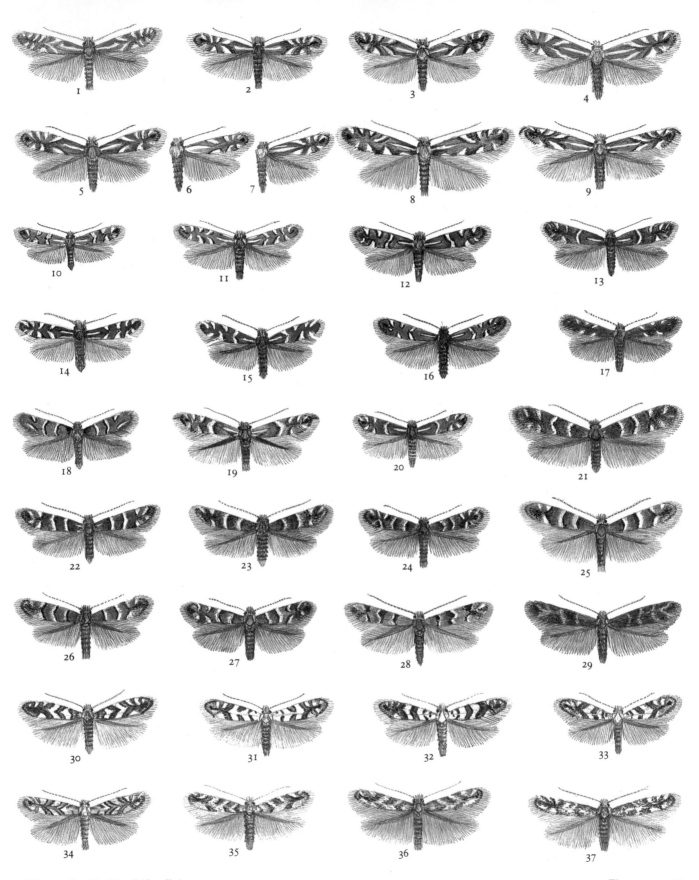

Plate 13: Gracillariidae Lithocolletinae

Figs 1–37, × 4½

Plate 14: Sesiidae

Figs 1–25, × 1½

1 *Sesia apiformis* (Clerck) ♂ Hornet Moth. *Page 371*

2 *Sesia apiformis* (Clerck) ♀ Hornet Moth. *Page 371*

3 *Sesia bembeciformis* (Hübner) ♂ Lunar Hornet Moth. *Page 372*

4 *Sesia bembeciformis* (Hübner) ♀ Lunar Hornet Moth. *Page 372*

5 *Paranthrene tabaniformis* (Rottemburg) ♂ Dusky Clearwing. *Page 373*

6 *Synanthedon spheciformis* ([Denis & Schiffermüller]) ♂ White-barred Clearwing. *Page 377*

7 *Synanthedon scoliaeformis* (Borkhausen) ♂ Welsh Clearwing. *Page 378*

8 *Synanthedon scoliaeformis* (Borkhausen) ♀ Welsh Clearwing. *Page 378*

9 *Synanthedon tipuliformis* (Clerck) ♂ Currant Clearwing. *Page 375*

10 *Synanthedon tipuliformis* (Clerck) ♀ Currant Clearwing. *Page 375*

11 *Synanthedon vespiformis* (Linnaeus) ♂ Yellow-legged Clearwing. *Page 376*

12 *Synanthedon vespiformis* (Linnaeus) ♀ Yellow-legged Clearwing. *Page 376*

13 *Synanthedon vespiformis* (Linnaeus) ab. *rufomarginata* Spuler ♀ Yellow-legged Clearwing. *Page 376*

14 *Synanthedon andrenaeformis* (Laspeyres) ♂ Orange-tailed Clearwing. *Page 380*

15 *Synanthedon flaviventris* (Staudinger) ♂ Sallow Clearwing. *Page 379*

16 *Synanthedon formicaeformis* (Esper) ♂ Red-tipped Clearwing. *Page 382*

17 *Synanthedon myopaeformis* (Borkhausen) ♂ Red-belted Clearwing. *Page 381*

18 *Synanthedon culiciformis* (Linnaeus) ♂ Large Red-belted Clearwing. *Page 383*

19 *Synanthedon culiciformis* (Linnaeus) ab. *flavocingulata* Spuler ♂ Large Red-belted Clearwing. *Page 383*

20 *Bembecia scopigera* (Scopoli) ♂ Six-belted Clearwing. *Page 385*

21 *Bembecia scopigera* (Scopoli) ♀ Six-belted Clearwing. *Page 385*

22 *Bembecia chrysidiformis* (Esper) ♂ Fiery Clearwing. *Page 387*

23 *Bembecia chrysidiformis* (Esper) ♀ Fiery Clearwing. *Page 387*

24 *Bembecia muscaeformis* (Esper) ♂ Thrift Clearwing. *Page 386*

25 *Bembecia muscaeformis* (Esper) ♀ Thrift Clearwing. *Page 386*

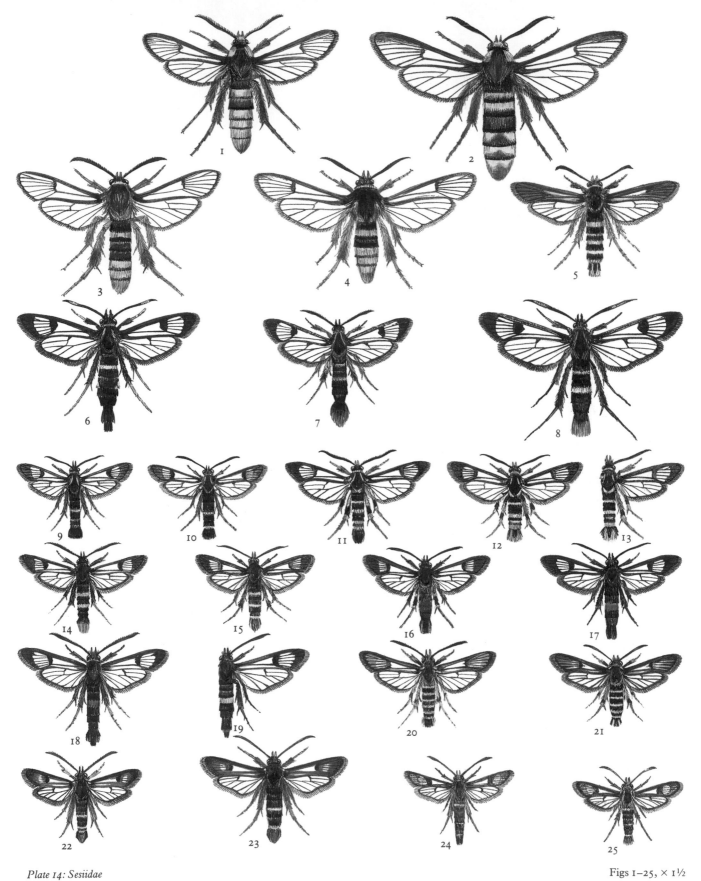

Plate 14: Sesiidae

Figs 1-25, × 1½

General Index

Principal entries are given in **bold type**. Plate references are shown as (B:2; 11:7). The index also includes references to figures in the text, as 66 (text fig. 32), and keys.

See separate index to host plants and other food substances, attractants and situations.

Index of Host Plants

and other food substances, attractants and situations